ACS SYMPOSIUM SERIES **537**

Polymers for Microelectronics

Resists and Dielectrics

Larry F. Thompson, EDITOR
AT&T Bell Laboratories

C. Grant Willson, EDITOR
The University of Texas at Austin

Seiichi Tagawa, EDITOR
Osaka University

Developed from a symposium sponsored
by the Division of Polymeric Materials:
Science and Engineering, Inc., of the American Chemical Society
and the Society of Polymer Science, Japan,
at the 203rd National Meeting
of the American Chemical Society,
San Francisco, California,
April 5–10, 1992

American Chemical Society, Washington, DC 1994

Library of Congress Cataloging-in-Publication Data

Polymers for microelectronics: resists and dielectrics / Larry F. Thompson, editor, C. Grant Willson, editor, Seiichi Tagawa.

    p.   cm.—(ACS symposium series, ISSN 0097–6156; 537)

"Developed from a symposium sponsored by the Division of Polymeric Materials: Science and Engineering, Inc., of the American Chemical Society at the 203rd Meeting of the American Chemical Society, San Francisco, California, April 5–10, 1992."

Includes bibliographical references and indexes.

ISBN 0–8412–2721–7

1. Microelectronics—Materials. 2. Polymers.

I. Thompson, L. F., 1944– . II. Willson, C. Grant, 1938– . III. Tagawa, Seiichi. IV. American Chemical Society. Division of Polymeric Materials: Science and Engineering. V. American Chemical Society. Meeting (203rd: 1992: San Francisco, Calif.) VI. Series.

TK7871.15.P6P628   1993
621.3815′31—dc20
                                    93–33955
                                          CIP

Foreword

THE ACS SYMPOSIUM SERIES was first published in 1974 to provide a mechanism for publishing symposia quickly in book form. The purpose of this series is to publish comprehensive books developed from symposia, which are usually "snapshots in time" of the current research being done on a topic, plus some review material on the topic. For this reason, it is necessary that the papers be published as quickly as possible.

Before a symposium-based book is put under contract, the proposed table of contents is reviewed for appropriateness to the topic and for comprehensiveness of the collection. Some papers are excluded at this point, and others are added to round out the scope of the volume. In addition, a draft of each paper is peer-reviewed prior to final acceptance or rejection. This anonymous review process is supervised by the organizer(s) of the symposium, who become the editor(s) of the book. The authors then revise their papers according to the recommendations of both the reviewers and the editors, prepare camera-ready copy, and submit the final papers to the editors, who check that all necessary revisions have been made.

As a rule, only original research papers and original review papers are included in the volumes. Verbatim reproductions of previously published papers are not accepted.

M. Joan Comstock
Series Editor

Contents

TOP-SURFACE IMAGING AND
DRY DEVELOPMENT RESISTS

ELECTRON-BEAM, X-RAY, AND PHOTORESISTS

INDEXES

Preface

POLYMERIC MATERIALS ARE WIDELY USED in the manufacture of electronic devices and systems. Because their structures can be tailored to provide a wide range of physical properties, polymers are the material of choice for many applications. Polymers with appropriate structures exhibit excellent dielectric constants and others high radiation sensitivity, two properties that are useful in the manufacture of electronic systems.

Radiation-sensitive polymers are used as the imaging materials in the lithographic process—the linchpin technology used to fabricate integrated circuits. In 1976, state-of-the-art electronic devices contained several thousand transistors with minimum features of 5–6 μm. Today, state-of-the-art electronic devices contain several million transistors with minimum features of less than 0.5 μm. Within 10 years, a new form of lithography will be required that routinely produces features of less than 0.25 μm. Short-wavelength (deep-UV) photolithography and scanning electron-beam, X-ray, scanning ion-beam, and projection electron beam lithography are the possible alternatives to conventional photolithography. However, each technology requires new resists and processes. When deep-UV photolithography is implemented, it will represent the first widespread use in manufacture of a lithographic technology that requires an entirely new resist technology. Because of the low brightness sources of the advanced lithographic technologies, polymers with exceptionally high radiation sensitivity are required. This sensitivity can be achieved with a chemical mechanism referred to as "chemical amplification."

Polymer dielectric materials have been in widespread use for several decades as insulators in printed wiring boards and as encapsulates in packaging. Currently, polymers are being used as the interlevel insulation materials in solid-state devices. Polyimides and fluorinated polymers have exceptional insulation capability and the required thermal and environmental stability to survive both the rigors of the manufacturing process and the thermal cycling of long-term use. This book presents a substantial body of new information on advanced polymer systems for these applications.

Acknowledgments

We thank Jennifer Gunn for her help in producing this book. We also thank the authors for choosing this venue for publishing their work and the ACS Division of Polymeric Materials: Science and Engineering, Inc., and the Society of Polymer Science, Japan, for sponsoring the meeting. We are pleased and proud to provide this volume to those interested in the current status of the rapidly expanding field of polymers for microelectronics.

LARRY F. THOMPSON
AT&T Bell Laboratories
600 Mountain Avenue
Murray Hill, NJ 07974

C. GRANT WILLSON
Current address:
Department of Chemical Engineering
CPE 3.474
The University of Texas at Austin
Austin, TX 78712–1062

SEIICHI TAGAWA
Current address:
Institute of Scientific and Industrial Research
Osaka University
8–1 Mihogaoka, Ibaraki
Osaka 567, Japan

August 12, 1993

CHEMICALLY AMPLIFIED RESISTS

Chapter 1

Chemical Amplification Mechanisms for Microlithography

E. Reichmanis, F. M. Houlihan, O. Nalamasu, and T. X. Neenan

AT&T Bell Laboratories, Murray Hill, NJ 07974

Continued advances in microelectronic device fabrication are trying the limits of conventional lithographic techniques. In particular, conventional photoresist materials are not appropriate for use with the new technologies that will be necessary for sub-0.5 μm lithography. One approach to the design of new resist chemistries involves the concept of chemical amplification, where one photochemical event can lead to a cascade of subsequent reactions that effect a change in solubility of the parent material. Generally, chemically amplified resists utilize photochemically generated acid to catalyze crosslinking or deprotection reactions. This article reviews the chemistries that have been evaluated for chemical amplification resist processes; acid generator, crosslinking, deprotection and depolymerization chemistry.

Significant advances are continually being made in microelectronic device fabrication, and especially in lithography, the technique that is used to generate the high resolution circuit elements characteristic of today's integrated circuits (*1*). These accomplishments have been achieved using "conventional photolithography" as the technology of choice. Incremental improvements in tool design and performance have allowed the continued use of 350–450 nm light to produce ever smaller features (*2*). However, the ultimate resolution of a printing technique is governed, at the extreme, by the wavelength of the light (or radiation) used to form the image, with shorter wavelengths yielding higher resolution (*3*). Additionally, the same basic positive photoresist consisting of a photoactive compound that belongs to the diazonaphthoquinone chemical family and a novolac resin has been in pervasive use since the mid 70's, and will likely be the resist of choice for several more years (*4, 5*).

Unfortunately, conventional photoresists are not appropriate for use with the new lithographic technologies that will be necessary for sub-0.5 μm lithography. The most notable deficiencies of the conventional novolac-quinonediazide resists are the sensitivity and absorption properties of the materials. For most resists, the quantum yield is significantly less than 1.0, and since the new lithographic tools in general have low brightness sources, high sensitivity resists are required. Additionally, the absorption of conventional photoresists is too high to allow uniform imaging through practical resist film thicknesses (~ 1 μm). Thus, no matter which technology becomes dominant after photolithography has reached its limit (0.3–0.5 μm), new resists and processes will be required, necessitating enormous investments in research and process development (*6*). The introduction of new resist materials and processes will also require a considerable lead-time, probably in excess of five years, to bring them to the performance level currently realized by conventional positive photoresists.

RESIST DESIGN REQUIREMENTS

The focus of this paper concerns the design of polymer/organic materials and chemistry that may prove useful in radiation sensitive resist films and relies heavily on a recent review on this subject (*7*). Such chemistry must be carefully designed to meet the specific requirements of each lithographic technology. Although these requirements vary according to the radiation source and device process requirements, the following are ubiquitous: sensitivity, contrast, resolution, etching resistance and purity (*8*). These properties can be achieved by careful manipulation of polymer structure, molecular properties and synthetic methods (*4*).

As mentioned above, sensitivity is a key issue that must be addressed in the development of resist materials. One approach to improving sensitivity involves the concept of chemical amplification (*9, 10*), which employs the photogeneration of an acidic species that catalyzes many subsequent chemical events such as deblocking of a protective group or crosslinking of a matrix resin (Figure 1). The overall quantum efficiency of such reactions is thus effectively much higher than that for initial acid generation. A chemically amplified resist is thus generally composed of three or more elements; i) a matrix polymer, ii) a photoacid generator, and iii) a moiety capable of effecting differential solubility between the exposed and unexposed regions of the film either through a crosslinking reaction or other molecular transformation. These elements may be either discrete molecular entities that are formulated into a multicomponent resist system[9,10] or elements of a single polymer (*11*).

To be effective in a chemically amplified resist formulation, the matrix polymer must i) exhibit solubility in solvents that allow the coating of uniform, defect free, thin films, ii) be sufficiently thermally stable to withstand the temperatures and conditions used with standard device processes, iii) exhibit no flow during pattern transfer of the resist image into the device substrate, iv) possess a reactive functionality that will allow a change in solubility after irradiation and v) have absorption characteristics that will permit uniform imaging through the thickness of a resist film. In general, thermally stable (> 150 °C), high glass transition temperature ($T_g > 90$ °C) materials with low absorption at the wavelength of interest are desired.

The photoacid generator should i) have sufficient radiation sensitivity to ensure adequate acid generation for good resist sensitivity (for photochemical reactions a quantum yield > 0.1 is desirable), ii) be free of metallic elements such as antimony or arsenic that are perceived to be device contaminants, iii) be fully compatible with the matrix resin to eliminate the possibility of phase separation, iv) be stable to at least 175 °C to avoid premature thermal generation of acid, v) be sufficiently acidic to effect the desired post-exposure reaction with high yield, and vi) for photochemical processes, have absorbance characteristics that are commensurate with uniform absorption of light through the thickness of the resist film (12).

If other additives are to be employed to effect the desired reaction, similar criteria apply. Specifically, they must be non-volatile, be stable up to at least 175 °C, possess a reactive functionality that will allow a change in solubility after irradiation, and have low absorbance.

ACID GENERATOR CHEMISTRY

While recent research regarding base catalyzed systems is now known (13, 14), the predominant chemistry associated with chemically amplified resists involves acidolytic reactions. The acid species is required for either crosslinking or deprotection reactions and is also often needed for depolymerization mechanisms. Acid generator chemistry will be discussed separately since any of the available materials might find application in a chemically amplified resist composition.

The dominant ionic photogenerators of acid are a class of materials called onium salts developed by Crivello and coworkers (9, 15–17). Typical examples are the diaryliodonium salts (15), 1, and triarylsulfonium salts, 2 (Figure 2) (18). When these salts are irradiated at wavelengths in the range of 200-300 nm, they undergo irreversible photolysis with rupture of a carbon iodine or carbon sulfur bond. Abstraction of a hydrogen atom from a surrounding "solvent", R-H, results in the formation of a protic acid (Scheme I) (15, 18).

Onium salts have several advantages as photochemical acid generators. They are thermally stable (typically > 150 °C) and may be structurally modified to alter their spectral absorption characteristics. A wide variety of acids may be photochemically generated from these materials, including such strong inorganic acids as hexafluoroarsenic and hexafluoroantimonic acids. Onium salts are also currently the only known source from which the strongest known organic acid, triflic acid, may be photogenerated.

There are many systems described in the literature concerning the photogeneration of acid from non-ionic compounds. Many of these involve the generation of sulfonic acids which are strong organic acids with reasonably low nucleophilicity. Houlihan et al have described photochemical acid generators based upon 2-nitrobenzyl esters (19, 20). These compounds photochemically generate acid through an intramolecular o-nitrobenzyl rearrangement, as shown in Scheme II. Nitrobenzyl esters have certain advantages as photochemical acid generators. The nitro group is a well known inhibitor for radical processes and as a result, secondary reactions occurring due to radical generation are minimized. Moreover, the nitrobenzyl esters described above are thermally stable, with typical stabilities approaching 200 °C. The thermal stability of the esters can be dramatically increased by the introduction of an electron withdrawing, sterically bulky group (Br, CF_3) at the other ortho position of the benzyl moiety.

A positive deep-UV photoresist system has been described by Ueno and coworkers that consists of 1,3,5-tris(methanesulfonyloxy)benzene, bisphenol-A

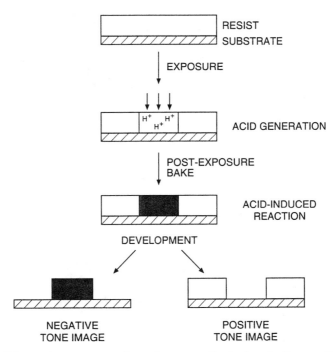

Figure 1. Schematic representation of a generalized chemically amplified resist process.

$$Ar_2I^+X^- \qquad Ar_3S^+X^-$$

$$\mathbf{1} \qquad\qquad \mathbf{2}$$

$$X^- = BF_4^-,\ PF_6^-,\ AsF_6^-,\ SbF_6^-,\ CF_3SO_3^-$$

Figure 2. Representative chemical structures of the diaryliodonium salts (**1**), and triarylsulfonium salts (**2**).

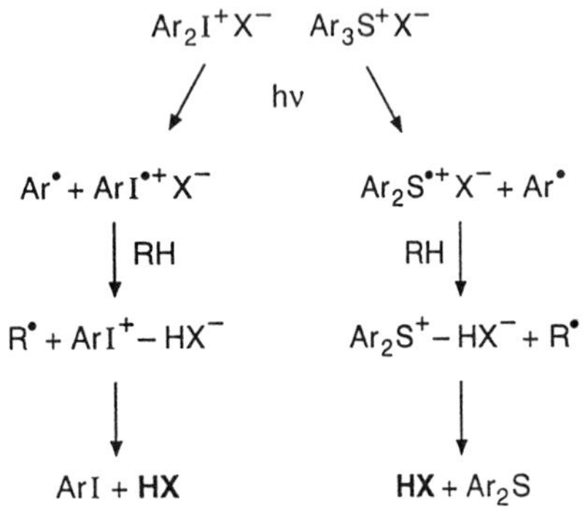

Scheme I

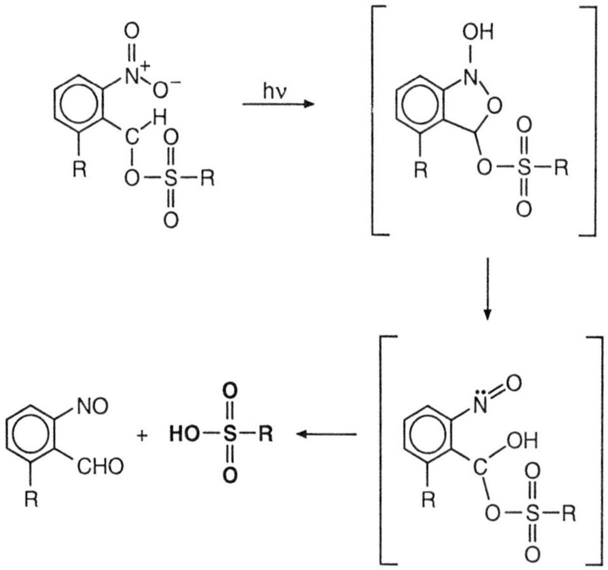

Scheme II

protected with *tert*-butoxycarbonyl groups as a dissolution inhibitor, and a novolac matrix polymer (*21*). This system is reported to generate methanesulfonic acid with a high quantum yield (number of acid moieties generated per photon absorbed). UV spectroscopic studies showed that the novolac resin strongly sensitized the substituted benzene towards acid generation, presumably via a charge-transfer intermediate.

Other chemistries that have been employed as photoacid generators in chemically amplified resist compositions include imino sulfonates (*22*), 4-nitrobenzenesulfonic acid derivatives (*23*), disulphone compounds (*24*), sulfonyl substituted diazomethanes (*25*), dibenzyl sulfones (*26*), and aryl-*bis*-trichloromethyl-s-triazines (*27*).

CROSSLINKING CHEMISTRY

Chemical amplification through acid catalyzed crosslinking for negative working resist applications has been achieved through various mechanisms. These include cationic polymerization, condensation polymerization, electrophilic aromatic substitution and acid catalyzed rearrangement. The acid species has been generated from either ionic materials, such as onium salts, or non-ionic precursors. While the simultaneous formation of radicals and radical cation intermediates from the photolysis or radiolysis of onium salts may be disadvantageous for deprotection and depolymerization chemistries, their simultaneous formation with a strong acid may actually enhance the crosslinking efficiency in select cases, thus improving the overall sensitivity of a negative resist.

Cationic Polymerization Mechanisms

The first chemically amplified resist systems to be developed were those based on the cationic polymerization of epoxy materials (*9*). One example of a resist material based on the above chemistry is a novolac-epoxy resin (*28*) formulated with an onium salt acid generator. Ito and Wilson (*10*) have demonstrated 1 μm resolution with a negative resist system comprised of a commercially available epoxy resin (Celanese Epi-Rez SU-8) based upon bisphenol-A and acid generators. For shorter wavelength exposure, more transparent materials are desirable. This criterion led to the design of styrene-allyl glycidyl ether (SAGE) copolymers for deep-UV lithographic applications (*29*).

Acid-catalyzed cationic polymerization is an attractive method that may be used in the design of resist materials as it not only provides excellent sensitivity owing to the chemical amplification mechanism, but also is insensitive to oxygen and trace amounts of water in the resist films. While these properties aid in the design of a negative resist that is not subject to environmental conditions and that possesses good process latitude, the mechanism generally may not be ideal for sub-micron lithography as features smaller than 1 μm are often subject to distortion due to solvent swelling of the irradiated regions.

Condensation Polymerization Mechanisms

Condensation polymerization mechanisms are probably the most prevalent in the design of chemically amplified negative resists and are the basis for the commercially available negative acting chemically amplified materials. Such resist systems generally consist of three essential components: i) a polymer resin with reactive site(s) (also called a binder) for crosslinking reactions (e.g. hydroxy functional polymer); ii) a radiation sensitive acid generator; and iii) an acid-activated crosslinking agent (*30-33*). Figure 3 depicts some of the alternative structures for the above components. The photogenerated acid catalyzes the reaction between the resin and crosslinking agent to afford a highly crosslinked polymer network that is significantly less soluble than the unreacted polymer resin. A post-exposure bake step prior to development is required to complete the condensation reaction as well as to amplify the crosslinking yield to enhance sensitivity and improve image contrast. The rate determining step for crosslinking in a system based upon poly(hydroxystyrene), a substituted melamine, and a photoacid generator is the formation of a carbocation from the protonated ester moiety (Scheme III) (*34*). Not surprisingly, the crosslinking efficiency (hence sensitivity and contrast) and resolution of these resists are a very strong function of post exposure bake parameters.

Sub-half-micron features could be resolved with deep-UV (*35*) and electron-beam (*36*) radiation with wide process latitude and high sensitivity using this chemistry. An example of the resolution capability is shown in Figure 4. Very sensitive X-ray and e-beam resist formulations based on similar chemistry using melamine and benzyl alcohol derivatives as crosslinking agents, formulated with onium salt photoacid generators in novolac or poly(hydroxystyrene) binders, have shown 0.2 μm resolution (*30, 31*).

An interesting system consisting of a novolac matrix resin, an onium salt photoacid generator and silanol compounds that act as dissolution promoters for novolac resins in aqueous base was described recently (*37, 38*). Compounds such as diphenylsilanediol (DPS) are readily soluble in aqueous base and may in fact increase novolac solubility in aqueous media by as much as a factor of 5. Upon exposure to light followed by post-exposure bake, acid catalyzed condensation of the silanol additive results in formation of a polysiloxane. While silanols are dissolution promoters, polysiloxanes are hydrophobic, aqueous-base insoluble resins that may act as dissolution inhibitors. Sufficient differential solubility is achieved between the exposed and unexposed areas of a resist film resulting in negative tone images. Polysiloxanes are alternate dissolution inhibitors that may be used in these processes (*39*).

Electrophilic Aromatic Substitution Mechanisms

Crosslinking via electrophilic aromatic substitution (*40, 41*) encompasses the concepts discussed in deprotection chemistry (vide infra) to afford a crosslinked network for negative resist applications. Photo-induced crosslinking was achieved in styrene polymers that are susceptible to electrophilic aromatic substitution by addition of an electrophile, in this case, a carbocation precursor and a photoacid generator. The photogenerated acid reacts with the latent electrophile during a post-exposure bake step to generate a reactive carbocation that reacts with an aromatic moiety in the matrix to result in a crosslinked network. The latent electrophile may be either an additive or a monomer that is copolymerized into

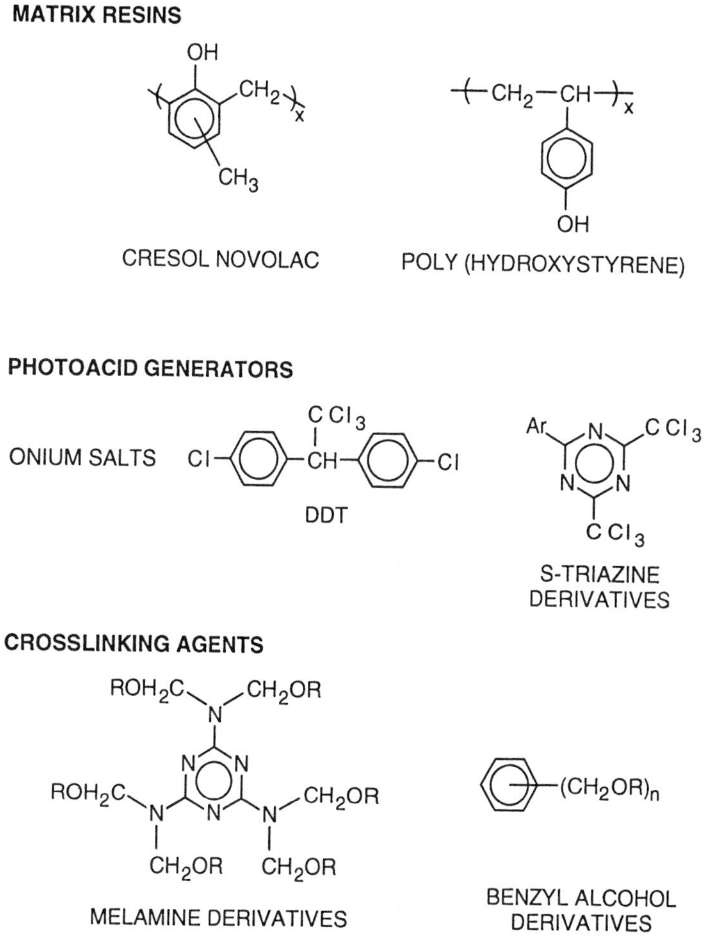

MATRIX RESINS

CRESOL NOVOLAC   POLY (HYDROXYSTYRENE)

PHOTOACID GENERATORS

ONIUM SALTS

DDT

S-TRIAZINE
DERIVATIVES

CROSSLINKING AGENTS

MELAMINE DERIVATIVES

BENZYL ALCOHOL
DERIVATIVES

Figure 3. Representative structures of photoacid generators, crosslinking agents and matrix resins used in acid catalyzed condensation polymerization resist mechanisms.

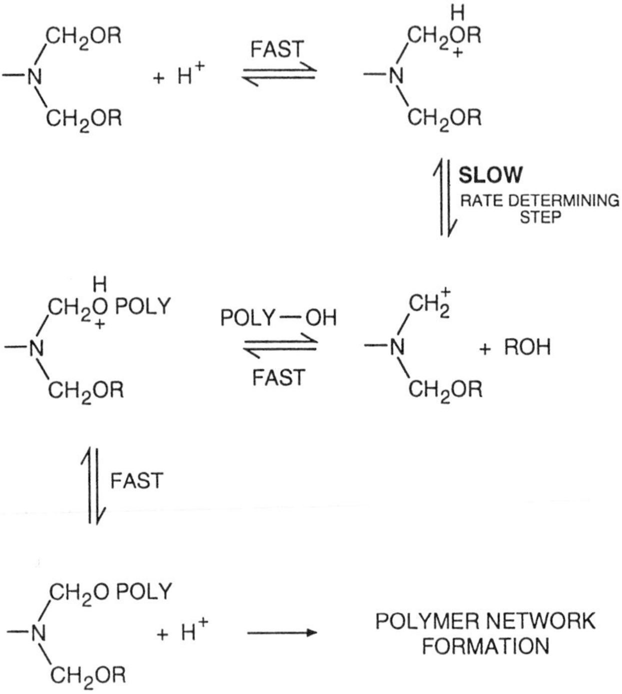

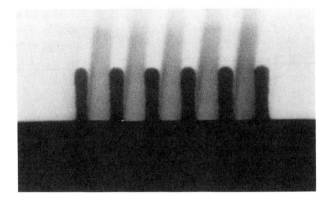

Scheme III

0.3 μm L/S

Figure 4. SEM micrograph depicting nominal 0.3 μm line/space images obtained in the negative, deep-UV resist, XP8843 available from the Shipley Co.

the polymer binder (Figure 5). In the former approach, dibenzyl acetate was added as a latent electrophile along with triphenylsulfonium hexafluoroantimonate to novolac or poly(vinylphenol) binders. Alternately, acetyloxymethylstyrene serves the same function when incorporated into poly(hydroxystyrene) as a comonomer.

DEPROTECTION CHEMISTRY

The pioneering work relating to the development of chemically amplified resists based on deprotection mechanisms was carried out by Ito et al. (*10*). These initial studies dealt with the catalytic deprotection of poly(*tert*-butoxycarbonyloxystyrene) (TBS) in which the thermally stable, acid-labile *tert*-butoxycarbonyl group is used to mask the hydroxyl functionality of poly(vinylphenol) (*42–44*). As shown in Figure 6, irradiation of TBS films containing small amounts of an onium salt, such as diphenyliodonium hexafluoroarsenate with UV light, liberates an acid species that upon subsequent baking catalyzes cleavage of the protecting group to generate poly(*p*-hydroxystyrene). While this reaction will take place at room temperature, it is much faster at 100 °C, requiring only a few seconds of baking. In the absence of an acidic species, the protected polymer undergoes no degradation during prolonged heating at 100 °C, and temperatures in excess of 200 °C are required to thermally initiate the deprotection reaction. Loss of the *tert*-butoxycarbonyl group results in a large polarity change in the exposed areas of the film. Whereas the substituted phenol polymer is a nonpolar material soluble in nonpolar lipophilic solvents, poly(vinylphenol) is soluble in polar organic solvents and aqueous base. This change in polarity allows formation of either positive or negative images, depending upon the developer. Nonpolar solvents such as a mixture of dichloromethane and hexane remove the unirradiated regions, generating a negative image, while an aqueous base developer such as tetramethylammonium hydroxide selectively removes the irradiated regions. These resists are sensitive to deep-UV and electron-beam irradiation and may be sensitized to longer wavelengths through the addition of appropriate mid- and near-UV dyes. TBS-onium salt resists have also been successfully used in the manufacture of integrated circuit devices owing to their high sensitivity and contrast (*45*). The high contrast behavior emanates from the exceptionally non-linear dissolution response as a function of radiation dose. For the TBS-onium salt resist described above in addition to other analogs (vide infra), the desired dissolution behavior is accomplished by the conversion of the hydrophobic t-butoxycarbonyl group to a hydrophilic hydroxyl group. This conversion, coupled with the fact that solubility in aqueous base is only achieved when > 95% of the protecting groups have been removed, results in a very high contrast resist.

Alternate protective groups and parent polymers have been utilized in the design of chemically amplified resists. Generally, thermally stable, acid labile substituents are desirable as protective groups for aqueous base soluble parent polymers. Some typical examples that have been employed include *tert*-butyl (*46, 47*), tetrahydropyranyl (*48–52*), and α,α-dimethylbenzyl (*49, 53, 54*). In situations where adequate moisture is expected to be present in the film, hydrolyzable groups such as trimethylsilyl have also been utilized (*55*). As mentioned above,

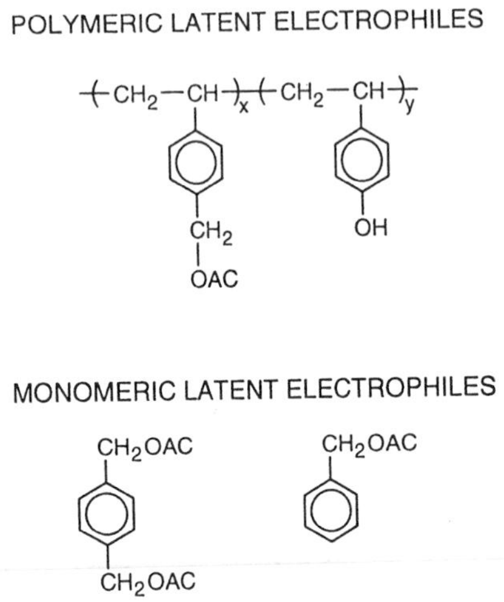

POLYMERIC LATENT ELECTROPHILES

MONOMERIC LATENT ELECTROPHILES

Figure 5. Schematic representation of latent electrophiles that may be used in the development of cross-linking chemically amplified resists.

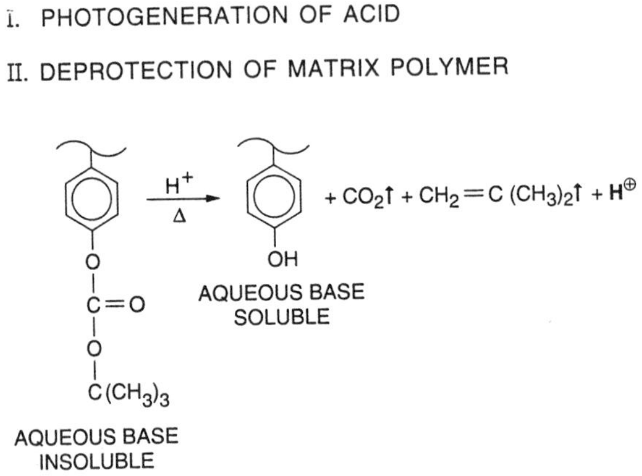

I. PHOTOGENERATION OF ACID

II. DEPROTECTION OF MATRIX POLYMER

Figure 6. Schematic representation of TBS based resist systems.

the parent polymer is typically an aqueous base soluble resin. Examples include
poly(hydroxystyrene) (*10*), poly(vinyl benzoic acid) (*56*) and poly(methacrylic
acid) (*53*).

The need for resist materials that do not undergo image deformation due to
flow during high temperature processing has stimulated efforts to design high
glass transition temperature (T_g) polymers. This interest has also been extended
to the development of high T_g chemically amplified materials. One of the first
such materials was an N-blocked maleimide/styrene resin (*57*). Alternatively, a
t-butoxycarbonyl group was used to block the hydroxyl functionality of N-(*p*-hy-
droxyphenyl) maleimide polymers and copolymers (*58*). The high T_g of these
materials generated images that were resistant to flow when heated at 200 °C for
up to 1 hour.

A matrix polymer used in a resist formulation currently undergoing development for use in manufacturing is poly(4-t-butoxycarbonyloxystyrene-sulfone)
(TBSS) (*59*). As in the case of TBS, the t-butoxycarbonyl moiety is used as the
acid labile protective group. The inclusion of sulfur dioxide into the backbone of
the polymer affords a high T_g that gives greater flexibility for processing.
Additionally, introduction of sulfur dioxide into similar polymers has effectively
improved their sensitivity to e-beam radiation due to C-S bond scission (*60*).
Negligible difference in sensitivity between TBS and TBSS was observed when
these polymers were used in conjunction with onium salt photoacid generator
materials. However, the resist exposure dose was reduced by as much as a factor
of 2.5 when a nitrobenzyl ester acid photogenerator was employed (*61*). In fact,
when exposed to x-ray irradiation, TBSS is an effective single component
chemically amplified resist (*11*). Presumably, radiation induced C-S bond scission
leads to generation of either sulfinic or sulfonic acid end groups that subsequently induce the deprotection reaction (Scheme IV). Figure 7 depicts images
that are readily obtained in TBSS formulated with a nitrobenzyl ester acid
generator upon 50 mJ/cm^2 exposure to 248 nm irradiation.

The concept of acid catalyzed deprotection may also be applied to resist
formulations utilizing a small molecule acting as a dissolution inhibitor for an
aqueous alkali soluble resin. When physically incorporated into an otherwise
soluble resin, an appropriately designed, hydrophobic species can effectively
limit the solubility of the matrix in aqueous alkali. After imaging, this inhibitor is
then converted to a hydrophilic substance, allowing selective dissolution of the
imaged regions. Materials that may effectively be used in such processes include
carbonates or ethers of phenols (*21, 48, 62, 63*), esters of carboxylic acids (*64,
65*), acetals (*66*) or orthocarboxylic acid esters (*66*). In one example, the t-butyl
ester of cholic acid (Figure 8) is used as a dissolution inhibitor for a phenol-for-
maldehyde matrix resin (*64*). When formulated with an onium salt, irradiation
generates a strong acid, which upon mild heating, liberates cholic acid. The
irradiated regions may then be removed by dissolution in aqueous base. One
motivating factor leading to the use of the cholate ester is that it is a large
molecule organic ester that undergoes a significant change in aqueous base
solubility (the solubility of sodium cholate in water is ~ 500 g/l while the esters
are insoluble) (*67*). This in turn can lead to a more effective dissolution
inhibition material. The dissolution inhibitor may also be combined with the acid

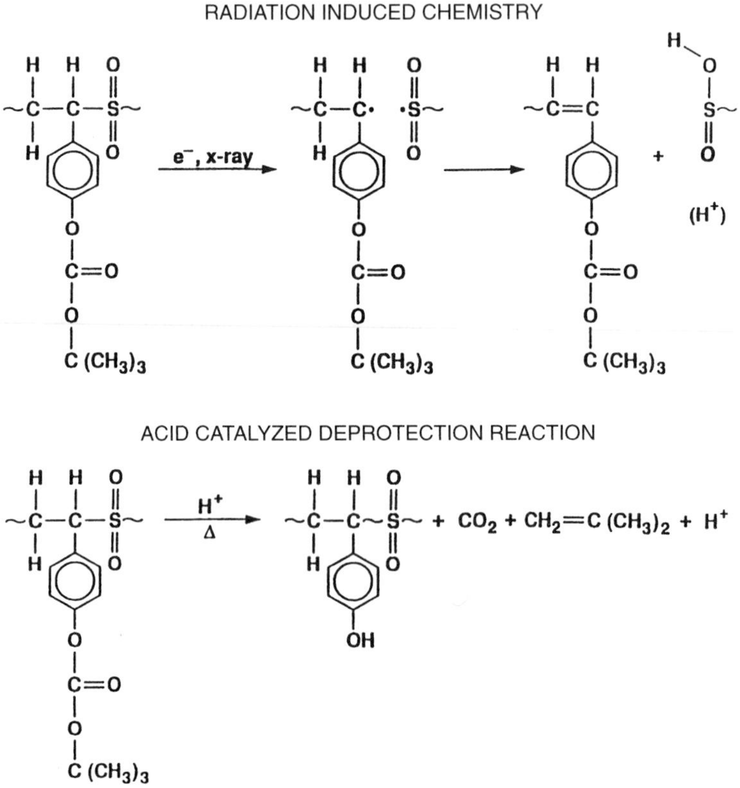
RADIATION INDUCED CHEMISTRY
e⁻, x-ray
(H⁺)
C (CH3)3
ACID CATALYZED DEPROTECTION REACTION
H⁺
Δ
+ CO2 + CH2=C (CH3)2 + H⁺
OH

Scheme IV

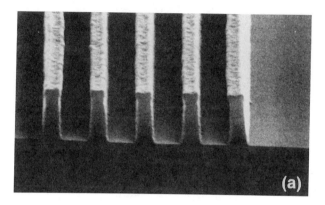

0.275 μm L/S

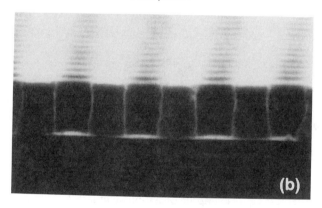

0.3 μm CONTACT HOLES

Figure 7. SEM micrograph depicting nominal 0.275 and 0.3 μm images obtained in TBSS formulated with a nitrobenzylester PAG upon 248 nm irradiation.

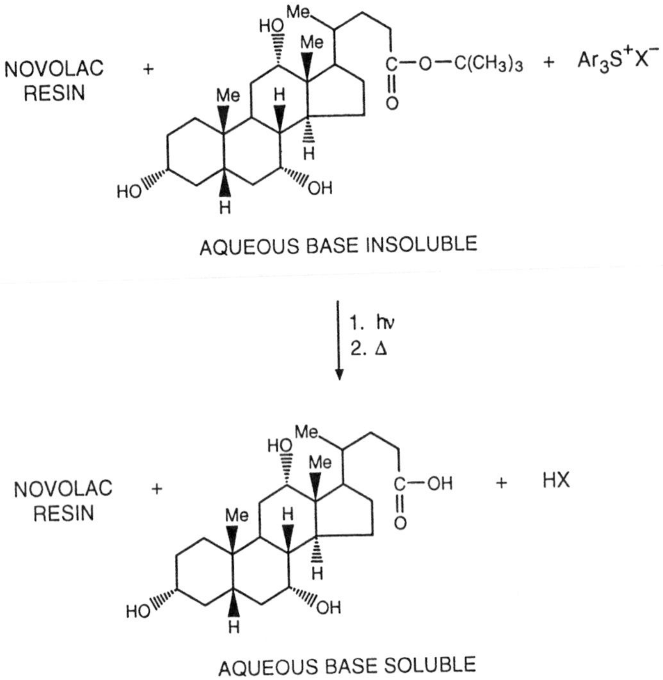

Figure 8. Structural representation of a t-butylcholate based chemically amplified resist.

generator functions into a single chemically amplified resist additive. Schwalm, et al developed a series of sulfonium salts which contained acid sensitive solublizing groups (Scheme V) (*68*). Upon irradiation, the onium salt generates acid which removes the protective groups from both the parent onium salt and irradiation products and renders them base soluble. Houlihan, et al (*69*) have devised a similar approach using 2-nitrobenzyl ester photoacid generating chemistry.

DEPOLYMERIZATION CHEMISTRY

Chemically amplified resists that act through a polymer depolymerization mechanism can be broadly divided into two classes: those that act through a thermodynamically induced depolymerization mechanism and those requiring catalytic cleavage of a polymer backbone. The former process depends upon the use of low ceiling temperature polymers that have been stabilized by suitable end capping. Introduction of a photocleavable moiety either at the end-cap or along the polymer backbone may then allow depolymerization to take place after irradiation and mild heating. A variant of this approach utilizes an end cap, or polymer chain that may be cleaved by photogenerated acid.

The first chemically amplified resists that operated by a thermodynamically induced depolymerization were based on polyacetals (or polyaldehydes) (*10, 44, 70*). These polymers have very low ceiling temperatures, but can be stabilized by end capping (*71*). Although PPA materials are sensitive self-developing resists, they have two major drawbacks: i) they liberate a volatile aldehyde during the exposure step which could be injurious to the optics of the exposure tool, and ii) they exhibit poor plasma etching resistance. Recently, it was found that contamination of the optical tool could be reduced by using poly(4-chlorophthalaldehyde) as the matrix resin in conjunction with an onium salt acid generator (*72, 73*). This resist system does not spontaneously depolymerize upon exposure to radiation, but requires a post-exposure bake step. The problem of plasma etching resistance has been addressed through the use of silylated PPA derivatives such as poly(4-trimethylsilylphthalaldehyde) (*72, 74*) which can form an SiO_2 barrier layer. Alternatively, PPA may be used as a dissolution inhibitor for a novolac which is itself more resistant to plasma etching (*75*).

The second type of chemically amplified depolymerization resist mechanism depends upon the incorporation of C-O bonds into the polymer backbone which can be cleaved by either hydrolysis or acidolysis. This concept was first advanced by Crivello, who proposed that polymers such as polycarbonates and polyesters could undergo photo-induced acid catalyzed hydrolysis reaction in polymeric film (*9, 76*). Although polymers could be designed to undergo catalytic chain cleavage in the presence of acid, such an approach depends upon the inclusion of stoichiometic amounts of water in the polymer film. Little further work was reported on this concept until recently, when a new system for dissolution inhibition was described based upon the hydrolysis of polysilyl ethers in a novolac resin (*24*).

Much of the work related to the design of chemically amplified resists that depolymerize upon catalytic cleavage of the polymer backbone has been done by Frechet et al. (*77–81*). In these systems, the polymer film depolymerizes through

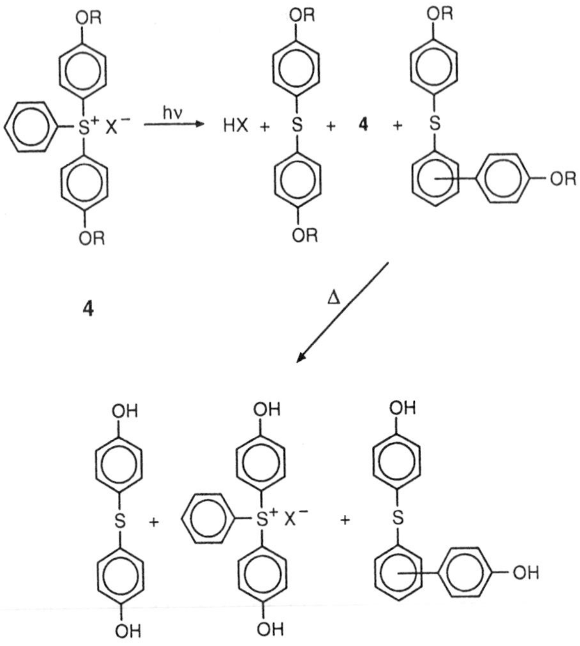

AQUEOUS ALKALI SOLUBLE PRODUCTS

R = ACID LABILE SUBSTITUENT

Scheme V

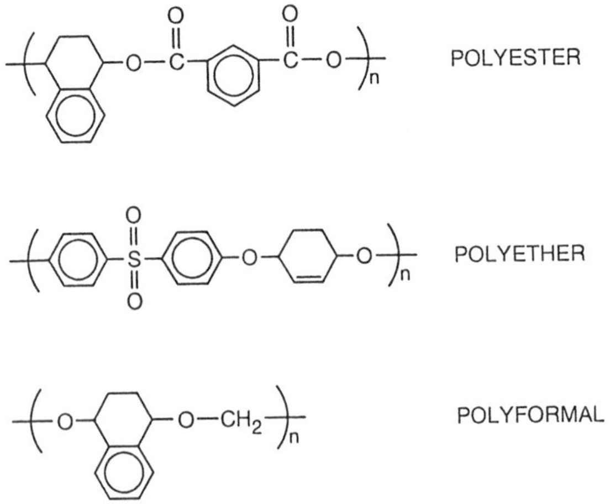

POLYESTER

POLYETHER

POLYFORMAL

Figure 9. Structural representation of a polyester, polyether, and polyformal employed in a chemically amplified resist.

thermally induced acid catalyzed cleavage of tertiary, secondary allylic, or secondary benzylic C-O bonds, to form a stable carbocation with at least one alpha hydrogen. This carbocation can then eliminate to form an alkene with concomitant regeneration of the acid catalyst. The first examples of such a resist were based upon polycarbonates which incorporated tertiary carbonate units along the polymer backbone (77). Typically these materials have a thermal stability of ~ 200 °C at which point they undergo decomposition to diols, carbon dioxide and dienes (77, 78). This decomposition may be accelerated by mild heating in the presence of photogenerated acid (78). Other similar carbonate resist systems have been described in which secondary benzylic and secondary allylic units are incorporated along the polymer chain (78). Of particular interest are the materials based upon 2-cyclohexene-1,4-diol, and a dihydroxy compound.

Polyesters (Figure 9) with tertiary, secondary allylic, and secondary benzylic units and polyethers (Figure 9) containing the latter two groups were also prepared and evaluated as resist materials when used with onium salt PAG materials (80). Selected resist systems formulated from these polymers and triphenylsulfonium hexafluoroantimonate undergo acidolytic cleavage to afford a neutral olefin or aromatic compound plus a diacid or diol. Polyformals containing secondary allylic and secondary benzylic C-O units capable of acidolytic cleavage have also been reported as effective polymers for chemically amplified resist applications (81). Polyformals undergo acidolysis to afford an aromatic compound, formaldehyde, and water. Because all these compounds are volatile, the resists systems made from these polymers and triphenylsulfonium hexafluoroantimonate are completely self-developing and have been reported to give lithographic sensitivities as high as 15 mJ/cm^2. Features obtained with these resist systems, however, tend to be rounded due to the low T_g of the polymer and possible plasticization by the released aromatic compounds before volatilization. Recently, other workers have reported a similar type of resist system based upon polyethers containing alkoxypyrimidine units (82). Resists made from these materials and triphenylsulfonium triflate exhibited sensitivities of 10 mJ/cm^2 upon irradiation at 250 nm. The products arising from acidolytic cleavage are pyrimidone and compounds containing both alkene and alcohol functionalities.

PROCESS CONSIDERATIONS

The inherent sensitivity associated with most chemically amplified resists emanates from the catalytic action of the acid during the post-exposure baking step. Typical turnover rates (catalytic chain length) for each acid molecule in a working resist formulation are in the 800-1200 range (83). The deprotection or crosslinking, and turnover rate are critically dependent on post-exposure bake temperature, time, and the method of bake itself.

A typical chemically amplified resist process involves conversion of the PAG molecule to a strong acid upon absorption of a photon and the rate of this reaction is fast, with the extent of reaction being governed by the quantum efficiency of the particular acid generator and exposure energy. The acid effects the desired reaction with a characteristic rate, which is a function of the acid concentration, the temperature and the diffusion rate of the acid in the polymer matrix (84–86). The diffusion rate in turn, depends on the acid structure, the temperature and the polarity of the polymer matrix. At room temperature, the rate of reaction is typically slow and it is generally necessary to heat the film to well above room temperature to increase the rate to acceptable levels. The acid (H^+) is regenerated and continues to be available for subsequent reaction, hence, the *amplification* nature of these resists. For this system to work satisfactorily, it would be necessary for the radiation generated acid concentration, (H^+), to remain constant. However, in most chemically amplified systems, undesired side reactions prematurely destroy the acid, i.e., reactions with contaminants such as water, ions or reactive sites on the polymer (*87–90*). The most notable contaminants are airborne amines that serve to prematurely neutralize the acid leading to a dependence on the post-exposure delay time (the time delay between the exposure and PEB process steps) (*88–90*). An understanding of these issues and the general structure property relationships of various PAG molecules and host polymers may lead to the design of a sensitive chemically amplified resist system that is insensitive to environmental issues.

CONCLUSION

The aim of this review has been to examine one approach to the design of new resist chemistries for microlithographic applications. Chemical amplification processes utilize a single photochemical event that leads to a cascade of subsequent reactions effecting a change in solubility of the parent material. The vast majority of materials that have been designed to date utilize a photochemically generated acid to catalyze subsequent crosslinking, deprotection or depolymerization reactions. Significant efforts have been expended in the design of acid generator chemistries compatible with device processing environments. The most notable examples include the onium salts and nitrobenzyl esters. Additionally, the matrix resin must exhibit suitable characteristics with the most common materials being derived from hydroxyphenols.

Since the conception of chemical amplification mechanisms for microlithographic applications approximately one decade ago, increasing attention has been given to such processes in that they provide advantages in terms of sensitivity and contrast with minimal increase in process complexity. Additionally, a given chemistry may find application to more than one lithographic technology. The original work in chemically amplified resists has spawned many research efforts to define chemistries appropriate for matrix materials and photogenerators of catalysts, primarily strong acids. There continue to be many challenges in the areas of both fundamental and applied materials chemistry as well as process engineering to both understand and develop new chemically amplified resists for use with the future lithographic technologies.

REFERENCES

1. "Electronic and Photonic Applications of Polymers", *ACS Advances in Chemistry Series* **218**, Bowden, M. J., Turner, S. R., Eds., ACS, Washington, D.C., 1988.
2. McCoy, J. H., Lee, W., Varnell, G. L., *Solid State Technology*, 1989, **32**(3), 87.

3. Thompson, L. F., Willson, C. G., Bowden, M. J., "Introduction to Microlithography", *ACS Symposium Series* **219**, ACS, Washington, D.C., 1983.
4. Reichmanis, E., Thompson, L. F., *Chemical Reviews*, 1989, **89**, 1273.
5. Willson, C. G., In "Introduction to Microlithography", *ACS Symposium Series* **219**, Thompson, L. F., Willson, C. G., Bowden, M. J., Eds., ACS, Washington, D.C., 1983, pp. 88-159.
6. Reichmanis, E., Thompson, L. F., In "Polymers in Microlithography: Materials and Processes", *ACS Symposium Series* **412**, Reichmanis, E., MacDonald, S. A., Iwayanagi, T., ACS, Washington, D.C., 1989, pp. 1-24.
7. Reichmanis, E., Houlihan, F. M., Nalamasu, O., Neenan, T. X., *Chem. of Mat.*, 1991, **3**, 394.
8. Thompson, L. F., Bowden, M. J., In "Introduction to Microlithography", *ACS Symposium Series* **219**, ACS, Washington, D.C., 1983, pp. 162-214.
9. Crivello, J. V., In "Polymers in Electronics", *ACS Symposium Series* **242**, Davidson, T., Ed., ACS, Washington, D.C., 1984, pp. 3-10.
10. Ito, H., Willson, C. G., In ref. 9, pp. 11-23.
11. Novembre, A. E., Tai, W. W., Kometani, J. M., Hanson, J. E., Nalamasu, O., Taylor, G. N., Reichmanis, E., Thompson, L. F., *Chem. of Mat.* 1992, **4**, 278.
12. Houlihan, F. M., Neenan, T. X., Reichmanis, E., Kometani, J. M., Thompson, L. F., Chin, T., Nalamasu, O., *J. Photopolymer Sci., and Technol.* 1990, **3**, 259.
13. Frechet, J. M. J., Cameron, J. F., *J. Amer. Chem. Soc.* 1991, **113**, 4252.
14. Winkle, M. R., Graziano, K. A., *J. Photopolym. Sci. Technol.*, 1990, **3(3)**, 419-422.
15. Crivello, J. V., Lee. J. L., Conlon, D. A., *Makromol, Chem.*, Makromol. Symp., 1988, **13/14**, 145.
16. Crivello, J. V., Lam, J. H. W., *Macromolecules*, 1977, **10**, 1307.
17. Crivello, J. V., Lam, J. H. W., *J. Polym. Sci., Polym. Chem. Ed.*, 1979, **17**, 977.
18. Dektar, J. L., Hacker, N. P., *J. Amer. Chem. Soc.*, 1990, **112**, 6004.
19. Houlihan, F. M., Shugard, A., Gooden, R., Reichmanis, E., *Macromolecules*, 1988, **21**, 2001.
20. Neenan, T. X., Houlihan, F. M., Reichmanis, E., Kometani, J. M., Bachman, B. J., Thompson, L. F., *Macromolecules*, 1990, **23**, 145.
21. Schlegel, L., Ueno, T., Shiraishi, H., Hayashi, N., Iwayanagi, T., *Chem. Mater.*, 1990, **2**, 299.
22. Shirai, M., Katsuta, N., Tsunooka, M., Tanaka, M., *Makromol. Chem.*, 1989, **190**, 2099.
23. Yamaoka, T., Omote, T., Adachi, H., Kikuchi, N., Watanabe, Y., Shirosaki, T., *J. Photopolym. Sci. Technol.*, 1990, **3**, 275.
24. Aoai, T., Aotani, Y., Umehara, A., Kokubo, T., *J. Photopolym. Sci. Technol.*, 1990, **3**, 389.
25. Pawlowski, G., Dammel, R., Lindley, C. R., Merrem, H-J, Röschert, H., Lingnau, J., *Proc. SPIE Adv. Resist Technol. and Proc. VII*, 1990, **1262**, 16.
26. Novembre, A. E., Hanson, J. E., Kometani, J. M., Tai, W. W., Reichmanis, E., Thompson, L. F., West, R. J., *Proc. Reg. Tech. Conf. on Photopolymers*, October 28-30, 1991, Ellenville, NY, 41-50.

27. Buhr, G., Dammel, R., Lindley, C. R., *Proc. Polym. Mat. Sci. Eng.*, 1989, **61**, 269.
28. Schlessinger, S. I., *Polym. Eng. Sci.*, 1974, **14**, 513.
29. Stewart, K. J., Hatzakis, M., Shaw, J. M., Seeger, D. E., Neumann, E., *J. Vac. Sci. Technol.*, 1989, **B7**, 1734.
30. Lingnau, J., Dammel, R., and Theis, J., *Solid State Technology*, 1989, **32(9)**, 105-112.
31. *ibid*, 1989, **32(10)**, 107-112.
32. Feely, W. E., *Eur. Patent Appl.*, 232, 972, 1980
33. Buhr, G., *U.S. Patent*, 4,189,323, 1980.
34. Berry, A. K., Graziano, K. A., Bogan, L. E. Jr., Thackeray, J. W., In "Polymers in Microlithography", *ACS Symposium Series*, 412, Reichmanis, E., MacDonald, S. A., Iwayanagi, T., Eds., ACS, Washington, D.C., 1989, pp. 87-99.
35. Thackeray, J. W., Orsula, G. W., Pavelchek, E. K., Canistro, D., *Proc. SPIE Advances in Resist Technology and Processing VI*, 1989, **1086**, 34.
36. Liu, H-Y, deGrandpre, M. P., Feely, W. E., *J. Vac. Sci. Technol.*, 1988, **B6**, 379.
37. Toriumi, M., Shiraishi, H., Ueno, T., Hayashi, N., Nonogaki, S., *J. Electrochem. Soc.*, 1987, **134**, 334.
38. Shiraishi, H., Fukuma, E., Hayashi, N., Tadano, K., Ueno, T., *Chem. Mater.*, 1991, **3**, 621.
39. Sakata, M., Ito, T., Yamashita, Y., *Japn. J. Appl. Phys.*, 1991, **30(11B)**, 3116.
40. Reck, B., Allen, R. D., Twieg, R. J., Willson, C. G., Matsuszczak, S., Stover, H. D. H., Li, N. H., Frechet, J. M. J., *Polym. Eng. Sci.*, 1989, **29**, 960.
41. Frechet, J. M. J., Matsuszczak, S., Stover, H. D. H., Willson, C. G., Reck, B., In "Polymers in Microlithography", *ACS Symposium Series* 412, Reichmanis, E., MacDonald, S. A., Iwayanagi, T., Eds., ACS, Washington, D.C., 1989, pp. 74-85.
42. Frechet, J. M. J., Eichler, E., Ito., H., Willson, C. G., *Polymer*, 1980, **24**, 995.
43. Ito, H., Willson C. G., Frechet, J. M. J., Farrall, M. J., Eichler, E., *Macromolecules*, 1983, **16**, 510.
44. Ito, H., Willson, C. G., *Polym. Eng. Sci.*, 1983, **23**, 1012.
45. Maltabes, J. G., et al., *Proc. SPIE Advances in Resist Technology and Processing VII*, 1990, **1262**, 2.
46. Conlon, D. A., Crivello, J. V., Lee, J. L., O'Brien, M. J., *Macromolecules*, 1989, **22**, 509.
47. Crivello, J. V., *J. Electrochem. Soc.*, 1989, **136**, 1453.
48. Hayashi, N., Hesp, S. M. A., Ueno, T., Toriumi, M., Iwayanagi, T., Nonogaki, S., *Proc. Polym. Mat. Sci. Eng.*, 1989, **61**, 417.
49. Frechet, J. M. J., Kallman, N., Kryczka, B., Eichler, E., Houlihan, F. M., Willson, C. G., *Polymer Bulletin*, 1988, **20**, 427.
50. Frechet, J. M. J., Eichler, W., Gauthier, S., Kryczka, B., Willson, C. G., In "The Effects of Radiation on High-Technology Polymers", *ACS Symposium Series* 381, Reichmanis, E., O'Donnell, J. H., Eds., ACS, Washington, D.C., 1989, pp. 155-171.
51. Hesp, S. A. M., Hayashi, N., Ueno, T., *J. Appl. Poly. Sci.*, 1991, **42**, 877-883.

52. Schlegel, L., Ueno, T., Shiraishi, H., Hayashi, N., Iwayanagi, T., *Microelectronic Eng.*, 1991, **14**, 227-236.
53. Ito, H., Ueda, M., Ebina, M., In "Polymers in Microlithography", *ACS Symposium Series* **412**, Reichmanis, E., MacDonald, S. A., Iwayanagi, T., Eds., ACS, Washington, D.C., 1989, pp. 57.
54. Ito, H., Ueda, M., *Macromolecules*, 1988, **21**, 1475.
55. Yamaoka, T., Nishiki, M., Koseki, K., Koshiba, M., "Proc. Regional Technical Conference on Photopolymers", Mid-Hudson Section SPE, Ellenville, New York, Oct. 30 - Nov. 2, 1988, pp. 27.
56. Ito, H., Willson, C. G., Frechet, J. M. J., *Proc. SPIE*, 1987, **771**, 24.
57. Osuch, C. E., Brahim, K., Hopf, F. R., McFarland, M. J., Mooring, A., Wu, C. J., *Proc. SPIE*, 1986, **631**, 68.
58. Turner, S. R., Ahn, K. D., Willson, C. G., In "Polymers for High Technology", *ACS Symposium Series* **346**, Bowden, M. J., Turner, S. R., Eds., ACS, Washington, D.C., 1987, pp. 200-210.
59. Tarascon, R. G., Reichmanis, E., Houlihan, F. M., Shugard, A., Thompson, L. F., *Polym. Eng. Sci.*, 1989, **29**, 850.
60. Bowden, M. J., Chandross, E. A., *J. Electrochem. Soc.*, 1975, **122**, 1370.
61. Houlihan, F. M., Reichmanis, E., Thompson, L. F., Tarascon, R. G., In "Polymers in Microlithography", *ACS Symposium Series* **412**, Reichmanis, E., MacDonald, S. A., Iwayanagi, T., Eds., ACS, Washington, D.C., 1989, pp.39-56.
62. McKean, D. R., MacDonald, S. A., Clecak, N. J., Willson, C. G., *Proc. SPIE*, 1988, **920**, 60.
63. Schlegel, L., Ueno, T., Shiraishi, H., Hayashi, N., Iwayanagi, T., *J. Photopolym. Sci. Technol.*, 1990, **3**, 281.
64. O'Brien, M. J., *Polym. Eng. Sci.*, 1989, **29**, 846.
65. O'Brien, J. M., Crivello, J. V., *Proc. SPIE*, 1988, **920**, 42.
66. Lingnau, J., Dammel, R., Theirs, J., "Proc. Regional Technical Conference on Photopolymers", Mid-Hudson Section, SPE, Ellenville, New York, Oct. 30 - Nov. 2, 1988, pp. 87-97.
67. Reichmanis, E., Wilkins, C. W., Jr., Chandross, E. A., *J. Vac. Sci. Technol.*, 1981, **19**, 1338.
68. Schwalm, R., *Proc. Polym. Mat. Sci. Eng.*, 1989, **61**, 278.
69. Houlihan, F. M., these proceedings.
70. Ito, H., Willson, C. G., "Proc. Regional Technical Conf. on Photopolymers", Mid-Hudson Section, SPE, Ellenville, New York, 1982, p. 331.
71. Vogl, O., *J. Polym. Sci., Polym. Chem. Ed.*, 1960, **46**, 261.
72. Ito, H., Ueda, M., and Schwalm, R., *J. Vac. Sci. Technol.*, 1988, **B6**, 2259.
73. Ito, H., Schwalm, R., *J. Electrochem. Soc.*, 1989, **136**, 241.
74. Ito, H., Ueda, N., Renaldo, A. F., *J. Electrochem. Soc.*, 1988, **136**, 245.
75. Ito, H., Flores, E., Renaldo, A. F., *J. Electrochem. Soc.*, 1988, **135**, 2328.
76. Crivello, J. V., "Proc. Regional Technical Conference on Photopolymers", Mid-Hudson Section, SPE, Ellenville, New York, 1982, p. 267.
77. Houlihan, F. M., Bouchard, F., Frechet, J. M. J., Willson, C. G., *Macromolecules*, 1986, **19**, 13.
78. Frechet, J. M. J., Eichler, E., Stanciulescu, M., Iizawa, T., Bouchard, F.,

Houlihan, F. M., Willson, C. G., In "Polymers for High Technology", *ACS Symposium Series* **346**, Bowden, M. J., Turner, S. R., Eds., ACS, Washington, D.C., 1987, pp. 138-148.

79. Frechet, J. M. J., Iizawa, T., Bouchard, F., Stanciulescu, M., Willson, C. G., Clecak, N., *Proc. Polym. Mat. Sci. Eng.*, 1986, **55**, 299.
80. Frechet, J. M. J., Stanciulescu, M., Iizawa, T., Willson, C. G., *Proc. Polym. Mat. Sci. Eng.*, 1989, **60**, 170.
81. Frechet, J. M. J., Willson, C. G., Iizawa, T., Nishikubo, T., Igarashi, K., Fahey, J., In "Polymers in Microlithography", *ACS Symposium Series* **412**, Reichmanis, E., Macdonald, S. A., Iwayanagi, T., Eds., ACS, Washington, D.C., 1989, pp. 100-112.
82. Inaki, Y., Horito, H., Matsumura, N., Takemoto, K., *J. Photopolym. Sci. Technol.*, 1990, **3**, 417.
83. McKean, D. R., Schaedeli, U., MacDonald, S. A., In "Polymers in Microlithography", *ACS Symposium Series* **412**, Reichmanis, E., MacDonald, S. A., Iwayanagi, T., Eds., ACS, Washington, D.C., 1989, pp. 27-38.
84. Schlegel, L., Ueno, T., Hayashi, N., Iwayanagi, T., *J. Vac. Sci. Technol. B.*, 1991, **9(2)**, 278.
85. Schlegel, L., Ueno, T., Hayashi, N., Iwayanagi, T., *Japn. J. Appl. Phys.*, 1991, **30(11B)**, 3132.
86. Nakamaura, J., Ban, H., Deguchi, K., Tanaka, A., *Japn. J. Appl. Phys.*, 1991, **30(10)**, 2619.
87. Houlihan, F. M., Reichmanis, E., Thompson, L. F., Tarascon, R. G., In "Polymers in Microlithography", *ACS Symposium Series* **412**, Reichmanis, E., MacDonald, S. A., Iwayanagi, T., Eds., ACS, Washington, D.C., 1989, pp. 39-56.
88. Nalamasu, O., Reichmanis, E., Cheng, M., Pol, V., Kometani, J. M., Houlihan, F. M., Neenan, T. X., Bohrer, M. P., Mixon, D. A., Thompson, L. F., *Proc. SPIE*, 1991, **1466**, 13.
89. MacDonald, S. A., Clecak, N. J., Wendt, H. R., Willson, C. G., Snyder, C. D., Knors, C. J., Deyoe, N. B., Maltabes, J. G., Morrow, J. R., McGuire, H. E., Holmes, S. J., *ibid.*, 2.
90. Nalamasu, O., Reichmanis, E., Hanson, J. E., Kanga, R. S., Heimbrook, L. A., Emerson, A. B., Baiocchi, F. A., Vaidya, S., *Proc. Reg. Tech. Conf.*, on Photopolymers, October 28-30, 1991, Ellenville, NY, 225-234.

RECEIVED January 15, 1993

Chapter 2

Synthesis of 4-(*tert*-Butoxycarbonyl)-2,6-dinitrobenzyl Tosylate

A Potential Generator and Dissolution Inhibitor Solubilizable through Chemical Amplification

F. M. Houlihan, E. Chin, O. Nalamasu, and J. M. Kometani

AT&T Bell Laboratories, Murray Hill, NJ 07974

A simple synthetic pathway was devised for 4-(*t*-butoxycarbonyl)-2,6-dinitro-benzyltosylate **1** starting from inexpensive 3,5-dinitrotoluic acid. This molecule incorporates both the *o*-nitrobenzylsulfonate moiety needed for a photogenerator of acid (PAG) and the *t*-butoxycarbonyl moiety needed for a dissolution inhibitor solubilizable by chemical amplification (DISCA). Preliminary evaluations show that the thermal stability of **1** towards thermal elimination of isobutene is almost identical to that of 2,6-dinitrobenzyl tosylate in a matrix containing *t*-butoxycarbonyl moieties (~160 °C). Resists made from 4% of either PAG **1** or 4-methoxycarbonyl-2,6-dinitrobenzyl tosylate (**2**) and poly(*t*-butoxycarbonyloxystyrene sulfone) gave lithographic performance (74 and 68 mJ/cm^2 respectively) similar to that of a resist formulated from 2,6-dinitrobenzyl tosylate. Since the absorbances of the three resists at 248 nm are similar (~0.36/μm) it can be inferred that the quantum yield for the 4-(alkoxycarbonyl)-2,6-dinitrobenzyl chromophore is similar to the value previously found for 2,6-dinitrobenzyl tosylate (0.16). The acidolytic behavior of PAG **1** was evaluated in a poly(hydroxystyrene) matrix where it was found that complete removal of the *t*-butyl group occurs with 100 mJ/cm^2 after post-exposure bake at 100 °C. These tests indicate that 4-(*t*-butoxycarbonyl)-2,6-dinitrobenzyltosylate has all the essential properties for a potential single component DISCA/PAG additive for use in chemically amplified deep UV resists based upon base soluble polymers such as polyhydroxystyrene derivatives.

Critical dimensions of integrated circuits (IC) have decreased in size to the subhalf-micron regime over the last few years. This has prompted the development of chemically amplified resists systems capable of operating with the limited photon flux available from deep UV exposure tools. These resists work by the amplification of a photochemical event that initiates a catalytic process in which as many as 1000 chemical changes are engendered in the system per photo-event (*1, 2*). One example of such a resist depends on the acidolytic deprotection of a *t*-butylcarbonate (*t*-BOC) (*3–5*) or *t*-butyl ester (*t*-Bu) (*6*) pendent group to produce a free pendent functionality that is soluble in aqueous

0097–6156/94/0537–0025$06.00/0

base (OH or CO_2H, respectively). The acidolysis process is initiated by irradiation of a photoacid generator (PAG) to produce a free protic acid, followed by post-exposure bake   (PEB) of the resist to achieve catalytic deprotection. Initially, PAGs for these systems were restricted to the onium salts developed by Crivello (7). More recently, we have developed a new class of non-ionic/non-metal containing PAG's based upon nitrobenzyl esters (8–10). Another strategy, which has been recently developed, involves resists employing a dissolution inhibitor solubilizable by chemical amplification DISCA. DISCA's are generally small hydrophobic molecules which act as dissolution inhibitors   and contain pendent groups susceptible to acidolysis (11–17). Examples of DISCA's include t-butyl esters of carboxylic acids and t-butyl carbonates of phenols (15). This class of resist typically consists of a base soluble polymer into which a DISCA and a PAG have been added. Exposure of the resist to radiation causes the hydrophobic DISCA to be converted with great quantum efficiency (because of the acidolysis chain process) to base soluble hydrophilic moieties. These wetting molecules engender a much faster rate of dissolution with aqueous basic developer for the exposed areas resulting in positive images. A concept called SUCCESS combines an onium salt PAG and the phenolic t-butyl carbonate based DISCA into one molecule (18). The thrust of this work is to report our initial efforts into synthesizing a molecule which would combine the function of a DISCA with a 2-nitrobenzyl ester based PAG. The target material chosen was 4-(t-butoxycarbonyl)-2,6-dinitrobenzyl tosylate 1. The chromophore was picked because of the convenient attachment point for the t-butyl group, and the starting material, 3,5-dinitrotoluic acid, was inexpensive. Tosylate was chosen as the protected acid because it offered a simple initial target which could easily be compared to other previously characterized 2-nitrobenzyl PAG's (8, 9). These investigations will center upon the elaboration of a synthetic pathway for 1, and an examination of its thermal stability.

Also, it will be shown that the absorbance and lithographic sensitivity engendered in a typical t-BOC based resist by 1 and that of a model compound without the DISCA feature, 4-(methoxycarbonyl)-2,6-dinitrobenzyl tosylate 2, are comparable to that found when using 2,6-dinitrobenzyl tosylate 3.

Furthermore, an IR investigation will demonstrate that the DISCA portion of 1 is undergoing effective acidolytic deprotection. The efforts towards using this material as a DISCA in a base soluble polymer resist matrix are a subject for future investigation and will be reported on at a later date.

EXPERIMENTAL

Materials characterization

Differential scanning calorimetry (DSC) and thermogravimetric analysis (TGA) data were obtained by using a Perkin-Elmer DSC-7 differential scanning calorimeter and a TGA-7 thermogravimetric analyzer interfaced with a TAC 7 thermal analysis controller and a PE-7700 data station. DSC and TGA samples were heated at a rate of 10 °C/min with purified N_2 gas flow of 20 cm^3/min.

Samples ranged in mass from 1.20 to 2.00 mg and were encapsulated in aluminum pans. Nuclear magnetic resonance spectra were obtained on a JEOL JMN-FX90Q Fourier transform spectrometer. IR spectra were obtained on a Digilab FTS-60 Fourier transform spectrometer. Mass spectra were obtained with a Hewlett Packard 5989A Mass spectrometer in the electron impact mode coupled with a 59980B particle beam and a 1050 Hewlett Packard Liquid Chromatography system. Elemental analyses were obtained by Galbraith Laboratories Inc., Knoxville, TN.

Material synthesis

Poly((4-*t*-butyloxycarbonyloxy)styrene sulfone) (PTBSS) was synthesized as previously described (*19*). A research sample of poly(4-hydroxystyrene) (PHS) (M_w = 12,300) was obtained from the Hoechst chemical company. 2,6-Dinitrobenzyl tosylate and 2,6-dinitrobenzyl 4-nitrobenzenesulfonate were made as previously described (*2, 8, 9*). All other chemicals were obtained from the Aldrich Chemical Company.

Synthesis of 4-carboxyl-2,6-dinitrobenzyl bromide. In a 50 mL pressure tube equipped with a Teflon valve were placed 2.23 g (9.76 mmol) of 3,5-dinitrotoluic acid and 0.50 mL, (9.8 mmol) of Br_2. The valve on the tube was closed and the contents frozen in liquid nitrogen. The tube was then evacuated (<1 mm Hg) and resealed. After thawing, the tube was placed in a wire cage, gradually heated to 150 °C in an oil bath over 45 min and maintained at this temperature for 24 h. Workup was done by cooling the tube to room temperature, freezing the contents in liquid nitrogen, and opening the tube. Upon warming, a large amount of HBr was evolved. The solid product was flushed with nitrogen overnight to remove HBr and unreacted bromine. The fused reaction mass was then broken up into a powder with a mortar and pestle, and dried under vacuum for 2 h. The dried solid was dissolved in acetone (50 mL), filtered through silica gel, and precipitated with pentane. After drying overnight under vacuum, crude 4-carboxyl-2,6-dinitrobenzyl bromide (7.36 g, 82% yield) was obtained. This material was not purified further but used directly to make 4-(*t*-butoxycarbonyl)-2,6-dinitrobenzyl bromide. ^1H NMR (dimethylsulfoxide d_6) 12.5 (broad s, 1H, OH carboxylic acid); 8.70 (s, 2H, CH nitroaryl); 4.87 (s, 2H, CH_2, benzyl) IR(NaCl, film): 1690, 1530, 1350cm^{-1}

Synthesis of 4-(t-butoxycarbonyl)-2,6-dinitrobenzyl bromide. Crude 4-carboxyl-2,6-dinitrobenzyl bromide (4.1 g 13 mmol), and 4-dimethylaminopyridine (DMAP) (0.146 g, 1.2 mmol) were suspended in 100 mL of dry CH_2Cl_2 under argon. To this was added with stirring *t*-butanol (1.2 mL, 13 mmol) followed by dicyclohexylcarbodiimide (DCC) (2.75 g, 13.3 mmol) which was added dropwise dissolved in 5.0 mL of CH_2Cl_2. The reaction mixture was left stirring for 2 h. Workup was done by filtration and, washing the filtrate with water (3 × 50.0 mL), 5% acetic acid (3 × 50.0 mL) and finally water (3 × 50.0 mL). The organic layer was then filtered through a short column of silica gel, dried over $MgSO_4$ and stripped of solvents under vacuum to give crude material. This crude material was dissolved in 30.0 mL of hot ethanol and allowed to cool. A small amount of black tar-like material deposited in the flask was removed by decanting. The solution was recrystallized by adding a small amount of water and cooling to 10 °C. This crude material was chromatographed over silica gel (ethyl acetate, hexane) to give 3.55 g (74% yield) of white crystals (mp: 85-87 °C). Anal. calcd for $C_{12}H_{13}O_6N_2Br$: C 39.91, H 3.63, N 7.76, Br 22.12. Found: C 40.13, H 3.67, N 7.58, Br 21.99. ^{1}H NMR(CDCl$_3$): 8.60(s, 2H, CH nitroaryl), 4.94(s, 2H, CH$_2$ benzyl), 1.64(s, 9H, CH$_3$ t-butyl). IR(NaCl film): 1720, 1550, 1350, 1300, 1150 cm^{-1}.

Synthesis of 4-(t-butoxycarbonyl)-2,6-dinitrobenzyl tosylate 1.   4-(t-Butoxycarbonyl)-2,6-dinitrobenzyl bromide (2.00 g, 5.54 mmol) was dissolved in 8 mL of dry acetonitrile and added to a solution of silver tosylate (4.13 g, 5.54 mmol) in 20 mL of dry acetonitrile under argon. The reaction mixture was then refluxed for 2 h. The workup was done by removal of acetonitrile under vacuum, redissolution in CH$_2$Cl$_2$, and filtration followed by recrystallization with Cl$_4$. After three recrystallizations 1.88 g of (75% yield) white crystals (mp: 149-151 °C) were obtained. Anal. calcd for $C_{19}H_{20}O_9N_2S$: C 50.44, H 4.46, N 6.19, S 7.09. Found: C 50.71, H 4.63, N 5.93, S 7.35 ^{1}H NMR(CDCl$_3$): 8.54(s, 2H, CH nitroaryl), 7.60(d,J = 9Hz,2H, CH tosylate *o* to SO$_2$) 7.31(d,J = 9Hz,2H, CH tosylate *o* to CH$_3$), 5.57(s, 2H, CH$_2$ benzyl), 2.45(s, 3H, CH$_3$ tosyl), 1.63(s, 9H, CH$_3$, t-butyl). IR (film NaCl) 1720, 1550, 1350, 1280, 980 cm^{-1}. MS: 396(M$^+$, —CH$_2$=C(CH$_3$)$_2$), 209(M$^+$, —OSO$_2$C$_6$H$_4$CH$_3$—CH$_2$=C(CH$_3$)$_2$—O), 166 (M$^+$, —OSO$_2$C$_6$H$_4$CH$_3$—C$_4$H$_9$—CO$_2$—O).

Synthesis of methyl ester of 3,5-dinitrotoluic acid.   A suspension was prepared consisting of 3,5-dinitrotoluic acid (55.60 g, 60.24 mmol) dissolved in absolute methanol (101 mL, 2.5 mole) containing 2.7 mL of H$_2$SO$_4$. The mixture was stirred under reflux for 5 h at which time most of the acid had reacted. workup was done by removing methanol under vacuum, adding water, and extracting with CH$_2$Cl$_2$ followed by washing of the organic layer with saturated NaHCO$_3$. After drying of the organic layer over MgSO$_4$, filtration, and removal of solvents, 40.47 g (69% yield) of yellowish crystals (mp: 80-82 °C) were obtained. Anal. calcd for $C_9H_8O_6N_2$: C 45.01, H 3.36, N 11.66. Found:C 45.05, H 3.36, N 11.50. ^{1}H NMR(CDCl$_3$): 8.59(s, 2H, CH nitroaryl), 4.04 (s, 3H, OCH$_3$), 2.64(s, 3H, CH$_3$). IR (film, NaCl): 1720, 1530, 1350, 1290, 1200, 1160 cm^{-1}. MS: 240(M$^+$), 223(M$^+$ —OH), 210(M$^+$— NO), 209(M$^+$— OCH$_3$), 193(M$^+$— NO$_2$— H).

Synthesis of 4-(methoxycarbonyl)-2,6-dinitrobenzyl tosylate.   3,5-Dinitrotoluic acid methyl ester (10.00 g, 41.66 mmol) was dissolved in 50 mL of CCl$_4$, to which was added in suspension N-bromosuccinimide (14.83 g, 83.32 mmol) and benzoyl peroxide (1.00 g, 4.12 mmol). The reaction mixture was refluxed for 17 h. Workup was done by filtering the reaction mixture through a short column of silica gel. After removal of solvent a crude product (13.47 g, 36% yield) was obtained which consisted of 35.5% 4-(methoxycarbonyl)-2,6-dinitrobenzyl bromide with the remainder being the starting material. This crude product (~15 mmol) was dissolved in 10 mL of dry acetonitrile and was added to a solution of silver tosylate (4.13 g, 14.8 mmol) in 40 mL of dry acetonitrile under argon. The reaction mixture was then refluxed for 2 h. The workup was done as in for **1**, followed by a recrystallization in hot CCl_4. In this way 6.01 g (52% yield) of pale yellow crystals (mp: 139 °C) were recovered. Anal. calcd for $C_{16}H_{14}O_9N_2S$: C 46.83, H 3.44, N 6.82, S 7.81. Found: C 46.61, H 3.39, N 6.72, S 7.43. ^1H NMR(acetone-d_6): 8.68(s, 2H, CH nitroaryl), 7.72(d, J = 8Hz, 2H, CH tosyl o to SO_2), 7.47(d, J = 8Hz, 2H, CH tosyl o to CH_3), 5.63(s, 2H, CH_2 benzyl), 4.01(s, 3H, OCH_3), 2.46(s, 3H, CH_3). IR (film NaCl) 1730, 1545, 1530, 1360, 1290, 1180, 980 cm^{-1}. MS: 239(M^+-$OSO_2C_6H_4CH_3$), 223(M^+-$OSO_2C_6H_4CH_3$-O)

Synthesis of t-butyl ester of 3,5-dinitrotoluic acid. A suspension was prepared consisting of 3,5-dinitrotoluic acid (5.00 g, 22.1 mmol) and DMAP (0.36 g, 2.9 mmol) in 65 mL of CH_2Cl_2. To this suspension was added under argon t-butyl alcohol (1.67 g, 22.1 mmol) followed by dicyclohexylcarbodiimide (4.56 g, 22.1 mmol). The reaction mixture was stirred overnight, filtered through a short SiO_2 column, and stripped of solvent under vacuum. The crude material was recrystallized from hot CCl_4 to give 5.01 of tan crystals (mp: 116 °C). Anal. calcd for $C_{12}H_{14}O_6N_2$: C 51.06, H 5.00, N 9.93. Found: C 51.10, H 4.94, N 9.94. ^1H NMR(acetone-d_6): 8.61(s, 2H, CH nitroaryl), 2.61(s, 3H, CH_3), 1.63(s, 9H, CH_3 t-butyl). IR (film, NaCl) 1710, 1540, 1365, 1310, 1150 cm^{-1}. MS: 267(M^+-CH_3), 252(M^+, -O-$C(CH_3)_3$), 209(M^+-O-$C(CH_3)_3$), 196(M^+- CH_2=$C(CH_3)_2$-NO) 180 (M^+- CH_2=$C(CH_3)_2$), 179 (M^+-$C(CH_3)_3$-NO_2), 160(M^+-CO_2-CH_2=$C(CH_3)_2$-O).

Lithography

Photoresist solutions for lithographic experiments consisted of 14 wt% of PTBSS dissolved in chlorobenzene with 4 mole% of PAG (relative to the polymer's t-BOC pendent groups). The solutions were filtered through a series of 1.0, 0.5, and 0.2-μm Teflon filters (Millipore, Inc.). Photoresists were spun (spin speed 3K) onto hexamethyldisilazane-primed oxidized silicon substrates and prebaked at 105 °C for 60 s. These films were ~1 μm thick and were measured with a Nanospec thickness gauge (Nanometrics Inc). Exposures were done using A GCA Laserstep® prototype deep UV exposure system with a NA = 0.35 lens and 5 × reduction optics. Proximity printing was used to obtain exposure response curves. The irradiated films were postexposure-baked(PEB) at 115 °C for 30 s. The resist films were developed with 0.18M tetramethylammonium hydroxide (TMAH) for 30 s. The absorbances of the resist films, spun on quartz disks, was determined using a Hewlett-Packard Model 8452A diode array UV spectrometer.

Infrared spectroscopy studies

A solution was prepared as above consisting of 20 mole% of **1** (relative to the 4-hydroxystyrene repeating unit of PHS) dissolved into a 14 wt% solution of PHS in cyclohexanone. The solution was spun onto an NaCl disk at 3K to give a

1 μm thick film. The coated NaCl disk was preexposure baked at 100 °C for 5 min, exposed (100, 200, 400 mJ/cm^2) with a Süss Model MA56A contact aligner equipped with a Lambda Physik excimer laser operating at 248 nm and post exposure-baked at 100 °C for 5 min. The IR spectra of the resist films were obtained during this process using the previously described Digilab instrument.

RESULTS AND DISCUSSION

Synthesis

The synthetic pathway to 1 (scheme 1) has the initial advantage of starting from the inexpensive 3,5-dinitrotoluic acid. The first step is a bromination of acid using molecular bromine. Previously, we have prepared many o-nitrobenzyl bromide PAG precursors from the corresponding toluene derivatives by employing N-bromosuccinimide (2, 8, 9) (NBS). While this procedure works moderately well with the methyl ester of 2,6-dinitrotoluic acid, unfortunately it does not give isolable products with 3,5-dinitrotoluic acid or its t-butyl ester. However, we successfully adapted the sealed tube bromination procedure devised by Fieser for 2,4,6-trinitrotoluene (20). However, attempts to extend this procedure to either the methyl or t-butyl ester of 3,5-dinitrotoluic acid led to bromination with deprotection of the ester group. The next step towards 1 was an esterification with t-butyl alcohol adapted from the general procedure described by Hassner (21) which is promoted by dicyclohexylcarbodiimide (DCC) using 4-dimethyl-aminopyridine (DMAP) as a catalyst. The last step, the reaction of the bromide with silver tosylate proceeds in the same manner as previously described for the synthesis of other o-nitrobenzyl PAG from o-nitrobenzyl bromide derivatives (2, 8, 9). The silver tosylate must, however, be pure as any tosic acid impurity may lead to deprotection of the t-butyl ester moiety during reaction.

Thermal stability

An important criteria for study in o-nitrobenzyl based PAG's is thermal stability. Scheme 2 shows the reaction which can lead to thermal generation of acid. Thermal generation of acid is undesirable for PAG applications as it leads to contrast degradation and even to loss of images in resists during PEB. Previously, we have established an empirical relationship between T_{min} (as measured by DSC from the exotherm minima for decomposition) Hammett σ and calculated E_s for o-nitrobenzyl tosylates (Figure 1). From this relationship, a predicted T_{min} of 238 °C is expected for both 1 and 2, knowing that σ_p for — CO$_2$R (R = alkyl or H) (22) is 0.45 and that σ' and E_s(calc) (22, 23) for NO$_2$ are 0.63 and —0.92, respectively.

DSC scans of 1, 2 and 3 (Figure 2) show that 2 has a T_{min} (224 °C) quite close to that predicted, while the T_{min} of 1 (204 °C) is somewhat lower similar to that previously observed[2] for 3. Figure 1 shows the placement of the new thermal stabilities on the empirical plot of T_{min}. An endotherm in the DSC (Figure 2) of 1 (163 °C) initially attributed to a melting transition was found by examination of the TGA (Figure 3) to correspond to a weight loss consistent with

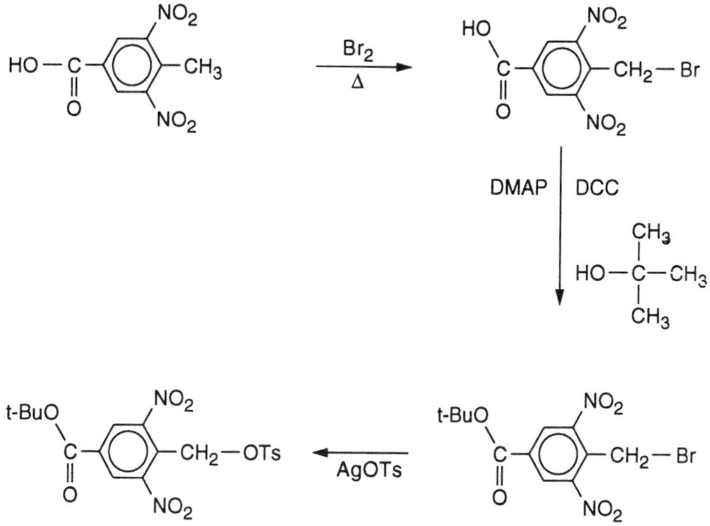

Scheme 1.

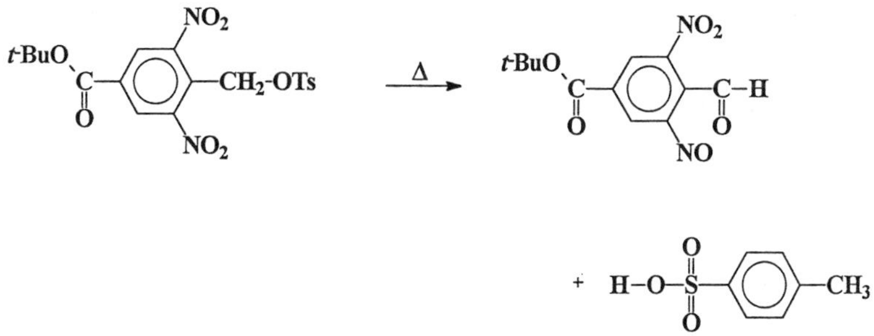

Scheme 2.

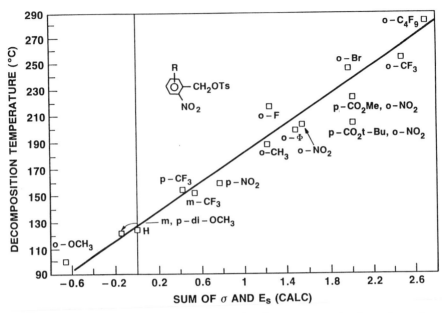

Figure 1. Plot of T_{min} versus $\Sigma\sigma' + \Sigma E_s$(calc) for the 2-nitrobenzyl tosylate derivatives.

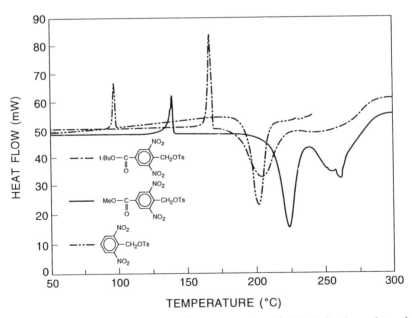

Figure 2. Comparative DSC's of 4-(*t*-butoxycarbonyl)-2,6-dinitrobenzyl tosylate **1**, 4-methoxycarbonyl-2,6-dinitrobenzyl tosylate **2**, and 2,6-dinitrobenzyl tosylate **3**.

loss of isobutene (14%). Scheme 3 shows the elimination reaction. The measurement of the isobutene elimination temperature by DSC (Figure 4) of the *t*-butyl ester of 3,5-dinitrotoluic acid **4** gives a much higher value (192 °C). Postulating that the tosylate moiety or impurities engendered by it are acting as a Lewis acid catalysts for the thermal elimination reaction, a DSC/TGA experiment was done with a 50/50 mixture of **4** and 2,6-dinitrobenzyl tosylate (Figure 5). This experiment lowers the stability of **4** to a value close (159 °C) to that found for **1**.

Since thermal elimination of isobutene occurs at 163 °C the T_{min} measured for **1** really corresponds to that of 4-carboxyl-2,6-dinitrobenzyl tosylate. However, the predicted T_{min} for this compound should have been ~238 °C and not the 204 °C value found. A DSC experiment with a 50/50 mixture of 3,5-dinitrotoluic acid and 2,6-dinitrobenzyl tosylate shows that the temperature of T_{min} is lowered to 192 ° from 200 °C (Figure 6). A possible explanation for the lower measured value of T_{min} is that the more polar matrix containing carboxylic acid moieties has lowered slightly the transition state for thermal generation of tosic acid.

A TGA of **1** with (5 mole%) and without added tosic acid (Figure 7) showed that the difference in decomposition temperature was ~70 °C. This should be a good estimate of the difference in relative thermal stabilities of **1** in actual exposed and unexposed resist areas. Therefore selective removal of *t*-butyl ester moieties in exposed areas should be possible providing care is taken in selecting a PEB temperature that does not decompose PAG in unexposed areas.

Lithographic sensitivity and absorbance

Compound **3** and 2,6-dinitrobenzyl 4-nitrobenzenesulfonate **5** are two examples of PAG's which have been successfully used in chemically amplified *t*-BOC based resist systems (*2*). Table 1 shows a comparison of the absorbance at 248 nm of a 6 mole% solutions of these materials and **1** and **2** in acetonitrile. As can be seen the absorbance in solution of compounds **1** and **2** are somewhat lower than that of **3** and considerably lower than that of 2,6-dinitrobenzyl 4-nitrobenzenesulfonate **5**. Table 1 also shows the absorbance per μm (ABS/μm) of **1**, **2** and **3** in a PTBSS polymer matrix. Since the ABS/μm of **1** and **2** are similar to **3** and the same acid is liberated in all three PAG's, it was expected (based upon our previously established relationship between the lithographic sensitivity and the product of the ABS/μm, ϕ, and catalytic chain length) (*9*) that the sensitivities of **1** and **2** should depend primarily on the relative values of the quantum yields.

The matrix polymer chosen for the lithographic experiments was PTBSS. Compound **2** being incapable of undergoing acidolysis, but having the same *o*-nitrobenzyl chromophore as **1**, was used to determine whether lithographic sensitivity was affected by the presence of a *t*-butyl ester in a PTBSS matrix. Table 1 summarizes the results of lithographic sensitivity and contrast for **1**, **2** and **3**.

The lithographic sensitivities of **1**, **2**, and **3** are similar. This indicates that no significant competition occurs between the acidolytic deprotection of **1** and PTBSS, and that the quantum yield for photogeneration of acid in **1** and **2** are only slightly lower than **3** (0.16) (*8*), namely ~0.11.

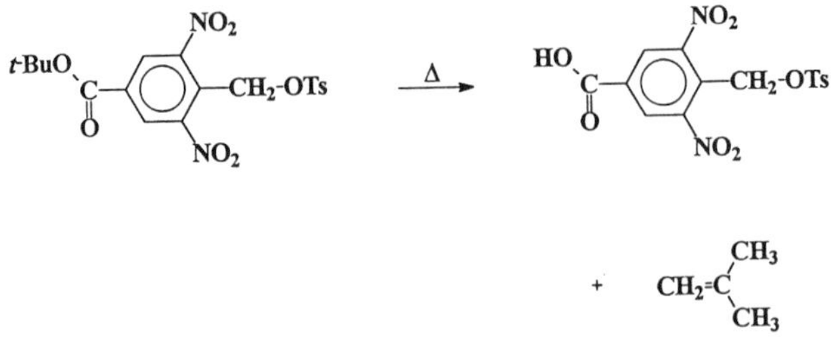

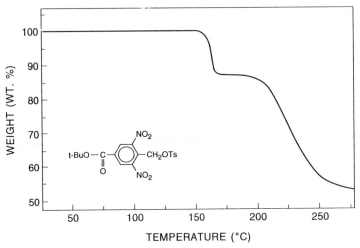

Scheme 3.

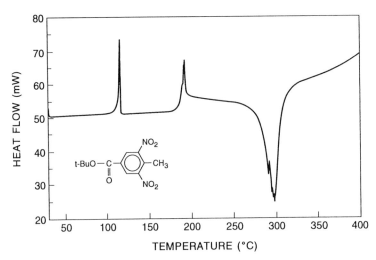

Figure 3. TGA of 4-(*t*-butoxycarbonyl)-2,6-dinitrobenzyl tosylate **1**.

Figure 4. DSC of the *t*-butyl ester of 3,5-dinitrotoluic acid **4**.

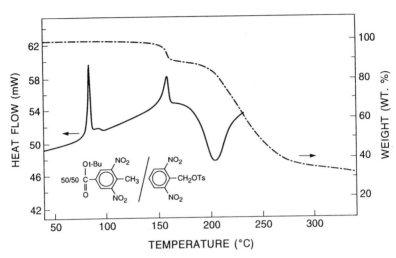

Figure 5. DSC and TGA of a 50/50 mixture of the *t*-butyl ester of 3,5-dinitrotoluic acid **4** and 2,6-dinitrobenzyl tosylate **3**.

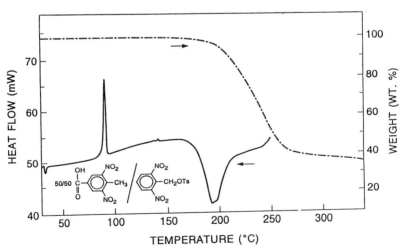

Figure 6. DSC and TGA of a 50/50 mixture of 3,5-dinitrotoluic acid and 2,6-dinitrobenzyl tosylate.

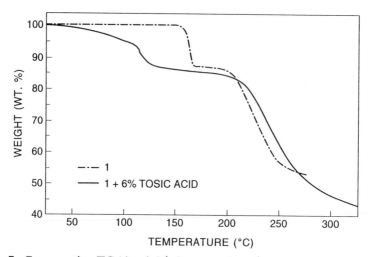

Figure 7. Comparative TGA's of 4-(*t*-butoxycarbonyl)-2,6-dinitrobenzyl tosylate
1 with and without added tosic acid.

Table 1. Solution absorbances of studied PAG's in acetonitrile, and absorbances,
lithographic sensitivity and contrast of PAG/PTBSS based resists

Compound	ABS 6 mole % CH$_3$CN	ABS/μm 4 mole % PBSS	Lithographic Sensitivity mJ/cm^2	Contrast
1	0.18	0.39	74	>10
2	0.18	0.38	68	>10
3	0.24	0.36	50	>10
5	0.37	–	–	–

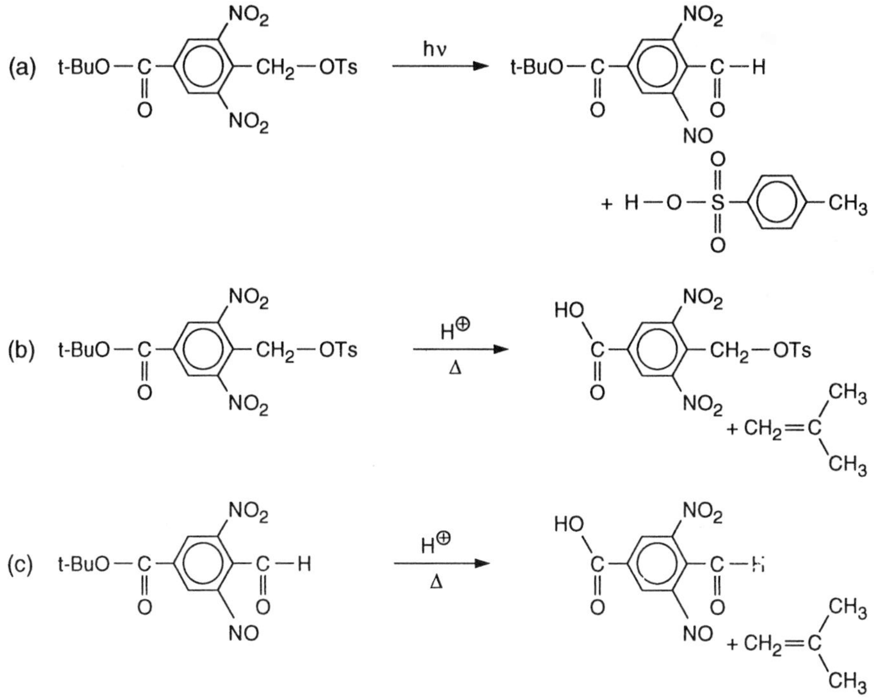

Scheme 4.

t-Butyl ester deprotection study by infrared spectroscopy

A primary requirement for a DISCA is that it undergoes effective acidolytic deprotection in order to impart maximum solubility in the exposed areas of a resist while remaining a hydrophobic molecule in unexposed areas. Scheme 4 outlines first the photogeneration of acid from **1**, then the acidolytic deprotection of unreacted PAG, and 4-(*t*-butoxycarbonyl)-2-nitro-6-nitrosobenzaldehyde photoproduct. This second step would ensure that no PAG derived hydrophobic moieties would remain in the exposed areas of a resist. IR spectroscopy was used to monitor whether this was occurring in the resist matrix. Poly(4-hydroxy-styrene) (PHS) was chosen as the matrix. In this study changes in the C-H stretch(*t*-Bu, 2981 cm^{-1}) were followed spectroscopically. Exposures were made at 100, 200 and 400 mJ/cm^2. It was found that PEB at 100 °C for 5 min was sufficient to completely deprotect **1** and its photo-product (Figure 8) even with an exposure of only 100 mJ/cm^2. In contrast, heating the PHS film containing **1** for as long as 20 minutes at 100 °C did not show any detectable decrease in the *t*-butyl IR frequency. This shows that **1** has the appropriate acidolytic behavior upon irradiation to be further evaluated as a DISCA candidate.

CONCLUSION

It has been found that 4-(*t*-butoxycarbonyl)-2,6-dinitrobenzyl tosylate **1** can be synthesized from a inexpensive and readily available commercial starting mate-

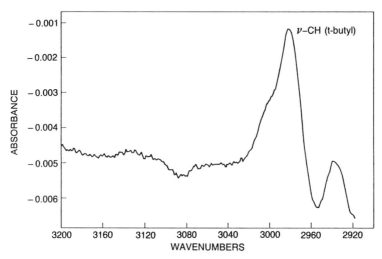

Figure 8. Infrared difference spectrum of a PHS film containing 20 mole% of 4-(*t*-butoxycarbonyloxy)-2,6-dinitrobenzyl tosylate **1**. The difference   spectrum was taken between the spectrum of the original film and that of the film after exposure to 100 mJ/cm^2 and PEB at 100 °C for 5 min. The positive band at 1981 cm^{-1} shows the loss of the *t*-butyl group.

rial and that it possesses the needed properties to be evaluated as a PAG/DISCA candidate both in terms of photogeneration of acid and ease of acidolysis of its pendent *t*-butyl ester group.

ACKNOWLEDGEMENTS

We thank R. A. Liselli, M. Cheng, and A. Timko for technical assistance. E. Reichmanis for technical discussion, and D. Mixon and M. Bohrer for a sample of PTBSS.

REFERENCES

1. McKean, D.; Schoendeli, U.; MacDonald, S.A. *Proc. Polym. Mater. Sci. Eng.* 1989, *60*, 45.
2. Neenan, T.X.; Houlihan, F.M.; Reichmanis, E.; Kometani, J.M.; Bachman, B.J.; Thompson, L.F. *Macromolecules,* **990**, *23*, 145.
3. Ito, H.; Willson, C.G. *Proc. Reg. Tech. Conf. on Photopolymers,* Mid-Hudson Section SPE, Nov. 8-10, 1982, Ellenville, NY, 331.
4. Fréchet, J.M.J.; Ito, H.; Willson, C.G. *Proc. Microcircuits Engineering 82,* Grenoble, France, **1982**, 260.
5. Ito, H.; Willson, C.G. *Polymers in Electronics,* ACS Symposium Series *242,* Davidson, T., Ed.; American Chemical Society, Washington DC, **1984**, 11.

6. Ito, H.; Willson, C.G.; Fréchet, J.M.J.; *Proc. SPIE,* **1987,** *771,* 24.
7. Crivello, J.V.; Lam, J.H.W. *Macromolecules,* **1977** *10,* 1307.
8. Houlihan, F.M.; Shugard, A.; Gooden, R.; Reichmanis, E. *Macromolecules,* **1988,** *21,* 2001.
9. Houlihan, F.M.; Neenan, T.X.; Reichmanis, E.; Kometani, J.M.; Chin, T. *Chem. Mater.,* **1991** *3,* 462.
10. Nalamasu, O.; Reichmanis, E.; Cheng, M.; Pol, V.; Kometani, J.; Houlihan, F.M.; Neenan, T.X.; Bohrer, M.P.; Mixon, D.A.; Thompson, L.F., *Proc. SPIE,* **1991,** *1466,* 13.
11. Schlegel, L.; Ueno, T.; Shiraishi, H.; Hayashi, N.; Iwayanagi, T. *Chem. Mater.,* **1990** *2,* 199.
12. Hayashi, N.; Hesp, S.M.A.; Ueno, T.; Toriumi, M.; Iwayanagi, T.; Nonogaki, S. *Proc. Polym. Mater. Sci. Eng.* **1989,** *61,* 417.
13. McKean, D.R.; McDonald, S.A.; Clecak, N.J.; Willson, C.G. *Proc. SPIE,* **1988,** *920,* 60.
14. Schlegel, L.; Ueno, T.; Shiraishi, H.; Hayashi, N.; Iwayanagi, T. *J. Photopolym. Sci. Technol.,* **1990,** *3,* 281.
15. O'Brien, M.J. *Polym. Eng. Sci.,* **1989,** *29,* 846.
16. O'Brien, M.J.; Crivello, J.V. *Proc. SPIE,* **1988** *920,* 342.
17. Lingnau, J.; Dammel, R.; Theis, J. *Proc. Regional Technical Conference on Photopolymers;* Mid-Hudson Section, SPE, Oct. 30-Nov. 2, 1988, Ellenville, NY, 87.
18. Schwalm, R. *Polym. Eng. Sci.,* **1989,** *61,* 278.
19. Tarascon, R.G., Reichmanis, E.; Houlihan, F.M.; Shugard,A.; Thompson, L.F. *Polym. Eng. Sci.,* **1989,** *29(3).* 850.
20. Fieser, L.F.; Doering, W.E.; *J. Am. Chem. Soc.* **1946,** *68,* 2552.
21. Hassner, A. *Tetrahedron Lett.,* **1978** *46,* 4475.
22. Hansch, C.; Leo, A. *Substituent Constants for Correlation Analysis in Chemistry and Biology* Wiley Interscience, 1979, p. 1-12 and 69.
23. *Steric Effects in Organic Chemistry,* Newman, M.S., Ed.; John Wiley & Son, Chapter 13, *Separation of Polar, Steric and Resonance Effects in Reactivity,* 1956, p. 556.

RECEIVED February 6, 1993

Chapter 3

Chemically Amplified Deep-UV Photoresists Based on Acetal-Protected Poly(vinylphenols)

Ying Jiang[1] and David R. Bassett

Technical Center, Union Carbide Chemicals and Plastics Company, Inc., 2000–2001 Kanawha Turnpike, South Charleston, WV 25303

A chemically amplified deep UV photoresist system based on acetal chemistry is reported. Acetal-protected poly(vinylphenols) were prepared either by free radical polymerization of the monomers or by chemical modification of poly(vinylphenol). In the presence of an acid as a catalyst, the polymers thermally decomposed to aqueous base soluble poly(vinylphenol) and some small molecules. Therefore, the resists were formulated with the acetal-protected polymers and a photoacid generator such as triphenylsulfonium hexafluoroantimonate. Positive-tone image could be resolved by exposing the resist film in deep UV region, post-baking, and developing in tetramethylammonium hydroxide solutions.

Since the concept of chemical amplification was introduced about ten years ago for resist application (1–2), significant advances have been made in designing resist systems utilizing photoacids and polymers containing acid-labile pendant groups (3–7). Most of these polymers were based on carbonate- or ether-protected poly(vinylphenols) or ester-protected polymeric acids (8, 9). Recently, systems based on tetrahydropyranyl- and furanyl-protected poly(vinylphenols) have also been reported (10). The basic idea of chemical amplification in these systems is to use a photo-generated acid (11, 12) to catalyze cleavage of pendant groups from protected polymers and therefore, to induce a change in solubility of the exposed area, such as from organic soluble poly(t-BOCstyrene) to aqueous base soluble poly(vinylphenol) (1–2). Therefore, dual-tone images of this type of photoresists consisting of polymers with acid-labile pendant groups and photoacid generators can be developed depending on developing conditions. These systems are attractive as deep UV resist materials because of their high sensitivities as the result of chemical amplification.

In this paper we report a new chemically amplified deep UV photoresist system based on acetal-protected poly(vinylphenols) and a photoacid generator.

[1]Current address: U.S. Surgical Corporation, 150 Glover Avenue, Norwalk, CT 06856

0097–6156/94/0537–0040$06.00/0
© 1994 American Chemical Society

The syntheses of monomers and polymers and their characterization as well as their acid-catalyzed deprotection reaction are discussed. Some initial results of imaging tests are also reported.

EXPERIMENTAL

Materials and Instrumentation

Methyl vinyl ether, 4-hydroxybenzaldehyde, potassium t-butoxide and methyltriphenylphosphonium bromide were purchased from Aldrich and used without purification. Vinyl phenyl ether from Polyscienece, Inc. was used as received. Poly(vinylphenols) were commercial samples from Hoechst Celanese. Pyridinium p-toluenesulfonate (PPTS) was prepared by mixing equal moles of pyridine and p-toluenesulfonic acid. After gently refluxing, the salt was purified by crystallization from methanol.

NMR spectra were recorded in deuterated chloroform or dimethyl formamide (DMF). FT-IR and UV spectra were recorded on a BioRad FTS-40 Fourier-Transform Infrared Spectrometer and a Hewlett Packard UV-visible spectrometer.

Films were spun on quarts or silicon wafers using a manual spinner (Solitec, Inc., Model 5100). Thickness measurements of dry films were made using a NANOSpec/AFT (Nanometrics, Model 010-0181). Radiation was done using a Xenon deep UV lamp (ORIEL, Model 66142); the ORIEL deep UV lamp was powered with an arc power supply (ORIEL, Model 8530).

Preparation of 4-(1-Methoxyethoxy)styrene

4-Hydroxybenzaldehyde (113 g) and 20 g of PPTS were suspended in 195 mL of methylene chloride in a pressure bottle with a magnetic stirring bar. Methyl vinyl ether (65.1 g) was condensed into a graduated flask and added to the pressure bottle in one portion. The reactor was then placed in a warm water bath with stirring and immediately the inside pressure was brought to about 10 psi. After three days of stirring, the reaction mixture became a clear solution. Thin layer chromatography (TLC) of the reaction mixture indicated that the reaction was complete. After the mixture was passed through a column packed with silica gel and the solvents were removed, 160 g of the product, 4-(1-methoxyethoxy)-benzaldehyde, were obtained and taken directly to the next step, Wittig reaction. The reaction gave 95.6% yield.

Methyltriphenylphosphonium bromide (389 g) and 126 g of potassium t-butoxide were added to 2 L of freshly distilled tetrahydrofuran (THF) under nitrogen protection. After 30 minutes of stirring, a solution of 160 g of 4-(1-methoxyethoxy)benzaldehyde prepared as above in 300 mL THF was added slowly. The reaction was stirred at room temperature overnight. Crashed ice (1 kg) was added to the reaction mixture with stirring. After two phases were separated, the aqueous layer was extracted with ethyl acetate twice. The combined organic phases were dried over meganesium sulfate and the solvents were removed under reduced pressure. The product was a colorless liquid, which was then further purified by vacuum distillation (b.p. = 69 °C/0.8 mmHg). In this way, 4-(1-methoxyethoxy)styrene was obtained in 71% yield (112 g).

Preparation of 4-(1-Phenoxyethoxy)styrene

4-Hydroxybenzaldehyde (13 g) was dissolved in 100 mL of freshly distilled THF under nitrogen. To this solution was then added 10 g of vinyl phenyl ether and 3 g of PPTS. After the mixture was stirred at room temperature for 24 hours, TLC analysis indicated that the reaction was not complete. Therefore, the reaction mixture was gently refluxed and closely followed by TLC. After refluxing for additional 24 hours, the reaction was complete. THF was removed under reduced pressure. Methylene chloride (200 mL) was added and the solution was washed with 2% tetramethylammonium hydroxide solution to remove unreacted 4-hydroxybenzaldehyde. After three times of washing, the solution seemed free of 4-hydroxybenzaldehyde. The solution was then washed with water and the organic phase was separated, dried, and concentrated. Slightly yellow liquid was obtained which, characterized by both FT-IR and NMR, was the desired product. The reaction yield was 60%.

A mixture was prepared, under nitrogen, consisting of 23.4 g of methyltriphenylphosphonium bromide and 8.5 g of potassium t-butoxide in 300 mL of freshly distilled THF. After 5 minutes of stirring, a solution of 10.5 g of 4-(1-phenoxyethoxy)benzaldehyde prepared as above in 100 mL THF was added slowly at room temperature. The reaction mixture was stirred at room temperature for 12 hours. The reaction was stopped by pouring the reaction mixture into 200 g of cold water. The organic phase was separated and the aqueous phase was extracted with ethyl acetate three times. The combined organic phases were dried over meganesium sulfate, filtered and concentrated. The product was purified by silica gel charomatography.

Free Radical Polymerization of 4-(1-Methoxyethoxy)styrene and 4-(1-Phenoxyethoxy)styrene

In a typical experiment, 100 g of 4-(1-methoxyethoxy)styrene were dissolved in 100 g of benzene. After 0.5 g of free radical initiator, 2,2'-azobis(isobutyronitril) (AIBN), was added, the solution was degassed with nitrogen for 30 minutes and then placed in an oil bath at 75 °C for 16 hours with stirring. The solution became very viscous which was diluted with THF. Precipitation of the polymer was done in a large amount of methanol. After the solid was filtered, dried in vacuo, 85.4 g of white powder was obtained, which was proven to be the desired polymer by FT-IR and NMR; poly[4-(1-methoxyethoxy)styrene].

Similarly, 4-(1-phenoxyethoxy)styrene could also be polymerized and purified using this experimental procedure. A series of reactions were carried out to investigate both solvent and initiator effects on polymerization and the results are summarized in Table 2.

Chemical Modification on Poly(vinylphenol)

In a high pressure reactor, 30 g of poly(vinylphenol) were dissolved in 380 mL of freshly distilled THF or suspended in dry methylene chloride. PPTS (6 g) was added under nitrogen. Methyl vinyl ether (200 mL) was condensed into a graduated flask and then added into the pressure reactor in one portion. After flushing with nitrogen briefly, the reactor was sealed and kept stirring at room temperture for three days. The polymer solution was precipitated in methanol. After brief drying, the polymer was further purified by dissolving in acetone and precipitating in a mixture of methanol and water. In this way, 39 g of polymer were obtained. FT-IR showed no absorption of the phenolic groups, indicating that the reaction had high yield of masking the OH groups by forming the acetal groups.

An attempt was also made to modify poly(vinylphenol) with vinyl phenyl ether under similar conditions described above. However, it was found that a significant amount of phenolic groups remained unreacted after a week of the reaction as indicated by free OH absorption in its FT-IR spectrum.

RESULTS AND DISCUSSION

Syntheses of Monomers and Polymers

The monomers were synthesized using a similar experimental procedure. Starting from 4-hydroxybenzaldehyde, the phenolic group was first blocked with the vinyl ethers to form the acetal-benzaldehydes; and then Wittig reaction converted the aldehyde group into the vinyl functionality. To demonstrate the reaction sequences, Scheme 1 illustrates the preparation of 4-(1-phenoxyethoxy)styrene and its free radical polymerization.

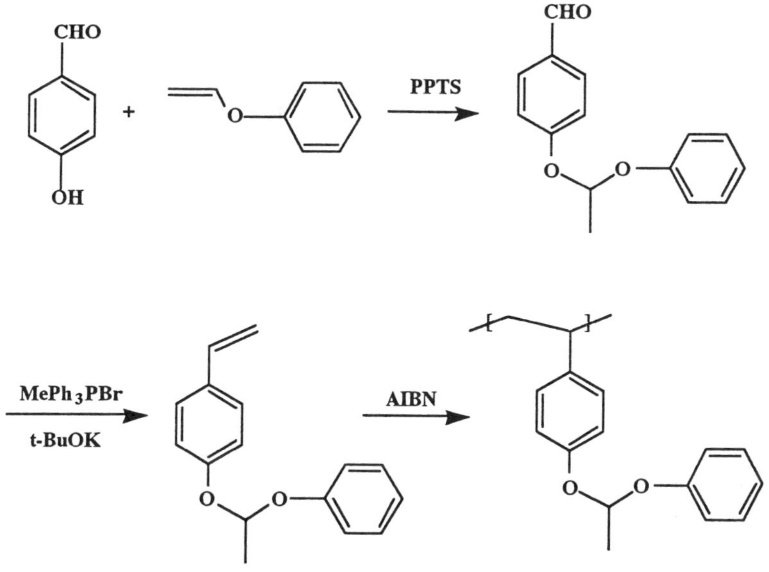

Scheme 1. Synthesis of 4-(1-Phenoxyethoxy)styrene and Its Polymerization

The nucleophilic addition of the phenolic functionality to vinyl phenyl ether was found to proceed satisfactorily with PPTS as the catalyst. Excess 4-hydroxybenzaldehyde was used in order to achieve maximum conversion of expensive vinyl phenyl ether. The unreacted 4-hydroxybenzaldehyde could be easily removed from the reaction mixture by washing with diluted aqueous base solutions and in the meantime, the catalyst, PPTS, was also removed by aqueous washing. The protected 4-hydroxybenzaldehyde was then taken to a high yield Wittig methylenation to afford the desired product.

4-(1-Methoxyethoxy)styrene was also synthesized following a similar reaction sequence discribed above. However, since methyl vinyl ether is a gas at room temperature, it had to be condensed through a cooling trap, weighed, and added into a pressure bottle, in which the reaction was carried out. Similarly for chemical modification of poly(vinylphenol) with methyl vinyl ether, same experimental procedure was carried out to handle the gaseous methyl vinyl ether.

Figure 1 shows the proton NMR spectra of 4-(1-phenoxyethoxy)-styrene and its polymer prepared by free radical polymerization. For the monomer's spectrum, a doublet located at 1.64 and 1.66 ppm was assigned to CH_3 of the ethoxy group and the proton of its CH group had a quartet located at 5.97 to 5.99 ppm; three protons of the double bond were located at 5.13 to 5.16; 5.59 to 5.97 and 6.64 to 6.69 ppm; two sets of aromatic protons were in the region of 6.93 to 7.34 ppm. FT-IR spectrum of the monomer showed a vinyl absorption at 1629 cm^{-1} which, after polymerization, disappeared completely as evidenced by the FT-IR of the polymer (Figure 2.)

For 4-(1-methoxyethoxy)styrene, its proton NMR was very similar to that of 4-(1-phenoxyethoxy)styrene except that a new peak of CH_3 of the methoxy group could be seen at 3.39 ppm.

The acetal-protected monomers were easily polymerized under free radical polymerization conditions. Table 1 summarizes some of the results of free radical polymerization of 4-(1-methoxyethoxy)styrene.

Careful analysis of FT-IR and NMR spectra of the polymers indicated that free radical polymerization did not induce any deprotection of the acetal group. The polymerization carried out in benzene at 70 °C produced a satisfactory result: high molecular weight polymer with narrower molecular weight distribution in high yield.

Similarly, free radical polymerization of 4-(1-phenoxyethoxy)-styrene was also investigated under different conditions as summarized in Table 2.

Solvents seemed to have a strong effect on the polymerization. Aromatic solvents, such as benzene and toluene, were suitable for the polymerization as the reactions resulted in higher yields and higher molecular weights; on the other hand, the polymerization carried out in a chlorocarbon, methylene chloride, produced lower molecular weight polymer with lower reaction yield. This is probably due to chain transfer effect of chlorocarbons. While AIBN seemed to be an excellent free radical initiator for this type of polymerization, benzoyl peroxide induced a poor polymerization reaction, resulting only oligomers with low reaction yield. A repeated polymerization with BPO produced similar results.

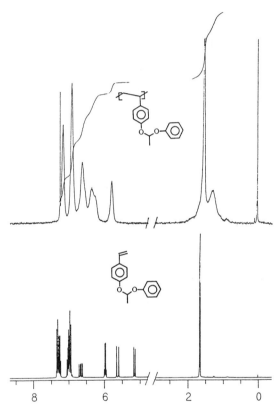

Figure 1. ¹H NMR Spectra of 4-(1-Phenoxyethoxy)styrene and Its Polymer.

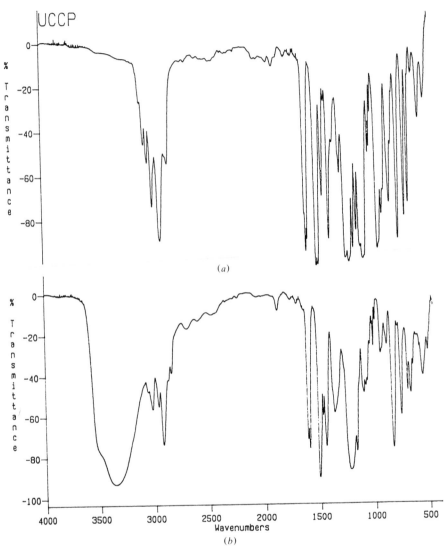

Figure 2. FT-IR Spectra of Poly[4-(1-phenoxyethoxy)styrene] Containing 5 wt% of Ph_3S^+ SbF_6; a) prebaked at 90 °C for 2 minutes and b) exposed to 10 mJ/cm2 of 254 nm radiation and postbaked at 110 °C for 2 minutes.

Table 1. Free Radical Polymerization of 4-(1-Methoxyethoxy)styrene[a]

Temp. (°C)	AIBN (mg)	Solv. (g)	Time (h)	Yield%	Mn[b]	Mw/Mn
85	21.4	Toluene (4)	48	83	23,100	2.9
75	46.3	Clbenzene (4.6)	24	95	26,700	3.7
70	40.0	Benzene (4)	16	96	61,000	2.4

a. Monomer: 4 g; all the reactions were carried out under nitrogen and the reaction mixtures were degassed with nitrogen for 30 minutes before heating.
b. GPC values with polystyrene standards.

^{1}H NMR spectrum of poly[4-(1-phenoxyethoxy)styrene] shown on the top of Figure 1 had much broader peaks compared to its monomer. New peaks located in the region of 0.9 to 2.1 ppms could be assigned to the polymer's backbone protons. While the methyl group's chemical shift remained almost unchanged at about 1.6 ppm compared to the monomer, the protons of the double bond disappeared completely after the polymerization.

Since free radical polymerizations resulted in polymers with rather broad molecular weight distributions, anionic polymerization of 4-(1-phenoxyethoxy)styrene was also attempted using n-butyllithium as the catalyst in order to prepare polymers with narrower molecular weight distributions. It was assumed that photoresists formulated with polymers with narrow molecular weight distributions would improve their lithographic performances, especially their resolution. Under our experimental conditions, no high molecular weight polymers were produced although in some cases the formation of a small amount of oligomers was observed.

Alternately, the acetal-protected polymers have also been prepared by chemical modification on poly(vinylphenol) by reacting the polymer with vinyl ethers using PPTS as the catalyst. While satisfactory results were obtained with chemical modification using methyl vinyl ether, the reaction with vinyl phenyl ether showed a low efficiency of blocking the phenolic groups as FT-IR studies indicated that the chemically modified polymer still showed some unprotected

Table 2. Free Radical Polymerization of 4-(1-Phenoxyethoxy)styrene[a]

Monomer (g)	Initiator (mg)	Solv. (g)	Yield (%)	Mn[b]	Mw/Mn
1.0	AIBN (10)	Benzene (1.5)	99.0	56,600	3.0
3.0	AIBN (30)	Toluene (3.0)	93.3	46,300	2.3
3.2	AIBN (32)	CH_2Cl_2 (3.2)	84.4	23,000	2.8
3.0	BPO[c] (30)	Toluene (3.0)	53.3	9,800	2.9

a. All the reactions were carried out under nitrogen for at 79–80 °C for 24 hours.
b. GPC values with polystyrene standards.
c. Benzoyl peroxide.

hydroxy groups. Lithographically, resists formulated from the methyl vinyl ether modified poly(vinylphenol) or the polymer prepared by polymerization of the monomer did not show much difference.

Design of Resist Materials

It is well known that acetals undergo acid-catalyzed hydrolysis and therefore, poly(vinylphenols) protected with acetal groups are potentially useful as chemically amplified deep UV photoresist materials if used with photoacid generators. The photoacid-catalyzed cleavage of pendant acetal groups in the exposed area results in a large change of polarity of the polymer: from a polymer soluble in an organic solvent to an aqueous base soluble poly(vinylphenol) and therefore, high quality imaging process is expected.

The system based on acetal-protected poly(vinylphenols) has the advantages that most chemically amplified photoresists have demonstrated, such as high sensitivity as the result of catalytic nature of deprotection reaction. However, the novel design of the system based on poly[4-(1-phenoxyethoxy)styrene] shows several other advantages also. The imaging chemistry of this system is illustrated in Scheme 2.

By introducing phenoxyethoxy as a protection group, the deprotection products are poly(vinylphenol), phenol, and acetaldehyde. Since phenol is not volatile under the lithographic conditions, the only mass loss as the result of deprotection is acetaldehyde, which is about 17% of the total mass of the polymer. Phenol molecules are evenly blended in poly(vinylphenol) film after the pendant protection group is cleaved and therefore, film loss of the system is greatly limited.

In addition, the phenoxy group, as a part of the protection group, is virtually a masked promoter. After the cleavage reaction, the deprotected compound, phenol from the phenoxy group, acts as a dissolution promoter since it is very soluble in aqueous base solutions. As the result, the sensitivity of the system should be improved.

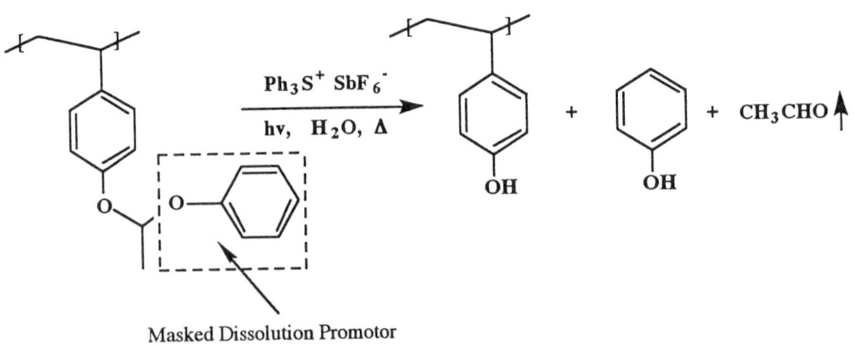

Scheme 2. Imaging Reaction for Poly[4(1-phenoxyethoxy)styrene]

Acid-catalyzed Deprotection: FT-IR and NMR Studies

Before imaging tests were carried out, the acid-catalyzed cleavage of the protected polymers was studied using FT-IR. In a typical experiment for solid-state photolysis studies, a solution of poly[4-(1-phenoxyethoxy)-styrene] containing 5 wt% of a photoacid generator, triphenylsulfonium hexafluoroantimonate, was spun on a NaCl pallet. A FT-IR spectrum was recorded as shown on the top of Figure 2. After the film was exposed to deep UV radiation and baked at 110 °C for 2 minutes, a FT-IR spectrum (the bottom of Figure 2) showed that the cleavage reaction had resulted in disappearance of the acetal structure on the polymer. From these spectra, it is seen clearly that the alkyl-oxygen streching band of the acetal at about 1060 cm^{-1} disappeared after the cleavage while formation of several new bands corresponding to the hydroxy groups could be observed: the phenolic O-H streching band at about 3370 cm^{-1} and deformation band at 1380 cm^{-1}.

Further study of the acid-catalyzed deprotection was carried out using NMR spectroscopy. In stead of using a solid-state film, a solution of poly[4-(1-phenoxyethoxy)styrene] in deuterated DMF was chosen to conduct the study. The reason to use DMF as the solvent was because both protected and deprotected polymers were soluble in DMF. After a trace of triflic acid was introduced into the polymer solution in a NMR tube and heated, gaseous molecules evolved from the solution, which were believed to be acetaldehyde. The solution was further heated at 110 °C for 2 minutes and by then gas evolution stopped. After cooling to room temperature, a proton NMR spectrum was taken, which is shown in Figure 3.

Compared to the NMR spectrum of the protected polymer (top, Figure 1), the chemical shifts of both CH$_3$ and CH of the acetal group disappeared completely after the deprotection as the result of evolving gaseous acetaldehyde. The new spectrum, Figure 3, showed a mixture of two components: broad chemical shifts for poly(vinylphenol) and two sets of sharp aromatic peaks for small molecular phenol. As expected, phenol remained with poly(vinylphenol) because it was not volatile under out experimental conditions.

Photolysis experiments carried out with thin films gave essentially same results as those with solution NMR study. 1H NMR spectrum of a reaction mixture after irradiation of a resist film provided evidence supporting our proposed imaging mechanism that only mass loss from the deprotection was acetaldehyde.

Imaging Experiments

The imaging tests were carried out using triphenylsulfonium hexafluoroantimonate or an organic compound developed in our laboratory as photoacid generators. The resists were formulated in either diglyme or PM acetate containing 3-8 wt% of photoacid generators and a small amount of a surfactant. After spin-coating onto a silicon wafer to 1mm thickness and solf-baking at 90 °C for 2 minutes, the wafer was exposed to UV radiation at 254nm through a mask in a contact printing process and a visible (latent) image was seen. After post-baked

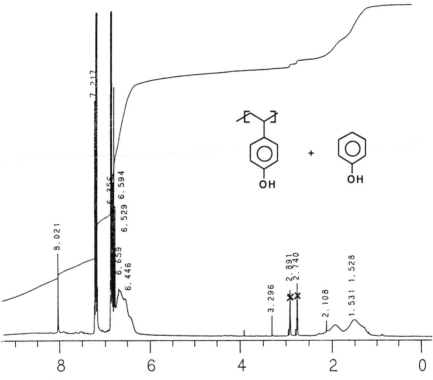

Figure 3. ^{1}H NMR Spectrum of Acid-catalyzed Cleavage of Poly[4-(1-pheno-xyethoxy)styrene]; Solution in DMF with Triflic Acid and Heated at 110 °C for 2 Minutes.

at 90-110 °C for 2 minutes and developed using dilute tetramethylammonium hydroxide solution or a commercial aqueous base developer, KTI 934, a positive image was resolved. Both resists formulated with poly[4-(1-methoxyethoxy)styrene] and poly[4-(1-phenoxyethoxy)styrene] had similar lithographic performances in terms of their resolution. However, resists formulated with poly[4-(1-phenoxyethoxy)styrene] demonstrated improved sensitivity.

The system had a sensitivity better than 15 mJ/cm2 and also good contrast. Non-swelling during development as the result of a large change of polarity of the polymers due to the acid-catalyzed deprotection contributed greatly to overall high quality of the imaging process. It is worthwhile to point out that the resist with poly[4-(1-phenoxyethoxy)-styrene] had very limited film loss as the result of the novel design of the polymer. Figure 4 shows a scanning electron micrograph of a positive image with 0.45 mm line/space obtained by the acid-catalyzed cleavage of the acetal-protected poly(vinylphenol).

CONCLUSIONS

This study has demonstrated that excellent photoresist materials incorporating chemical amplification can be obtained using polymers operating on the basis of acetal chemistry. A variety of acetal groups are available for designing novel

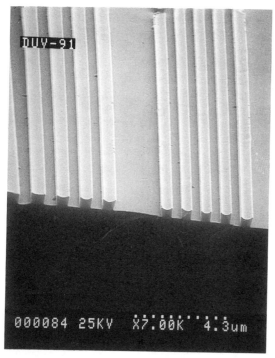

Figure 4. Scanning Electron Micrograph of Positive Image of the Acetal-protected Poly(vinylphenol); 15 mJ/cm^2 of 254 nm Radiation and Baked at 95 °C for 2 Minutes.

resist materials to satisfy different requirements, for instance, the carefully designed chemistry in poly[4-(1-phenoxyethoxy)styrene] can greatly limit film loss and improve sensitivity. The concept of using a masked dissolution promoter as a part of protection group can be further extended to other polymers besides poly(vinylphenol) or to new designs for dissolution inhibitors.

ACKNOWLEDGMENTS

Support from Texas Instruments for imaging tests and from KTI Chemicals is greatly acknowledged. The authors also thank every one from the Photoresist R & D group of Union Carbide Corp.

REFERENCES

1. Ito, H.; Willson, C. G.; Frechet, J. M. J.; *Digest of Technical Papers of 1982 Symposium on VLSI Technology*, **1982**, p. 86.
2. Frechet, J. M. J.; Ito, H.; Willson, C. G.; *Proc. Microcircuit Engineering*, **1982**, 260.
3. Frechet, J. M. J.; Eichler, E.; Willson C. G.; Ito, H.; *Polymer*, **1983**, 24, 995.
4. Ito, H.; Willson C. G.; Frechet, J. M. J.; U.S. Patent No. 4,491,628, 1985.
5. Houlihan, F. M.; Reichmanis, E.; Thompson, L. F.; Eur. Patent Appl. No. 89301556.0, 1989.
6. Ito, H.; Ueda, M.; *Macromolecules*, **1988**, 21, 1475.
7. Conlon, D. A.; Crivello, J. V.; Lee, J. L.; O'Brien, M. J.; *Macromolecules*, **1989**, 22, 509.
8. Ito, H.; Willson, C. G.; Frechet, J. M. J.; *Proc. SPIE*, **1987**, 771, 24.
9. Ito, H.; Ueda, M.; Ebina, M.; in *"Polymers in Microlithograph"*, Reichmanis, E.; MacDonald, S. A.; Iwayanagi, T.; Eds., ACS Symposium Series No. 412, ACS, Washington, D. C. 1989, p. 57.
10. Hesp, S. A. M.; Hayashi, N.; Ueno, T.; *J. Appl. Polym. Sci.*, **1991**, 42, 877.
11. Crivello, J. V.; Lam, J. H. W.; *J. Polym. Sci., Polym. Chem. Ed.*, **1979**, 17, 977.
12. Crivello, J. V.; Lee, J. L.; Conlon, D. A.; *Makromol. Chem., Makromol. Symp.*, **1988**, 13/14, p. 145.

RECEIVED May 7, 1993

Chapter 4

Novel Analytic Method of Photoinduced Acid Generation and Evidence of Photosensitization via Matrix Resin

N. Takeyama, Y. Ueda, T. Kusumoto, H. Ueki, and M. Hanabata

Sumitomo Chemical Company, Ltd., Osaka Research Laboratory, 3-1-98, Kasugadenaka, Konohana-ku, Osaka City, Osaka 554, Japan

A novel analytic method of photoinduced acid generation was developed by which the relative efficiency of photoinduced acid generation can be easily determined. For one halogenated photoacid generator, a photosensitization mechanism via matrix resin was suggested.

Recently chemical amplification type photoresists have been developed. In the present paper, we described a novel analytic method of photoacid generation. This method is very simple and useful for investigation of photoacid generators.

EXPERIMENTAL

Materials

In this study, we chose a typical negative type chemically amplified resist (1), which consists of three components, i.e. polyhydroxystyrene (PHS), photoacid generator (PAG) and a crosslinker (Figure 1). Poly(4-hydroxy)styrene (PHS) was a commercial sample (MARUKA LYNCUR-M;Mw = 4150) from MARUZEN PETROCHEMICAL CO., LTD. Isocyanuric acid tris(2,3-dibromopropyl)ester (TBI) and 2-phenylsulfonylacetophenone (PSAP) were purchased from TOKYO KASEI KOGYO CO., LTD. and used without purification. Trichloromethyl-s-triazine (TCT) was purchased from MIDORI KAGAKU CO., LTD. and used without purification. Hexamethoxy methyl melamine (HMM) was synthesized according to the literature procedure (2) and purified by column chromatography.

Measurements

The equipment used in detecting photogenerated acic is shown schematically in Figure 2. The generated acid was detected by the conduct meter (TOA DENPA KOGYO CO., LTD. ; Cell No. CG-201PL; Cell Const. = 1.005). A

0097–6156/94/0537–0053$06.00/0

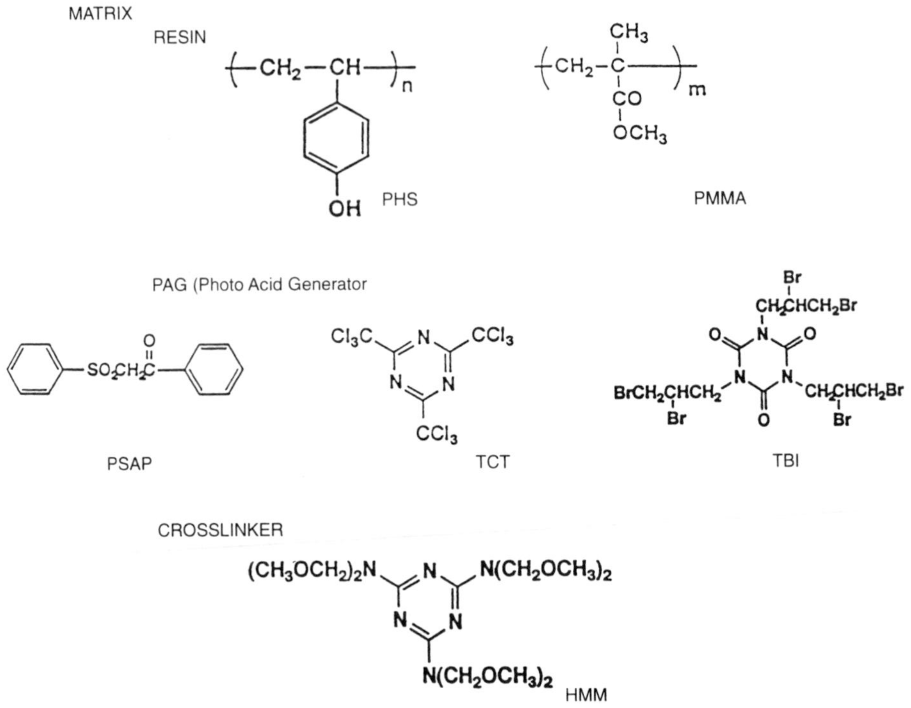

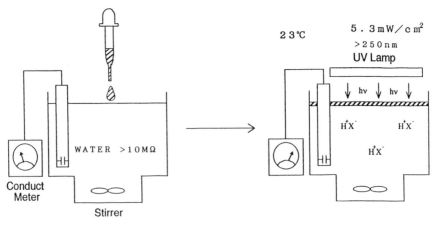

Figure 1. Structure of samples used in this experiment

Figure 2. Equipment and measurements in this experiment

dimethyl carbitol solution of the resist components was dropped in the water (the resistance greater than 10 MW). A thin film containing resin and PAG, was formed on the water surface after solvent evaporation. The acid formed after UV irradiation moved to the water phase rapidly, resulting in an increase in conductivity.

In this study, the film thickness (~ 2 mm) was controlled by the concentration of the sample solution, and was uniform. The film was placed on a glass slide and its tickness determinened.by Nanospec. The acidic species in the water phase were confirmed by ion chromatography.

RESULTS AND DISCUSSION

(a) PSAP

PSAP absorbs between 200 and 300 nm (Figure 3); on irradiation, benzene sulfonic acid is generated. For a film of PSAP about 0.5 mm thick (white points in Figure 4), photoacid generation increased with increasing irradiation time. However in the case of a PHS film containing 5 wt% PSAP, acid generation decreased in comparison with the single PSAP film (black points in Figure 4). We attribute this difference to the strong absorption band of PHS hindering light absorption by PSAP. As expected, PHS did not produce acidic species on irradiation.

(b) TCT

TCT has a small absorption band centered at approximately 275 nm (Figure 3). On irradiation of TCT, HCl generation occured, both with and without the PHS matrix. The PHS matrix caused a strong increase in photoacid generation (Figure 5). The amount of TCT was the same in each film. The film thickness of the single TCT film was about 0.5 mm.

(c) TBI

TBI does not absorb as wavelength gerater than 250 nm. Consequently, there was no HBr production from the single TBI film. However, in a film containing TBI, HBR was generated (Figures 3 and 6).

(d) TBI in a PMMA matrix

PMMA does not absorb at wavelength above 230 nm, so PMMA seemed to be a neutral matrix for this test. In a PMMA film containing only TBI, there was no HBr production. However, HBr generation increased in matrices with increasing PHS content (Figure 7), i.e. the presence of PHS led to photoinduced acid generation.

(e) 3 component resist system

In the typical negative type resist containig HMM, TBI and PHS, the initial rate of HBr production was similar both with and without HMM (Figure 8).

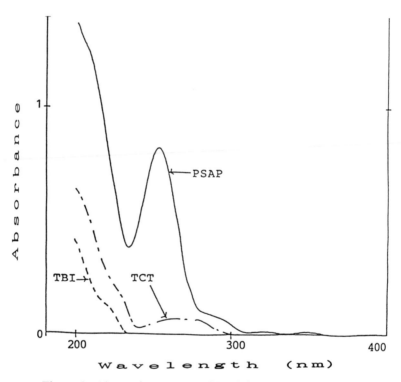

Figure 3. Absorption spectra of PAG in a PMMA matrix

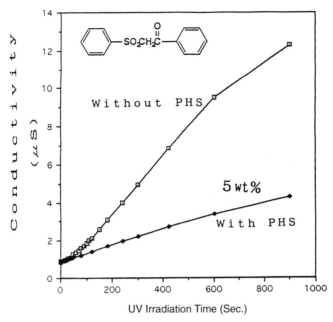

Figure 4. Time profiles of photoinduced acid generation of PSAP with and without PHS matrix

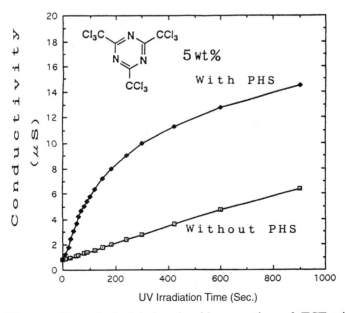

Figure 5. Time profiles of photoinduced acid generation of TCT with and without PHS matrix

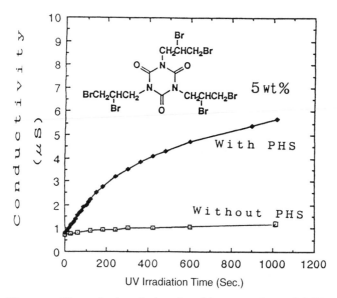

Figure 6. Time profiles of photoinduced acid generation of TBI with and without PHS matrix

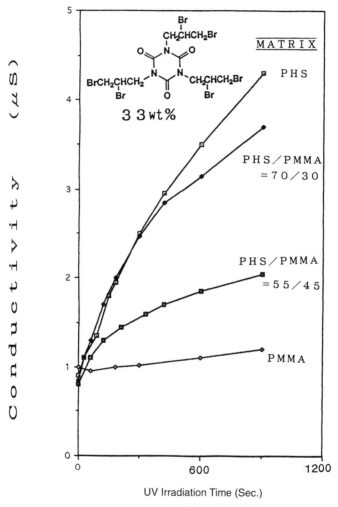

Figure 7. Time profiles of photoinduced acid generation of TBI as a function on PHS content in a PMMA matrix

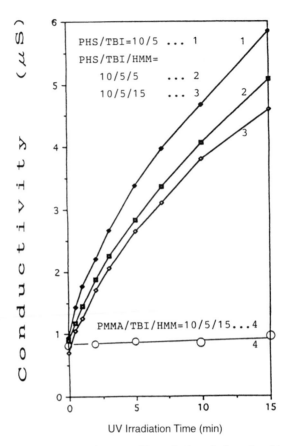

UV Irradiation Time (min)

Figure 8. Effect of HMM on time profiles of photoinduced acid generation of TBI in a PMMA and a PHS matrix

There was no HBr production from the PMMA film containing TBI and HMM. HMM did not hinder the photoinduced acid generation.

The mechanisms by which acids are photogenerated from TCT and PSAP are known (Figure 9). (3, 4) On the other hand, in the case of TBI which has no absorption band above 250 nm, HBr generation occurs in matrices containing PHS. We propose a photosensitization of TBI by PHS. During Deep UV irradiation, molecules of TBI, sensitized by excited PHS molecules, produced H + Br-. The quantity of HBr is a function of the light intensity, how many photons are absorbed by PHS, the efficiency of sensitization and the efficiency of TBI decomposition to generate HBr. This sensitization mechanism does not seem to be an electron transfer via a charge transfer complex between PHS and TBI because there was no clear charge transfer absorption band between PHS and TBI (Figure 10). Photosensitization may occur via a chemical reaction between excited state PHS and ground state TBI, or a direct electron transfer from excited PHS to TBI.

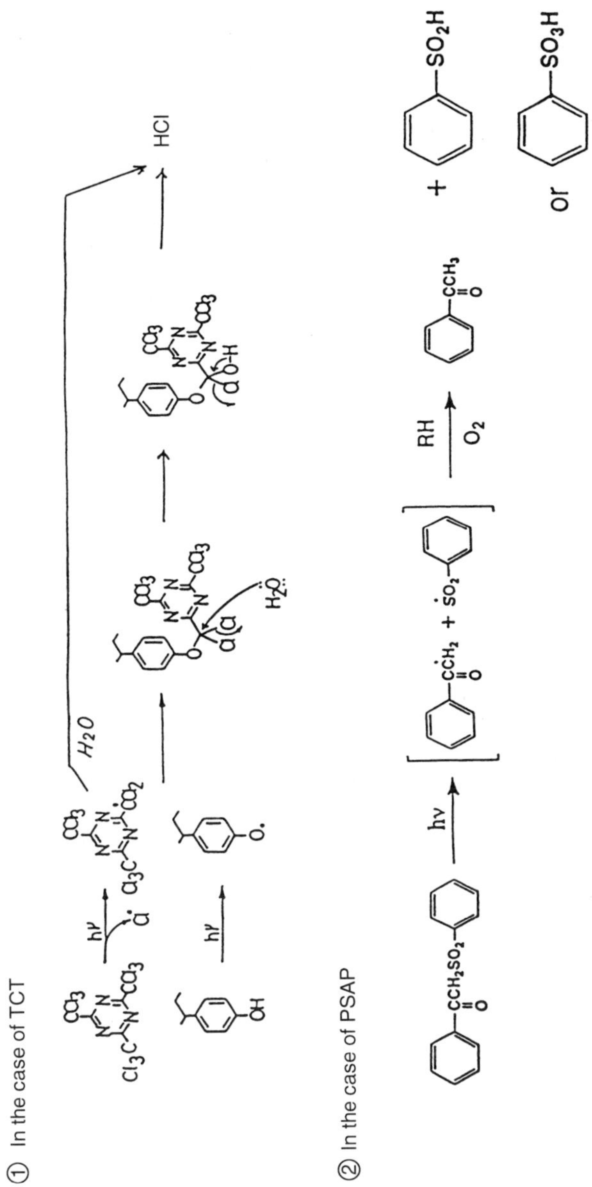

Figure 9. Mechanisms of photoinduced acid generation of TCT and PSAP

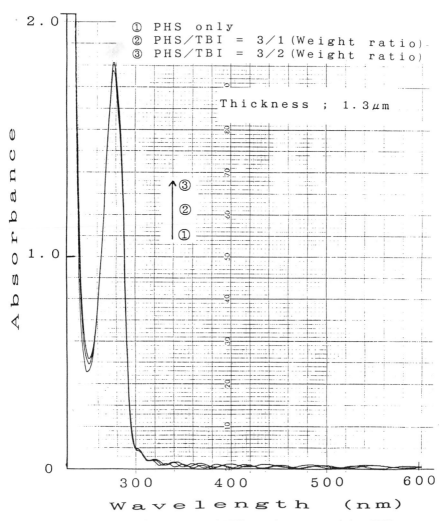

Figure 10. Absorption spectra of PHS membrane containing TBI

CONCLUSION

A novel analytic method of photoinduced acid generation was developed. This mehod is limited to noncharged type photoacid generators. The relative efficiency of photoinduced acid generation was easily detected using this method. And the analytic limitation of this method was 10-7 molar order of H + . In the photoinduced acid generation mechanism from TBI, aphoto sensitization mechanism via the PHS matrix resin was suggested.

REFERENCES

1. Lingnau, J.; Dammel, R.; Theis, J. Solid State Technology, 1989, 32 (9), 105.; ibid, 1989, 32 (10), 107.
2. Tomita, B. J. Polym. Sci., Pholm. Chem. Ed., 1977, 15, 2347.
3. Ito, T.; Sakata, M.; Yamasita, Y.; et al. "Polymers for Microelectronics", 1990, VCM publ., New York, pp293.
4. Tsunooka, M.; Yanagi, H.; Kitayama, M.; Shirai, M. J. Photopolym. Sci. Techol., 1991, 239, 4.

RECEIVED May 19, 1993

Chapter 5

Acid-Catalyzed Dehydration

A New Mechanism for Chemically Amplified Lithographic Imaging

H. Ito, Y. Maekawa[1], R. Sooriyakumaran[2], and E. A. Mash[3]

IBM Research Division, Almaden Research Center, 650 Harry Road, San Jose, CA 95120–6099

New chemical amplification resist systems based on acid-catalyzed dehydration reactions are described. Intramolecular dehydration of poly[4-(2-hydroxy-2-propyl)styrene] results in the change of the polarity from a polar to a nonpolar state, allowing swelling-free negative imaging with alcohol as a developer. The pendant α-methylstyrenic moieties generated by dehydration further undergo dimerization, resulting in concomitant crosslinking, which disallows positive imaging. The *tertiary*-alcohol resist provides an extremely high sensitivity. Polystyrene with a pendant *secondary* alcohol group has been also prepared and evaluated as a resist material. Its unexpectedly high sensitivity is a result of crosslinking through ether linkage produced *via* intermolecular dehydration. The O-alkylation product is then converted to a linear dimer of styrene. Lithographic evaluations and mechanistic studies are presented.

Chemical amplification (*1*) based on radiation-induced acidolysis is a very attractive approach to dramatically increasing radiation sensitivities of resist systems for use in short wavelength lithographic technologies (deep UV < 300 nm, electron beams, and x-ray radiations). Acid-catalyzed deprotection (de-esterification) of polymer pendant groups to change a polarity of repeat units and the solubility of polymer has been successfully employed in the design of the tBOC dual tone resist (*2*), which allowed manufacture of 1-megabit dynamic random access memory devices by deep UV lithography (*3*). The polarity change from a nonpolar to a polar state as embodied in the tBOC resist provides positive imaging with aqueous base or alcohol or swelling-free negative imaging with nonpolar organic solvents (Scheme I). Such an acid-catalyzed deprotection mechanism to photochemically generate a base-soluble functionality has attracted a great deal of attention and is currently pursued by many research groups for the design of aqueous base developable, positive resist systems for replacement of novolac/diazoquinone resists (*4*).

A polarity reversal has been also utilized in the chemical amplification scheme (Scheme I). One such system employs thermal deesterification and acid-catalyzed rearrangement of poly(2-cyclopropyl-2-propyl 4-vinylbenzoate)(*5*).

---

[1]Current address: Department of Chemistry, University of Wisconsin, Madison, WI 53706
[2]Current address: IBM General Technology Division, East Fishkill Facilities, Hopewell Junction, NY 12533
[3]Current address: Department of Chemistry, University of Arizona, Tucson, AZ 85721

0097–6156/94/0537–0064$07.00/0
© 1994 American Chemical Society

We have been also interested in inducing a reverse polarity change from a polar to a nonpolar state for lithographic imaging (Scheme I) and reported that pinacol rearrangement, which is acid-catalyzed, of pendant *vic*-diol to ketone (Scheme II) cleanly and facilely occurs in polymeric films (*6, 7*). Pinacol rearrangement of small *vic*-diols to ketones or aldehydes in phenolic resins provides aqueous base developable, negative, chemical amplification resists with high sensitivity and high resolution (*6–8*). Other examples of the reverse polarity change that have been reported include photochemical transformation of pyridinium ylides to 1,2-diazepines (*9*), acid-catalyzed condensation of dissolution-promoting silanol to dissolution-inhibiting siloxane in a novolac resin (*10*), photochemically-induced partial etherification and esterification of poly(4-hydroxystyrene) (*11–13*). We believe that polarity alteration through clean structural transformation could offer high resist contrasts, swelling-free development in a negative mode, and some versatile applications. Therefore, we have continued to exploit the reverse polarity change in the resist design and report in this paper sensitive resist systems based on acid-catalyzed dehydration of pendant alcohol to olefin (Scheme III) as well as a different pathway in dehydration of pendant *secondary* alcohol.

EXPERIMENTAL

Materials

4-(2-Hydroxy-2-propyl)styrene [4-vinylphenyl dimethyl carbinol, α,α-dimethyl(4-vinyl)benzyl alcohol] and 4-(1-hydroxyethyl)styrene [4-vinylphenyl methyl carbinol, α-methyl(4-vinyl)benzyl alcohol] were synthesized according to Scheme IV by reacting 4-vinylphenylmagnesium chloride with acetone and acetaldehyde, respectively, purified by column chromatography, and subjected to radical polymerization with benzoyl peroxide (BPO) or 2,2'-azobis(isobutyronitrile) (AIBN) in tetrahydrofuran (THF) at 60 °C. The monomer (*14*) and polymer syntheses (*15, 16*) can be found in the literature. In our polymerizations the monomer concentration was kept low (6 mL THF/g monomer) and the conversions were also kept relatively low (~70 %) to avoid gelation. The unreacted monomers were removed by column chromatography and the polymers were purified by precipitation in water. Simple precipitation procedures alone did not remove the unreacted monomers.

Triphenylsulfonium hexafluoroantimonate and trifluoromethanesulfonate (triflate) employed as deep UV acid generators in our formulations were synthesized according to the literature (*17, 18*). Propylene glycol monomethyl ether acetate and cyclohexanone were used as our casting solvents.

Lithographic Evaluation and Imaging

Resist films were spin-cast onto Si wafers for imaging experiments, NaCl plates for IR studies, and quartz discs for UV measurements and then baked at 80-130 °C for 2 min. The deep UV exposure systems employed in these studies

1. polarity change   : nonpolar ⟶ polar

2. polarity reversal   : nonpolar ⟶ less polar
                                    ↓
                               more polar

3. reverse polarity change : polar ⟶ nonpolar

Scheme I. Three modes of polarity alteration.

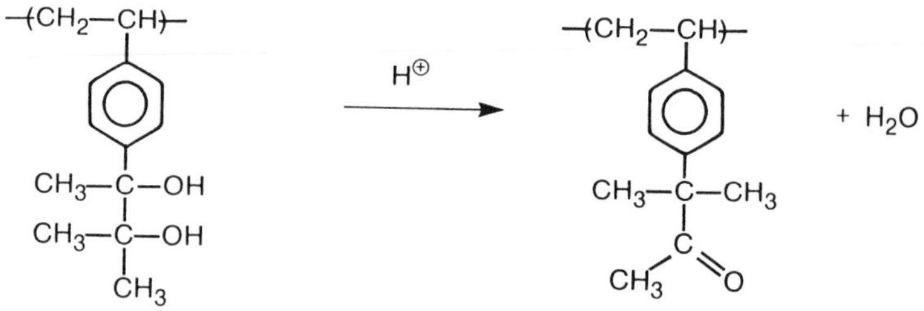

Scheme II. Pinacol rearrangement of polymeric *vic*-diol.

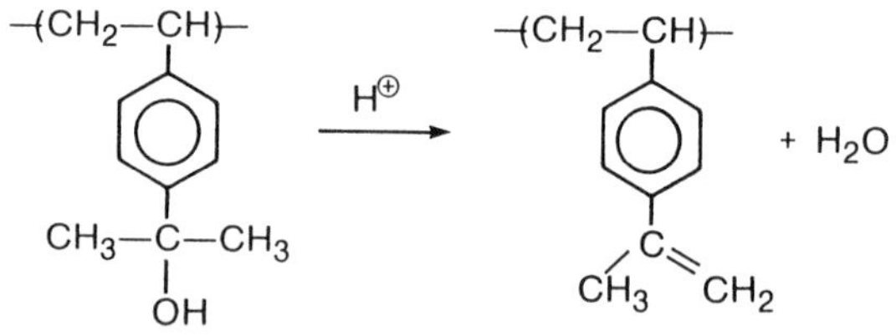

Scheme III. Acid-catalyzed intramolecular dehydration.

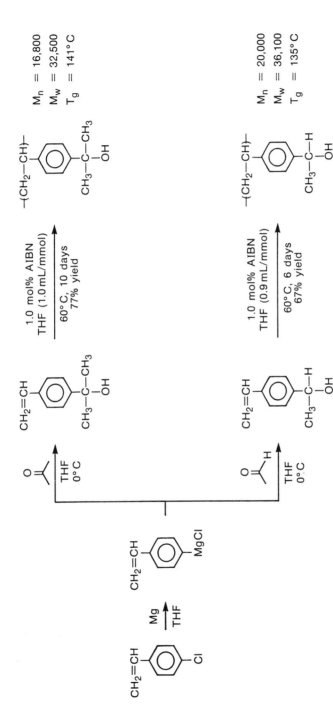

Scheme IV. Syntheses of polystyrenes bearing *tertiary* and *secondary* alcohol.

were an Optical Associate, Inc. apparatus and a Ultratech X-248E 1x stepper. The developer solvent for negative imaging was isopropanol (IPA).

Measurements

Molecular weight determination was made by gel permeation chromatography (GPC) using a Waters Model 150 chromatograph equipped with 6 μStyragel columns at 30 °C in THF. Thermal analyses were performed on a Du Pont 910 at 10 °C/min for differential scanning calorimetry (DSC) and on a Perkin Elmer TGS-2 at a heating rate of 5 °C/min for thermogravimetric analysis (TGA) under nitrogen atmosphere. IR spectra of the resist films were obtained with an IBM IR/32 FT spectrometer using 1-mm thick NaCl discs as substrates. UV spectra were recorded on a Hewlett-Packard Model 8450A UV/VIS spectrometer using thin films cast on quartz plates. NMR spectra were obtained on an IBM Instrument NR-250/AF spectrometer. Film thickness was measured on a Tencor alphastep 200.

RESULTS AND DISCUSSION

Polymer Syntheses and Characterization

The preparation and polymerization of the styrenes bearing the *tertiary* (*15*) and *secondary* (*16*) alcohol structures have been reported in the literature. Fréchet incorporated the *tertiary* alcohol structure into polystyrene by Friedel-Crafts acetylation followed by a Grignard reaction (*19*). The *tertiary* alcohol was then acetylated and used as the site for living cationic graft polymerization of isobutene (*19*). The same author also reported synthesis of the *tertiary* alcohol monomer by the same method as ours and described preparation of polymers of styrenes bearing pendant *tertiary* reactive sites and their use as rubbers, coatings, and lubricants in his recent U. S. Patent (*20*). When the polymerizations of these monomers are run at high monomer concentrations and/or to high conversions, the reaction mixtures tend to gel presumably due to chain transfer reactions. However, low monomer concentrations ($\sim$ 6 ml THF/g monomer) and medium conversions ($\sim$ 70 %) provided good, soluble polymers (M_n = 17,000-23,000, M_w = 33,000-40,000, and M_w/M_n = 1.7-2.0).

NMR spectra of the alcohol polymers in THF-d_8 are presented in Figures 1 (^{1}H) and 2 (^{13}C). The strong absorptions at ca. 1.4 ppm are due to the pendant methyl groups and the resonances near the upper field THF absorption (1.5-2.0 ppm) due to the backbone methylene and methine protons. The aromatic protons resonate between 6.3 and 7.2 ppm. The proton resonance at 4.3 ppm in Figure 1a can be assigned to the hydroxyl group of the *tertiary* alcohol. The two proton resonances at ca. 4.5 ppm in Figure 1b are due to the hydroxyl and methine groups of the *secondary* alcohol moiety. ^{13}C NMR spectra of the polymers presented in Figure 2 are straightforward, indicating that the polystyrenes with the pendant alcohol groups made by radical polymerization are free of foreign structures and the unreacted monomers (after column chromatography).

The *tertiary* and *secondary* alcohol polymers exhibit their glass transitions at 141 °C (M_n = 16,800 and M_w = 32,500) and at 135 °C (M_n = 20,000 and

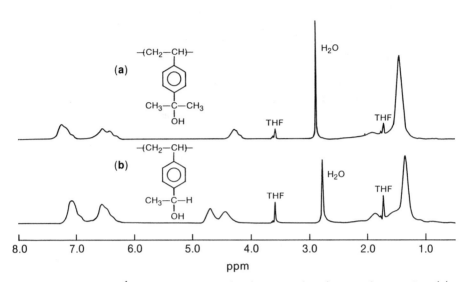

Figure 1. 250 MHz ^{1}H NMR spectra of polystyrene bearing pendant *tertiary* (a) and *secondary* alcohol (b) in THF-d$_8$.

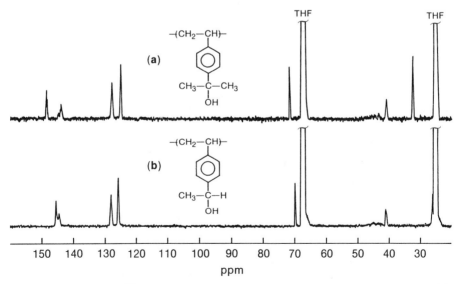

Figure 2. 62.9 MHz ^{13}C NMR spectra of polystyrene bearing pendant *tertiary* (a) and *secondary* alcohol (b) in THF-d$_8$.

M_w = 36,100), respectively, according to DSC analyses. TGA curves of the alcohol polymers are presented in Figure 3. The polymer powders were first heated to 100 °C on a TGA apparatus to remove absorbed water, cooled to room temperature, and then heated in a normal fashion to investigate thermal events. Whereas the *secondary* alcohol exhibits only a gradual weight loss between 300 and 400 °C, followed by main chain scission, the *tertiary* alcohol polymer undergoes a sharp weight loss of 11 % at ca. 200 °C, which corresponds to a quantitative loss of water. However, although formation of olefinic double bonds in the heated *tertiary* alcohol polymer was detected by IR spectroscopy (*21*), the thermal dehydration appears to be significantly affected by moisture. A *tertiary* alcohol sample further dried *in vacuo* at 100 °C for 1 hr did not exhibit any distinct sharp weight loss below 370 °C. Thus, the thermal dehydration of the *tertiary* alcohol seems to be a complex phenomenon and is currently under investigation.

Tertiary Alcohol System

We formulated a resist by mixing our very first *tertiary* alcohol polymer (M_n = 7,700 and M_w = 11,500) with 4.75 wt% of triphenylsulfonium hexafluoroantimonate, which was subjected to preliminary lithographic evaluation. A deep UV sensitivity curve is presented in Figure 4 for this formulation. The resist provided a high contrast ($\gamma > 6$) and an extremely high sensitivity with a 0.79-μm-thick film becoming completely insoluble in IPA at < 0.1 mJ/cm^2 of 254 nm radiation. The resist was imaged using an Ultratech 248 nm 1x stepper. Under our processing conditions, however, the resist was very much overexposed even at 0.4 mJ/cm^2, which is the lowest dose the stepper can provide. Although small features were fused due to overexposure, Figure 5 demonstrates good printability of the resist. The undercut profile resulted from the high optical density (OD) of the resist film (OD = 0.72/μm at 248 nm). The polymer itself had a high UV absorption (OD = 0.55/μm at 248 nm) due to the effect of the benzoyl end group derived from BPO on the UV absorption of low molecular weight polymers (M_n = 7,700) in this case (*22*) and also due to contamination with the unreacted monomer (no column chromatography performed).

The good preliminary results encouraged us to continue to work on the *tertiary* alcohol resist. We prepared a polymer with higher molecular weights (M_n = 16,800 and M_w = 32,500) using AIBN as the initiator and purified the polymer by column chromatography to remove the unreacted monomer. The use of AIBN in preparation of low molecular weight polymers results in reduction of sensitivity due to the poisoning effect of the CN group attached to the polymer end (*22*). The destructive CN effect was relatively minor in this case due to relatively high molecular weights of the polymer and could be accepted because the sensitivity of the resist was excessively high. In order to further reduce the sensitivity to a workable range, we decreased the concentration of the acid generator from 4.75 to 1.5 wt%. At the same time we replaced hexafluoroantimonate with non-metallic triflate.

In Figure 6 are presented IR spectra of the *tertiary* alcohol resist containing 1.5 wt% of triphenylsulfonium triflate before and after UV exposure. The film

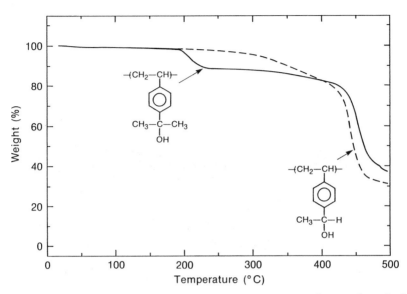

Figure 3. TGA curves of polystyrene with pendant *tertiary* and *secondary* alcohol (heating rate: 5 °C/min).

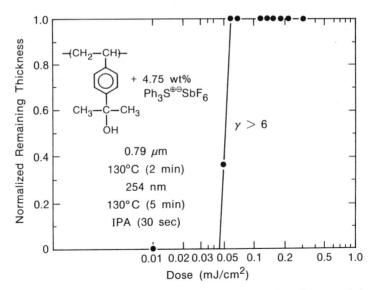

Figure 4. Deep UV sensitivity curve of *tertiary* alcohol resist containing 4.75 wt% $Ph_3S^{+-}SbF_6$

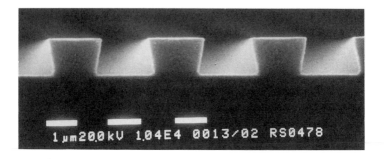

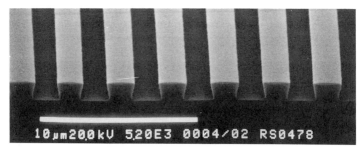

Figure 5. Scanning electron micrograph of negative images projection-printed at 0.4 mJ/cm^2 of 248 nm radiation in preliminary *tertiary* alcohol resist containing 4.75 wt% Ph$_3$S^{+-}SbF$_6$.

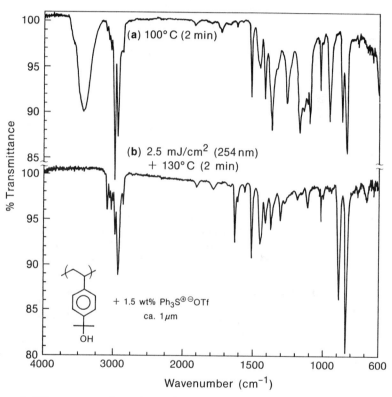

Figure 6. IR spectra of *tertiary* alcohol resist before and after deep UV exposure.

baked at 100 °C for 2 min exhibits a strong OH absorption at 3500 cm^{-1} (Figure 6a), which completely disappears upon exposure to 2.5 mJ/cm^2 of 254 nm radiation followed by postbake at 130 °C for 2 min (Figure 6b), indicating that the acid-catalyzed dehydration of the *tertiary* alcohol is very facile. The appearance of a new sharp peak at ca. 1627 cm^{-1} suggests generation of conjugated olefinic double bonds, which is accompanied by a strong absorption at 886 cm^{-1} characteristic to a vinylidene structure. Thus, the IR studies suggest that the *tertiary* alcohol undergoes acid-catalyzed dehydration to an α-methylstyrene structure as shown in Scheme III and that the olefinic double bond of α-methylstyrene does not enter into cationic polymerization appreciably. As the loss of water does not occur at 130 °C in the absence of acid, the dehydration reaction observed upon postbake is acid-catalyzed.

The formation of the conjugated structure is also suggested by UV spectroscopic studies. In Figure 7 are presented UV spectra of a 0.92-μm-thick film of the *tertiary* alcohol resist containing 1.5 wt% of triphenylsulfonium triflate before and after 254 nm exposure. The resist film is highly transparent in the 250 nm region with an OD of 0.254/μm at 248 nm (OD of the polymer itself is 0.184/μm at 248 nm). Postbaking the film at 130 °C for 2 min after exposure to 1.0 mJ/cm^2 of 254 nm radiation resulted in a dramatic increase in its UV absorption with its OD reaching 2.5, which indicated that the resist was highly sensitive and that the acidolysis product had a conjugated structure.

Deep UV sensitivity curves of the *tertiary* alcohol resist (ca. 1.0 μm thick) containing 1.5 wt% of triphenylsulfonium triflate are presented in Figure 8. The resist film was prebaked at 100 °C, exposed to 254 nm radiation, and postbaked at 100 °C for 2 min. The film thickness was measured after postbake (Δ) and after development with IPA for 1 min (·). The exposed area begins to lose its thickness at ~0.5 mJ/cm^2 upon postbake and the maximum shrinkage is attained at about 0.6 mJ/cm^2, which corresponds to the quantitative loss of water (11 %). The small shrinkage associated with the dehydration chemistry may be advantageous over the large thickness loss typically observed in deprotection chemistries (the tBOC resist could lose as much as 45 % of its thickness upon postbake). The exposed film begins to become insoluble in IPA at ca. 0.5 mJ/cm^2 and full retention of its thickness is achieved at 0.6 mJ/cm^2 with a high γ of 8.7. The resist is still very sensitive even at a low sulfonium salt loading of 1.5 wt%. In Figure 9 is presented a scanning electron micrograph of negative 1-μm line/space patterns contact printed in the *tertiary* alcohol resist at 3.4 mJ/cm^2. Bake temperatures were 80 °C and development was performed for 1 min in IPA. In spite of the high sensitivity, the resist solution is stable at room temperatures for at least 6 months.

The reverse polarity change from a polar alcohol to a greasy olefin (Scheme III) thus allows negative imaging with use of a polar alcohol as a developer. However, positive-tone imaging with xylenes as a developer was unsuccessful, suggesting concomitant crosslinking involving the α-methylstyrene structure. A small amount of crosslinking, which may not be detected by IR and UV spectroscopies, could render the exposed regions insoluble. The *tertiary* alcohol resist began to become soluble in xylenes at ca. 1 mJ/cm^2 and lost ca. 60 % of its thickness in the exposed regions upon development with xylenes at 2–5

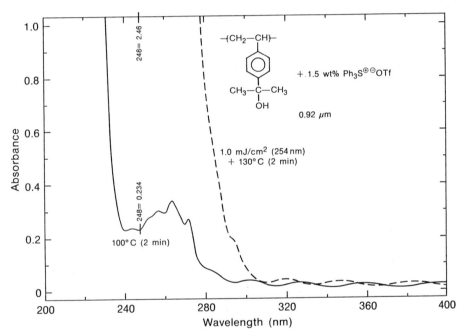

Figure 7. UV spectra of *tertiary* alcohol resist before and after deep UV exposure.

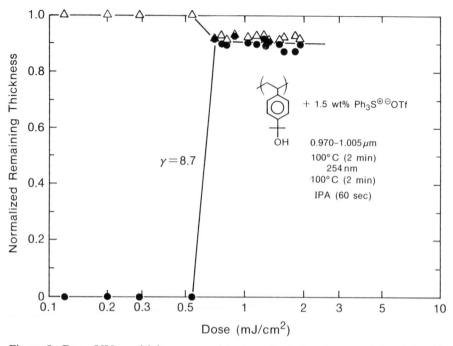

Figure 8. Deep UV sensitivity curves of *tertiary* alcohol resist containing 1.5 wt% $Ph_3S^{+-}OTf$.

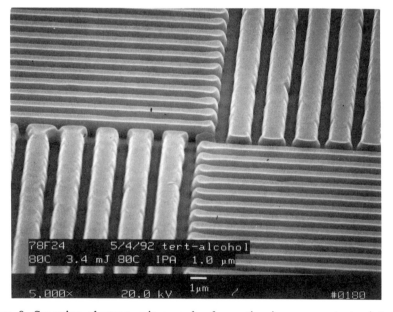

Figure 9. Scanning electron micrograph of negative images contact-printed at 3.4 mJ/cm^2 in *tertiary* alcohol resist containing 1.5 wt% $Ph_3S^{+-}OTf$.

mJ/cm^2 and then the fraction insoluble in xylenes increased slightly to ca. 55 % when the exposure dose was raised to 9 mJ/cm^2.

In order to further investigate the imaging mechanism and side reactions, we followed model reactions in solution with NMR and isolated and identified model products in some cases. The results are summarized in Scheme V. When the *tertiary* alcohol polymer was treated with trifluoromethanesulfonic acid (triflic acid, TfOH) in methanol at room temperature, a methyl ether was the sole product as the ^{13}C NMR spectra in acetone-d$_6$ in Figure 10 indicate. When CDCl$_3$ was used as the solvent for the reaction of the *tertiary* alcohol polymer with triflic acid, ^{1}H NMR indicated predominant formation of polystyrene with a *p*-isopropenyl group (ca. 67 % at room temperature after ca. 1 hr) without any precipitation. The solution became slightly cloudy and the olefin proton resonances decreased in intensity upon standing at room temperature for 1 day, suggesting some crosslinking. The benzylic *tertiary* carbocation generated by dehydration exclusively reacts with the solvent methanol or undergoes in CDCl$_3$ β-proton elimination to form α-methylstyrene. The *tertiary* alcohol is consumed rapidly and α-methylstyrene is the predominant product (61 %) at an early stage (2 hrs) as the ^{1}H NMR spectra in Figure 11 indicate. A set of resonances centered at 5.3 ppm and a strong signal at 2.2 ppm in Figure 11b are due to α-methylstyrene. Although the spectrum of the mixture (Figure 11b) looks very complex, the rest of the resonances can be all assigned to α-methylstyrene dimers, which are produced by reaction of the carbocation onto the β-carbon of α-methylstyrene followed by β-proton elimination (linear dimers) and by electrophilic aromatic substitution of the dimeric carbocation onto the penultimate benzene ring (cyclic dimer, indane) (Scheme V). A second set of resonances centered at 5 ppm is due to the vinylidene protons, the singlet at 2.9 ppm to the methylene protons, and the singlet at 1.3 ppm to the methyl protons of the linear dimer with an external double bond, which amounted to 14 % after 2 hrs. The second linear dimer bearing a trisubstituted olefin structure exhibits its olefinic proton resonance at 6.2 ppm (very small in Figure 11b) and its methyl resonances at 1.6 ppm and was a very minor product (1 % after 2 hrs). The second major product observed after 2 hrs was the cyclic dimer (24 %), as evidenced by relatively strong methyl proton resonances at 1.8, 1.4, and 1.1 ppm and an AB quartet centered at 2.4 ppm due to the magnetically non-equivalent methylene protons of the indane ring. The product distribution simply changed without generating new structures upon standing, according to NMR analyses, as summarized in Scheme V. α-Methylstyrene was consumed (from 61 % after 2 hrs to 39 % after 5 hrs) and converted to the dimers upon standing. The linear dimer with the trisubstituted olefin structure was still a minor product (6 %) after 5 hrs. The other linear dimer and the cyclic dimer amounted to 27-28 % each after 5 hrs. The cyclic dimer was isolated by column chromatography from a similar but separate experiment and subjected to NMR analyses. Thus, α-methylstyrene is the major product in the acid-catalyzed dehydration of α,α-dimethylbenzyl alcohol but can undergo further reactions to linear and cyclic dimers. In contrast to the solution acidolysis at room temperature, however, treatment of the *tertiary* alcohol with triflic acid at 100 °C for 2 min in bulk, which mimics the postbake

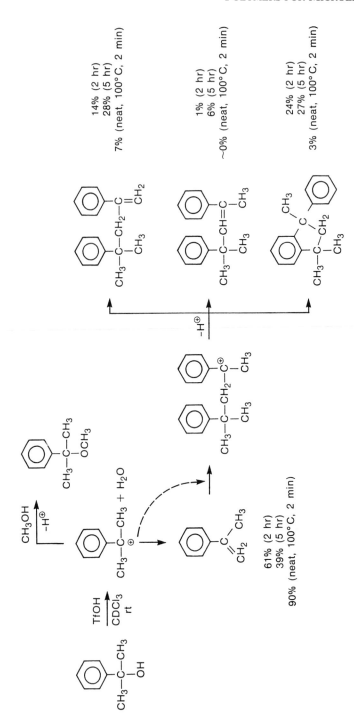

Scheme V. Pathways in acid-catalyzed dehydration of *tertiary* alcohol.

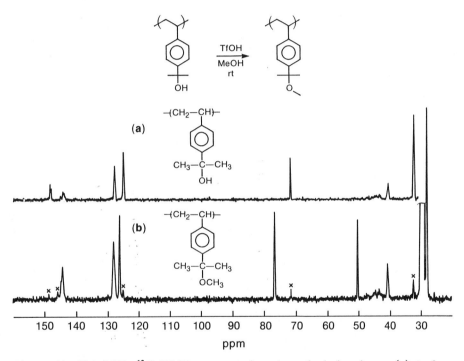

Figure 10. 62.9 MHz ^{13}C NMR spectra of *tertiary* alcohol polymer (a) and a polymer produced by treatment with triflic acid in methanol (b) in acetone-d$_6$

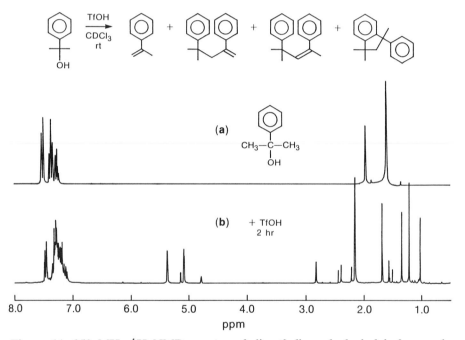

Figure 11. 250 MHz ^{1}H NMR spectra of dimethylbenzyl alcohol before and after addition of triflic acid at room temperature in CDCl$_3$.

conditions of the resist more closely, produced α-methylstyrene almost exclusively (90 %).

It is therefore concluded based on the spectroscopic studies that the *tertiary* alcohol polymer undergoes acid-catalyzed intramolecular dehydration to form an α-methylstyrene structure and that this reverse polarity change from a polar to a nonpolar state is primarily responsible for the negative imaging. However, crosslinking via dimerization through the α-methylstyrene structure may contribute to the negative imaging mechanism and disallow positive-tone imaging.

Secondary Alcohol System

A *secondary* carbocation is significantly less stable than a *tertiary* carbocation. However, encouraged by an extremely high sensitivity exerted by the *tertiary* alcohol resist, we decided to investigate a corresponding *secondary* alcohol polymer, which we expected to offer a decent sensitivity. Deep UV sensitivity curves of the *secondary* alcohol resist (1 μm thick) containing 1.5 wt% of triphenylsulfonium triflate are presented in Figure 12. The film was prebaked at 100 °C for 2 min, exposed to 254 nm radiation, and postbaked at 100 °C for 2 min. The film thickness was measured after postbake (Δ) and after development with IPA ($\cdot$). In contrast to the *tertiary* alcohol system, no thickness loss was observed upon postbake below 3 mJ/cm^2. However, the resist became insoluble in IPA at a low dose of 1 mJ/cm^2 and the 100 % thickness retention was achieved at $\sim$1.2 mJ/cm^2 with a high γ of 6.8. Thus, the *secondary* alcohol resist is only slightly less sensitive than the *tertiary* counterpart. In Figure 13 is presented a scanning electron micrograph of negative images contact-printed at 3.6 mJ/cm^2 in a 1.0-μm-thick film of the *secondary* alcohol resist containing 1.5 wt% of triphenylsulfonium triflate. The prebake and postbake temperatures were 80 °C and IPA was used as the developer.

In contrast to the *tertiary* alcohol system, however, IR measurements did not detect any significant spectral changes at the imaging dose, which was consistent with observation of no shrinkage on postbake. An IR spectrum of the resist film exposed to a very high dose of 20 mJ/cm^2 and postbaked at 130 °C for 2 min is presented in Figure 14 together with a spectrum of a prebaked film. The absorptions at 1700-1800 cm^{-1} are due to a residual casting solvent (cyclohexanone free and hydrogen-bonded with the polymer). Although shrinkage of the OH absorption and appearance of a conjugated olefinic double bond (at 1630 cm^{-1}) are clearly observed, the dehydration reaction is only partial even at the very high dose. The resist film, initially very transparent with an OD of 0.266/μm at 248 nm, became intensely opaque upon postbake after exposure to 3 mJ/cm^2. It is not clear at the moment whether or not only a small amount of conjugated structures could cause such a large increase in the UV absorption.

The *secondary* alcohol system provides an unexpectedly high resist sensitivity although the degree of expected dehydration is very small. In order to elucidate the imaging mechanism, we carried out model reactions in solution and analyzed the products by NMR. When α-methylbenzyl alcohol was treated with triflic acid in CDCl$_3$ at room temperature for 1 hr, the alcohol was completely consumed and almost quantitatively converted to an enantiomeric mixture of

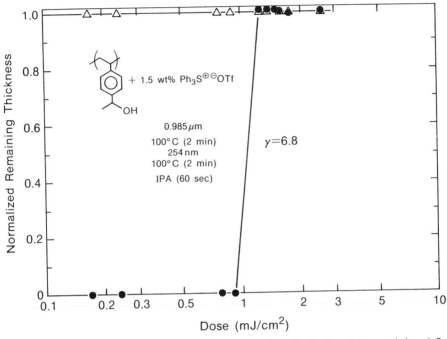

Figure 12. Deep UV sensitivity curves of *secondary* alcohol resist containing 1.5 wt% $Ph_3S^{+-}OTf$.

Figure 13. Scanning electron micrograph of negative images contact-printed at 3.6 mJ/cm² in *secondary* alcohol resist containing 1.5 wt% $Ph_3S^{+-}OTf$.

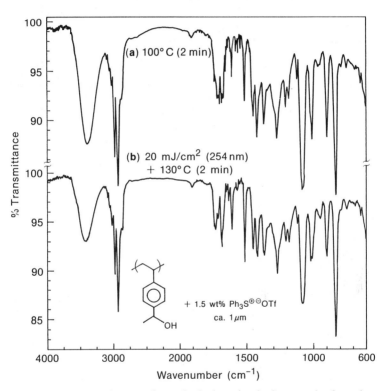

Figure 14. IR spectra of *secondary* alcohol resist before and after deep UV exposure.

di(α-methylbenzyl) ether through intermolecular dehydration as the ^{1}H NMR spectra in Figure 15 demonstrate. Styrene and its oligomers were also observed in a minute amount. In a separate experiment the ether dimer was isolated by column chromatography, a ^{13}C NMR spectrum of which is exhibited in Figure 16a. Heating the reaction mixture at 60 °C for 30 min resulted in conversion of the ether dimer to monomeric and dimeric styrenes (Figure 15b). It is interesting to note that one of the enantiomers is consumed faster than the other. Two sets of quartets centered at ca. 5.6 and 6.75 ppm are due to the β- and α-protons of styrene. Thus, in contrast to the rapid intramolecular dehydration of the *tertiary* alcohol to form α-methylstyrene, the *secondary* alcohol initially undergoes acid-catalyzed intermolecular dehydration to an ether dimer, which then splits off styrene under more harsh conditions (Scheme VI). Upon further standing at 60 °C for 1 day, the ether dimer was destroyed nearly completely and the styrene monomer was also completely consumed as Figure 15c indicates. The final product shown in Figure 15c has been confirmed to be a linear dimer of styrene (Scheme VI), ^{13}C NMR spectrum of which is presented in Figure 16b along with the spectrum of the ether dimer. Very small amounts of a cyclic dimer (indane) and oligomers were produced as well. Acid-catalyzed dimerization of substituted styrenes has been extensively studied by Higashimura et al. (*23–25*) and they showed that oxo acids or their derivatives including triflic acid produced efficiently and selectively linear unsaturated dimers in nonpolar solvents at a high temperature (70 °C), which supports our result.

It is concluded that the negative imaging of the *secondary* alcohol resist is primarily based on crosslinking through ether linkage produced *via* intermolecular dehydration. A further reaction to form the linear unsaturated dimer structure could also contribute to crosslinking. Thus, since the intermolecular dehydration to form the ether dimer and further reactions to styrenic dimers are not significant at the imaging dose, the reverse polarity change plays only a small role in the imaging of the *secondary* alcohol resist system, and therefore the resist seems to suffer from swelling during development.

SUMMARY

1. Acid-catalyzed intramolecular dehydration of a pendant *tertiary* alcohol structure to an olefinic structure occurs very efficiently and quantitatively in the solid polymer film, which results in a reverse polarity change from a polar to a nonpolar state, allowing highly sensitive negative imaging with alcohol as a developer.

2. The α-methylstyrene structure produced in the exposed region through intramolecular dehydration of the *tertiary* alcohol polymer could further undergo acid-catalyzed dimerization to linear unsaturated dimes and a cyclic dimer, which would result in minor crosslinking and could contribute to negative imaging, disallowing positive imaging.

3. The corresponding *secondary* alcohol resist system provides an unexpectedly high sensitivity while acid-catalyzed dehydration in the polymer film is slow, which is due to crosslinking through ether linkage produced via intermolecular dehydration.

4. The ether dimer could further react with acid to split styrene, which dimerizes to a linear unsaturated dimer.

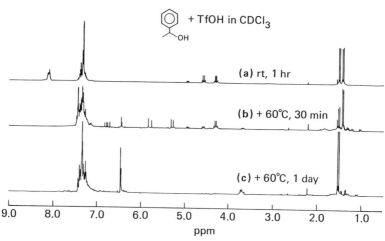

Figure 15. 250 MHz ^{1}H NMR spectra of methylbenzyl alcohol treated with triflic acid for 1 hr at room temperature (a), subsequently heated at 60 °C for 30 min (b), and further heated at 60 °C for 1 day (c) in CDCl$_3$.

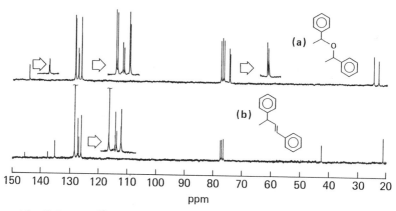

Figure 16. 62.9 MHz ^{13}C NMR spectra of enantiomeric mixture of ether dimer isolated (a) and linear unsaturated styrene dimer (b) in CDCl$_3$.

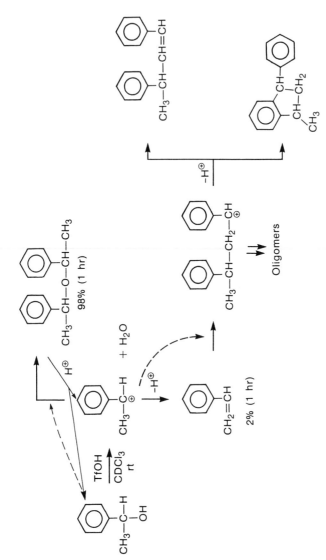

Scheme VI. Pathways in acid-catalyzed dehydration of *secondary* alcohol.

5. Thus, the imaging mechanisms for the *tertiary* and *secondary* alcohol systems are much different. The former is based on the reverse polarity change induced by intramolecular dehydration and the latter relies on crosslinking through intermolecular dehydration.

ACKNOWLEDGMENT

The authors thank N. Clecak for his Ultratech stepper exposure of one of the resists, G. May for his NMR maintenance, H. Truong for her thermal analyses, and S. Ferline for her GPC work. Prof. J. Fréchet kindly brought his recent patent article to our attention, which is gratefully acknowledged.

REFERENCES

1. Ito, H.; Willson, C. G. *Polym. Eng. Sci.* **1983**, *23*, 1012.
2. Ito, H.; Willson, C. G. In *Polymers in Electronics*; Davidson, T., Ed.; Symposium Series 242; American Chemical Society: Washington, D. C., **1984**, p. 11.
3. Maltabes, J. G.; Holmes, S. J.; Morrow, J.; Barr, R. L.; Hakey, M.; Reynolds, G.; Brunsvold, W. R.; Willson, C. G.; Clecak, N. J.; MacDonald, S. A.; Ito, H. *Proc. SPIE* **1990**, *1262*, 2.
4. Ito, H. In *New Aspects of Radiation Curing in Polymer Science and Technology*; Fourssner, J. P.; Rabek, J. F., Eds.; Elsevier: London, **1993**; Vol. 4, Chapter 11, pp. 237–359, and references therein.
5. Ito, H.; Ueda, M.; England, W. P. *Macromolecules* **1990**, *23*, 2589.
6. Sooriyakumaran, R.; Ito, H.; Mash, E. A. *Proc. SPIE* **1991**, *1466*, 419.
7. Ito, H.; Sooriyakumaran, R.; Mash, E. A. *J. Photopolym. Sci. Technol.* **1991**, *4*, 319.
8. Uchino, S.; Iwayanagi, T.; Ueno, T.; Hayashi, N. *Proc. SPIE* **1991**, *1466*, 429.
9. Schwalm, R.; Böttcher, A.; Koch, H. *Proc. SPIE* **1988**, *920*, 21.
10. Ueno, T.; Shiraishi, H.; Hayashi, N.; Tadano, K.; Fukuma, E., Iwayanagi, T. *Proc. SPIE* **1990**, *1262*, 26.
11. Ito, H.; Schildknegt, K.; Mash, E. A. *Proc. SPIE* **1991**, *1466*, 408.
12. Ito, T.; Sakata, M.; Yamashita, Y. *J. Photopolym. Sci. Technol.* **1991**, *4*, 403.
13. Nakamura, K.; Sekimoto, N.; Yamada, T.; Tomascewski, G. *J. Photopolym. Sci. Technol.* **1991**, *4*, 415.
14. Kunitomo, T.; Tanimoto, S.; Oda, R. *Kogyo Kagaku Zasshi* **1965**, *68*, 1976.
15. Anda, K.; Iwai, S. *J. Polym. Sci., Part A* **1969**, *7*, 2414.
16. An, Y.; Ding, M. *Chem. Abstr.* **1985**, *103*, 142466a.
17. Crivello, J. V.; Lam, J. H. W. *J. Org. Chem.* **1978**, *43*, 3055.
18. Miller, R. D.; Renaldo, A. F.; Ito, H. *J. Org. Chem.* **1988**, *53*, 5571.
19. Jiang, Y.; Fréchet, J. M. J. *Polym. Preprints* **1989**, *30(1)*, 127.
20. Fréchet, J. M. J. *U. S. Patent* 5,084,522, **1992**.
21. Anda, K.; Nishiwaki, T.; Iwai, S. *Kobunshi Kagaku* **1972**, *29*, 118.
22. Ito, H.; England, W. P.; Lundmark, S. B. *Proc. SPIE* **1992**, *1672*, 2.
23. Higashimura, T.; Hiza, M.; Hasegawa, H. *Macromolecules* **1979**, *12*, 217.
24. Higashimura, T.; Hiza, M.; Hasegawa, H. *Macromolecules* **1979**, *12*, 1058.
25. Hiza, M.; Hasegawa, H.; Higashimura, T. *Polym. J.* **1980**, *12*, 379.

RECEIVED December 30, 1992

Chapter 6

An Alkaline-Developable Positive Resist Based on Silylated Polyhydroxystyrene for KrF Excimer Laser Lithography

Eiichi Kobayashi, Makoto Murata, Mikio Yamachika, Yasutaka Kobayashi, Yoshiji Yumoto, and Takao Miura

Japan Synthetic Rubber Company, Ltd., 100 Kawajiri-cho, Yokkaichi-shi, Mie 510, Japan

A chemically amplified, positive-working resist system based on silylated polyhydroxystyrene has shown its capability for application to quarter micron lithography. The present paper describes the recent improvement in the resist performance achieved through studies on polymer characteristics and process conditions. Possible measures to suppress a peculiar problem of positive-working chemical amplification systems, i.e., formation of T-shaped profile, is also presented. Although the development is still on the way, the silylated polyhydroxystyrene based resist system shows excellent properties on resolution capability, sensitivity and process latitude.

KrF excimer laser lithography is one of the most promising technologies for production of devices with sub-half micron design rule owing to its theoretical capability in resolution (*1*). Chemical amplification (CA) technique, which can reduce the energy required for imaging, seems to be effective to achieve the desirable through-put against the relatively weak photointensity of the laser (*2*). Thus, many CA type resist systems have been proposed and extensively investigated since 1983 (*3–5*). We also have been investigating a CA type positive-working resist system which is based on silylated polyhydroxystyrene (SiPHS) and photoacid generator (PAG) (*6, 7*). Figure 1 shows the CA scheme of the SiPHS resist system. Upon exposure a PAG releases an acid, and catalytic cleavage of the protecting silyl groups takes place. This catalytic mechanism contributes to raise the resist sensitivity beyond the limitation of quantum yield. We chose the trimethylsilyl group as the protecting silyl group because it was found to leave rapidly under acidic environment even at room temperature and was quite stable under neutral dry condition (*8*). The deprotection at room temperature is advantageous for the resist because its sensitivity does not depend much on post-exposure bake (PEB) temperature. Lithographic images can be obtained with a low temperature PEB treatment or even without PEB treatment (*9*).

On the other hand, several recent papers point out curious profile abnormality for some CA type positive-working resists which prohibits us from

0097–6156/94/0537–0088$06.00/0

bringing these resist systems up to practical use; i.e., T-shaped pattern profile due to formation of surface skin which is insoluble in the developer (*10–12*). It seems to be important to acquire superior resist performance to adjust (1) structure of the protecting silyl group, (2) protecting ratio, what we call the silylation ratio, of the SiPHS, (3) molecular weight of the SiPHS and (4) process conditions such as soft bake and PEB temperature. In this paper, presented are the possible measures for avoiding the T-shaped pattern profile in terms of polymer characteristics and process conditions.

EXPERIMENTAL

Characterization of SiPHS

Molecular weight of the polymer was determined by Gel Permeation Chromatography. A Toso HLC-8020 was employed to measure polystyrene-equivalent molecular weight.

Protecting degree of polyhydroxystyrene, defined as the silylation ratio which is the number of the protected hydroxyl groups against all the hydroxyl groups of the polymer was calculated from 1H-NMR spectrum recorded on a JEOL EX90A by comparing the area of methyl protons of trimethylsilyl group with the area of aromatic protons.

Softening point of the polymer was measured on a Mettler FP 800 thermosystem at a heating rate of 2.5 °C/minute.

Glass transition temperature of the polymer was measured on a Du Pont model 910 under inert atmosphere at a heating rate of 10 °C/minute.

Measurement of dissolution rate

Dissolution rate of the polyhydroxystyrene films and SiPHS resist films, spin-coated and soft baked on Si wafers, were determined by using a Perkin-Elmer Development Rate Monitor (DRM).

Evaluation

A resist solution which consists of SiPHS as the base resin and triphenylsulfonium triflate as the photoacid generator was spin-coated on a Si wafer, and baked for 2 minutes. Lithographic imaging was performed on KrF Excimer Laser Steppers, Canon FPA-3000EX1 (NA0.45) or NIKON NSR 1755EX8. Then PEB was applied for 2 minutes, followed by development in a 2.38wt% aqueous solution of tetramethyl-ammonium hydroxide (TMAH).

RESULTS AND DISCUSSION

Effects of Mw and the silylation ratio

1. Molecular weight (Mw) of polyhydroxystyrene. It is generally important to maximize the difference in dissolution rate a between the exposed and the unexposed regions of a resist films to obtain high lithographic performance. Thus, polyhydroxystyrene, which is produced upon desilylation reaction, seems to be favorable to have higher dissolution rate in our resist system. In order to compare the dissolution behavior of polyhydroxy-styrene polymers varying in Mw, we have synthesized the polymers whose Mw are ranging from about 2×104 to 6×104, and have monitored their dissolution behavior as shown in Figure 2. Unexpectedly, the dissolution rate has been found to be almost independent of Mw within this Mw range (Similar result was also obtained by Long et al. (13)). This result suggests that the difference in Mw would not cause drastic change of resist performance at least in this Mw range. Although further investigation on Mw influence to lithographic imaging was carried out, no significant difference was observed as expected.

2. Silylation ratio of the polymer. It is necessary to decide the silylation ratio to balance the thermal stability with solubility of the resist films. Softening point of SiPHS polymers are decreased as the silylation ratio increased (as is shown below).

Silylation ratio of SiPHS (%)	Softening point (°C)
100	120
83	180
70	> 200
30	> 200

From these results, silylation ratio of SiPHS is needed to be less than 70% in order to ensure sufficient thermal stability. On the other hand, at low silylation ratio, significant loss of residual film thickness was observed due to high solubility of the polymer in the TMAH developer. Figure 3 shows the residual film thickness in the unexposed regions of the resist films based on SiPHS with various silylation ratio after development. More than 30% of protected hydroxystyrene groups seem to be necessary to keep the loss of film thickness within 10%. Therefore we chose the silylation ratio between 30% and 70% at first stage. Further investigation was focused on the dissolution behavior of the resist film. Figure 4 shows the dissolution behavior of the exposed resist films consisting of SiPHS varying in silylation ratio in the TMAH developer. The applied PEB temperature was 90 °C for 2 minutes. A curious induction period was observed at the initial stage of development and it became longer as the silylation ratio increased. This induction indicates that the resist surface is less soluble than the inside of the resist film, so that the longer induction of the highly silylated polymer is likely to have relation to the T-shaped profile. Dissolution rate could be no longer monitored with our DRM system when the silylation ratio was over 60%.

Prevention of the T-shaped profile

1. Mechanism for surface skin formation. Although the mechanism of surface skin formation in this system has not been clarified yet, the result of DRM

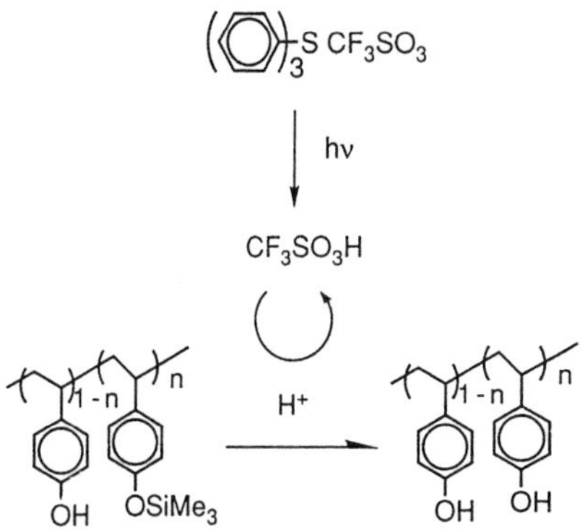

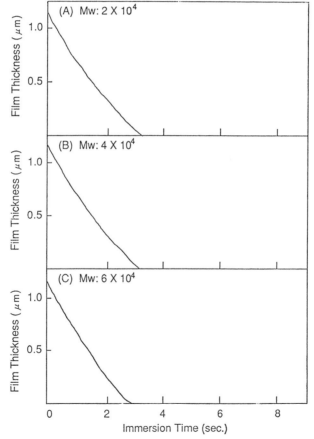

Figure 1. Chemical amplification scheme of the silylated polyhydroxystyrene deep UV resist.

Figure 2. Dissolution behavior of polyhydroxy-styrene films in 2.38% aqueous solution of tetramethylammonium hydroxide.

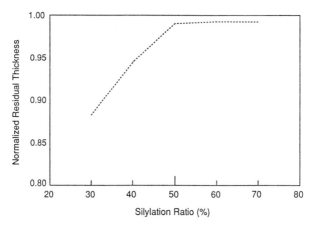

Figure 3. Dependence of residual film thickness in the unexposed regions of SiPHS resists on the silylation ratio.

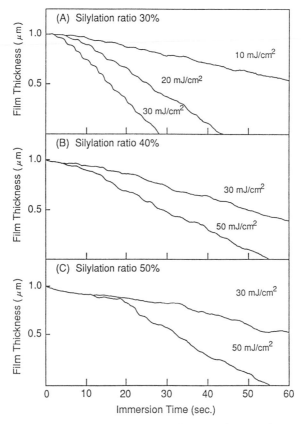

Figure 4. Dissolution behavior of exposed resist films in 2.38% aqueous solution of tetramethyl-ammonium hydroxide.

measurements suggests that optimization of the silylation ratio is a most effective method to prevent the surface skin formation. Thus we have further investigated characteristics of the SiPHS with various silylation ratio from the view point of glass transition temperature of the polymer. Figure 5 shows the result of DSC measurements. The two distinctive peaks appeared around 140 °C and 90 °C can be assigned to Tg of hydroxystyrene blocks and fully protected silyloxystyrene blocks, respectively. The lower Tg of the polymer is similar to that of polystyrene. Other peaks appeared between them are probably corresponding to the Tg of random block structures made up with hydroxystyrene and silylated hydroxystyrene segments (Figure 6). The peaks around 90 °C are not observed in the DSC curves for the polymers below the silylation ratio of 40%, while for more than 50% silylation, transition peak of low Tg segments became greater as the silylation ratio increased. The appearance of low Tg segments seems to be well agreed with the appearance of induction period observed in the DRM measurement. Thus we would like to propose one possible mechanism for surface skin formation as described below.

The low Tg segments are supposed to soften at 90 °C and thus PAG and/or photo induced acid seems to be able to move rapidly through those soft segments in the resist film during soft bake or PEB treatment. The 60 or 70% silylated resist films show high sensitivities compared with less silylated ones (Figure 7), so that we assume the induced acid could migrate faster through the low Tg softened segments and the catalytic chain length of deprotecting reaction becomes longer in these highly silylated resist films. Besides, Chatterjee et al. have reported that spin coating of resist depletes PAG concentration at the resist surface when the PAG has poor solubility in the polymer (*14*). This depletion effect can be stronger for the resist with highly silylated polymer because solubility of PAG is likely to decrease against the silylation ratio. Since PAG and/or the induced acid could migrate faster through the low Tg softened segments, PAG depletion towards inside of the resist film seems to be accelerated in a highly silylated resist system. These two factors for the highly silylated SiPHS, i.e., low compatibility with PAG and low Tg segments would cause the acid depletion at the resist surface resulting in the aqueous-alkaline insoluble surface skin layer.

2. Prevention of surface skin formation. The results of DRM and DSC studies suggest that lowering the silylation ratio can be an effective approach to solve the surface skin problem, because lowering the ratio makes the induction period shorter probably due to reduction of the depletion effect and decrease in softened segments as shown in Figure 4. However, since it is necessary to keep the silylation ratio above a certain level in order to achieve a high contrast resist, there should exist some limitation for this approach.

Another factor which decides the polymer condition in the resist is the baking temperature and it was found that the induction period could also be controlled by adjusting PEB temperature. Figure 8 shows the characteristic curves for the 30, 50 and 70% silylated SiPHS based resist films with and without PEB treatment. Without PEB can be regarded as lowering the PEB temperature down to room temperature 25 °C. For the 30%-silylated resist, the film

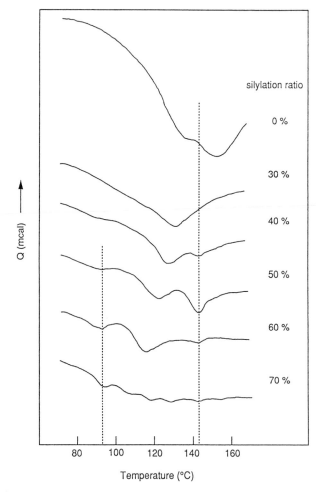

Figure 5. DSC curves for SiPHS with various silylation ratio.

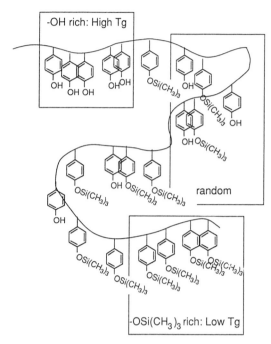

Figure 6. Schematic drawing of the structure of SiPHS.

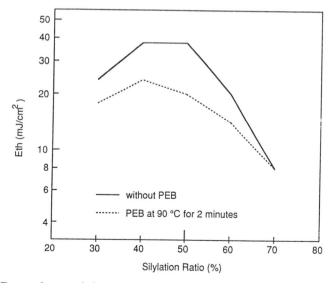

Figure 7. Dependence of the optimum sensitivity for 0.4 μm line patterning on the silylation ratio of SiPHS.

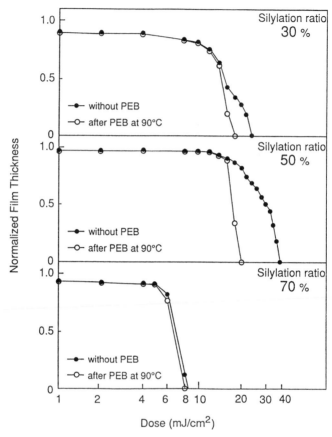

Figure 8. Characteristic curves for SiPHS resists.

gradually lost its thickness as the exposure dose increased, indicating absence of the insoluble surface in both cases with and without PEB due to the low silylation ratio. Effect of PEB temperature was typical when the silylation ratio was 50%. Without PEB was obtained a similar curve to those of 30%-silylated one, while sudden loss of the film occured in the case of PEB at 90 °C, which was regarded as the formation of surface inhibition layer. Lowering PEB temperature was no longer effective for the 70%-silylated resist, resulting in the T-shaped profile in any case. Figure 9 shows the dissolution behavior of the exposed 60%-silylated resist film in the aqueous base developer. The induction period was apparently decreased by lowering PEB temperature. This result seems to indicate the restraint of acid migration by lowering PEB temperature. We have confirmed that even the resist film consisting of 50%-silylated polyhydroxystyrene can be used without surface skin problem under low temperature PEB treatment.

As summarized in Figure 10, both lowering the silylation ratio and PEB temperature are effective to prevent the T-shaped profile. In order to achieve

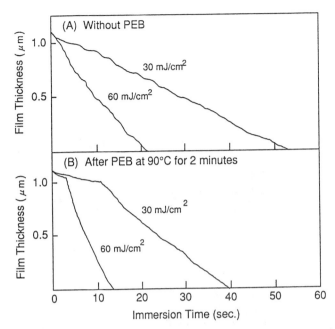

Figure 9. Dissolution behavior of exposed 60%-silylated SiPHS resist film in 2.38% aqueous solution of tetramethylammonium hydroxide.

high resist performance, combination of the 50%-silylated polymer and 50 °C PEB treatment was revealed to be preferable.

EVALUATION OF THE RESIST

Under the optimized conditions in both materials and process described above, the SiPHS resist has shown excellent performance. Figure 11 shows an imaging result of a SiPHS based resist. 0.28 μm line and space pairs along with 0.30 μm contact holes are formed in a 1.0 μm thick resist film on a silicon wafer with mask linearity down to 0.30 μm line and space pairs. The focus budget for 0.30 μm line and space pairs is about 1.0 μm. Figure 12 shows thermal stability of a 30 μm square pattern. No deformation was observed even after a baking treatment at 150 °C for 2 minutes.

SUMMARY

A positive deep-UV resist consisting of SiPHS was further investigated. Through the DRM and DSC studies, both lowering the silylation ratio of SiPHS and lowering PEB temperature were found to be effective to avoid the formation of T-shaped profile. The resist was capable of imaging 0.28 μm line and space pairs along with 0.30 μm contact holes in the optimized conditions.

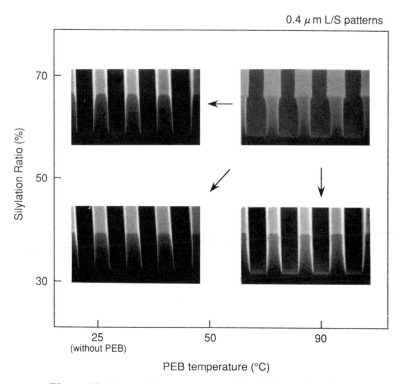

Figure 10. Preventive measures of the T-shaped profile.

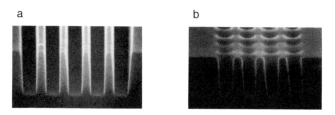

Figure 11. Resolution capability of SiPHS based resist. a, 0.28 μm line and space pairs and b, 0.30 μm holes.

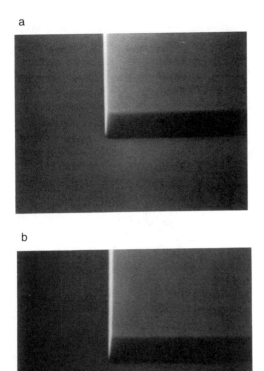

Figure 12. Thermal stability of SiPHS based resist. a, as developed; b, baked at 150 °C for 2 minutes.

ACKNOWLEDGEMENTS

The authors would like to thank both Canon Inc. and Nikon corporation for their kind offers of stepper demonstrations. The authors also would like to thank Japan Synthetic Rubber Co., Ltd. for supporting the present study.

REFERENCES

1. Jain, K. *Excimer Laser Lithography*; SPIE Optical Engineering press, 1990.
2. Ito, H.; Willson, C. G. ACS Symposium Series.No. 242; American Chemical Society: Washington, DC, 1984.
3. Reichmanis, E.; MacDonald, S. A.; Iwayanagi, T., Eds.; ACS Symposium Series No. 412; American Chemical Society: Washington, DC.
4. Ueno, T.; Shiraishi, H.; Hayashi, N.; Tadano, K.; Fukuyama, E.; Iwayanagi, T. *Proc. SPIE.* **1990,** 1262, 26.

5. Yamaoka, T.; Nishiki, N.; Koseki, K.; Koshiba, M. *Polym. Eng. Sci.* **1989**, 29, 856.
6. Murata, M.; Takahashi, T.; Koshiba, M.; Kawamura, S.; Yamaoka, T. *Proc. SPIE.* **1990**, 1262, 8
7. Kobayashi, E.; Murata, M.; Yamachika, M.; Kobayashi, Y.; Yumoto, Y.; Miura, T.; *Proc. PMSE.* **1992**, 66, 47.
8. Murata, M.; Kobayashi, E.; Yumoto, Y.; Miura, T.J. Photopolym. Sci. Technol. 1991, No. 3, 509.
9. Murata, M.; Kobayashi, E.; Yamachika, M.; Kobayashi, Y.; Yumoto, Y.; Miura, T. *J. Photopolym. Sci. Technol.* **1992**, No. 1, 79.
10. Schwartzkopf, G.; Niazy, N. N.; Das, S.; Surendran, G.; Covington, J. B. *Proc. SPIE.* **1991**, 1466, 26.
11. MacDonald, S. A; Clecak, N. J.; Wendt, H. R.; Willson, C. G.; Snyder, C. D.; Knors, C.J.; Deyoe, N.B.; Maltabes, J.G.; Morrow, J. R.; McGuire, A. E.; Holmes, S. *J. Proc. SPIE.* **1991**, 1466, 2.
12. Pawlowski, G.; Przybilla, K.; Spiess, W.; Wengenroth, H.; Roschert, H. *J. Photopolym. Sci. Technol.*, **1992**, 5, No. 1, 55.
13. Long, T.; Rodriguez, F.; *Proc. SPIE.* **1991**, 1466, 188.
14. Chatterjee, S.; Jain, S.; Lu, P. H.; Kahanna, D. N.; Potvin, R. E.; McCaulley, J. A.; Rafalko, *J. Proc. Regional Technical Conference on Photopolymers* Mid-Hudson Section SPE; Ellenville, NY, 1991 Oct. 28-30, 239.

RECEIVED December 30, 1992

Chapter 7

A Test for Correlation between Residual Solvent and Rates of N-Methylpyrrolidone Absorption by Polymer Films

W. D. Hinsberg[1], S. A. MacDonald[1], C. D. Snyder[2], H. Ito[1], and R. D. Allen[1]

[1]IBM Research Division, Almaden Research Center, 650 Harry Road, San Jose, CA 95120
[2]IBM Corporation, 5600 Cottle Road, San Jose, CA 95119

Resist systems based on chemical amplification can exhibit very high radiation sensitivity, but this is accompanied in most cases by an extreme susceptibility to adventitious basic airborne contaminants. Recent work in our laboratory has shown that the rate of uptake of such contaminants by thin spin-cast films can vary widely depending on the structure of the polymer. In the present study we probe whether those variations are attributable to differences in levels of residual casting solvent entrapped in the films. Radiochemical methods are used to quantify solvent retention and to measure the rate of uptake of N-methylpyrrolidone, a representative organic contaminant. A series of polymer films of differing structure are examined. In this series the residual solvent levels range from 0.05 to 21 wt %. The measured rates of absorption of NMP vapor are not correlated with residual solvent content. However, small changes in polymer structure cause a large change in the rate of NMP uptake.

Resist systems based on acid-catalyzed chemical amplification (CA) can exhibit very high radiation sensitivity, but this is accompanied in most cases by an extreme susceptibility to adventitious airborne contaminants which interfere with the resist chemistry (1, 2). Airborne amines (1, 2) and N-methylpyrrolidone (3) (NMP) have been reported to cause image degradation characterized by linewidth shifts and, in the case of positive-tone CA resists, by formation of a thin, poorly-soluble "skin" on the resist film. These basic substances are damaging even at extremely low vapor concentrations on the order of 15 ppb (1, 2). The practical application of such CA resists is simplified if the interfering substances can be rigorously excluded from the resist film. Two means of achieving this are by application of a protective overcoat (2) or by purification of the enclosing atmosphere using activated carbon filtration (1).

If it is assumed that an absorbed contaminant degrades the performance of a CA resist by simple acid-base neutralization of the catalytic species, then the degree of degradation should be related to the relative basicities of the contami-

0097–6156/94/0537–0101$06.00/0

nating species versus the resist components, the catalytic chain length, and the rate at which the contaminant enters the resist film. Recent work at our laboratory has demonstrated that the rate of contaminant uptake varies sharply as the structure of the polymer comprising the film is changed (4). In that study, the rate of uptake of radiolabeled NMP vapor by spin-cast polymer films varied by a factor of 40 depending on the polymer structure. No simple relation linking polymer structure and NMP uptake was evident in these data.

By understanding the causes of this variation, it may prove possible to design resist materials with improved tolerance toward airborne basic contaminants. We know from past work that the amount of residual casting solvent in thin resist films varies significantly depending on the characteristics of the polymer and the solvent, and it is well-accepted that residual solvent can influence diffusion processes in polymer films. Therefore, one possible factor influencing contaminant uptake is that the films studied have differing levels of retained casting solvent.

Polymer permeation by vapors and gases is generally treated in terms of the *solution diffusion* model (5), in which the vapor is first sorbed onto the surface of the film in a rapid equilibration step and then slowly diffuses through the polymer (Figure 1). The rate of uptake is determined in part by the ease with which the contaminant diffuses in the film. It is well-known that the mobility of small molecules in polymeric materials can be strongly affected by the concentration of solvent in the polymer matrix (6) Presumably the solvent acts to disrupt interactions between polymer chains and thereby enhances motion within the matrix. Recent reports from other workers have described residual solvent effects in lithographic materials. In one study, significant differences in image profiles were observed with one chemically amplified deep-UV resist depending on the choice of casting solvent (7) Of particular note was the appearance of a poorly-soluble surface skin when the resist film was cast from diglyme. This effect was attributed to the influence of residual solvent on acid diffusion within the film. Another group has provided more direct evidence that residual casting solvent influences mobility in chemically amplified resists (8). Figure 2 shows data from their study. In Figure 2a is plotted the acid diffusion range in two resist films of identical composition which were post-apply baked at different temperatures. The film baked at the higher temperature exhibits a much decreased diffusion range, attributed to a lower level of residual solvent. Figure 2b shows a similar result for resist films of different polymer composition which were post-apply baked using identical conditions. The different compositions retain casting solvent to different degrees. Again the acid diffusion range is much lower in the film with less solvent residue.

The goal of the present study is to establish experimentally whether the differences in contaminant uptake such as those seen in our previous study (4) can be similarly attributed to variations in levels of residual casting solvent. Since the final shape of a lithographic relief image is determined by a large set of interdependent variables, it is difficult to characterize the effect of residual solvent on susceptibility to airborne contamination in an unambiguous way by SEM examination of resist patterns. In our prior work on NMP uptake (4), films

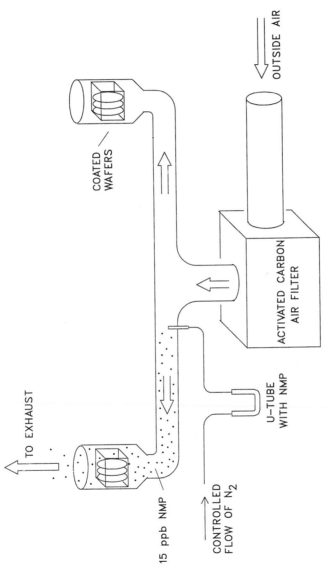

Figure 1. Schematic diagram of polymer permeation by a vapor.

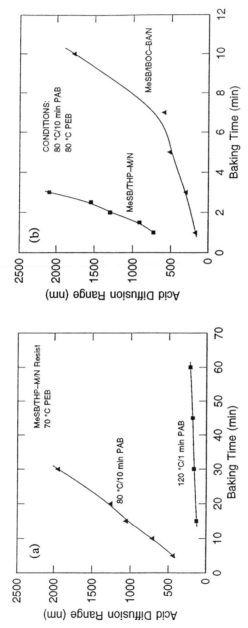

Figure 2. Influence of (a) post-apply bake temperature and (b) polymer structure on acid diffusion range. These data are taken from reference 8.

were prepared by spin-casting from various solvents (diglyme, cyclohexanone, toluene or propylene glycol monomethyl ether acetate (PGMEA), depending on solubility characteristics of the polymer). In the present work we have selected for study a series of polymers soluble in a single casting solvent (PGMEA) but with widely varying structural characteristics. The methodology used here is first, to accurately quantify residual casting solvent in films of this set of polymers, and second, to test for a correlation between the level of residual solvent and the uptake rate of NMP, a representative organic contaminant.

EXPERIMENTAL

Quantitation of Residual Casting Solvent

A radiochemical method was used to determine solvent retention after a typical spin-coat/post-apply bake cycle. Carbonyl-^{14}C-propylene glycol monomethyl ether acetate (PGMEA*) casting solvent was prepared by reaction of propylene glycol monomethyl ether with carbonyl-^{14}C-acetyl chloride in pentane solution in the presence of triethylamine (Figure 3), followed by an aqueous extractive workup, drying over $MgSO_4$, and distillation. The final material (98.7% purity by GC analysis, with pentane the only volatile impurity) had a specific activity of 57.7 $\mu Ci/mmol$.

A series of polymer solutions were prepared by combining an aliquot of PGMEA* with a non-labeled PGMEA solution of the polymer. The final solutions had specific activities of ca. 5 $\mu Ci/mmol$ PGMEA. These solutions were applied as ca. 1 μm thick films to tared 25 mm diameter silicon wafers, weighed to ±0.01 mg and post-apply baked on a hotplate at 100 °C for 300 seconds. After a final weighing the films were stripped with 5 ml unlabeled PGMEA and rinsed twice with Scintiverse II liquid scintillation cocktail (LSC) (Fisher Scientific). The combined washings were assayed for ^{14}C using a Packard Tri-Carb 460 Liquid Scintillation System. All analyses were performed in duplicate; results from duplicate runs generally agreed within 5 %, and at worst agreed within 10 % of their average value.

Quantitation of NMP Uptake

The apparatus shown in Figure 4 was employed for this measurement. Its construction and use have been previously described (*1, 4*). In the present work the airborne substance was methyl-^{14}C-N-methylpyrrolidone, mixed into the purified airstream to a final concentration of 15 ppb. A series of polymer solutions identical to those prepared above but containing only unlabeled PGMEA were applied as ca. 1 μm films to 125 mm dia silicon wafers and post-apply baked on a hotplate at 100 °C for 300 seconds. Immediately after coating, the wafers were immersed in the ^{14}C-NMP-doped airstream (for a period of 60 minutes unless otherwise specified). The films were then stripped using 5-7 ml of unlabeled PGMEA, twice rinsed with LSC solution, and the combined washings assayed for ^{14}C content as above. Results on duplicate wafers typically agreed within 10%.

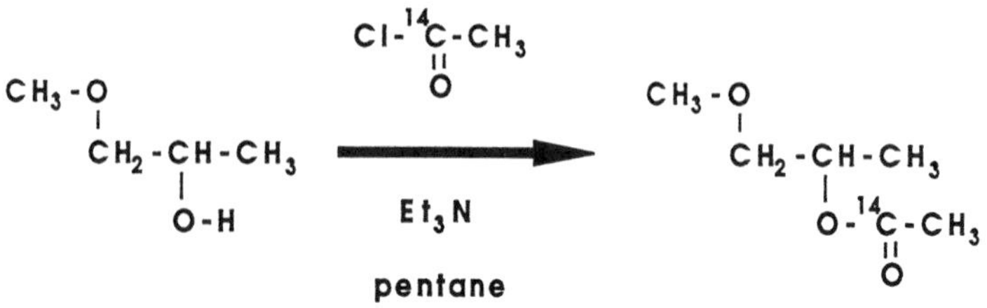

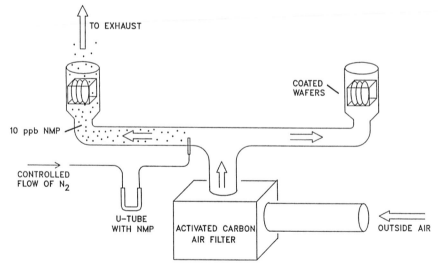

Figure 3. Synthesis of carbonyl-^{14}C-propylene glycol monomethyl ether acetate.

Figure 4. Schematic diagram of the air handling apparatus used in this study. (Reprinted with permission from reference 4. Copyright 1992 Society of Photo-Optical Instrumentation Engineers.)

RESULTS AND DISCUSSION

Table 1 summarizes the residual solvent contents for a series of polymer films of interest in microlithography, tabulated on a weight percent basis. Depending on the properties of the polymer, the residual solvent contents vary from 0.05 to 21 wt %. The general trend is that the level of solvent residue is related to the overall polarity of the polymer, with polymers of similar structure exhibiting nearly identical solvent levels.

One striking aspect of these data is that the amount of residual solvent can be substantial. In the case of the m-cresol novolac film, about one-fifth of the film on a weight basis is retained solvent. We have found in further work on the cresol novolac system that (1) the addition of 23 wt. percent of a typical 5-substituted diazonapthoquinone to the novolac film does not significantly change its retention of PGMEA; (2) the solvent level does not decrease significantly with more extended bake times as long as 30 minutes, suggesting that the PGMEA is relatively strongly bound to the polymer; and (3) that otherwise identical films formulated using ethyl cellosolve acetate as casting solvent retain similar proportions of solvent after baking. These observations demonstrate that residual casting solvent can be a major component of films of DNQ-novolac photoresists, though its presence and its role in influencing resist properties is often disregarded.

NMP absorption measurements indicate that the nature of the casting solvent has relatively small influence on NMP uptake. A comparison of NMP absorption by polymer films cast from PGMEA with that of otherwise identical films cast from other solvents reveals similar rates of uptake. For example, 1.0 μm films of cresol-novolac (post-apply baked as above) absorb 165 ng NMP in 15 minutes when cast from a diglyme solution, and 176 ng NMP when cast from PGMEA solution. Using data of Ouano (9), the estimated solvent content of the two films is comparable. Qualitatively similar uptake results are obtained for epoxy-novolac films cast from cyclohexanone or from PGMEA.

Figure 5 displays the relative absorption of NMP vapor for a series of polymers, plotted as a function of the residual (PGMEA) solvent content. There is no apparent correlation between solvent content of the film and its propensity to absorb NMP vapor. For example, the residual PGMEA content of P(para-TBOCST) is 1/35th that of the cresol novolac film, yet its rate of NMP uptake is about 3.5 times that of novolac. Similarly, compare the relative solvent contents and NMP absorption rates of P(para-TBOCST) and P(TBMA-MMA-MAA). This provides further support that residual solvent has relatively small influence on NMP uptake.

Recall that all films in this study and in our previous work were subjected to identical processing conditions. That, in combination with the evidence presented here on the role of residual casting solvent, leads to the conclusion that polymer structural characteristics are the dominant factor controlling contaminant absorption. A closer examination of the data of Figure 5 support this conclusion:

• The residual PGMEA contents of films of P(para-HOST) and cresol-novolac are nearly identical. The structural features of both polymers are quite similar

Table 1. Measured Residual Solvent Levels for a Series of Spin-Coated
Polymer Films

POLYMER	STRUCTURE	RESIDUAL SOLVENT (WT %)
Epoxy Novolac		0.05
P(meta-TBOCST)		0.5
P(para-TBOCST)		0.6
P(TBMA-MMA)		5.9
P(TBMA-MMA-MAA)		8.6
P(3-Me-4-HOST)		12.8
P(para-HOST)		20.4
Cresol Novolac		21.1

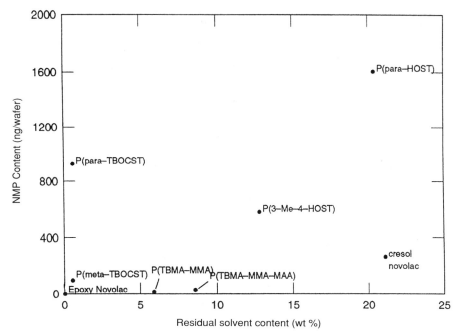

Figure 5. Plot of amount of NMP absorbed in one hour by a series of thin polymer films versus their residual casting solvent contents.

(Table 1)—in fact the nominal repeat units are isomers. However their rates of NMP uptake differ sharply, with the value for P(para-HOST) being six times that for cresol-novolac.

• In an even closer comparison, the levels of solvent residue in P(para-TBOCST) and P(meta-TBOCST) are identical, and their structures differ only in the position of substitution on the aromatic ring. However, their rates of NMP uptake differ by over a factor of seven.

Clearly, polymer structure strongly influences NMP uptake. In addition to altering the chemical properties, polymer structure can also influence such physical properties as polarity, molecular packing, and thermal properties. The data at hand are insufficient to reach a conclusion regarding the relative importance of these effects. A more extensive examination of structural effects has been undertaken to explore this relationship.

CONCLUSIONS

We have presented evidence that residual casting solvent is relatively unimportant in determining the propensity of a polymer film to absorb NMP. Films cast from different solvents, but otherwise identical in composition and processing, exhibit similar NMP absorption rates. Though residual solvent levels can vary widely depending on the characteristics of the polymer, rates of absorption of NMP vapor by the films are unrelated to the amount of retained solvent. We have provided evidence that polymer structural changes, rather than variations in residual solvent content, are responsible for the large changes in NMP uptake rates previously reported.

REFERENCES

1. S. A. MacDonald, N. J. Clecak, H. R. Wendt, C. G. Willson, C. D. Snyder, C. J. Knors, N. B. Deyoe, J. G. Maltabes, J. Morrow, A. E. McGuire, S. J. Holmes, *Proc. Soc. Photo-Opt. Instr. Eng.*, **1466**, 2-12 (1991).
2. O. Nalamasu, E. Reichmanis, M. Cheng, V. Pol, J. M. Kometani, F. M. Houlihan, T. X. Neenan, M. P. Bohrer, D. A. Mixon, L. F. Thompson, *Proc. Soc. Photo-Opt. Instr. Eng.*, **1466**, 13-25 (1991).
3. O. Nalamasu, M. Cheng, A. G. Timko, V. Pol, E. Reichmanis, L. F. Thompson, *J. Photopolym. Sci. Tech.*, 4(3), 299-317 (1991).
4. W. Hinsberg, S. MacDonald, N. Clecak and C. Snyder, *Proc. Soc. Photo-Opt. Instr. Eng.*, **1672**, 24-32 (1992).
5. T. deV. Naylor, in *Comprehensive Polymer Science*, G. Allen and J. Bevington, eds., Pergamon Press, New York, 1989; Chap. 20.
6. G. S. Park, in *Diffusion in Polymers*, J. Crank and G. S. Park, eds., Academic Press, New York, 1968; Chap. 5.
7. H. Kikuchi, N. Kurata, K. Hayashi, *J. Photopolym. Sci. Technol.*, 4(3), 357-360 (1991).
8. L. Schlegel, T. Ueno, N. Hayashi and T. Iwayanagi, *J. Vac. Sci. Technol.*, 9(2), 278-289 (1991).
9. A. Ouano, in *Macromolecular Solutions*, R. Seymour and G. Stahl, eds., Pergamon Press, Elmsford, NY, 1982, p. 208.

RECEIVED December 30, 1992

Chapter 8

Dissolution Rates of Copolymers Based on 4-Hydroxystyrene and Styrene

C.-P. Lei, T. Long, S. K. Obendorf, and F. Rodriguez

Olin Hall, Cornell University, Ithaca, NY 14853

Copolymers were synthesized by free-radical polymerization using 4-acetoxystyrene and styrene in various ratios. These polymers then were hydrolyzed to the corresponding 4-hydroxystyrene (phenol) copolymers. . The dissolution rates of the copolymers (using laser interferometry) decreased with a decrease in hydroxyl group content in aqueous developers (sodium hydroxide, potassium hydroxide, and a tetramethylammonium hydroxide based commercial developer). While an increase in pH value enhances the dissolution, an increase in cation size of these hydroxides at constant pH decreases the dissolution rate. The polymers become essentially insoluble in aqueous developers when the content of the monomer containing the hydroxyl group is less than 70 mole%. The dissolution rate of P(4HS/S) in organic solvents, methyl isobutyl ketone (MIBK), isopropyl alcohol (IPA), and their mixtures, increases as the styrene content increases. However, in IPA, dissolution rate goes through a maximum and then drops down with styrene content greater than 50 mole%. In mixtures of these two solvents, the polymer dissolves faster than it does in either pure solvent.

For many years, novolak polymers have been widely used in commercial photoresists because they have the properties of dissolution inhibition on alkali development and possess reasonable etch resistance (1). In alkali developers, they exhibit minimal swelling, a major problem which is often encountered with organic developers. However, novolaks have other inherent problems, such as low molecular weights, broad molecular weight dispersities, and difficulties in the control of structure during synthesis. Especially, novolak has high absorptivities in the wavelength of 240–300 nm. That means it will strongly absorb the 248 nm wavelength light generated by KrF excimer lasers and result in poor profiles (2, 3).

Poly(4-hydroxystyrene) [P4HS] has been suggested as a replacement candidate due to its similar phenolic functionalities to novolak polymers. P4HS

0097–6156/94/0537–0111$06.00/0
© 1994 American Chemical Society

provides better structure control and possible higher molecular weight in the synthesis process. In addition, it offers adequate thermal stability and etch resistance as well as high transparency in DUV range. However, in a comparison of P4HS with novolak in aqueous developers, Hanrahan and Hollis (4) concluded that the dissolution speed of this polymer in aqueous base is unacceptably high in comparison with commercial resist products.

In the present work, an effort has been made to improve the understanding of the P4HS dissolution and to approach the goal of finding a suitable replacement for the novolaks. Copolymers of 4-hydroxystyrene and styrene were synthesized in order to evaluate the influence of composition on polymer dissolution rates. The dissolution rates in both aqueous and organic developers were measured by laser interferometry. The influence of the reduction in phenolic hydroxyl groups of P4HS on dissolution behavior is also discussed.

EXPERIMENTAL

Polymer Synthesis

The two monomers used in the polymer synthesis were 4-acetoxystyrene (4AOS) (courtesy of M. T. Sheehan of Hoechst Celanese Company), and styrene, obtained from Eastman Kodak Company. Benzoyl peroxide, purchased from Aldrich Chemical Company, was used as the initiator. In order to obtain various polymer compositions, the copolymers of 4AOS with 0 to 60 mol% styrene were synthesized by free radical polymerization. Each copolymer of P(4AOS/S) was hydrolyzed with more than five times the potassium hydroxide (KOH) to hydroxyl mole ratio needed to obtain P(4HS/S), the copolymer of 4HS with styrene. A series of P(4HS/S) were thus formed with a styrene content from 0 to 60 mol%. Complete hydrolysis of the P(4AOS/S) copolymers was confirmed using infrared analysis by the loss of the carbonyl peak at 1760 cm^{-1} concurrent with the appearance of a strong hydroxyl band at 3200-3600 cm^{-1}.

Gel Permeation Chromatography

All (polystyrene equivalent) molecular weights were determined using the method of size exclusion chromatography by gel permeation chromatography (GPC). The instrument used was a GPC system produced by Waters Associates with Du Pont's Zorbax columns. THF was used as the eluting solvent. Polystyrene standards, obtained from Polymer Laboratories Limited, were used for calibration. The number-average molecular weights of both the P(4AOS/S) and the P(4HS/S) were in the range of 18,000–22,000 and their polydispersities were around 4 (Table 1). After hydrolysis, the molecular weights of P(4HS/S) would be expected to decrease due to the cleavage of the acetoxyl groups. Greater GPC molecular weights for the P(4HS/S) copolymers than expected may result from a difference in the interaction of the hydroxyl groups with column packing. Molecular weights were measured again after the polymers were spun on wafers and prebaked. This was done to ensure that no artifacts were introduced by the processing steps.

Table 1.   Molecular weights (polystyrene equivalent) and polydispersities
$(M_n M_w)$ for P(4AOS/S) and P(4HS/S)

Styrene (wt.%)	Content mole%	Polymer	Mn	Mw	Mw/Mn
0	0	P(4AOS/S)	22,10	77,400	3.50
		P(4HS/S)	17,700	92,800	5.12
5	7.62	P(4AOS/S)	22,300	92,300	4.14
		P(4HS/S)	20,000	98,800	4.89
10	14.8	P(4AOS/S)	22,600	92,600	4.10
		P(4HS/S)	17,600	97,800	5.30
12.5	18.3	P(4AOS/S)	20,600	75,000	3.65
		P(4HS/S)	19,600	99,300	5.06
15	21.7	P(4AOS/S)	21,100	81,400	3.87
		P(4HS/S)	18,900	94,300	4.73
20	28.2	P(4AOS/S)	20,500	75,000	3.66
		P(4HS/S)	19,700	74,700	3.80
25	34.3	P(4AOS/S)	22,100	80,900	3.66
		P(4HS/S)	19,000	94,200	4.97
30	40.2	P(4AOS/S)	19,500	73,300	3.82
		P(4HS/S)	17,900	76,100	4.26
40	51.1	P(4AOS/S)	18,200	73,400	4.04
		P(4HS/S)	17,300	75,000	4.33
50	61.1	P(4AOS/S)	19,000	76,800	4.03
		P(4HS/S)	18,300	83,200	4.54

Laser Interferometry

In preparing the samples, each polymer of P(4HS/S) was first dissolved in methyl isobutyl ketone (MIBK) at 10% by weight. The solution was spin-coated onto a clean, 3-inch diameter, oxide-coated silicon wafer. The spinning speed and time were set at 1,500 rpm for 60 seconds to obtain about $0.6 \sim 1.0$ μm thick films. After spin coating on wafers, these polymer films were baked in an oven at the temperature of 160 °C for one hour and then cooled down gradually in the oven for about 30 minutes. All the measurements of P(4HS/S) dissolution rates were carried out at room temperature (23 ± 1 °C) using a laser interferometer apparatus (Spectra Physic Model 102-4, wavelength 632.8 nm). Details of the procedure have been described previously (*5, 6, 7*). The wafer coated with a thin film of polymer was placed in a transparent beaker filled with a developing solvent which was stirred by a magnetic stirrer. The laser beam was directed horizontally towards a vertically placed wafer with an incident angle of 10 degrees. The reflected beam was collected by a silicon photocell and the signals were recorded as a function of time by a chart recorder with adjustable chart speed. The periodicity of the reflected light intensity was used to calculate the rate of dissolution (*5, 6*).

RESULTS AND DISCUSSION

Glass Transition Temperatures

All glass transition temperatures of P(4AOS/S) and P(4HS/S) were determined by differential scanning calorimetry (DSC). Except for the acetoxy homopolymer, the Tg of the P(4AOS/S) copolymers tend to decrease regularly from 120 °C to 101 °C as the styrene content increases (Figure 1). After hydrolysis, the Tg of P(4HS/S) decreases regularly up to a styrene content of 30 mol%. Above 30 mol%, there is a steep drop in the Tg (Figure 1). Such a drop in Tg may be indicative of a major change in copolymer configuration and interaction. This copolymer composition roughly corresponds to the range in which copolymers become alkali insoluble. It will be shown that dissolution rates of copolymers with more than 30 mol% styrene are close to zero even at pH values as high as 13 (8 g/l NaOH) and 13.2 (MF-319).

Dissolution of P(4HS / S) in Aqueous Base

P(4HS/S) copolymers were developed in aqueous solutions of sodium hydroxide (NaOH) and potassium hydroxide (KOH). The pH values of these solutions were measured with a conventional glass electrode after an approximately two minute immersion time. All quoted values should be viewed with caution since the glass electrodes which were used for convenience are not always an absolute measure of [OH$^-$]. However, they do represent an internally consistent method of comparison. Some results of dissolution rate measurements are shown in Figure 2. It is known that P4HS is soluble in strong aqueous base but that polystyrene does not dissolve in aqueous solution. Thus, the decrease in dissolution rate of P(4HS/S) copolymer with the increase of the styrene content was predictable. In either base, a fairly uniform sinusoidal pattern was obtained during film dissolution, indicating that the polymer dissolution rate did not change much with time.

The NaOH solution with pH = 13 yields a much higher dissolution rate for P(4HS/S) than that with pH = 12.8. The fact that a difference in pH of only 0.2 causes such a large difference in dissolution rates has been reported before for novolaks and other phenolic polymers. Several authors have reported dissolution rates for novolaks that increased as much as 8-fold for pH changes of a few tenths (8, 9). Our observation matches in general the dissolution behavior of P4HS previously reported by Long and Rodriguez (10). However, there seems to be a limiting copolymer composition for solubility of P(4HS/S) in aqueous base with pH near 13. The dissolution rate of P(4HS/S) with styrene content more than 30 mole% is close to zero even when the pH value is as high as 13.

In addition, it is obvious that at a similar pH value P(4HS/S) dissolves faster in NaOH solution than it does in KOH solution. The decrease in dissolution rate with the increase in the cation size of aqueous base developers is not an unexpected phenomena. It has been demonstrated by other authors in the dissolution study of novolak polymers with metal hydroxides as developers (8, 11, 12).

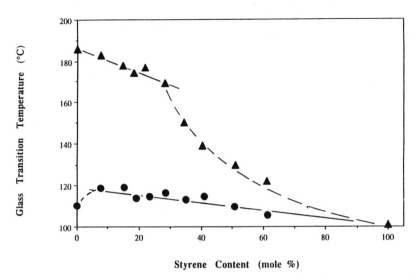

Figure 1. Glass transition temperatures of copolymers generally decrease as the styrene content increases. Copolymers with acetoxy derivative (triangles) and with 4-hydroxy styrene (circles).

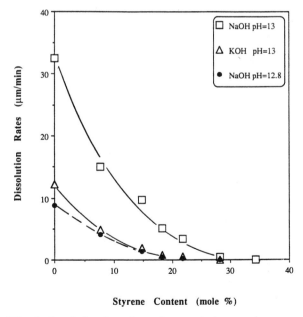

Figure 2. Dissolution behavior of copolymers in inorganic aqueous alkali.

In some commercial formulations, surfactants are added to developers in order to improve wetting. Two common surfactants were examined as additives to sodium hydroxide. Triton X-100 has the formula (where n is about 10):

$$CH_3-\underset{\underset{CH_3}{|}}{\overset{\overset{CH_3}{|}}{C}}-CH_2-\underset{\underset{CH_3}{|}}{\overset{\overset{CH_3}{|}}{C}}\text{—————}(OCH_2CH_2)_n-OH$$

When this non-ionic surfactant is added to a sodium hydroxide solution of pH = 12.9, the pH does not change. The dissolution rate is depressed, but increased amounts of the surfactant do not further depress the rate (Figure 3). An incidental observation can be made from these data. That is, the dissolution rate does decrease somewhat for coated wafers that have been held 4 days at room temperature after prebaking regardless of the medium in which they are dissolved. The reason for the rate depression by Triton X-100 is not clear. The formation of an adsorbed layer at the aqueous-polymer interface is likely. Raising the concentration of surfactant in the bulk of the liquid would not necessarily increase the thickness of the adsorbed layer, hence the lack of any cumulative effect.

A common ionic surfactant was also tested. Sodium lauryl sulfate actually decreases the pH of the sodium hydroxide solution from its initial value of 12.8 down to 12.7. The almost linear increase in dissolution rate with surfactant concentration (Figure 4) is not far from the expected "salt" effect noted by Huang and others (8,13). In a separate study, we found that addition of 0.05 M sodium chloride to a sodium hydroxide solution of pH = 12.8 was enough to increase the dissolution rate of P4HS polymers by a factor of 2 to 3. As a comparison, a 0.05 M solution of sodium lauryl sulfate (FW = 288) corresponds to 1.5 weight% which increases the rate only by a factor of about 1.5 (Figure 4). The lower pH observed upon the addition of the sodium lauryl sulfate may explain the suppressed salt effect in this case.

Microposit Developer MF-319 is a tetramethylammonium hydroxide (TMAH) based developer. The initial dissolution rates for P(4HS/S) in MF-319 are shown in Figure 5. The decrease in dissolution rate of P(4HS/S) with the increase of styrene content is found to be similar to the dissolution behavior in NaOH. It is also observed that the styrene content of the polymer must be less than 30% in order for the polymer to be soluble in MF-319. In addition, the dissolution rate of P(4HS/S) in MF-319 consistently decreases with time from the initial rate (Figure 6). Special thick polymer films (around 1.5 μm) were made to examine the change in dissolution rate during development. It was found that the polymer dissolves faster in the beginning and then slows down as the wafer surface is approached. The dissolution rate near the starting point can be twice as fast as the asymptotic value that is reached before the end point for a very thick film. When a thin film is tested, the initial rate is the same as that for the thick film, but the rate decreases all the way to the asymptotic value of the thick film. Thus it would seem that the effect is due to a change in structure at

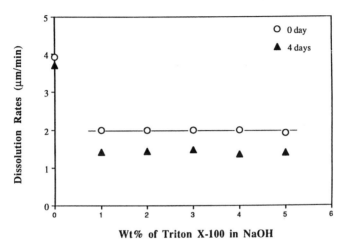

Figure 3. Decrease in dissolution rate from addition of surfactant (Triton X-100).

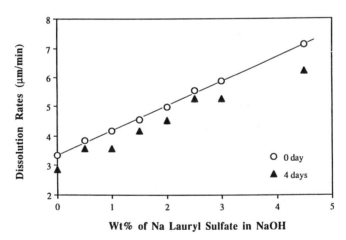

Figure 4. Increase in dissolution rate from addition of surfactant (Sodium lauryl sulfate).

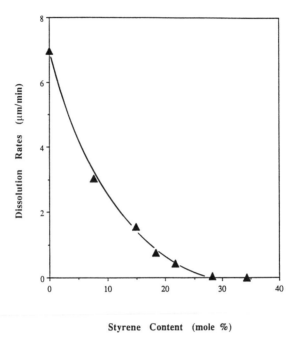

Styrene   Content   (mole %)

Figure 5. Dissolution behavior of copolymers in a TMAH-based, commercial developer (Shipley Microposit MF-319).

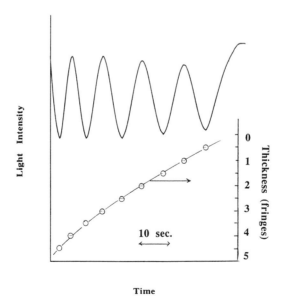

Time

Figure 6. Dissolution trace of copolymer with 18.3 mol% styrene in Microposit developer.

the polymer-solvent interface and not in the bulk of the resist layer. If residual solvent were the cause, one would expect less solvent near the surface compared to the bulk and the dissolution rate would accelerate with time rather than slow down. The buildup of a transition or gel layer seems to be the cause. Although the swelling of positive photoresists in alkali developing has been observed before (*11*), the exact mechanism and the reason for the change in dissolution remains unknown.

The "salt" effect also operates with TMAH. In experiments with a low molecular weight P4HS, dissolution rates were measured for various combinations of developers (Table 2). First of all, it is seen that the cation size effect continues when tetraethylammonium hydroxide is used in place of TMAH with a corresponding 6-fold decrease in rate. When TMACl, the chloride salt of TMAH, is added to TMAH, the rate increases. When the same salt is added to sodium hydroxide, the rate decreases. This last case is equivalent to adding NaCl to a TMAH solution whereby the salt effect is overcome by the cation size effect.

Dissolution of P(4HS / S) in Organic Solvents

MIBK, IPA, and their mixtures were also used as developers for P(4HS/S). While P4HS is soluble in both MIBK and IPA, polystyrene dissolves rapidly in MIBK but not at all in IPA. The solubilities of P(4HS/S) copolymers in these organic solvents were therefore of interest for study.

The dissolution behavior of P(4HS/S) in both MIBK and IPA are shown in Figure 7. As expected, the dissolution rate in MIBK increases with the increase in styrene content. In IPA, the dissolution rate also increases as the styrene content increases until it reaches a maximum and then drops off with styrene content greater than 50 mole%. This presents additional evidence for the effects of hydroxyl groups on the dissolution of P(4HS/S) copolymers. When the copolymer contains enough hydroxyl groups to dominate the dissolution, a higher hydroxyl content gives a higher dissolution rate. On the other hand, when hydroxyl groups are no longer in majority, the polymer tends to be insoluble in IPA like polystyrene.

It is usually difficult to predict the dissolution behavior in solvent mixtures. In some cases, a small amount of nonsolvent added to the solvent increases the dissolution rate significantly (*5, 6*). In this study, two solvents, MIBK and IPA, were mixed in different weight ratios to study the dissolution behavior. The

Table 2. Dissolution Rates for P4HS of Molecular Weight = 6,500

Developer (pH = 12.8)	Dissolution rate, μm/min
Sodium hydroxide	7.37
Tetramethylammonium hydroxide	0.14
Tetraethylammonium hydroxide	0.024
Sodium hydroxide + 0.1 M TMACl	1.55
TMAH + 0.1 M TMACl	0.64

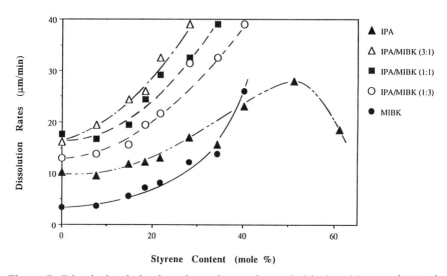

Figure 7. Dissolution behavior of copolymers in methyl isobutyl ketone (MIBK), isopropyl alcohol (IPA), and mixtures.

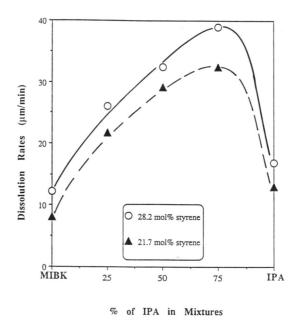

Figure 8. Copolymers dissolve faster in mixtures of MIBK and IPA than in either solvent alone.

results also are presented in Figure 7. It is obvious that dissolution rate of P(4HS/S) in mixtures increases with the increase of styrene content. In each mixture, P(4HS/S) dissolves faster than it does in either pure MIBK or pure IPA. This is true for every P(4HS/S) copolymer. Moreover, in mixtures, the dissolution rate goes through a maximum as the IPA ratio increases (Figure 8).

The rationalization has been put forward before (*14, 15*) that a smaller molecule might penetrate the bulk of the resist rapidly and preplasticize the bulk of the polymer film even when the small molecule is not a solvent for the polymer. IPA might act in this way since there is a modest difference in molecular size (FW = 60 for IPA compared to 100 for MIBK). Neutral water should be even more effective than IPA as a preplasticizer. In Long's experiments (*15*), acetone absorption by poly(vinyl acetate) was greatly accelerated by the presence of water vapor. Cooper (*14*) observed a doubling of the dissolution rate of poly(methyl methacrylate) when 10% water was added to methyl ethyl ketone. In the present work, the accelerating effect of water is much more pronounced (Figure 9). The solubility limit of water in MIBK is about 1.9% at 20 °C (*16*).

CONCLUSIONS

In alkaline developers (NaOH, KOH solutions, and Microposit Developer MF-319), the dissolution rate of P(4HS/S) decreases as the styrene content increases. In any one developer, higher pH values yield higher dissolution rates. At similar pH values, dissolution rates decrease with an increase in cation size. These observations are consistent with literature results. However, when the

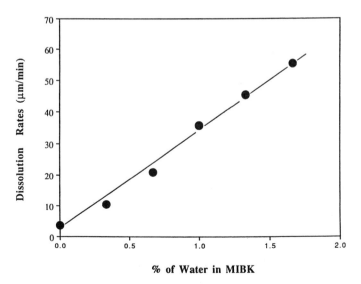

% of Water in MIBK

Figure 9. Accelerated dissolution of copolymers in MIBK by addition of water (5 mole% styrene in copolymer).

hydroxyl (phenol) group content is less than 70 mole%, P(4HS/S) becomes nearly insoluble even at a developer pH value as high as 13. In the TMAH based commercial developer, a change in dissolution rates was found during development. That is, the dissolution rate decreases with development time.

In organic developers (MIBK, IPA, as well as their mixtures), the dissolution rate of P(4HS/S) increases as the styrene content increases. However, in IPA, dissolution rate of P(4HS/S) goes through a maximum and then drops down with styrene content greater than 50%. In mixtures of MIBK and IPA, P(4HS/S) dissolves faster than it does in pure MIBK or pure IPA.

ACKNOWLEDGEMENTS

The authors wish to acknowledge partial financial support from the Office of Naval Research, the College of Human Ecology, the Department of Textiles and Apparel, and the School of Chemical Engineering. The 4-acetoxystyrene was a gift from the Hoechst Celanese Company. The cooperation of Cornell's National Nanofabrication Facility was useful also.

REFERENCES

1. Mckean, D.R., Hinsberg, W.D., Sauer, T.P., Willson G., Vicari, R., and Gordon, D.J., J. Vac. Sci. Technol. B 8 (6), 1466 (1990).
2. Pawlowski, G., Sauer, T., Dammel, R., Gordon, D.J., Hinsberg, W., McKean, D., Lindley, C.R., Merrem, H.-J., Roschert, H., Vicari, R., and Willson, C.G., Advances in Resist Technology and Processing VII, SPIE, 1262, 391 (1990).
3. Przybilla, K., Roschert, H., Spie, W., Eckes, Ch., Chatterjee, S., Pawlowski, G., and Dammel, R., Advances in Resist Technology and Processing VIII, SPIE, 1466, 174 (1991).
4. Hanrahan, M.J., and Hollis, K.S., Advances in Resist Technology and Processing IV, SPIE, 771, 128 (1987).
5. Groele, R.J., Ph.D Thesis, Cornell University, Ithaca, New York (1988).
6. Krasicky, P.D., Groele, R.J., and Rodriguez, F., Chem. Eng. Comm., 54, 279 (1987).
7. Rodriguez, F., Krasicky, P.D., and Groele, R.J., Solid State Tech., 28 (5), 125(1985).
8. J. P. Huang, E. M. Pearce, A. Reiser, and T. K. Kwei, in, E. Reichmanis, S. A. MacDonald, and T. Iwayanagi (eds), Polymers in Microlithography, ACS Symp. Series 412, ACS, Washington, DC, 1989, Chapter 22, p. 364.
9. T-F. Yeh, H-Y. Shih, A. Reiser, M. A. Toukhy, and B. T. Beauchemin, Jr., J. Vac. Sci. Tech. B., 10, 715 (1992).
10. Long, T., and Rodriguez, F., Advances in Resist Technology and Processing VIII, SPIE, 1466, 188 (1991).
11. Arcus, R.A., Advances in ResistTechnology and Processing III, SPIE, 631, 124 (1986).
12. Hinsberg, W.D., and Gutierrez, M.L., Processing Kodak Microelectronics Seminar, Interface 1983, San Diego (1983).

13. M. K. Templeton, C. R. Szmanda, and A. Zampini, Proc. SPIE 771, 136 (1987).
14. W. J. Cooper, P. D. Krasicky, and F. Rodriguez, J. Appl. Polym. Sci., 31, 65 (1986).
15. F. A. Long and L. J. Thompson, J. Polym. Sci., 14, 321 (1954).
16. A. J. Papa and P. D. Sherman, Jr., "Ketones" in Encycl. Chem. Tech., 13, Wiley, New York, 1981, p. 894.

RECEIVED December 30, 1992

Chapter 9

Synthesis and Polymerizations of N-(tert-Butoxy)maleimide and Application of Its Polymers as a Chemical Amplification Resist

Kwang-Duk Ahn and Deok-Il Koo

Polymer Chemistry Laboratory, Korea Institute of Science and Technology, P.O. Box 131, Cheongryang, Seoul 130–650, Korea

N-(tert-Butoxy)maleimide, t-BuOMI was synthesized as a new kind of a protected acid-labile monomer. Its radical copolymerizations were performed and the thermal deprotection behavior of its copolymers were investigated. t-BuOMI was readily copolymerized with styrene derivatives (X-St) to give copolymers, P(t-BuOMI/X-St) in high conversions. The t-BuOMI units of the tert-butyl (t-Bu) protected copolymers were converted into the N-hydroxy-maleimide (HOMI) units by heating at about 280 °C releasing 2-methylpro-pene and the facile deprotection of the side-chain t-Bu groups resulted in a large polarity change in the polymer structure. The deprotected copolymers, P(HOMI/X-St) have very high glass transition temperatures higher than about 250 °C and showed solubilities in aqueous base solutions whereas the protected polymers are soluble only in organic solvents. Acidolytic deprotec-tion of P(t-BuOMI/St) was observed at 130 °C or lower temperatures in the presence of catalytic acids. Resist solutions of P(t-BuOMI/St) containing triphenylsulfonium salts as a photoacid generator were prepared and the films were imagewise exposed to 260 nm light by a contact mode. The films were post-exposure baked at 130 °C and were followed by development with 2.38 wt% TMAH solution to obtain positive tone images.

The design and synthesis of protected polymers having acid-labile groups are considered to be of challenging research for obtaining imageable polymers which can serve high sensitivity resist materials for fabricating microelectronic devices (1). Adequately protected polymers bring a remarkable change in polarity, thereof considerable differentiation in solubilities, when the side-chain protect-ing groups of polymers are deprotected by thermolysis or acidolysis. Those protected polymers are best suited for the application as chemically amplified resists to achieve high sensitivity (2).

A typical acid-labile protecting group is tert-butoxycarbonyl (t-BOC) group which is known to be readily deprotected in the presence of catalytic acids. Poly[p-(t-butoxycarbonyloxy)styrene] has been investigated in detail as a proto-type t-BOC protected polymer and successfully applied in the manufacture of VLSI devices (3, 4). In this case, the acid-labile t-BOC group is utilized to

0097–6156/94/0537–0124$06.00/0

protect the phenol functionality of poly(p-hydroxystyrene) which is then regenerated to give a large polarity change after deprotection of t-BOC side-chains.

One of required properties placed on resist polymers is thermal stability of patterned resist images for the use in advanced lithographic processes such as plasma etching and ion implantation. Thus the resist polymers which have high glass transition temperature (T_g) above 200 °C have been searched. One of common synthetic techniques to improve thermal stability of resist polymers is the incorporation of a maleimide unit into copolymers (5, 6). Recently we reported the synthesis and polymerizations of a new t-BOC protected maleimide monomer, N-(*tert*-butyloxycarbonyl)maleimide, t-BOCMI (7). t-BOC protected maleimide copolymers were found to be applicable as deep UV resist materials having superior thermal stability along with high sensitivity by chemical amplification (8). Upon deprotection of the t-BOC groups of the copolymers, the deprotected maleimide copolymers exhibit very high thermal stability with T_gs of about 250 °C because of the maleimide (MI) backbone structure. The above-mentioned t-BOC protected polymers are readily deprotected by loss of carbon dioxide and isobutylene to corresponding polymers with phenol or maleimide functionality, thermally at about 150 to 180 °C and acidolytically at about 100 °C in the presence of catalytic acids.

The previously mentioned t-BOC protected polymers are considered to have rather low thermal deprotection temperatures at about 150 to 180 °C for practical application (9). Therefore our objective was to make acid-labile polymers based on MI backbone structures which have protecting groups rather than t-BOC units to obtain much higher deprotection temperatures. Now we report the synthesis and polymerizations of N-(*tert*-butoxy)maleimide (t-BuOMI) as a novel protected maleimide monomer while thermal deprotection of its copolymers occurs at higher than 200 °C. Characteristic properties of t-BuOMI copolymers along with the capability of facile deprotection to N-hydroxymaleimide (HOMI) structure at higher temperatures were investigated and resist applicaton of the copolymers in the deep UV region are also discussed.

EXPERIMENTAL

Materials and Instruments

Furane, N-hydroxymaleimide (HOMI), styrene (St), p-methylstyrene (MeSt), and p-chlorostyrene (ClSt) were purchased from Aldrich Chemical Co. Maleic anhydride and hydroxylamine hydrochloride were obtained from Kanto Chemical Co. p-Trimethylsilylstyrene (SiSt), p-(*tert*-butoxycarbonyloxy)styrene (t-BOCSt), and p-acetoxystyrene (AcOSt) were kindly donated by Korea Kumho Petrochemical Co., Eastman Kodak Co., and Hoechst Celanese Corp., respectively. The styrene monomers (X-St) and solvents were purified by distillation with standard procedures. Other chemicals were purified by conventional methods. The radical initiators, N,N'-azobis(isobutyronitrile) (AIBN) and dicumyl peroxide (DCP) were purchased from Aldrich and used after recrystallization. As photoacid generators (PAG), triphenyl sulfonium hexafluoroantimonate and triflate were prepared according to the known procedures or obtained by kind donation. The resist solvent, cyclohexanone was purchased from Aldrich and used without purification. The aqueous base developer, 2.38 wt% tetramethylammonium hydroxide (TMAH) solution was kindly donated from Hoechst Korea Ltd. and Tokyo Ohka Co. Proton NMR spectra were taken on a JEOL Model PMX-60 SI (60MHz) spectrometer or a Varian Gemini 300MHz spectrometer in deuteriochloroform using TMS as an internal standard. Carbon-13 NMR spectra were also obtained with a Varian Gemini spectrometer in deuteriochloroform. Infrared spectra were recorded on an Polaris FT-IR spectrophotometer of Mattson Instrument Co. and UV absorption spectra with a Shimadzu UV-240 spectrophotometer, and mass spectra with a JEOL JMS-DX 303 spectrometer. Elemental analysis was done with a Perkin-Elmer Model 240C elemental analyzer. Thermal analysis was carried out on a Du Pont Model 910 DSC and Model 951 TGA at a heating rate of 10 °C per min under nitrogen atmosphere. Solution viscosities of polymers were measured with a Cannon-Fenske viscosity tube (No. 50) or an Ubbelohde viscometer tube mounted to an automatic measuring apparatus of Schott-Gerate GMBH at 25 °C in THF as a solvent.

Preparation of N-(tert-Butoxy)maleimide, t-BuOMI

The furan/N-hydroxymaleimide (furan/HOMI) adduct i. e., N-hydroxy-3,6-epoxy-1,2,3,6-tetrahydrophthalimide, was prepared by a reaction of 3,6-epoxy-1,2,3,6-tetrahydrophthalic anhydride and hydroxylamine in a high yield of 81% according to the known procedure (10). The melting point of the adduct is 185 °C [lit., (10) 187 °C]. To a solution of the furan/HOMI adduct (80.00 g, 0.44 mol) in 500 ml of methylene chloride in a pressure reactor added were 1.0 ml of concentrated sulfuric acid and 200 ml of isobutylene, and the mixture was stirred for 8 days at 110 °C. After finishing the reaction, the insoluble impurities were filtered off and the filtrate was washed with distilled water to remove acid residue, and the volatiles were evaporated under reduced pressure. The yellow powdery N-tert-butoxy adduct (furan/t-BuOMI), i.e., N-(tert-butoxy)-3,6-epoxy-1,2,3,6-tetrahydrophthalimide was obtained in a yield of 82.2% (86.10 g) with mp of 145 °C and used for the next procedure. furan/t-BuOMI: ^{1}H NMR (60MHz, CDCl$_3$) (ppm) 1.30 (s, 9H, t-Bu), 2.70 (s, 2H, 2 -CO-C$\underline{H}$-), 5.20 (s, 2H, 2 -O-C$\underline{H}$-), 6.30 (s, 2H, 2 = C$\underline{H}$-); IR (KBr) (cm^{-1}) 2990 (t-Bu), 1790 and 1720 (imide), 1370 (t-Bu), 1190 and 1150 (ether).

In a sublimation apparatus 10.44 g of the furan/t-BuOMI adduct was placed and pyrolyzed at 140-150 °C for 1h under reduced pressure. The solid product was obtained as a white crystal by sublimation in a yield of 74.6% (5.55 g). Recrystallization from a solution of methylene chloride and n-hexane (1:10 by vol) gave 4.26 g (yield 57.3%) of the desired monomer t-BuOMI in needle crystal with mp 92 °C. t-BuOMI: ^{1}H NMR (60MHz, CDCl$_3$) (ppm) 1.30 (s, 9H, t-Bu), 6.56 (s, 2H, 2 = C$\underline{H}$-); IR (KBr) (cm^{-1}) 3100 (olefinic CH), 2980 (t-Bu), 1730 (imide), 1370 (t-Bu), 1150 (ether); ^{13}C NMR (300MHz, CDCl$_3$) (ppm) 27.11 (Me), 85.73 (t-Bu), 132.35 (C = C), 167.85 (carbonyl of MI); MS 169.40 (M), 154

(M − CH$_3$, 3), 112 (1), 57 (2-methyl propene, 100). Anal. Calcd for C$_8$H$_{11}$NO$_3$: C, 56.79; H, 6.55; N, 8.28. Found: C, 56.60; H, 6.53; N, 8.14.

Polymerization

All the polymerizations were carried out in ampoules sealed after freeze-thaw cycles. The t-BuOMI monomer and the styrene comonomers (X-St) were copolymerized in 1 to 1 molar feed ratio in dioxane using AIBN as a radical initiator. The radical polymerizations were conducted under conditions described in Table I. The copolymers were obtained by precipitating into methanol, whereas the homopolymers P(t-BuOMI) were precipitated into a methanol-water solution. The structure of obtained polymers was fully characterized by spectroscopy and the thermal deprotection of the side-chain t-butyl groups was investigated by TGA and DSC analysis. As a representative copolymerization, t-BuOMI and styrene was copolymerized according to the following procedure to obtain P(t-BuOMI/St). To a pyrex ampoule placed were 3.38 g (0.02 mol) of t-BuOMI, 2.09 g (0.02 mol) of styrene and 131.3 mg (0.80 mmol, 2 mol%) of AIBN in 5.50 ml of dioxane. The copolymerization was proceeded for 3h at 55 °C and instantaneous polymer formation was observed at the initial stage. The product was diluted with dioxane and precipitated into 2 L of methanol. The white powdery alternating copolymer P(t-BuOMI/St) was obtained in a conversion of 90% (4.90 g) after drying in vacuo at 40 °C. The inherent viscosity of the copolymer was determined to be 1.18 dl/g in THF at 25 °C.

Photoimaging of P(t-BuOMI / St)

Resist solutions by 20 wt% in cyclohexanone were prepared by dissolving P(t-BuOMI/St) with various molecular weights and sulfonium salt such as triphenylsulfonium hexafluoroantimonate (TPSHFA) or triflate (TPSOTf). The spun films with 2000 to 3000 rpm on HMDS-treated silicon wafers were soft-baked at 90 °C for 1 min on a hot plate and imagewise exposed to deep UV by a contact mode. The exposure of the resist films was made on a Hybralign Series 400 Exposure Systems of Optical Associates Inc. equipped with a light source of a 500W short arc Hg-Xe lamp and optics tuned to 260 nm with light intensity of 15mW/cm^2. Then the exposed resist films underwent post-exposure bake (PEB) at a given temperature on a hot plate followed by development with commercial 2.38 wt% TMAH aqueous solution such as NMD-3 and AZ 700MIF developer for 1 min.

RESULTS AND DISCUSSION

Synthesis of N-(tert-Butoxy)maleimide

A maleimide monomer, N-(*tert*-butoxy)maleimide, t-BuOMI was prepared by means of a retro-Diels-Alder reaction which is a useful procedure to synthesize reactive maleimide derivatives (*10*) as described in Scheme I.

Table I. Radical Copolymerizations of N-(tert-Butoxy)maleimide

P(t-BuOMI/X-St)[a]	AIBN[b] (mol%)	M/S[c] (g/ml)	time (hr)	conver- sion (%)	inherent viscosity[d]
P(t-BuOMI)	0.5	1.00	48	84	0.06
P(t-BuOMI)[e]	1	—	24	77	0.25
P(t-BuOMI)[e]	1	—	3	56	0.17
P(t-BuOMI/St)	1	1.00	3	89	1.02
P(t-BuOMI/St)	2	1.00	3	90	1.18
P(t-BuOMI/St)	4	0.50	2	90	0.75
P(t-BuOMI/St)	6	0.33	3	92	0.59
P(t-BuOMI/t-BOCSt)	4	1.00	3	79	1.00
P(t-BuOMI/SiSt)	4	0.62	2	79	0.86
P(t-BuOMI/AcOSt)	2	1.00	2	84	1.23
P(t-BuOMI/MeSt)	2	1.00	2	87	0.83
P(t-BuOMI/ClSt)	2	1.00	2	82	0.45
P(HOMI/St)[f]	4	1.00	3	69	0.21
P(HOMI/SiSt)[g]	2	1.00	5	71	0.24

[a]All the t-BuOMI copolymers have alternating structures. Copolymerizations were carried out in 1 to 1 molar feed ratio at 55°C in dioxane: St, styrene;   t-BOCSt, p-(t-butoxycarbonyloxy)styrene; SiSt, p-trimethylsilylstyrene;   AcOSt, p-acetoxystyrene; MeSt, p-methylstyrene; ClSt, p-chlorostyrene; HOMI, N-hydroxymaleimide. [b]The mol% of the initiator AIBN to the total amounts of two monomers used. [c]M/S is the ratio of the total weight of two monomers to the volume of dioxane solvent. [d]Inherent viscosities in dl/g were measured at a concentration of 0.20 g/dl in THF at 25°C. [e]Homopolymerization with DCP as an initiator in bulk at 110°C. [f]P(HOMI/St) was obtained by copolymerization of HOMI and styrene in 1 to 1 molar feed ratio. [g]P(HOMI/SiSt) was obtained by copolymerization of HOMI and SiSt in 1 to 1 molar feed ratio.

The furan/HOMI adduct was prepared in a quantative yield by two steps, firstly, a Diels-Alder reaction and secondly a reaction with hydroxylamine. The *tert*-butyl (t-Bu) group was introduced to the furan/HOMI adduct by a reaction with isobutylene in a pressure reactor and the resulting furan/t-BuOMI adduct was obtained as a yellow powder in a high yield. The *tert*-butoxy (t-BuO) adduct was pyrolyzed at about 145 °C and the desired monomer, t-BuOMI was collected by sublimation in a yield of 75%. Recrystallization gave the colorless needle crystal of t-BuOMI with melting point of 92 °C. In an NMR spectrum, t-BuOMI shows only two singlet peaks at 6.56 ppm for two olefinic protons and at 1.30 ppm for nine protons of the t-Bu groups. The structure of t-BuOMI was confirmed by ^{1}H and ^{13}C NMR, IR, mass spectra, and elemental analysis.

Polymerization

The radical copolymerizations of the t-BuOMI monomer with various comonomers were carried out and the results are summarized in Table I. t-BuOMI was readily copolymerized with styrene derivatives (X-St) such as styrene (St), *p*-(*tert*-butoxycarbonyloxy)styrene (t-BOCSt), *p*-trimethylsilylstyrene (SiSt), *p*-acetoxystyrene (AcOSt), *p*-methylstyrene (MeSt), and *p*-chlorostyrene (ClSt) in high conversions in the presence of a radical initiator within three hours (Scheme II). However, the homopolymerization of t-BuOMI were rather sluggish and usually obtained were the low molecular weight polymers with some amounts of the free *N*-hydroxymaleimide (HOMI) unit by adventitious deprotection of t-Bu groups during the polymerization, particularly, in solution. This homopolymerization behavior of t-BuOMI is ascribable to the bulky side-chain of the t-Bu group along with 1,2-disubstitution. In contrast to the poor homopolymerization behavior of the t-BOC protected MI monomer, t-BOCMI (*7*), the new monomer t-BuOMI rendered homopolymers in quite high conversions above 77% by a polymerization at 110 °C in bulk with dicumyl peroxide (DCP) as a radical initiator, but still in low molecular weights.

Molecular weights of the t-BuOMI copolymers were controlled by the amount of the initiator and dioxane solvent used in copolymerizations. To obtain a large amount of copolymers suitable molecular weights, some of the copolymerizations were conducted by using large quantities of the AIBN initiator and the copolymers were used for characterizing the resist properties. The copolymerizations of t-BuOMI with styrene monomers were performed in 1 to 1 molar feed ratio and the obtained copolymers, poly(t-BuOMI-co-X-St), i.e., P(t-BuOMI/X-St) were confirmed to have 1 to 1 molar composition ratio from the proton NMR spectral analyses. Thus the polymers P(t-BuOMI/X-St) are expected to be of alternating structure because it is well known that nearly alternating structure is formed when an electron-poor monomer (t-BuOMI) and an electron-rich monomer (styrene) are copolymerized (*11*).

Thermal Deprotection and Structural Change

In thermogravimetric analysis (TGA), the copolymer P(t-BuOMI-alt-St), i.e., P(t-BuOMI/St) was found to undergo rapid thermal deprotection of the t-Bu

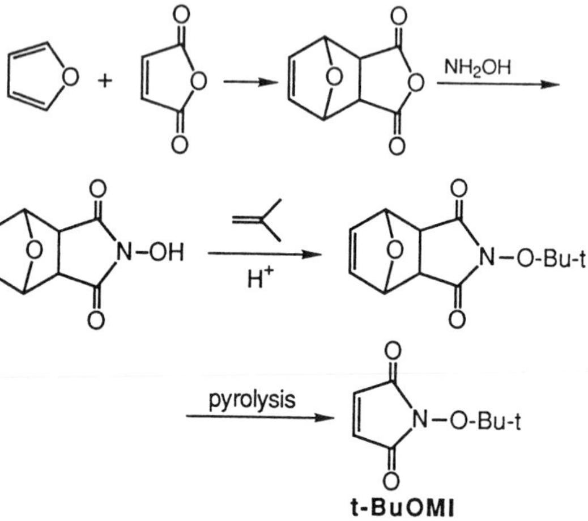

t-BuOMI

Scheme I.

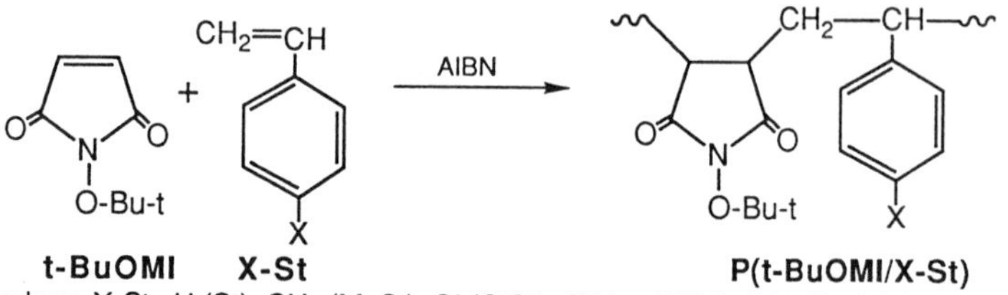

t-BuOMI   X-St   **P(t-BuOMI/X-St)**

where X-St : H (St); CH$_3$ (MeSt); Cl (ClSt); SiMe$_3$ (SiSt); OAc (AcOSt);
O-CO$_2$-Bu-t (t-BOCSt).

Scheme II.

groups at about 280 °C to yield poly(N-hydroxy- maleimide-alt-styrene), P(HOMI/St) by removing 2-methylpropene as shown in Figure 1. P(t-BuOMI/SiSt) and P(t-BuOMI/MeSt) are deprotected at higher temperatures to the corresponding copolymers P(HOMI/SiSt) and P(HOMI/MeSt), respectively, and the thermograms are compared with P(t-BuOMI/St) in Figure 1. The protected copolymers, P(t-BuOMI/X-St) are converted into P(HOMI/X-St) by thermal deprotection of t-Bu groups above 270 °C as shown in Scheme III.

The weight loss of the alternating copolymer P(t-BuOMI/St) in a TGA thermogram of Figure 1 was estimated to be of 21% which is the same amount as the theoretically calculated weight loss due to the release of 2-methylpropene from the copolymer and is listed in Table II. The DSC thermograms of P(t-BuOMI/St) in Figure 2 reveal the glass transition (T_g) at 200 °C and an endothermic event corresponding to the deprotection of t-Bu groups at 283 °C (T_{dp}) in the first run, and T_g of the deprotected polymer P(HOMI/St) at 245 °C and onset decomposition at 340 °C with some kinds of reactions due to HOMI units in the second run. In DSC measurements, the first run was recorded to 300 °C, then the same sample was cooled down to room temperature and the second run was conducted to the main-chain decomposition. Therefore T_g of the protected P(t-BuOMI/St) is 200 °C before deprotection and the observed glass transition temperature in the second run corresponds to that of the deprotected polymer and was confirmed to be identical to that of P(HOMI/St) obtained by the direct radical copolymerization of the corresponding monomers HOMI and St. T_gs of P(t-BuOMI/AcOSt) and P(t-BuOMI/ClSt) were found to be of 210 °C whereas P(t-BuOMI/MeSt) and P(t-BuOMI/SiSt) did not show T_g.

In the particular case, the copolymer of t-BuOMI and t-BOCSt, P(t-BuOMI/t-BOCSt) exhibits two-step thermal deprotection behavior by DSC and TGA analysis as shown in Figure 3, since the copolymer has two kinds of protecting groups in both the repeating units. In the first step the deprotection of t-BOCSt units to p-hydroxystyrene (HOSt) units occurs at 185 °C measured by DSC. P(t-BuOMI/t-BOCSt) is converted to P(t-BuOMI/HOSt) evolving isobutylene and carbon dioxide. In the second deprotection at 252 °C, P(t-BuOMI/HOSt) is converted to P(HOMI/HOSt) which does not show T_g before its main-chain decomposition. The two-step deprotection behavior of P(t-BuOMI/t-BOCSt) is also confirmed by mass loss in a TGA analysis in Figure 3 and the amount of mass loss in each step is described in Table II.

The thermal deprotection temperatures (T_{dp}) of the t-Bu protected polymers P(t-BuOMI/X-St), T_gs of the deprotected polymers P(HOMI/X-St), and the onset decomposition temperatures (T_{dc}) of the deprotected polymers were measured in nitrogen atmosphere and are summarized in Table II. The amounts of weight loss of the protected copolymers during the thermal deprotection measured by TGA agreed well with the calculated amounts. All the t-BuOMI copolymers are deprotected at about 270 °C or above and the deprotected polymers P(HOMI/X-St) having HOMI units showed high T_gs of about 245 °C or no T_g observed.

In the case of the homopolymer P(t-BuOMI), the deprotection was observed at somewhat low temperature of 247 °C and a smaller amount of weight loss than the theoretical amount was found due to adventitious deprotection during

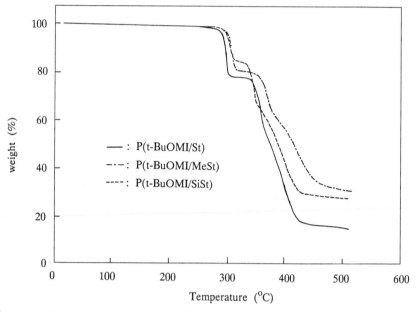

Figure 1. TGA thermograms of P(t-BuOMI/St), P(t-BuOMI/SiSt) and P(t-BuOMI/MeSt) in a nitrogen stream at the heating rate of 10 °C/min.

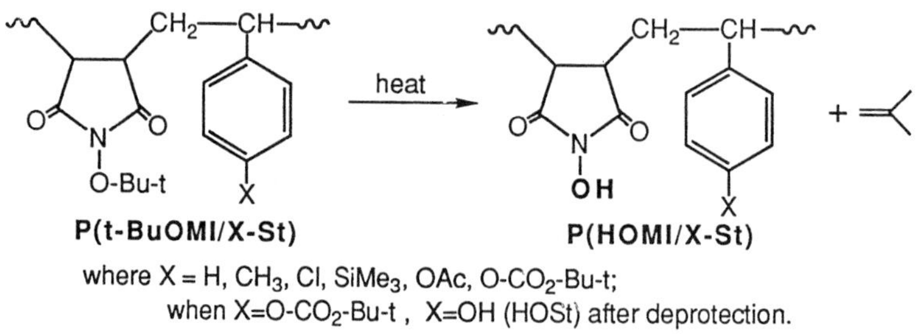

where X = H, CH₃, Cl, SiMe₃, OAc, O-CO₂-Bu-t;
when X=O-CO₂-Bu-t , X=OH (HOSt) after deprotection.

Scheme III.

Table II. Thermal Properties of the t-BuOMI Copolymers

solvent	P(t-BuOMI/St)	P(HOMI/St)*
acetone	+ +	−
chloroform	+ +	−
hexane	−	−
toluene	+ +	−
chlorobenzene	+ +	−
anisole	+ +	−
cyclohexanone	+ +	−
methyl isobutyl ketone	+ +	−
2-ethoxyethyl acetate	+ +	−
N,N-dimethylformamide	+ +	+ +
tetrahydrofuran	+ +	+
dioxane	+ +	+
methanol	−	−
0.7 N KOH (aq)	−	+ +
1.0 N NaOH (aq)	−	+ +
0.3 N TMAH**(aq)	−	+ +

Remark: + +, very soluble; +, soluble; −, insoluble.
*The alternating copolymer P(HOMI/St) was obtained by thermal deprotection of P(t-BuOMI/St) at 280°C.
**Tetramethylammonium hydroxide (2.38 wt%).

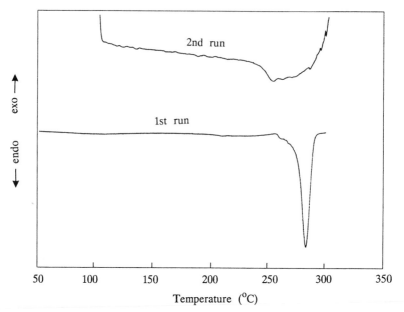

Figure 2. DSC analysis of P(t-BuOMI/St) in a nitrogen stream at the heating rate of 10 °C/min: the first run for the deprotection of t-Bu side-chains and the second run for the deprotected polymers.

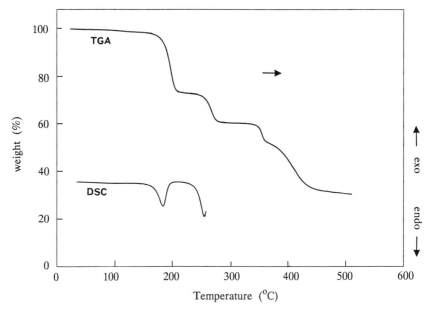

Figure 3. Thermograms of TGA and DSC for the two-step deprotection of P(t-BuOMI/t-BOCSt).

the homopolymerization. The observed mass loss of the obtained P(t-BuOMI) was 21 to 30% depending on the polymerization conditions whereas the theoretically calculated amount is 33%. In addition, P(t-BuOMI) revealed no observable T_g even after deprotection in the second run of DSC. As a similar polymer, Crivello and his coworkers (*12*) investigated the deprotection of poly(4-*tert*-butoxystyrene) to poly(p-hydroxystyrene) thermally and acidolytically, and positive image formation was achieved. The tendency of facile, complete thermal deprotection of t-Bu groups is more prominent in the case of t-BuOMI copolymers when compared with that of poly(4-*t*-butoxystyrene).

The thermal deprotection behavior was easily followed by infrared spectral change using a film of P(t-BuOMI/St) as shown in Figure 4. The starting protected polymer has absorption bands at 2980 and 1370 cm^{-1} for t-Bu groups, at 1790 and 1730 cm^{-1} for imide carbonyls and at 1190 cm^{-1} for ethers, whereas the deprotected polymer P(HOMI/St) obtained by heating above 285 °C shows absorption bands at 3200 cm^{-1} for imino groups, 1780 and 1710 cm^{-1} for imide carbonyl groups, and 1220 cm^{-1} for *N*-hydroxy groups due to the deprotection of the side-chain t-Bu groups. Furthermore the characteristic absorption peak at 1370 cm^{-1} due to symmetric bending of methyl groups in t-Bu groups disappeared in the deprotected copolymer P(HOMI/St). The IR spectrum of the deprotected polymer was found to be identical to that of P(HOMI/St) which was directly obtained by a radical copolymerization of the corresponding monomers.

Solubility Change by Deprotection

All the t-BuOMI copolymers were white powders having a good film forming property. The t-BuO protected copolymers show considerable change in solubility after deprotection due to the large polarity change. P(t-BuOMI/St) is very soluble in common organic solvents such as acetone, chloroform, toluene, anisole, and DMF, but insoluble in aqueous alkaline solutions and methanol as described in Table III. Instead, the deprotected polymer P(HOMI/St) is soluble in aqueous base solutions, dioxane, and DMF, but insoluble in common organic solvents. Other t-BuOMI copolymers are also showed the similar solubility behavior before and after deprotection. The solubility of the deprotected polymers in aqueous base solutions is of the utmost importance for practical application as positive resist materials.

Lithographic Evaluation of P(t-BuOMI / St)

The t-BuO protected MI polymers, P(t-BuOMI/X-St) appeared to have very low absorption in deep UV region like the t-BOC protected MI polymers (*8*). P(t-BuOMI/St) has an absorbance of 0.15/μm at 248 nm and an UV absorption spectrum is showed in Figure 5. P(t-BuOMI/SiSt) was found to have an absorbance of 0.13/μm at 248 nm. This low absorption in deep UV region is quite comparable to the tetrahydropyranyl-protected methacrylate copolymers for resist application (*13*). The acidolytic deprotection of t-Bu groups of P(t-BuOMI/St) was detected at 130 °C or lower temperatures in the presence of p-toluenesulfonic acid as shown in Figure 6 that indicates capability of the

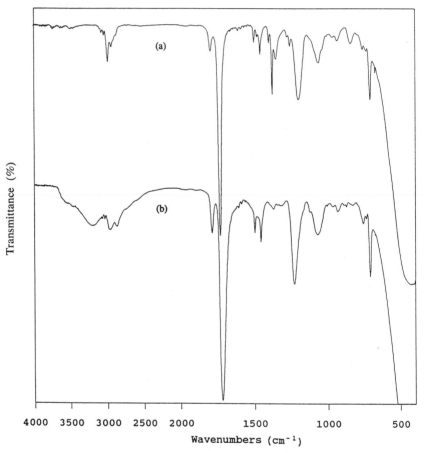

Figure 4. Infrared spectral change of P(t-BuOMI/St) before (a) and after (b) thermolysis.

Table III. Solubility of P(t-BuOMI/St) and P(HOMI/St)

P(t-BuOMI/X-St)[a]	P(HOMI/X-St)[a]	weight loss[b](%) found	calc.	T_{dp}[c] °C	T_g[d] °C	T_{dc}[e] °C
P(t-BuOMI)	P(HOMI)	21-30	33	240-247	–	335
P(t-BuOMI/St)	P(HOMI/St)	21	21	283	245	340
P(t-BuOMI/t-BOCSt)	P(t-BuOMI/HOSt)[f]	25	26	185	–	–
	P(HOMI/HOSt)	13	14	252	275	340
P(t-BuOMI/SiSt)	P(HOMI/SiSt)	15	16	300	–	335
P(t-BuOMI/AcOSt)	P(HOMI/AcOSt)	16	17	270	–	340
P(t-BuOMI/MeSt)	P(HOMI/MeSt)	19	20	290	250	340
P(t-BuOMI/ClSt)	P(HOMI/ClSt)	18	18	270	–	330
P(HOMI/St)	–	–	–	–	245	340
P(HOMI/SiSt)	–	–	–	–	–	335

[a]After deprotection of t-Bu groups of the original copolymers P(t-BuOMI/X-St), t-BuOMI units are converted into *N*-hydroxymaleimide (HOMI) units in the deprotected copolymers P(HOMI/X-St). [b]Measured in wt % by TGA and theoretical calculation. [c]T_{dp} is the deprotection temperature measured in the first run of DSC (cf. Fig. 2). [d]T_g is the glass transition temperature of the deprotected copolymer measured in the second run of DSC (cf. Fig. 2). [e]T_{dc} is the onset decomposition temperature of main-chains measured by TGA. [f]t-BOCSt units are converted to *p*-hydroxystyrene (HOSt) units by releasing carbone dioxide and 2-methylpropene (ref. 3).

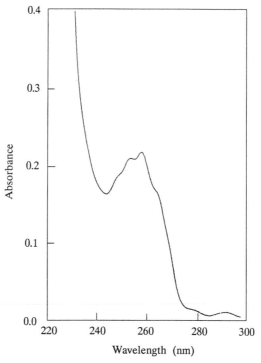

Figure 5. An UV absorption spectrum of P(t-BuOMI/St) in 1.4 um thick film.

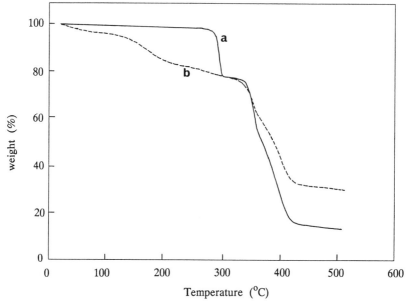

Figure 6. Comparison of TGA thermograms of (a) P(t-BuOMI/St) and (b) P(t-BuOMI/St) containing 10 wt% of p-toluenesulfonic acid.

polymer as a chemically amplified resist. It is evident that the acidolytic depro-tection slowly proceeds even at temperatures below 130 °C in the solid state despite of its high thermal deprotection temperature of 283 °C.

In the preliminiary resist evaluation of P(t-BuOMI/St) containing onium salts as a PAG, positive patterns were obtained by deep UV or electron beam irradiation followed by post-exposure bake (PEB) and development with aqueous base solutions. Two kinds of triphenylsulfonium salts, hexafluoroantimonate (TPSHFA) and triflate (TPSOTf) were used more than 15 wt% to generate proper positive tone images even high temperature PEB treatment at about 150 °C. The resist films of P(t-BuOMI/St) containing TPSHFA were imagewise exposed to 260 nm light by a contact mode and followed by PEB treatment at 130 °C for 1min, and developed with 2.38 wt% TMAH solution for 1min. Thus the SEM photographs of positive tone images were obtained as shown in Figure 7. The inherent viscosity of the polymer was 1.0 dl/g for (a) and 0.5 dl/g for (b)

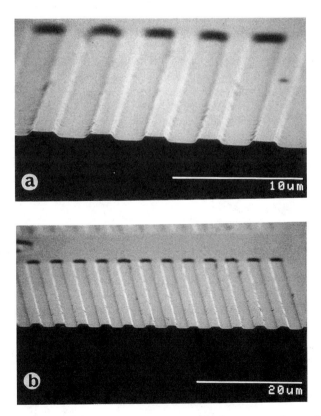

Figure 7. SEM photographs of coded 2.2 μm L/S patterns formed in 0.6 μm thick resist of P(t-BuOMI/St) containing TPSHFA: exposure to 260 nm light for 5 sec by a contact mode; PEB at 130 °C for 1 min. (a) 20 wt% TPSHFA, (b) 16.7 wt% TPSHFA.

in Figure 7. The sensitivities of the resists are not enough compared with the known chemical amplification resists (2, 13). The pK_a values of N-hydroxysuccinimides are reported to be as high as 6 to 7 due to the strongly acidic N-hydroxy functionality (14) and are higher than those values of the imino protons of succinimides and phenolic protons which are known to be about 10. However, the high deprotection temperatures of the t-Bu protected HOMI copolymer, namely P(t-BuOMI/St), rendered rather low sensitivity as a resists material compared with t-BOC protected polymers.

CONCLUSIONS

As a new protected monomer, N-(tert-butoxy)maleimide (t-BuOMI) was synthesized by a retro-Diels-Alder reaction in a high yield and the radical copolymerizations of t-BuOMI with styrene monomers (X-St) were carried out to obtain a new kind of t-BuO protected polymers based on the maleimide structure. The protected maleimide copolymers with styrene monomers, P(t-BuOMI/X-St) were completely converted into the free N-hydroxymaleimide copolymers P(HOMI/Y-St) by thermal deprotection of the side-chain t-Bu groups at about 270 to 300 °C. The deprotected polymers P(HOMI/X-St) have very high T_gs at 245 °C or above along with good solubilities in aqueous base solutions. The acidolytic deprotection of t-Bu groups from P(t-BuOMI/St) was occurred at 130 °C or lower temperatures in the presence of catalytic acids despite of its original high deprotection temperature. t-BuO protected maleimide copolymers were found to possess specific requirements such as alkaline solubility, high T_g, low UV absorption and facile deprotection for application as thermally stable, sensitive deep UV resist materials based on the chemical amplification concept. The full accounts of synthesis and polymerizations of t-BuOMI and further evaluation of its polymers as chemical amplification resists will be published elsewhere.

ACKNOWLEDGMENT

The authors are deeply grateful to Korea Ministry of Science and Technology for the financial support on the research project of advanced resist materials.

REFERENCES

1. Willson, C. G.; Ito, H.; Fréchet, J. M. J.; Tessier, T. G.; Houhihan, F. M.; J. Electrochem. Soc. 1986, 133, 181.
2. Reichmanis, E.; Houlihan, F. M.; Nalamasu, O.; Neenan, T. X. Chem. Mater. 1991, 3, 394; references cited therein.
3. Ito, H.; Willson, C. G. In Polymers in Electronics; ACS Symposium Series, No. 242; Davidson, T., Ed.; American Chemical Society: Washington, DC, 1984; p 11.
4. Maltabes, J. G.; Holmes, S. J.; Morrow, J. R.; Barr, R. L.; Hakey, M.; Reynolds, G.; Brunsvold, W. R.; Wilson, C. G.; Clecak, N.; MacDonald, S.; Ito, H. Proc. SPIE 1990, 1262, 2.

5. Turner, S. R.; Arcus, R. A.; Houle, C. G.; Schleigh, W. R. *Polym. Eng. Sci.* **1986**, *26*, 1096.
6. Turner, S. R.; Ahn, K.-D.; Willson, C. G. In *Polymers for High Technology: Electronics and Photonics*; ACS Symposium Series, No. 346; Bowden, M. J.; Turner, S. R., Eds.; American Chemical Society: Washington, DC, 1987; p 200.
7. Ahn, K.-D.; Lee, Y. H.; Koo, D.-I. *Polymer* **1992**, *33*, 4851.
8. Ahn, K.-D.; Koo, D.-I.; Kim, S.-J. *J. Photopolym. Sci. Technol.* **1991**, *4*, 433.
9. Brunsvold, W.; Conley, W.; Crockatt, D.; Iwamoto, N. *Proc. SPIE* **1989**, *1086*, 357.
10. Narita, M.; Teramoto, T.; Okawara, M. *Bull. Chem. Soc. Jpn.* **1971**, *44*, 1084.
11. Turner, S. R.; Anderson, C. C.; Kolterman, K. M. *J. Polym. Sci., Polym. Lett.* **1989**, *27*, 253.
12. Conlon, D. A.; Crivello, J. V.; Lee, J. L.; O'Brien, M. J. *Macromolecules* **1989**, *22*, 509.
13. Taylor, G. N.; Stillwagon, L. E.; Houlihan, F. M.; Wolf, T. M.; Sogah, D. Y.; Hertler, W. R. *Chem. Mater.* **1991**, *3*, 1031.
14. ames, D. E.; Grey, T. E. *J. Chem. Soc.* **1955**, 631.

RECEIVED December 30, 1992

Chapter 10

Acid-Sensitive Pyrimidine Polymers for Chemical Amplification Resists

Yoshiaki Inaki, Nobuo Matsumura, and Kiichi Takemoto

Department of Applied Fine Chemistry, Faculty of Engineering, Osaka University, Suita, Osaka 565, Japan

Pyrimidine bases such as uracil and thymine are known to form photodimers with exposure to UV light (1). It seemed interesting to explore the photochemical reaction of the pyrimidine bases to the photolithographic process in microelectronics fabrication technology.

We have studied the synthesis and application of a series of polymers containing pendent pyrimidine bases, and their application to negative-type photoresists, as well as polymers containing pyrimidine photodimers in the main chain and their applications to positive-type photoresists (2–8).

Nucleic acid bases exist as mixtures of two or more rapidly interconvertible isomers of tautomeric forms. Tautomers, at least in principle, can be separated at low temperatures where the rate of interconversion is low. The keto-enol equilibrium is the classic example of tautomerism. The enol is present in small amount, since it is usually less stable than the keto form. Uracil (1) is one of the nucleic acid bases, that exists as mixture of the tautomeric keto and enol forms (Scheme 1). The ratio of the enol tautomer to the keto tautomer of uracil, however, is small. The existence of the enol form of Scheme 1 is the basis for referring to uracil as dihydroxypyrimidine.

The enol form is readily formed from the keto tautomer by virtue of the fact that hydrogen atoms attached to nitrogen atoms that are immediately adjacent to carbonyl groups are acidic. Although the enol tautomer of uracil is difficult to isolate, 2,4-dialkoxypyrimidine that is thought to be a tautomeric form of uracil, can be easily prepared from dichloropyrimidine (4) (Scheme 2). The 2,4-dialkoxypyrimidine is very sensitive to acids forming the more stable uracil.

Di-*tert*-butoxypyrimidine (2) is a stable and water insoluble compound. In the presence of a catalytic amount of an acid this compound rearranges to give uracil which is soluble in alkaline aqueous solution (Scheme 1). In this reaction, a proton attacks the nitrogen atom and the alkyl group leaves to give the alkyl cation and the keto tautomer. The driving force for this reaction is the tautomerism of uracil.

0097–6156/94/0537–0142$06.75/0

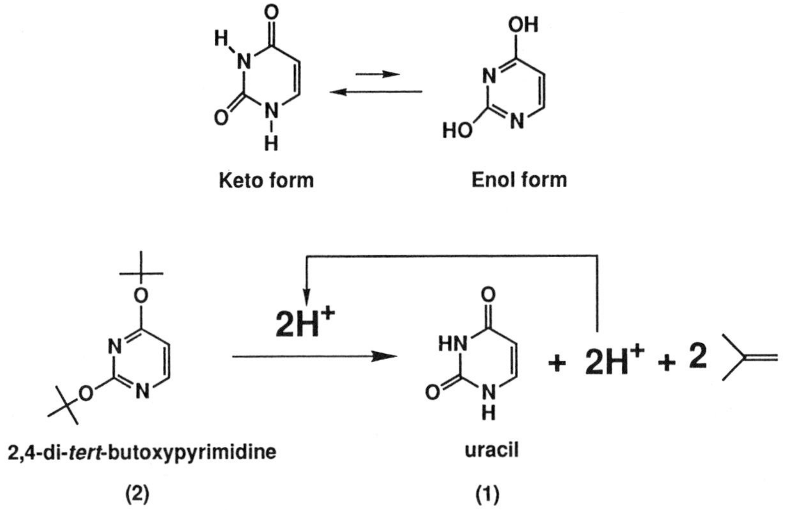
Keto form
Enol form
2H+
+ 2H+ + 2
2,4-di-*tert*-butoxypyrimidine
(2)
uracil
(1)

Scheme 1.

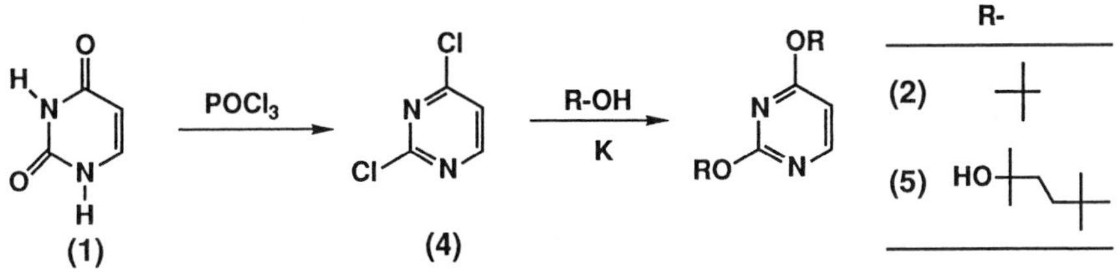
POCl3
R-OH
K
(1)
(4)
R-
(2)
(5)

Scheme 2.

This paper deals with the application of the tautomerism of uracil to a chemical amplification photoresist. Polyethers containing 2,4-dialkoxypyrimidine units (3) were prepared. These polymers were found to be very sensitive to an acid and to be applicable to chemical amplification photoresists (Scheme 3).

EXPERIMENTAL

Materials

Preparation of the model compound (Scheme 2).   2,4-Di-*tert*-butoxypyrimidine (2) was prepared by the reaction of potassium *t*-butoxide with 2,4-dichloropyrimidine (4) which was obtained by the reaction of uracil (1) and phosphorus oxychloride. Another model compound (5) was prepared from the potassium salt of 2,5-dimethyl-2,5-hexanediol with 2,4-dichloropyrimidine.

2, 4-Di-tert-butoxypyrimidine (2).   Uracil (1) was reacted with phosphorus oxychloride to give 2,4-dichloropyrimidine (4) (*9*). Potassium (0.78 g, 20 mmol) was added to *tert*-butanol (20 ml) and reacted under reflux for 1 hour to give a potassium *tert*-butoxide solution. To the solution, 2,4-dichloropyrimidine (1.5 g, 10 mmol) was added at room temperature, and the mixture was stirred under reflux for 1 hour. After the reaction, the precipitated KCl was removed by filtration, and the solvent was removed by evaporation. The residue was dissolved in diethyl ether, and washed with 30% aqueous KOH and dried over sodium sulfate. After evaporation of the solvent, the product was obtained by distillation under reduced pressure. Yield 1.1 g (49%). IR: 1580 and 1155 cm^{-1}. ^{1}H-NMR (CDCl$_3$, TMS, δ): 1.4 (18H, s), 6.0 (1H, d), and 7.9 (1H, d).

2,4-Di-(1,1,4,4-tetramethyl-4-hydroxybutoxy)pyrimidine (5).   Potassium (0.78 g, 20 mmol) was added to a solution of 2,5-dimethyl-2,5-hexanediol (2.8 g, 20 mmol) in dioxane (20 ml), and the solution was refluxed until potassium completely reacted. The reaction mixture was cooled to room temperature, 2,4-dichloropyrimidine (1,5 g, 10 mmol) was added slowly to the solution and the mixture was refluxed for 1 hr. The solution was filtered to remove KCl and the solvent was removed by evaporation. From the residue, excess 2,5-dimethyl-2,5-hexanediol was removed by sublimation at 60 °C under reduced pressure to give a glassy product. Yield 2.23 g (60%). The product was identified by IR and ^{1}H-NMR spectra. IR: 1580, and 1560 cm^{-1}. ^{1}H-NMR (CDCl$_3$, TMS, δ): 1.0(12H, s), 1.4(12H, s), 1.9(4H, t), 2.2(4H, t), 6.1(1H, d), and 7.9(1H, d).

Preparation of the polymers (Scheme 4)

Polyethers (3) containing dialkoxypyrimidine units in the main chain were prepared from dichloropyrimidine (4) with various diols. Potassium (0.78 g, 20 mmol) was added to a solution of a diol (11 mmol) in dioxane (50 ml), and the mixture was refluxed until potassium completely reacted. After the reaction mixture was cooled to room temperature, 2, 4-dichloropyrimidine (1.5 g, 10 mmol) was added to the mixture. The mixture was again heated to reflux for 20 min, and the solvent was removed by evaporation. Water (100 ml) was added to the residue, and the mixture was stirred overnight. The polymer was isolated as a solid by filtration. The polymers were purified by dissolution in chloroform and precipitation with methanol. The results of the polymerization are tabulated in Table 1. IR (3c): 1580 cm^{-1}. ^{1}H-NMR (CDCl$_3$, TMS, δ) (3c): 5.3(4H, s), 6.35(1H, d), 7.4(4H, b), and 7.9(1H, d).

Measurements

IR spectra of polymer films spin-coated on undoped silicon wafers were recorded on a JASCO IR-810 infrared spectrophotometer. UV spectra were recorded on a JASCO U-best-30 UV/Vis spectrophotometer. TGA and DSC analysis were carried out using a SEIKO SSC/580DS TGA/DSC instrument at a heating rate of 10.0 deg/min. in nitrogen, from 50 to 300 °C.

The molecular weight of the polymers was determined by gel permeation chromatography using Toyo Soda HLC CP8000 with a thermostated column: Cosmosil GPC100 and GPC300 with chloroform as the eluent, and a UV detector operating at 254 nm. The molecular weight was obtained by compairing retention times to those of polystyrene standards.

Acid hydrolysis of the polymer in solution

To a solution of polymer (0.1 g) in chloroform (5 ml), trifluoromethanesulfonic acid (3c: 0.1 g, 3d-e: 0.01 g) was added, and the mixture was stirred overnight at room temperature. The precipitate was separated by a centrifugation, and the product (3c: 0.05 g, 3d: 0.02 g, 3e: 0.04 g) was analyzed by IR and NMR spectra.

Photoacid-catalyzed reaction in solid film

Two-component resist solutions were prepared from polymer (0.09 g), triphenylsulfonium trifluoromethanesulfonate (0.01 g), and chloroform (1 ml) as a solvent. The resist solutions were spin-coated onto silicon wafers and baked in an oven at 100 °C (at 60 °C for 3e) for 10 min. The resist films were exposed by monochromatic light at 250 ± 1 nm using a JASCO CRM-FA spectro-irradiator equipped with a 2 kW Xenon-Arc lamp and with a exposed energy integrator. The films were then heated at 60 °, 100 °, or 150 °C for 10 min, and the IR spectra were measured.

Solubility measurements

The resist solutions of polymer (0.09 g) with photoacid generator (triphenylsulfonium trifluoromethanesulfonate, 0.01 g) in chloroform (1 ml) were cast on a quartz plate, and heated in an oven at 100 °C (at 60 °C for 3e) for 10 min. The exposed film was post-exposure baked in an oven at 100 °C for 10 min, and developed in tetramethylammonium hydroxide (TMAH) aqueous solution (0.1%) for 15 seconds. The film thickness was measured by UV spectra at 270 nm.

Sensitivity measurements

The resist solution of polymer **3d** (0.09 g) with photo acid generator (triphenylsulfonium trifluoromethanesulfonate, 0.01 g) in ethylene glycol monomethyl ether acetate (0.9 g) was spin-coated (1000 rpm) onto silicon wafers and baked on a hot-plate at 100 °C to give a 0.35 μm film. In a case of polymer **3e**, the resist solution of polymer (0.09 g) with photo acid generator (triphenyl-sulfonium trifluoromethanesulfonate, 0.01 g) in chloroform (0.9 g) was spin-coated (500 rpm) onto silicon wafers and baked on a hot-plate at 100 °C to give a film of 0.45 μm thickness. The resist films were exposed using a Hg-Xe lamp with a 251.5 nm interference filter and post-exposure baked at 100 °C for 60 seconds. The film thickness was measured using a Sloan DEKTAK 3030 after development in a 0.1% TMAH aqueous solution for 15 seconds.
A positive image was obtained with contact optical exposure at 250 nm (6 mJ/cm^2) and development for 15 seconds in a 0.1% TMAH aqueous solution.

EB sensitivity

In the case of EB resist, exposures were carried out using a point electron beam system (ELIONIXJELS-3300) at 20 kV. After exposure, the film was baked 100 °C for 5 minutes (**3d**) or 1 minute (**3e**), and developed in 0.1% TMAH aqueous solution for 15 seconds to give the positive image.

RESULTS AND DISCUSSION

Preparations of Polymers

Polyethers (**3**) containing dialkoxypyrimidine units in the main chain were prepared from dichloropyrimidine (**4**) with various diols (Scheme 4). Table 1 shows the result of the polymerization, molecular weight of the polymers, and elementary analysis for five polymers, prepared from alicyclic and aromatic primary, secondary, and tertiary diols.

Table 1. Yield and Analytical Data of Polyethers Containing Pyrimidine

(3)	Yield (%)	Mol. Wt $\times 10^3$	Anal (Calcd.) H	C	N	Anal (Found) H	C	N
a	86	8.5	7.32	65.43	12.72	7.14	63.71	12.26
b	86	3.0	6.29	62.49	14.57	6.13	58.76	13.87
c	96	2.3	4.71	67.28	13.08	4.83	65.21	11.81
d	95	2.7	5.82	69.41	11.56	5.84	68.51	11.17
e	99	1.3	6.71	71.09	10.36	7.14	70.65	8.47

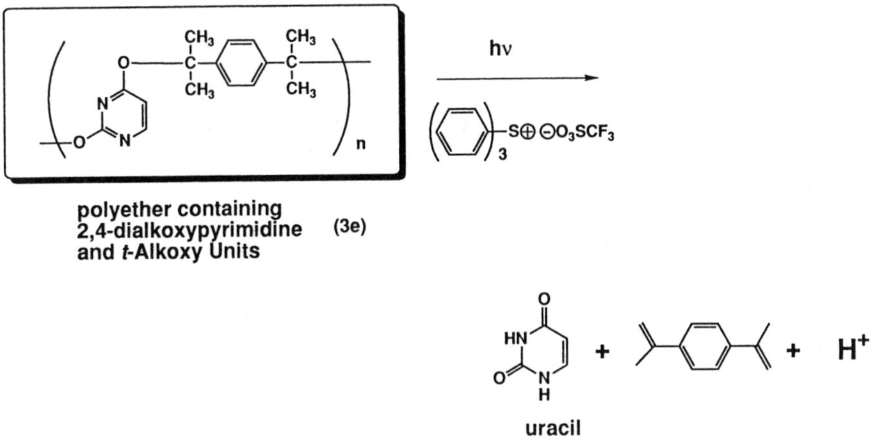

polyether containing
2,4-dialkoxypyrimidine   (3e)
and *t*-Alkoxy Units

uracil

Scheme 3.

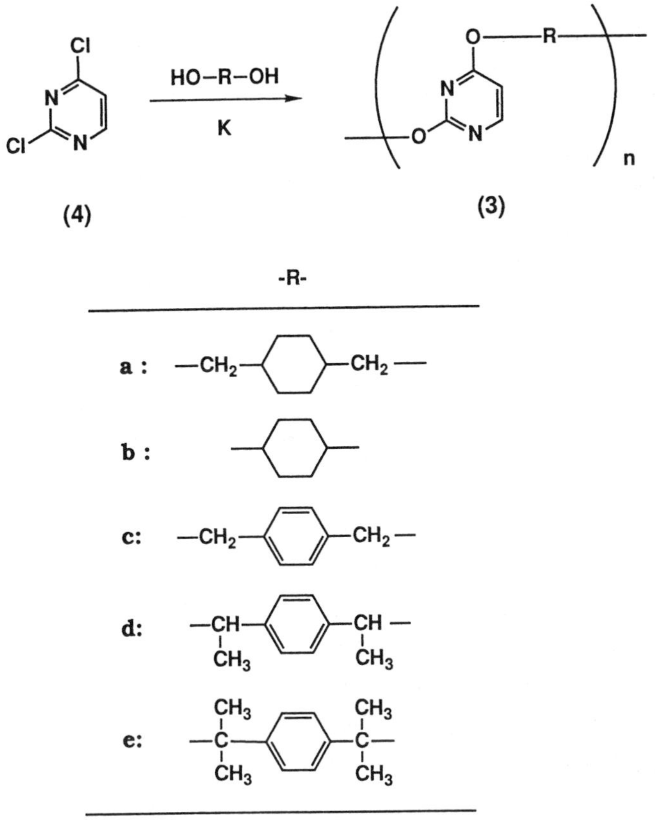

(4)   (3)

-R-

a :   —CH₂— ⬡ —CH₂ —

b :   — ⬡ —

c: —CH₂— ⬡ —CH₂—

d: —CH— ⬡ —CH —
       |CH₃ |CH₃

e: —C— ⬡ —C—
       |CH₃ |CH₃
     CH₃   CH₃

Scheme 4.

Thermal analysis of the polymers

DTA and TGA curves for the polymer 3c-e are shown in Figures 1–3. Polymer 3c was stable to 300 °C. However, polymer 3e decomposed at around 200 °C, and the residue after heating was identified by IR spectrum as uracil. Tg of this polymer was about 120–130 °C. These polymers are soluble in chloroform and insoluble in water.

Tautomerization by an acid in solution

Acid decomposition of the 2,4-dialkoxypyrimidine derivatives was studied in solution by addition of an acid. Figure 4 shows the UV spectra of 2,4-di-*tert*-butoxypyrimidine (2) in acetonitrile. Addition of acid (CF_3SO_3H) caused change of UV spectrum as shown in the Figure 4. The spectrum after the reaction was found to be the same as the spectrum of uracil in the presence of acid. This result suggested that the 2,4-dialkoxypyrimidine compounds can easily tautomerize to uracil by addition of acid.

Addition of a small amount of the acid to the solution of polymers in chloroform caused precipitation of a solid that was identified as uracil from its IR and NMR spectra. In the case of the polymer (3c) prepared from the primary diol, however, it was difficult to rule out the possibility of a rearrangement shown in Scheme 5. The attack of proton on the nitrogen at N3 of pyrimidine a may cause tautomerization of the pyrimidine base to give b. At this step, the attack of another proton on the nitogen at N1 of pyrimidine b may cause tautomerization of the pyrimidine base to give c. However, the leaving cation can attack the unprotonated nitrogen of the pyrimidine b to give the N-substituted uracil derivative d. The product d could not be identified by IR spectra, but was identified by NMR spectra as the N-substituted uracil derivative. This fact suggests that in addition to the main elimination reaction, the primary polyether underwent rearrangement as a side reaction.

Tautomerization of the model compound in solid state

The model compound (5) with triphenylsulfonium trifluoromethanesulfonate (10 wt%) in the solid film was irradiated at 250 nm. Photolysis of the model compounds was carried out in the solid film by exposure to monochromic UV light from a spectro-irradiator. After post-baking of this film, IR spectra were observed as shown in Figure 5. The absorbance at 1580 cm^{-1} assigned to the C = N and C = C stretching vibrations of alkoxypyrimidine (10) decreased with an increase in energy dose. At the same time, a new peak at 1700 cm^{-1} assigned to the C = O streching vibration of uracil increased. The spectrum after 75 mJ/cm^2 irradiation was the same as that of uracil suggesting a complete reaction. In Figure 6, the relative absorbances of 1580 and 1710 cm^{-1} were plotted against exposure dose. In this figure, the decrease in 1580 cm^{-1} absorbance occurrs in parallel with the decrease of the 1710 cm^{-1} band.

Photoacid-catalyzed reactions of the polymers in the solid film

Photolysis of the 2,4-dialkoxypyrimidine polymer (3) in the presence of a photo-acid generator was carried out in the solid film. Thin films of these

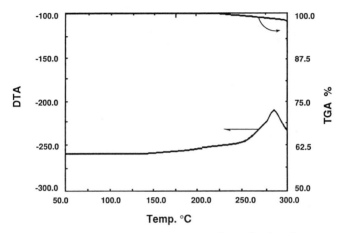

Figure 1. DTA and TGA curves for polyether **3c**.

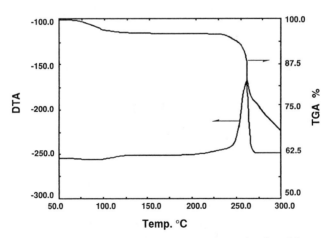

Figure 2. DTA and TGA curves for polyether **3d**.

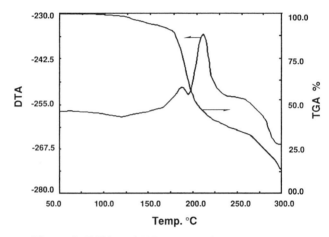

Figure 3. DTA and TGA curves for polyether **3e**.

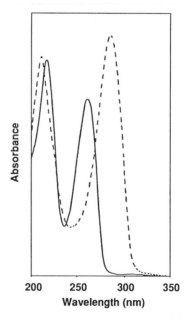

Figure 4. UV spectra of 2,4-di-*tert*-butoxypyrimidine (**2**) in acetonitrile. Dotted line: without acid. Solid line: with CF_3SO_3H.

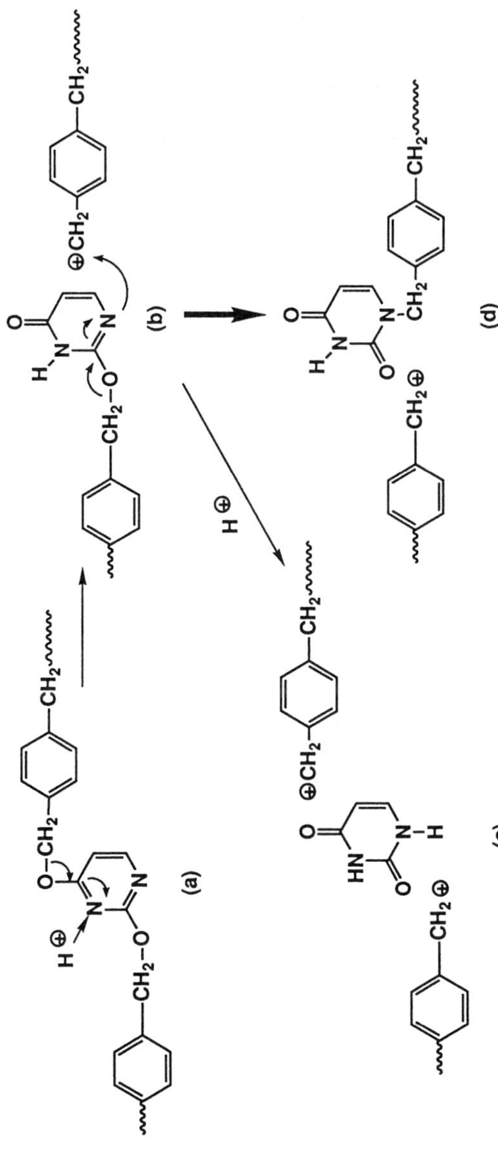

Scheme 5.

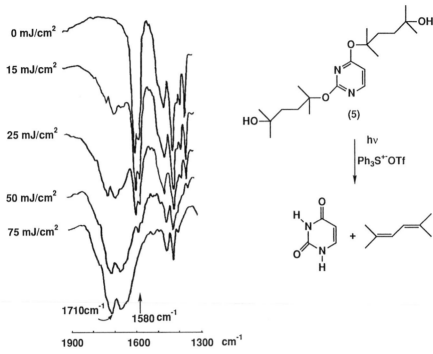

Figure 5. IR spectra of the model compound 5 with triphenylsulfonium trifluoromethanesulfonate (10 wt%) in solid film. Exposure: 250 nm. PEB: 100 °C for 10 min.

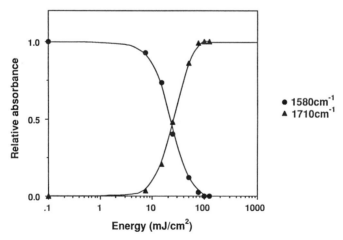

Figure 6. Acid decompositions of the model compound 5 with triphenylsulfonium trifluoromethanesulfonate (10 wt%) in solid film. Exposure: 250 nm. PEB: 100 °C for 10 min.

polymers were obtained from their chloroform solutions by spin coating on silicon substrates and prebaking at 100 °C for 5 minutes. The films were exposed to the monochromatic light (250 nm) until the reaction was complete. The reaction was followed by IR spectrophotometer at 1580 cm^{-1} and 1710 cm^{-1}, which are the absorption bands characteristic of the 2,4-dialkoxypyrimidine and the regenerated uracil, respectively (Figures 7–9).

The changes of the IR spectra of the polymers **3a** and **3b** were small, because these polymers were prepared from primary and secondary alicyclic diols. On the other hand, the change of IR spectra of the polymers **3c**, **3d**, and **3e** was remarkable because these polymers have a benzyloxy group.

Figure 7 shows IR spectral change of polymer **3c** obtained from a primary alcohol in the presence of 10% acid generator. The peak at 1580 cm^{-1} of the original unirradiated polymer (**i** in Figure 7) is assigned to alkoxypyrimidine. After excess exposure of UV light at 250 nm in the presence of acid generator and post baking at 100 °C, the peak at 1580 cm^{-1} disappeared, and new peak at 1710 cm^{-1} appeared (**ii** in Figure 7). The peak at 1710 cm^{-1} is assigned to uracil. Therefore, this change in IR spectrum was caused by the tautomerization of the alkoxypyrimidine to uracil. However, postbaking at 150 °C was necessary for the complete conversion to uracil (**iii** in Figure 7). Without the acid generator, this polymer was stable even after exposure to UV light and postbaking at 200 °C.

Figure 8 shows the IR spectra of the polymer **3d** prepared from the secondary alcohol, after no irradiation (**i** in Figure 8), 25 mJ/cm^2 (**ii** in Figure 8), and 100 mJ/cm^2 irradiation at 250 nm followed by postbaking at 100 °C (**iii** in Figure 8). After 100 mJ/cm^2 irradiation, this polymer decomposed completely. In the case of polymer **3e** from the tertiary alcohol, however, the reaction was complete with 20 mJ/cm^2 irradiation (at 250 nm) and postbake at 60 °C (Figure 9).

These IR spectra of polymer **3c-e** on silicon wafer in the presence of triphenylsulfonium triflate during irradiation at 250 nm indicated tautomeric change of the alkoxypyrimidine to uracil. From the results of IR spectra, sensitivities of these polymers were calculated (Figure 10). The highest sensitivity was obtained for the polymer **3e**. For the complete reaction of the polymer **3e**, the UV dose of 10 mJ/cm^2 and a post-bake at 60 °C are sufficient. Polymer **3d** from the secondary diol shows a sensitivity of 100 mJ/cm^2. However, the sensitivity of polymer **3c**, prepared from the primary diol was very low even after postbaking at 100 °C.

Excellent sensitivity of polymer **3e** with *tert*-alcohol units suggests the acid catalyzed reaction as shown in Scheme 6.

Development

Polyethers containing 2,4-dialkoxypyrimidine in the main chain were hydrophobic and insoluble in water. Uracil, however, is soluble in an alkaline aqueous solution. Therefore, the solubility of the polyether changes following photoacid tautomeriszation. Figure 11 shows UV spectra of the polymer **3d** in the presence of 10 wt% photoacid generator in the solid film on a quartz plate, prepared by casting from chloroform solution.

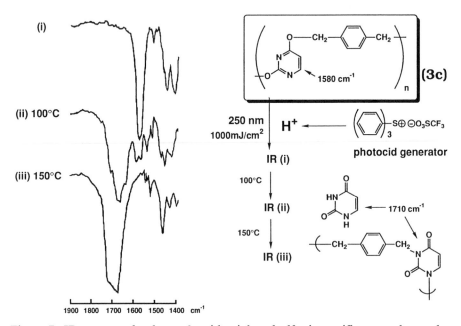

Figure 7. IR spectra of polymer **3c** with triphenylsulfonium trifluoromethanesulfonate (10 wt%) in solid film. Exposure: 250 nm.

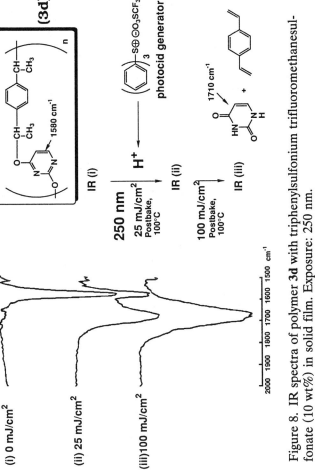

Figure 8. IR spectra of polymer **3d** with triphenylsulfonium trifluoromethanesulfonate (10 wt%) in solid film. Exposure: 250 nm.

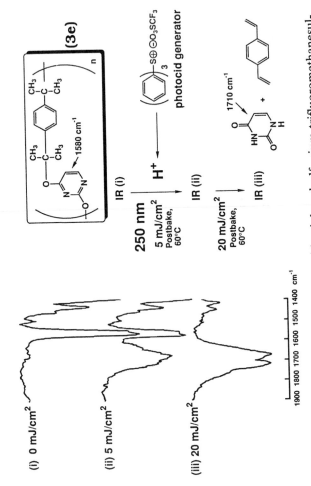

Figure 9. IR spectra of polymer **3e** with triphenylsulfonium trifluoromethanesulfonate (10 wt%) in solid film. Exposure: 250 nm.

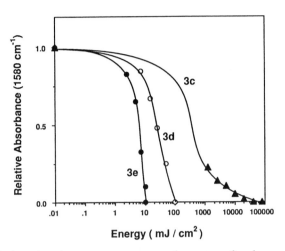

Figure 10. Relative absorbance versus exposed energy of polymers (**3c-e**), post-baked at 100 °C (**3c, d**) and 60 °C (**3e**), photoacid generator 10 wt%.

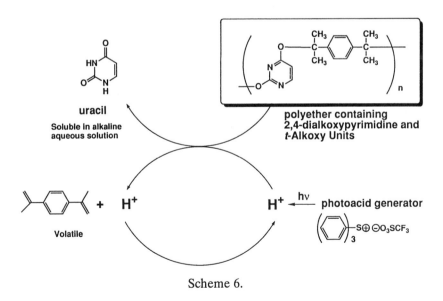

Scheme 6.

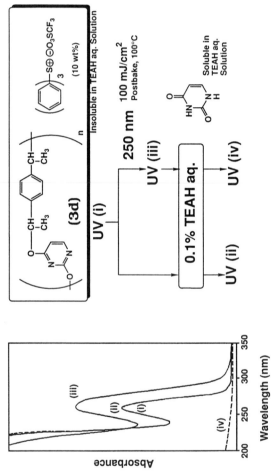

Figure 11. UV spectra of polymer **3d** in the presence of 10 wt% photoacid generator in solid film on a quartz plate.

The original polymer film after prebaking has the spectrum (i) in Figure 11. Spectrum (ii) is for the polymer film after development of the unirradiated polymer in 0.1% triethylammonium hydroxide aqueous solution. No decrease in absorbance suggestes insolubility of polymer **3d** in the alkaline aqueous solution. After exposure to UV light and postbaking, spectrum (i) in Figure 11 changed to spectrum (iii), corresponding to the reaction of pyrimidine to uracil. After development of (iii) in Figure 11, products were completely dissolved in the developer as shown by spectrum (iv). These results suggest that polymer **3d** can be used as positive type chemical amplification photoresist.

Figure 12 shows UV spectra of polymer **3e** in the presence of a 10 wt% photoacid generator in the film spin-coated on a quartz plate: spectrum (i in Figure 12) is the original polymer film after prebaking and (ii in Figure 12) the polymer film after development by 0.1% aqueous triethylammonium hydroxide. No decrease in absorbance suggested insolubility of polymer **3e** in the alkaline aqueous solution. After exposure to UV light (100 mJ/cm^2 at 250 nm) and postbaking, spectrum (i in Figure 12) changed to spectrum (iii). Development of the 100 mJ/cm^2 postbaked film gives spectrum (iv in Figure 12). In this case, however, slight residue after development was observed and is possibly due to contamination of the polymer. These results suggest that polymer **3e** can be used as a highly sensitive positive type chemical amplification photoresist after some improvement of preparation method of the polymer.

Sensitivity of the Resist

Polyethers containing 2,4-dialkoxypyrimidine were evaluated as a chemical amplification photoresist. Figure 13a shows relative film thickness after development against exposed energy for the polymer **3d** containing photoacid generator (10 wt% triphenylsulfonium triflate). A positive image on silicon wafer was obtained from contact exposure (600 mJ/cm^2) and post-expose bake at 100 °C for 10 min.

Figure 13b shows relative film thickness after development against exposed energy for the polymer **3e** containing photoacid generator (10 wt% triphenylsulfonium triflate). In this case, only 4 mJ/cm^2 exposure caused complete reaction to give products which were soluble in alkaline aqueous solution. Therefore, this system is a highly sensitive chemical amplification photoresist.

Electron-Beam Radiation of Polymers

Polyethers **3d-e** were evaluated as chemical amplification electron-beam resists. Figure 14a shows relative film thickness after development against exposed energy for the polymer **3d** containing acid generator (10 wt% triphenylsulfonium triflate) after development with 0.1% aqueous triethylammonium hydroxide. Sensitivity of this polymer as EB resist was about 300 μC/cm^2. A positive image was obtained by EB irradiation using polymer **3d** where line & space was 0.5 μm (Figure 15).

Figure 14b is a sensitivity data for polymer **3e** with 10 wt% photoacid generator. Sensitivity of this resist was about 7 μC/cm^2. A positive image was

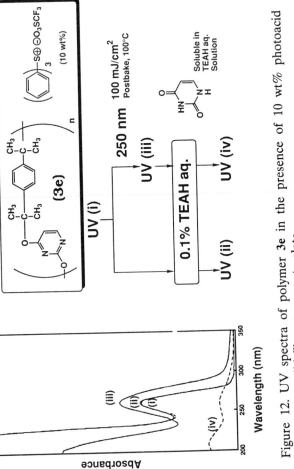

Figure 12. UV spectra of polymer 3e in the presence of 10 wt% photoacid generator in solid film on a quartz plate.

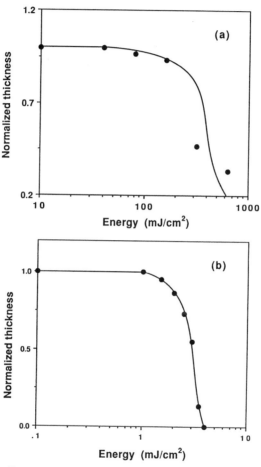

Figure 13. Normalized thickness against exposed energy at 250 nm for (a) polymer **3d** and (b) polymer **3e**. Photoacid generator: 10 wt% triphenylsulfonium triflate. Prebake: 100 °C for 1 min. Postbake: 100 °C for 1 min. Development: 0.1% TMAH aqueous solution for 15 sec.

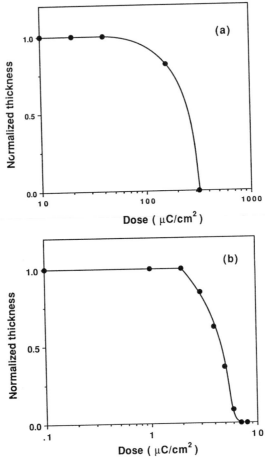

Figure 14. Normalized thickness versus EB(20kV) dose for (a) polymer **3d** and (b) polymer **3e**. Photoacid generator: 10 wt% triphenylsulfonium triflate. Prebake: 100 °C for 1 min. Postbake: 100 °C for 5 min (polymer **3d**) and 1 min. (polymer **3e**). Development: 0.1% TMAH aqueous solution for 15 sec.

Figure 15. Scanning electron micrograph of positive images in polymer **3d** containing 10 wt% of triphenylsulfonium triflate, exposed to 100 $\mu C/cm^2$ of electron beam(20kV), postbaked at 100 °C for 10 min, and then developed with TMAHaq(0.1%) for 15 sec. Line & space: 0.5 μm.

also obtained for polymer **3e** with 10 wt% photoacid generator. In this case, sensitivity was high (10 μC/cm^2) and postbaking temperature was low (70 °C for 30sec). However, slight residue after development (0.1% TMAHaq) was observed. Further preparation of the polyether containing tertiary alkyl unit is in progress.

CONCLUSION

Tautomerism of uracil was applied to a chemical amplification photoresist system. Polymers containing 2,4-dialkoxypyrimidine derivatives, an enol form of uracil, were highly sensitive to acid, and rearranged to uracil with photo generated acid. The sensitivity of the polymer containing tertiary alkyl unit after irradiation at λ = 250 nm was about 4 mJ/cm^2. In conclusion, we found that polyethers of a pyrimidine containing tertiary alcohol were highly sensitive deep-UV, and EB chemical amplification resists that gave positive images.

ACKNOWLEDGMENTS

The authors wish to acknowledged Manufacturing Development Laboratory, Mitsubishi Electric Corp., for photolithographic evaluation of the materials.

REFERENCES

1. Wang, S. Y. Photochemistry and Photobiology of Nucleic Acids ; Academic: New York, 1976.
2. Inaki, Y.; Moghaddam, M. J.; Takemoto, K. In Polymers in Microlithography; Reichmanis, E.; MacDonald, S.; Iwayanagi,T., Eds, ; ACS Symposium Series No. 412; American Chemical Society: Washington, DC, 1989; p 303.
3. Moghaddam, M. J.; Hozumi, S.; Inaki, Y.; Takemoto, K. J. Polymer Sci. Polymer Chem. Ed., 1988, 26, 3297.
4. Moghaddam, M. J.; Hozumi, S.; Inaki, Y.; Takemoto, K. Polymer J. 1989, 21, 203.
5. Moghaddam, Kanbara, K.; M. J.; Hozumi, S.; Inaki, Y.; Takemoto, K. Polymer J. 1990, 22, 369.
6. Moghaddam, M. J.; Inaki, Y.; Takemoto, K. Polymer J, 1990, 22, 468.
7. Inaki, Y.; Horito, H.; Matsumura, N.; Takemoto, K. J. Photopolym. Sci. Technol., 1990, 3, 417.
8. Horito, H.; Inaki, Y.; Takemoto, K. J. Photopolym. Sci. Technol., 1991, 4, 33.
9. Bhat, C. C.; Munson, H. R. In Synthetic Procedures in Nucleic Acid Chemistry; Zorbach, W. W.; Tipson, R. S. Eds.; Interscience: New York, 1968, Vol. 1, p 83.
10. Tsuboi, M.; Kyougoku, Y. In Synthetic Procedures in Nucleic Acid Chemistry; Zorbach, W. W.; Tipson, R. S. Eds.; Interscience: New York, 1973, Vol. 2, p 215.

RECEIVED May 17, 1993

Chapter 11

Methacrylate Terpolymer Approach in the Design of a Family of Chemically Amplified Positive Resists

R. D. Allen[1], G. M. Wallraff[1], W. D. Hinsberg[1], L. L. Simpson[2], and R. R. Kunz[3]

[1]IBM Research Division, Almaden Research Center, 650 Harry Road, San Jose, CA 95120–6099
[2]System Technology Division Development Laboratory, IBM Corporation, 1701 North Street, Endicott, NY 13760
[3]Lincoln Laboratory, Massachusetts Institute of Technology, 244 Wood Street, Lexington, MA 02173

A new family of chemically amplified positive resists based on methacrylate terpolymers has been developed. The three different monomers each perform a separate function in the terpolymer. These resists were originally designed for use in printed circuit board (PCB) fabrication. The flexibility of this approach in the design of positive resists has recently been demonstrated in the development of several new integrated circuit (IC) positive resists for deep UV (248 nm) and deep, deep UV (193 nm) lithography. These advances demonstrate that resists for wide application can be designed from a common platform of materials technology.

Chemically amplified (CA) resists have achieved recent prominence as materials useful in submicron imaging. In an effort to extend this imaging chemistry to applications requiring comparatively thick films (5–25 μm), we have developed a family of acrylic polymers which provide high performance, aqueous developing positive photoresists (*1–3*). The polymer design is quite flexible and general, as evidenced by the wide array of positive-imaging polymers which have been produced. We describe here the methacrylate polymer chemistry used to design CA positive photoresists, and present illustrative examples of members of this new resist family. Included are discussion of two positive CA resists for printed circuit board (PCB) lithography, a positive resist for deep-UV (248 nm) lithography and a positive-tone, single layer resist for 193 nm lithography, all of which are based on methacrylate terpolymers.

EXPERIMENTAL

All polymers in this study were synthesized by free radical solution polymerization. Inhibitors may be removed by column purification through neutral or basic alumina, or by distillation from calcium hydride, although experience dictates that for most methacrylate monomers, inhibitor levels are too low to interfere

0097–6156/94/0537–0165$06.00/0

with polymerization. Typically, the acrylic monomers are charged into a multi-necked, roundbottom flask at a concentration of 20-35% solids, followed by the polymerization solvent (e.g., tetrahydrofuran, 2-butanone, etc.), and the polymerization initiator AIBN (azo-isobutyronitrile). The polymerization is typically carried out under a nitrogen blanket and allowed to proceed to high conversion (80-90%). The polymer is isolated by precipitation into a non-solvent, and dried under vacuum at elevated temperature. More detail regarding polymer synthesis, characterization and lithographic processing can be found in a recent paper (1).

RESULTS AND DISCUSSION

The photoresist requirements for integrated circuit (IC) and printed circuit board (PCB) technologies differ substantially. Integrated circuit CA resists are typically based on aromatic polymers, due to their high resistance to plasma processes used in IC fabrication, but are unsuitable for PCB applications due to inadequate mechanical properties and high cost. Our design goal was to develop a positive CA photoresist compatible with existing PCB expose/develop tool sets which also met the following resist requirements: 1) the resist is developable in aqueous carbonate solutions; 2) is photoimageable in the mid-UV region (near 365 nm); 3) forms tough films with good adhesion to copper; 4) is chemically stable as an etching resist; and 5) the formulation (polymer plus photopackage) can be readily prepared in relatively large quantities at low cost.

The use of methacrylate (and/or acrylate) monomers as building blocks for new polymer materials offers great latitude in designing materials with controlled properties. Methacrylate monomers with a wide range of functionality are readily available at low cost. Coupled with clean copolymerization behavior, this allows the ready synthesis of acrylic polymers with well controlled physical and chemical properties. For these reasons, methacrylate monomers were used as building blocks for our CA resist for PCB applications.

The synthetic flexibility and demonstrated performance (e.g., photospeed and contrast) of the positive PCB resists based on functional methacrylate polymers provided the incentive to extend the methacrylate terpolymer design concepts to the submicron lithography arena. The methacrylate terpolymers have been modified in attempts to provide thin film, high resolution resists suitable for exposure at 248 and 193 nm. This work demonstrates that high performance and low materials cost are not mutually exclusive.

Polymer Design

Initially, a photoresist formulation based on poly(t-butyl methacrylate) (TBMA) homopolymer, described in the patent literature (4) and elsewhere (5), was evaluated. This polymer combines an acrylic backbone with a pendant protecting group capable of undergoing the acid-catalyzed cleavage reaction necessary for chemical amplification. The photoresist formulation consisted of 90% polymer and 10% (wt) triphenylsulfonium hexafluoroantimonate (TPSSb). Although thin, spin-coated films of formulated resist showed excellent photospeed (2.5-3.0 mJ/cm^2), the exposed pattern could not be developed unless the developing solution contained a minimum of 25 volume percent of isopropyl alcohol. In addition, solvent-cast films were optically cloudy, due to poor solubility of sulfonium salts in the non-polar polymer matrix (*6, 7*) and very brittle. For these reasons, the TBMA homopolymer-based resist is not suited for PCB applications.

To improve onium salt solubility and mechanical performance, the polymer structure was modified through the incorporation of methyl methacrylate (MMA). By substituting an MMA-TBMA (1/1) random copolymer for PTBMA, we expected that the resist mechanical properties would improve at the expense of the lithographic properties. In fact, we found that not only had the resist mechanical properties improved, but the photospeed (1.0-1.5 mJ/cm^2) and contrast had increased as well. Although the reasons for the improvement are not completely clear, the increased phase compatibility of the onium salt with the MMA-modified polymer is likely to be important, as increased polymer/onium salt compatibility has been shown to result in greater efficiency of photoacid generation. A solvent-cast resist film containing the onium salt TPSSb (10 wt.%) was optically clear, indicating improved phase compatibility of the salt in the copolymer. We have demonstrated that in these copolymer systems, useful differential solubility can be achieved with low levels of the acid-labile component. This feature is critical in our resist design, as it allows the incorporation of substantial amounts of other monomers to control various resist properties. Unfortunately, like the PTBMA homopolymer resist, imaged resist films of MMA-TBMA copolymers do not develop in purely aqueous alkaline solutions. Again, isopropyl alcohol is required as a cosolvent to achieve image development.

The strong influence of carboxylic acid functionality on aqueous solubility suggests an approach to attaining aqueous developability in this resist family. A series of MMA-TBMA-MAA terpolymers (Figure 1) with MAA concentrations ranging from 5%-33% by weight were prepared. Within this relatively narrow composition range, it was found that a full spectrum of aqueous dissolution behavior exists. At low MAA content, even highly exposed films are insoluble in aqueous base. At high MAA content, unexposed films have an unacceptably high development rate. Fortunately, within this range of behavior there exists a compositional window where the exposed films rapidly dissolve in aqueous base and the unexposed areas remain completely insoluble. This optimum range corresponds to 10-20% (mole) MAA content. Within the optimum compositional range, nonlinear dissolution behavior is obtained in combination with good resist photospeed. Table I shows a representative list of terpolymers of MMA-TBMA-MAA of different composition and behavior. The molecular weights are moderate to high in all cases, and the glass transition temperatures increase with methacrylic acid content. The materials listed here exhibit a range of dissolution properties, from non-developing resists (even at very high exposure doses) when the methacrylic acid content is 6.5% and below, to rapid unexposed thinning at methacrylic acid contents above 30% (mole). Intermediate compositions show acceptable development rates in the exposed regions at low doses, and excellent dissolution inhibition in the unexposed regions. Figure 2 shows a photoresist dissolution curve obtained via QCM (quartz crystal microbalance) for a low MAA content terpolymer resist. These dissolution properties contribute to the very high contrast of this resist. For example, measured values of gamma exceed

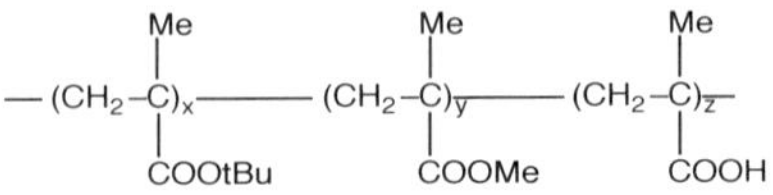

Figure 1. Chemical structure of **MMA-TBMA-MAA** random terpolymers.

Table I. Characteristics of Methacrylate Terpolymers

MMA–TBMA–MAA[A] (charged)	MOLE % COOH	$M_N{}^B$ (x 10^{-3})	MWD[B]	$T_g{}^C$ (°C)
1/1/0	0	40.5	2.3	111
1/1/0.1	6.5	39.0	2.1	120
1.1/1/0.4	20.5	39.5	2.1	154
1/1/0.5	25	37.0	2.1	160
1/1/0.75	34	36.0	2.2	–

A: wt %
B: GPC (polystyrene standard)
C: DSC 10 °C/min

10 with a threshold dose of 6 mJ/cm^2 in a visible-light sensitized resist formulation. The nonlinear effect of methacrylic acid concentration on exposed dissolution rate has also been demonstrated in non-CA resists by Schwalm (8).

The polymer mechanical properties are essentially unchanged from those of MMA-TBMA copolymers, while the adhesion to copper substrates is markedly improved. This is presumably due to the interaction of the carboxylic acid groups with the substrate surface.

Liquid Applied PCB Resist

This acrylic-based, chemically amplified positive resist appears well suited for the imaging of printed circuit boards. Detailed discussion of photoacid generator/sensitizer package for a visible laser direct imaging resist containing the methacrylate terpolymer appears elsewhere (2, 3). To evaluate the resist performance in a manufacturing environment, a large quantity of resist was prepared: the terpolymer, a PAG, and several processing additives were dissolved in propylene glycol methylether acetate to about 20% solids. This material was used as a print-and-etch resist for several weeks in a small, quick turn-around PCB manufacturing facility. The resist performed well with much better resolution than conventional dry film photoresist. A demonstration of this capability was the successful imaging of a PC mother board, shrunk to the size of a business card, by reducing all the features by a factor of 5 (minimum feature size of 25μm or 1/1000 inch). The resist process gave excellent first pass yields which were dominated by particulate defects resulting from solvent coating in a non-clean room processing environment. Resolution capability of this resist was further demonstrated by spin-coating on silicon wafers and exposing with a Perkin-Elmer 500 projection system in the UV-4 mode. Fine-line images produced in this manner are shown in Figure 3.

Dry Film PCB Resist

We have found that the resist chemistry discussed above is quite versatile and can be readily modified to achieve a given set of desired properties while maintaining good lithographic performance. For example, a simple modification provided a resist suitable for dry film application.

Typically, the PCB industry applies photoresist to the copper-clad substrate as dry films, which are laminated using a heated roller. Dry films are especially useful in PCB technology for protecting plated vias or through-holes during wet etching. The thermal properties of the resist are paramount in the design of a dry film resist material. Lamination temperatures of 100 °C to 120 °C are desirable. Experience dictates that the glass transition temperature (T$_g$) of the resin should be ca. 30-50 degrees below the desired lamination temperature.

The resist materials described earlier can be prepared as dry films which are tough and flexible, and are not prone to cracking or physical delamination. However, the MMA-TBMA-MAA terpolymers optimized for aqueous dissolution have T$_g$'s in the 140-150 °C range. Lamination of such dry films requires elevated temperatures, at or near the decomposition temperature of the resist.

Dry film lamination of the MMA-TBMA-MAA resist is impractical. To lower the glass transition (and thus the lamination temperature) of these

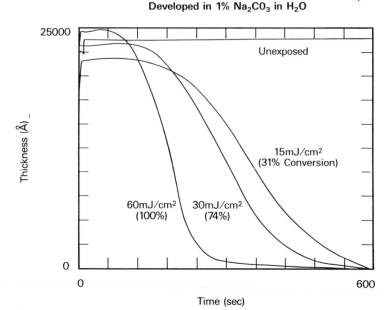

Figure 2. QCM aqueous development curves for a visible laser photoresist formulation based on a low MAA content methacrylate terpolymer, exposed at 514 nm to a variety of doses. Conversions are calculated from FTIR data.

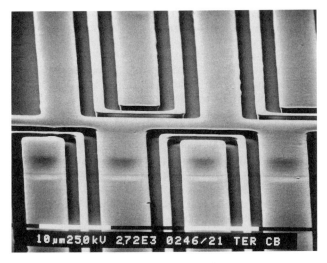

Figure 3. 1.25 micron images printed in MMA-TBMA-MMA (1.1/1.0/0.4), with iodonium triflate photoacid generator, using anthracene derivative sensitizer, exposed with PE-500 projection system in UV-4 mode.

materials, longer side-chain methacrylates were substituted into the MMA-TBMA-MAA terpolymer for MMA. Ethyl methacrylate (EMA), butyl methacrylate (BMA), and octyl methacrylate (OMA) were terpolymerized with TBMA and MAA in a variety of compositions. Resists prepared from these polymers had lower T_g's, but were too hydrophobic to dissolve in aqueous carbonate, even at high exposure doses. For example, a tetrapolymer of OMA-MMA-TBMA-MAA (1.0/1.0/1.0/0.5) had a glass transition at 117 °C, transferred at 140 °C, but showed little or no aqueous solubility.

A more useful approach was to introduce acrylate monomers into the resist through copolymerization. For example, a simple replacement of MAA with acrylic acid (AA) at the appropriate composition lowers the terpolymer T_g by 30-40 °C, without degrading its desirable imaging or dissolution characteristics. In practice, the desirable lamination temperature is no greater than 120 °C, so a further lowering of the T_g is required. Incorporation of ethyl acrylate (EA) into the polymer is one way to achieve this. The homopolymer of EA has a T_g below -20 °C and EA copolymerizes readily with methacrylates. Further, the solubility contribution of EA is similar to MMA (same C/O ratio) so a direct substitution of EA for MMA should not strongly perturb the dissolution properties. To test this, a resist was formulated from an EA-MMA-TBMA-AA tetrapolymer of composition 0.75/0.75/1.0/0.35. The resist showed excellent imaging characteristics, formed very tough films, had a T_g of 83 °C, and transferred to copper at 120 °C without decomposition.

The imaging capabilities of the tetrapolymer dry film resist were demonstrated by spin-coating the formulated material (from solution with 10% iodonium salt) onto Mylar to give a 1.0 μm film. This dry film was then transferred onto a silicon wafer at 100 °C and imagewise exposed at 254 nm (contact exposure). Resolution patterns of 2 μm spaces were opened after development as shown in Figure 4. In practice, films approaching 1 mil (25 μm) in thickness were required to achieve the mechanical strength necessary to 'tent' (protect) Plated-Through-Holes (PTH's) during wet-etching.

Integrated Circuit Deep-UV Resists

The polymers described above are produced in a single synthetic step which is very reproducible. The facile synthetic route, coupled with the low cost of starting materials, makes the production of these materials relatively inexpensive. While this attribute is key to the design of a PCB resist, the increasing sensitivity of integrated circuit (IC) production costs to the price of resist makes this class of materials equally attractive for IC fabrication using deep-UV lithography.

Another intriguing aspect of the methacrylate terpolymer CA resist polymers is their airborne contaminant absorption properties relative to more conventional CA resist materials (*9*). N-methylpyrrolidone (NMP) has been reported to result in serious image degradation of CA positive resists when present in very low levels in the air in resist processing environments (*10*). This is a very serious problem, as NMP is ubiquitous in IC fabrication lines, where it is used in resist stripping operations and as a solvent for polymer dielectrics (polyimides). A quantitative method was developed to measure the uptake of NMP in thin polymer films (11). It was found that the MMA-TBMA-MAA terpolymers absorb an order of magnitude less NMP than conventional materials used in CA resists. Table II contains representative data demonstrating the unusual resistance of these methacrylate polymers to NMP uptake. Enhanced environmental stability of these materials is anticipated, and is an attractive feature in positive CA resist materials for IC applications.

Resists based solely on acrylate/methacrylate polymers are not thought to be well suited for IC applications due to their poor resistance (vs. aromatic polymers) to the plasma etch processes used in semiconductor manufacturing for pattern transfer. The recent introduction of high density plasma tools which show dramatically improved etching discrimination eases somewhat the required etch resistance of resists for semiconductor manufacturing. Several modifications of the terpolymer chemistry discussed above have been made to increase the resistance of the acrylic materials to plasma etching. The most common means of attaining increased plasma etch resistance is by incorporating a large proportion of aromatic function in the polymer structure. The PCB resist design concepts enumerated earlier, in which three classes of monomers each serving a separate function are incorporated into the polymer, are equally applicable to the design of IC resists. For example, substitution of aromatic monomers for MMA provides increased etch resistance. Taylor and coworkers (12, 13) recently reported on copolymers of benzyl methacrylate and THP-protected methacrylic acid as an aqueous processable, dry etch stable DUV resist. Resolution was limited by large scale acid diffusion in this system. We have prepared and evaluated terpolymers of benzyl methacrylate, t-butyl methacrylate and methacrylic acid (Figure 5) as a potential hydroxystyrene-free positive resist. DUV resists based on these polymers are developable in dilute metal-ion-free developers, and are capable of submicron imaging using 248 nm radiation at a photospeed of 5 mJ/cm². Figure 6 shows a positive-tone image contact printed in a resist based on a terpolymer containing 60% benzyl methacrylate, 30% TBMA and 10% MAA, with 2% triphenylsulfonium triflate as the photoacid generator. Using a non-optimized process, 1.25 μm features were printed in a 1.25 μm resist film at a dose of 5.0 mJ/cm² and a develop time of 1 minute in 0.1N TMAH. Perhaps a more promising method for increasing etch resistance is the addition of a phenolic resin to the MMA-TBMA-MAA terpolymer. Blends of these chemically dissimilar materials result in one-phase, miscible behavior. The methacrylate terpolymer in this case behaves as a polymeric dissolution inhibitor, with the unique property of adjustable inhibition power through change in terpolymer composition. Positive DUV resists based on this new design concept will be discussed in future papers.

The methacrylate terpolymer chemistry has potential extendibility to 193 nm exposure technologies. The MMA-TBMA-MAA terpolymer is optically transparent at 193 nm, with an absorbance of 0.13/μm, and the addition of 5 wt. % of the photoacid generator di-t-butylphenyl iodonium triflate increases the absorbance to only 0.4/μm. An absorbance spectrum for a 3.75μm film is shown in Figure 7. We are currently investigating the viability of such a single layer resist (SLR) for 193 nm projection lithography. We hope to answer questions

Figure 4. Contact print of dry film terpolymer, transferred from mylar to silicon wafer at 120 °C, using thumb pressure, then exposed, baked and developed.

Table II. Airborne NMP Uptake of Polymer Films

POLYMER	AMOUNT NMP ABSORBED (ng/wafer)
poly(4–hydroxystyrene)	1650
poly(4–t–BOC–styrene)	1083
poly(MMA–co–TBMA–co–MAA)	52

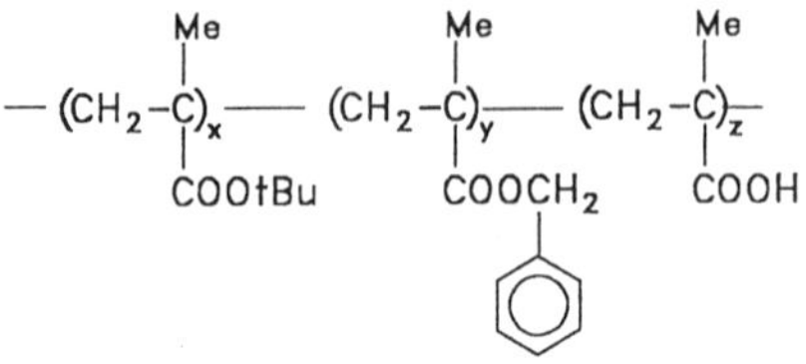

$$-(CH_2-\underset{\underset{COOtBu}{|}}{\overset{\overset{Me}{|}}{C}})_x \quad (CH_2-\underset{\underset{COOCH_2}{|}}{\overset{\overset{Me}{|}}{C}})_y \quad (CH_2-\underset{\underset{COOH}{|}}{\overset{\overset{Me}{|}}{C}})_z$$

Figure 5. Chemical structure of BMA-TBMA-MAA terpolymer.

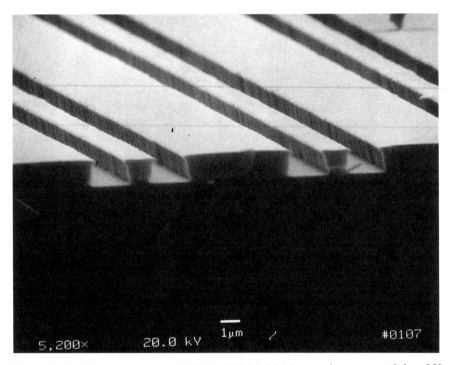

5,200×   20.0 kV   1 μm   #0107

Figure 6. Positive-tone image of BMA-TBMA-MAA terpolymer containing 2% triphenylsulfonium triflate.

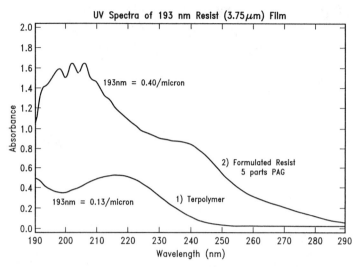

Figure 7. UV absorbance spectra of MMA-TBMA-MAA terpolymer with and without photoacid generator (diphenyliodonium triflate). Film thickness is 3.75 microns.

such as depth of focus and ultimate resolution capability using this resist. The resist is very fast at 193 nm exposure, requiring only 8 mJ/cm^2, and has encouraging resolution capability at this early stage of research, demonstrating sub-0.5 micron resolution at an NA of 0.22. The resist is useful for the characterization of new 193 nm exposure systems and has been demonstrated to have 0.2 micron resolution in a resist thickness of 0.75 microns with an 0.5 NA optical system. Figure 8 shows 0.325, 0.35 and 0.375μm features printed using 193 nm projection exposure (NA 0.5) in the MMA-TBMA-MAA terpolymer resist film containing 1% diphenyliodonium triflate. The optimization of imaging properties and the design of terpolymers with improved plasma etch resistance are the subjects of current research. Workers at Fujitsu have recently disclosed a means of gaining etch stability with non-aromatic and 193 nm transparent materials by the incorporation of alicyclic units (*14*).

SUMMARY

This work demonstrates that CA resist chemistry can find practical application to IC packaging. A series of CA resists specifically targeted for circuitization applications has been prepared and examined in this study. These relatively thick resists are based on polymers incorporating t-butyl methacrylate as the acid-labile functionality, methacrylic acid for aqueous developability and good adhesion, and methyl methacrylate to attain the desired mechanical properties. Both the polymer and the imaging chemistry are flexible and can be readily modified for a specific application. Examples include both liquid apply and dry film photoresists for PCB applications, and deep-UV resists for IC fabrication.

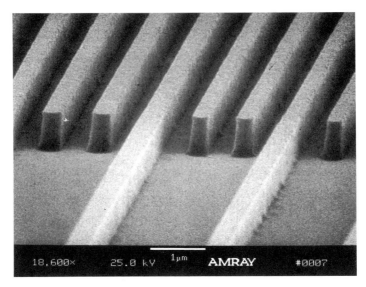

Figure 8. SEM micrographs of sub-0.4 micron features, printed at 193 nm in the MMA-TBMA-MAA terpolymer with 1% photoacid generator.

Additionally, we have demonstrated for the first time very fast, high resolution, aqueous processable, single layer resists for 193 nm lithography using the methacrylate terpolymer design approach.

ACKNOWLEDGEMENTS

The authors from IBM would like to thank members of the IBM Austin "Fast Lane" PCB manufacturing facility (B. Carpenter, M. McMaster and J. LaTorre) for their efforts in process development of the liquid-apply PCB resist, and Hoa Truong, Rani Saxena and Quan Ly of IBM Almaden for their help in polymer synthesis and characterization. The Lincoln Laboratory portion of this work was sponsored by the Defense Advanced Research Projects Agency.

REFERENCES

1. Allen, R. D.; Wallraff, G. M.; Hinsberg, W. D. and Simpson L. L.; *J. Vac. Sci. Tech.*, **1991**, B9(6), 3357.
2. Allen, R. D.; Wallraff, G. M.; Hinsberg, W. D.; Willson, C. G.; Simpson, L. L.; Weber, S. E. and Sturtevant, J. L.; *Proc. ACS Polym. Mat. Sci. Engr.*, **1992**, (66), 49.
3. Wallraff, G. M.; Allen, R. D.; Hinsberg, W. D.; Willson, C. G.; Simpson, L. L.; Weber, S. E. and Sturtevant, J. L.; *Jo. Imag. Sci. Tech.*, **1992**, 36(5), 468.
4. Ito, H.; Willson, C. G. and Frechet J.; US Patent 4,491,628 (1985).
5. Ito, H. and Ueda, M.; *Macromolecules*, **1988**, 21, 1475-1482.
6. McKean, D. R.; Schaedeli, U. P. and MacDonald S. A.; *J. Polym. Sci.: Part*

A: *Polym. Chem.*, **1989**, 27, 3927; Allen, R. D.; Schaedeli, U. P.; McKean, D. R. and MacDonald S. A.; *Proc. ACS Polym. Mat. Sci. Engr.*, **1989**, 61, 185.

7. McKean, D. R.; Allen, R. D.; Kasai, P. H.; Schaedeli, U. P. and MacDonald, S. A.; *SPIE Advances in Resist Technology and Processing IX*, **1992**, 1672, 94.
8. Schwalm, R.; *J. Electrochem. Soc.*, **1989**, 136(11), 3471.
9. Hinsberg, W. D.; MacDonald, S. A.; Snyder, C. D.; Ito, H. and Allen R. D.; *Proc. ACS Polym. Mat. Sci. Engr.*, **1992**, 66, 52.
10. Nalamasu, O.; Cheng, M.; Timko, A.; Pol, V.; Reichmanis, E.; Thompson, L.; *J. Photopolymer. Sci. Tech.*, **1991**, 4, 299.
11. Hinsberg, W. D.; MacDonald, S. A.; Clecak, N. J.; Snyder, C. D.; *SPIE Advances in Resist Technology and Processing IX*, **1992**, 1672, 24.
12. Taylor, G.; Stillwagon, L.; Houlihan, F.; Wolf, T.; Sogah, D. and Hertler, W.; *Chem. Mater.*, **1991**, 3, 1031-1040.
13. Taylor, G.; Stillwagon, L.; Houlihan, F.; Wolf, T.; Sogah, D. and Hertler, W.; *J. Vac. Sci. Technol.*, **1991**, B 9(6), 3348.
14. Kaimoto, Y.; Nozaki, K.; Takechi, S. and Abe, N.; *SPIE Advances in Resist Technol. and Process. IX*, **1992**, 1672, 66.

RECEIVED February 25, 1993

TOP-SURFACE IMAGING AND
DRY DEVELOPMENT RESISTS

Chapter 12

Surface-Imaging Resists Using Photogenerated Acid-Catalyzed SiO$_2$ Formation by Chemical Vapor Deposition

Masamitsu Shirai and Masahiro Tsunooka

Department of Applied Chemistry, College of Engineering, University of Osaka Prefecture, Sakai, Osaka 593, Japan

Silicon oxide formation at the near surface of UV irradiated polymers using a chemical vapor deposition (CVD) method was studied. Copolymers of 1,2,3,4-tetrahydro-1-naphthylideneamino p-styrenesulfonate (NISS) and either methyl methacrylate, isopropyl methacrylate, or benzyl methacrylate were used. Water sorption from the atmosphere occurred at the irradiated polymer surface because of the formation of p-styrenesulfonic acid units. When the irradiated polymer films were exposed to the vapor of tetraalkyl orthosilicates, silicon oxide was formed at the irradiated surface of the polymer films. No silicon oxide was formed at unirradiated areas. The silicon oxide formation rate was strongly affected by the structure of the methacrylate units in the polymers, the hydrolytic reactivity of the silicon compounds, the concentration of the silicon compounds in the vapor phase, and the number of photochemically formed p-styrenesulfonic acid units in the polymers. After UV irradiation and subsequent exposure to the vapor of alkoxysilanes, the polymer films showed good etching resistance to oxygen plasmas.

In the microlithographic process the use of deep UV light to provide higher resolution causes new problems due to decreased depth of focus and increased substrate reflectance. Surface imaging resist systems are expected to reduce or eliminate these difficulties. Several approaches have been tried to accomplish the surface imaging system. One approach involves the selective functionalization of the exposed or unexposed areas of photoresists using the vapor of inorganic halides or organometallic compounds. Taylor and co-workers (1, 2) reported a process in which a bisarylazide/isoprene crosslinkable polymer system was irradiated with UV light and, in a subsequent step, treated with the vapor of inorganic halides such as SnCl$_4$, SiCl$_4$, or (CH$_3$)$_2$SiCl$_2$. In this system reaction products containing Si or Sn could be formed at the near surface of polymer films. Pattern development for microlithography was achieved using oxygen reactive ion etching (O$_2$ RIE). In an alternative scheme, the surface functionalization of polymer films bearing photochemically formed phenolic -OH groups using hexamethyldisilazane was studied by MacDonald and co-workers (3). Coopmans and Roland (4) described the gas-phase silylation of a

0097–6156/94/0537–0180$06.00/0

diazonaphthoquinone/phenolic matrix resin, yielding a negative tone image by O_2 RIE. Another approach to designing surface imaging systems involves the selective formation of polysiloxane or metal oxides at the irradiated surface of polymer films. Follett and co-workers (5) have reported the plasma-developable electron-beam resists. In this system dichlorodimethylsilane selectively diffused into the irradiated areas of poly(methyl methacrylate) film and this step was followed by hydrolysis of the chlorosilane upon exposure to water vapor, resulting in polysiloxane network formation. Hydrophobic polymers such as chlorine-containing polystyrenes underwent oxidation to form hydroxyl and carbonyl groups when irradiated in air with 248.4- or 193-nm light. Since the irradiated regions sorbed water from the atmosphere, the treatment of the irradiated films with gaseous $TiCl_4$ led to the formation of a TiO_2 layer at the surface. This system gave a negative tone image by oxygen plasma etching (6–9).

Figure 1 shows the surface imaging system using photogenerated acid-catalyzed SiO_2 formation by the chemical vapor deposition (CVD) method. Upon irradiation with UV light the surface of polymers having imino sulfonate units becomes hydrophilic because of the formation of sulfonic acid. Water sorption from the atmosphere occurs at the top surface of the irradiated films. When the irradiated surface is exposed to the vapor of alkoxysilanes, a silicon oxide network is formed at the near surface of the polymers. No silicon oxide network is formed at unirradiated areas

$$n\,Si(OR)_4 + 2n\,H_2O \xrightarrow{H^+} n\,SiO_2 + 4n\,ROH$$

because the photochemically formed sulfonic acid units are necessary for the silicon oxide formation by hydrolysis and subsequent polycondensation reactions of alkoxysilanes (10, 11). This system gives a negative tone image by oxygen plasma etching.

Polymers bearing pendant 1,2,3,4-tetrahydro-1-naphthylideneamino p-styrenesulfonate (NISS) units were synthesized and they were utilized in the surface imaging resist system using CVD method (12, 13). The NISS units in the polymers can form p-styrenesulfonic acid units upon UV irradiation as shown in Scheme I (14). This paper describes the polymer synthesis, the photochemistry of NISS units in polymers, the water sorption to the irradiated polymer films, the formation of silicon oxide networks at the irradiated polymer surface, and the etching resistance to oxygen plasma of the polymer films which are irradiated and subsequently exposed to the vapor of alkoxysilanes.

EXPERIMENTAL

Preparation of 1,2,3,4-tetrahydro-1-naphthylideneamino p-styrenesulfonate (NISS): To a solution of 1-tetralone oxime (13.7g, 0.085 mol) in pyridine (32 mL) below 10 °C was slowly added p-styrenesulfonyl chloride (19g, 0.094 mol), which was freshly prepared from potassium p-styrenesulfonate and PCl_5 according to the method reported by Iwakura et al (15). The reaction was continued with

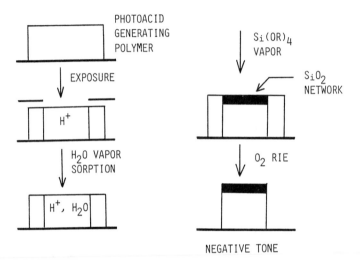

Figure 1. Surface imaging scheme using photogenerated acid-catalyzed SiO$_2$ formation by chemical vapor deposition.

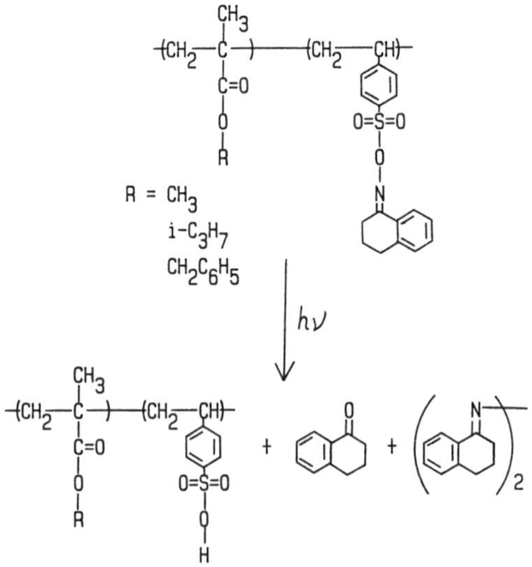

Scheme I.

vigorous stirring below 20 °C for 3 h. The reaction mixture was poured into 360 mL of ice-cold 5% HCl and extracted with chloroform. After washing the chloroform layer thoroughly with water, the chloroform layer was dried over K_2CO_3 and evaporated under reduced pressure. The oily residue was continuously extracted with hot n-heptane, which yielded white crystals upon cooling. After recrystallization from n-heptane 6.4 g (23%) of pure product was obtained; mp 80-82 °C. IR (KBr): 1370 and 1180 cm^{-1} (O-S-O). ^{1}H-NMR (CDCl$_3$, 270 MHz) δ 1.85 (q, 2H, CH$_2$), 2.73 (t, 2H, CH$_2$), 2.48 (t, 2H, CH$_2$), 5.45-5.90 (dd, 2H, J = 11, 19Hz, CH$_2$ =), 6.77 (dd, 1H, CH =), 7.17-8.00 (m, 8H, aryl H). Anal. Calcd for $C_{18}H_{17}NO_3S$; C, 66.06; H, 5.19; N, 4.28. Found C, 66.27; H, 5.39; N, 3.85.

Preparation of copolymers

The copolymers were prepared by the photochemically initiated copolymerization of NISS and methacrylates with azobis(isobutyronitrile) (AIBN) as an initiator at 29 °C by irradiation with UV light at wavelengths above 350 nm. The flux was 4 mW/cm^2. The photodecomposition of NISS did not occur by the light at wavelength above 350 nm. The concentrations of total monomer and AIBN in benzene were 4.5 mol/L and 1.6×10^{-2} mol/L, respectively. The sample solution was degassed under vacuum by repeated freeze-thaw cycles before polymerization. The content of NISS units in polymers was determined by measuring the absorbance at 254 nm in CH$_2$Cl$_2$. The molar extinction coefficient of the NISS units was estimated to be equal to that of the model compound 1,2,3,4-tetrahydro-1-naphthylideneamino p-toluenesulfonate, ϵ being 15300 L/mol · cm at 254 nm in CH$_2$Cl$_2$ at room temperature. The characteristics of poly(methyl methacrylate-co-NISS) (PMANI), poly(isopropyl methacrylate-co-NISS) (PPRNI), and poly(benzyl methacrylate-co-NISS) (PBZNI) are shown in Table I. Although the polymers could be obtained by the conventional thermally initiated copolymerization of these monomers with AIBN, they showed wide molecular weight distributions (Mw/Mn > 3.5).

Reagents

Tetramethyl orthosilicate (TMOS), tetraethyl orthosilicate (TEOS), tetra-n-propyl orthosilicate (TPOS), methyltrimethoxysilane (MTMOS), and methyltriethoxysilane (MTEOS) were reagent grade and used without purification.

Water sorption

A laboratory-constructed piezoelectric apparatus was used to measure water sorption in the polymer films (*16, 17*). The AT-cut quartz crystal with gold electrodes (Webster Electronics, WW1476) had a resonance frequency of 10.000 MHz. With this crystal, a frequency shift of 1 Hz corresponded to a mass change of 0.84 ng.

Polymers were deposited onto the quartz crystal (1.2 cm diameter) by casting from chloroform solution. The area coated with the polymer film was usually 0.19 cm^2. The quartz crystal was placed in the middle of the sealed glass vessel which had a quartz window for UV irradiation. An aqueous solution of inorganic salts (2M-KNO$_3$, saturated NaBr, or saturated CaCl$_2$) was placed at the bottom of the vessel to control its humidity at a constant temperature. Irradiation of the polymer films on the quartz crystal through the quartz window of the vessel was carried out with 254-nm light. The intensity of the incident light determined with a chemical actinometer (potassium ferrioxalate (*18*)) was 0.1 mJ/(cm$^2 \cdot$ s) at 254 nm.

Deposition of SiO$_2$

The polymer films (8.8 × 22 mm) were prepared on glass plates (8.8 × 50 mm) by casting from chloroform solution and drying under vacuum at room temperature. The polymer weight on the glass plate was usually 2×10^{-4}g (thickness ≈ 1 μm). After exposure with 254-nm light, the glass plate coated with polymer film was placed at the center of a 500 mL of glass vessel which had gas-inlet and -outlet valves. Water (50 mL) was placed at the bottom of the vessel and equilibrated for 10 min prior to introduction of the vapor of Si(OR)$_4$ or CH$_3$Si(OR)$_3$. Relative humidity in the vessel was 95%. During the SiO$_2$ network formation nitrogen gas (50 mL/min) flowed through a bubbler which contained liquid Si(OR)$_4$ or its homologues. The bubbler and reaction vessel were placed in a thermostatic oven at 30 °C. The amounts of SiO$_2$ formed at the near surface of the polymer were determined from the difference between the weight of the sample plate before and after exposure to the vapor of the silicon compounds.

Etching with an oxygen plasma

Oxygen plasma etching was carried out at room temperature using a laboratory-constructed apparatus where the oxygen plasma was generated using two parallel electrodes and RF power supplies. The typical etching conditions were as follows: 20W power (13.56MHz), power density of 1.0 W/cm^2, 30-125 mTorr, and oxygen flow of 1 sccm. The self-biasing voltage was -40 V.

RESULTS AND DISCUSSION

Photoreaction of Imino Sulfonates

It has been reported that upon irradiation with UV light the cleavage of -O-N=bonds in imino sulfonates and the subsequent abstraction of hydrogen atoms from residual solvent in the polymer film or from polymer molecules leads to the formation of sulfonic acid, azines and ketones (*14*). The change in the absorption spectrum of PMANI(C) film upon irradiation with 254-nm light is shown in Figure 2. The absorbance at 254 nm decreased with irradiation time due to the cleavage of -O-N=bonds of NISS units in the polymers and an isosbestic point was observed at 305 nm. The slight increase in absorbance above 300 nm is due to the formation of tetralone azine. In our system photolytic decomposition of NISS units was complete after exposure of 250 mJ/cm^2. The

Table I. Polymer Properties

Polymer	R[a]	M_n $\times 10^{-4}$	M_w/M_n	NISS Content in Polymer (mol %)	T_g[b] (°C)	T_d[c] (°C)
PMANI(A)	CH_3	7.0	3.0	25	–	–
PMANI(B)	CH_3	10.0	2.5	15	–	132
PMANI(C)	CH_3	11.5	2.0	8	114	–
PPRNI(A)	$i\text{-}C_3H_7$	13.3	2.1	32	–	–
PPRNI(B)	$i\text{-}C_3H_7$	13.1	1.9	15	105	149
PBZNI(A)	$CH_2C_6H_5$	17.0	2.2	30	–	–
PBZNI(B)	$CH_2C_6H_5$	15.5	2.3	18	83	147

[a] See Scheme I.

[b] Glass transition temperature.

[c] Decomposition temperature of NISS units in polymers

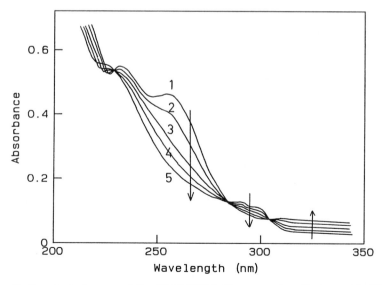

Figure 2. Spectral change of the PMANI(C) film upon irradiation with 254-nm light. Exposure dose (mJ/cm^2): (1) 0, (2) 25, (3) 87, (4) 150, (5) 250. Film thickness: 0.4 μm.

quantum yields (Φ) for the photolysis of NISS units in the polymer films at 254 nm were 0.33, 0.31, 0.36, 0.26, and 0.38 for PMANI(A), PMANI(B), PMANI(C), PPRNI(B), and PBZNI(B), respectively. They were not strongly dependent on either structure of the methacrylate unit or the NISS content of the polymers.

WATER SORPTION

The weight fractions of the sorbed water were 4.7, 3.1, 2.3, 0.3, and 0.1wt% for PMANI(A), PMANI(B), PMANI(C), PPRNI(B), and PBZNI(B), respectively, at 95% relative humidity. For the PMANI series of polymers, a plot of weight fraction of water sorbed vs. mole fraction of NISS gave a linear line. The value extrapolated to 0% of NISS fraction was 1.1% which was in good agreement with the value (1.05%) measured for poly(methyl methacrylate). The introduction of NISS units into the polymers enhanced their water sorption ability. The more hydrophobic nature of the methacrylate ester groups decreased the water sorption ability.

Figure 3 shows the relationship between irradiation time and water sorbed into the irradiated PMANI(A) film at various relative humidities. The weights of polymer films cast on the quartz crystal were 1000 ng. If the polymer is assumed to have a density of 1 g/cm^3, the film thickness is 51 nm and the absorbance of the film at 254 nm is < 0.1. The relative humidities in the vessel were adjusted to be 95, 58, and 32%. Water sorption began the moment that the polymer film was irradiated with 254-nm light. It increased with irradiation time and more gradually increased after irradiation until sorption equilibrium was established. Sorbed water was removed, when the sample was placed under a dry nitrogen atmosphere. The sorption and desorption processes were completely reversible.

Figure 4 shows the relationship between water sorbed and the weight of the irradiated polymer film (area = 0.19 cm^2) on the quartz crystal. The water sorption was measured after a 7 min irradiation (42 mJ/cm^2) and after equilibration for 60 min. The water sorption for all polymers examined increased with increasing polymer weight and reached a constant value at high polymer weight. The maximum amounts of sorbed water were 270, 80, 44, 30, and 20 ng for PMANI(A), PMANI(B), PMANI(C), PPRNI(B) and PBZNI(B), respectively. Although the water sorption increased with increasing styrenesulfonic acid units formed photochemically, the hydrophobic nature of the methacrylate ester groups decreased the water sorption. A saturation phenomenon for the water sorption was observed above ca. 1000 ng (thickness = 51 nm) for the irradiated polymers. This is not due to the inability of light to penetrate below that thickness, since the absorbance of the polymer film was < 0.1 at 254 nm.

Upon photolysis of the NISS units, 1-tetralone and 1-tetralone azine, which have poor affinity for water, could be formed in addition to p-styrenesulfonic acid units. The sorption of water at the surface of the irradiated polymer films may induce the aggregation of 1-tetralone and/or 1-tetralone azine in the layer, which may act as "hydrophobic barrier" to prevent further diffusion of water molecules. To test this hypothesis, irradiated PMANI(B) was exposed to the vapor of a mixed solvent of water and methanol (60/20, v/v), the latter having good affinity for 1-tetralone and its azine. The vapor sorption was observed to

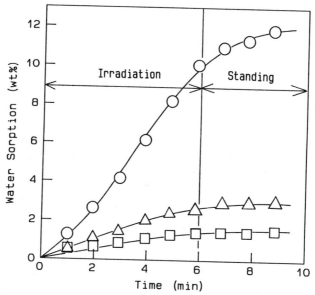

Figure 3. Relationship between irradiation time and water sorption to PMANI(A) film at 25 °C. Relative humidity: (O) 95, (△) 58, (□) 32%. Light intensity: 0.1 mJ/(cm^2 · s). Polymer weight: 1000 ng.

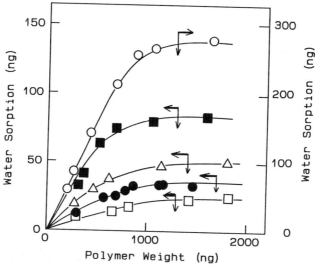

Figure 4. Water sorption at 95% RH into polymer films irradiated with 42 mJ/cm^2 of 254-nm light at 25 °C. Polymer: (O) PMANI(A), (■) PMANI(B), (△) PMANI(C), (●) PPRNI(B), (□) PBZNI(B).

increase linearly with increasing polymer weight. No saturation phenomenon was observed.

SILICON OXIDE DEPOSITION

In the presence of water and strong acids, the hydrolysis and subsequent polycondensation reactions of $Si(OR)_4$ or its homologues lead to the formation of silicon oxide or polysiloxane networks, which is well known as the sol-gel process for the silica glass formation (10, 11). When the irradiated polymer films bearing NISS units were exposed to the vapor of $Si(OR)_4$ or $CH_3Si(OR)_3$ at 30 °C, SiO_2 or polysiloxane was formed in the near surface region of the films, which was confirmed by IR analysis (transmission method). Irradiated PMANI(A) film exposed to TMOS vapor showed new peaks at 1080 (Si-O-Si), 940 (Si-OH), and 3300 cm^{-1} (Si-OH). The presence of Si-OH peaks suggests that the polycondensation reaction of $Si(OH)_4$ does not occur completely and silanol groups still remain in the film. No SiO_2 formation was observed in the unirradiated areas of the polymer films, since the styrenesulfonic acid units formed photochemically are essential to catalyze the hydrolysis of the silicon compounds.

Figure 5 shows the relationship between irradiation time and SiO_2 formation in PMANI series polymers. SiO_2 formation increased with increasing irradiation time and decreased in the order PMANI(A) > PMANI(B) > PMANI(C). This means that the SiO_2 formation rate is proportional to the number of sulfonic acids formed photochemically in the polymer matrix. It has been reported that the rate of hydrolysis of tetraalkyl orthosilicates in solution is first order in acid concentration under conditions of excess water (19, 20). As shown in Figure 4, the amount of water initially sorbed into the irradiated polymer surface was much smaller than the amount of water that was needed to form the observed amount of SiO_2 by the hydrolysis of TMOS at the polymer surfaces. Thus water must be supplied from the atmosphere during SiO_2 network formation.

Figure 6 shows the influence of methacrylate structure on SiO_2 formation at the irradiated polymer surface. The NISS content of the three polymers was almost the same. SiO_2 formation decreased in the order PMANI(B) > PPRNI(B) > PBZNI(B), the same trend observed in water sorption experiments. Thus the SiO_2 formation rate is strongly affected by the amount of water sorbed in the irradiated polymer surface.

Figure 7 shows the effect of silicon compound structure on SiO_2 or polysiloxane formation in irradiated PMANI(B) film. The formation rate decreased in the order MTMOS > TMOS > MTEOS > TEOS > TPOS. SiO_2 formation using TPOS was negligibly small under present conditions. It has been reported that the pseudo-first-order rate constant (k_{obs}) for the hydrolysis of alkoxysilanes using p-toluenesulfonic acid as a catalyst decreased in the order MTEOS ≈ MTMOS > TMOS > TEOS ≈ TPOS (13). Furthermore the concentration (V_c) of the alkoxysilanes in the vapor phase decreased in the order MTMOS > TMOS > MTEOS > TEOS > TPOS, which is consistent with their boiling points (13). Thus it was confirmed that the SiO_2 or polysiloxane formation rate in the irradiated polymer surface was determined by both k_{obs} and V_c values.

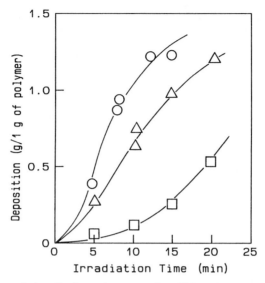

Figure 5. Effect of irradiation time on the SiO$_2$ deposition using TMOS.
Polymer: (○) PMANI(A), (△) PMANI(B), (□) PMANI(C). CVD treatment
time was 4 min at 30 °C. Relative humidity: 95%. Polymer weight: 2×10^{-4}g.
Surface area: 1.9 cm^2.

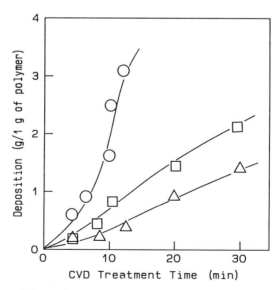

Figure 6. Effect of the polymer structure on the SiO$_2$ deposition using TMOS at
30 °C. Polymer: (○) PMANI(B), (□) PPRNI(B), (△) PBZNI(B). Exposure dose:
62 mJ/cm^2. Relative humidity: 95%. Polymer weight: 2×10^{-4}g. Surface area:
1.9 cm^2.

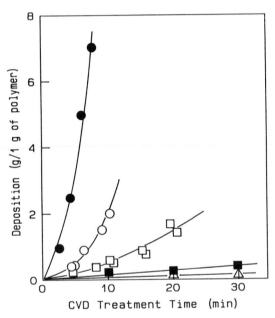

Figure 7. Effect of silicon compound structure on the SiO$_2$ deposition in irradiated PMANI(B) film at 30 °C. Silicon compound: (●) MTMOS, (○) TMOS, (□) MTEOS, (■) TEOS, (△) TPOS. Exposure dose: 62 mJ/cm^2. Polymer weight: 2×10^{-4}g. Surface area: 1.9 cm^2. Relative humidity: 95%.

SiO$_2$ formation using TMOS vapor increased with increasing polymer film thickness and showed a saturation at thicknesses above ca. 400 and 800 nm for PMANI(A) and PMANI(C), respectively. This roughly corresponds to the penetration depth of 254-nm light. This means that TMOS diffused into the polymer film and the hydrolysis and subsequent polycondensation reactions occurred beneath the film surface and at the film-air interface. The depth of the SiO$_2$ formation layer was beyond the depth of the water sorption layer ($\approx$ 50 nm) of the irradiated polymer films (see Figure 4). The methanol liberated during the hydrolysis of TMOS in the film may help the diffusion of water in the film by destroying the "hydrophobic barrier" of the photochemically formed 1-tetralone and/or 1-tetralone azine as discussed above for the sorption of the vapor of a mixture of water and methanol by the irradiated polymers.

OXYGEN PLASMA ETCHING

Figure 8 shows the effect of CVD treatment time on the oxygen plasma etching of the irradiated PMANI(A) film. The etching rate of PMANI(A) was 0.1 μm/min in the present etching conditions. The typical curve for the etching with oxygen plasma was observed for the PMANI(A) film which was irradiated and subsequently exposed to the vapor of MTEOS for 15 min. The etching rate can be divided into three different regions that occur in sequence: initial rapid region, very slow region, and rapid region which was almost equal to the etching rate for PMANI(A) film. The period where the etch resistance to oxygen plasma can be observed increased with CVD treatment time of the irradiated films. The initial rapid region corresponds to the process that polymers are etched to make SiO$_2$ layer acting as etch barrier to oxygen plasma. In the very slow region SiO$_2$ layer is working as a good etch barrier, where the etching rate was 100 times slower than that of PMANI(A) film. The latter rapid region shows the etching of PMANI(A) layer after the removal of SiO$_2$ layer by oxygen plasma etching.

Figure 9 shows the effect of alkyl group of methacrylate units of the polymers on the etching using oxygen plasma. The difference in the etching rates of the unirradiated PMANI(A), PPRNI(A), and PBZNI(A) was not large. The period where the etching resistance to oxygen plasma was observed increased in the order PBZNI(A) < PPRNI(A) < PMANI(A), the same trend observed in SiO$_2$ formation rate at the surface of the film.

CONCLUSIONS

Polymers bearing 1,2,3,4-tetrahydro-1-naphthylideneamino p-styrenesulfonate (NISS) unit, which forms p-styrenesulfonic acid upon UV irradiation, were synthesized by the copolymerization of NISS and either methyl methacrylate, isopropyl methacrylate, or benzyl methacrylate. When the UV-irradiated polymer films were exposed to the vapor of tetraalkyl orthosilicates or their homologues under humid conditions at 30 °C, SiO$_2$ or polysiloxane networks were formed at the surface. The rate of SiO$_2$ or polysiloxane formation was strongly dependent on the water sorption ability of the polymers, the number of the photochemically formed p-styrenesulfonic acid units, the hydrolytic reactivity of

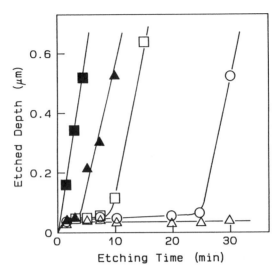

Figure 8. Effect of CVD treatment time on the oxygen plasma etching of the PMANI(A) films which were irradiated with dose of 171 mJ/cm^2 and subsequently exposed to the vapor of MTEOS at 30 °C and 58% RH. CVD treatment time: (■) 0, (▲) 10, (□) 13, (○) 15, and (△) 20.

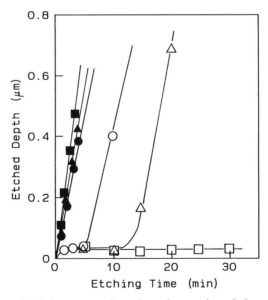

Figure 9. Effect of alkyl groups of methacrylate units of the polymers on the etching using oxygen plasma. The sample films were irradiated (104 mJ/cm^2) and subsequently exposed to the vapor of MTEOS for 10 min at 30 °C and 95% RH. Open and filled symbols mean after and before CVD treatments, respectively. Polymer: (□, ■) PMANI(A), (△, ▲) PPRNI(A), and (○, ●) PBZNI(A).

the silicon compounds and the concentration of silicon compounds in the vapor phase. The polymer films obtained after UV irradiation and subsequent exposure to the vapor of tetraalkyl orthosilicates or their homologues showed a good etching resistance to oxygen plasma. We are convinced that the preliminary results point out the potential for application of this system as surface imaging resists.

REFERENCES

1. Taylor, G. N.; Stillwagon, L. E.; Venkatesan, T. *J. Electrochem. Soc.* **1984**, 131, 1658.
2. Wolf, T. M.; Taylor, G. N.; Venkatesan, T.; Kraetsch, R. T. *J. Electrochem. Soc.* **1984**, 131, 1664.
3. MacDonald, S. A.; Ito, H.; Hiraoka, H.; Willson, C. G. *Proceedings of the Regional Technical Conference on Photopolymers*, New York; Mid-Hudson Section, SPE: Ellenville, New York, **1985**; p 177.
4. Coopmans, F.; Roland, B. *Proc. SPIE* **1986**, 631, 34.
5. Follett, D; Weiss, K.; Moore, J. A.; Steckl, A, J.; Liu, W. T. *The Electrochemical Society Extended Abstracts*; The Electrochemical Society: Pennington, NJ, 1982; Vol. 82-2, Abstract 201, p 321.
6. Stillwagon, L. E.; Vasile, M. J.; Baiocchi, F. A.; Silverman, P. J.; Taylor, G. N. *Microelectron. Eng.* **1987**, 6, 381.
7. Taylor, G. N.; Nalamasu, O.; Stillwagon, L. E. *Microelectron. Eng.* **1989**, 9, 513.
8. Nalamasu, O.; Taylor, G. N. *Proc. SPIE* **1989**, 1086, 186.
9. Taylor, G. N.; Nalamasu, O.; Hutton, R. S. *Polymer News* **1990**, 15, 268.
10. Bradley, D. C. *Chem. Rev.* **1989**, 89, 1317.
11. Hench, L. L.; West, J. K. *Chem. Rev.* **1990**, 90, 33.
12. Shirai, M.; Hayashi, M.; Tsunooka, M. *J. Photopolym. Sci. Technol.* **1991**, 4, 235.
13. Shirai, M.; Hayashi, M.; Tsunooka, M. *Macromolecules* **1992**, 25, 195.
14. Shirai, M.; Masuda, T.; Ishida, H.; Tsunooka, M.; Tanaka, M. *Eur. Polym. J.* **1985**, 21,781.
15. Iwakura, Y.; Uno, K.; Nakabayashi, N.; Chiang, W. Y. *J. Polym. Sci.* **1967**, 5, 3193.
16. Sauerbrey, G. *Z. Phys.* **1959**, 155, 206.
17. Alder, J. F.; McCallum, J. J. *Analyst* **1983**, 108, 1291.
18. Murov, S. L. *Handbook of Photochemistry*, Dekker, New York, **1973**; pp. 119-128.
19. Aelion, R.; Loebel, A.; Eirich, F. *J. Am. Chem. Soc.* **1950**, 72, 5705.
20. Smith, K. A. *Macromolecules* **1987**, 20, 2514.

RECEIVED December 30, 1992

Chapter 13

Polysilphenylenesiloxane Resist with Three-Dimensional Structure

K. Watanabe, E. Yano, T. Namiki, and Y. Yoneda

Fujitsu Laboratories, Ltd., 10–1 Morinosato-Wakamiya, Atsugi 243–01, Japan

A newly designed organosilicon resist, three-dimensional structure polysilphenylenesiloxane (TSPS), had been developed for use as a high-resolution negative bi-level electron-beam or KrF-excimer laser resist. The TSPS molecule is structured as a rigid three-dimensional mesh consisting of a silphenylenesiloxane core surrounded by functional groups. The advantages of such a structure are lower degree of swelling, a high oxygen-reactive ion etching resistance, and a high softening temperature. A 0.075 μm line-and-space pattern is well-defined after electron beam exposure. TSPS can also be used as a deep-UV resist. The addition of 10% of DMPA (2,2-dimethoxy-2-phenyl acetophenone) makes TSPS's deep-UV sensitivity about 20 times that of TSPS alone. A 0.25 μm pattern can be delineated with TSPS/novolak bi-level resist systems using a KrF-excimer laser stepper (NA = 0.37).

As the minimum VLSI pattern size is now in the sub-half micron region, conventional singlelayer resists can no longer satisfy all VLSI fabrication requirements. Due to the excessive absorbance or the light reflected from the patterned substrates, controlling the pattern profile and line width is difficult with a singlelayer resist.

Bi-level resist systems (1), consisting of a thin top imaging resist layer and a thick bottom layer, have many advantages over singlelayer resist systems, such as planarization of substrate topography, reduction of the light reflected from the patterned substrate and the ability to fabricate fine patterns with high aspect ratios.

To be suitable for fine bi-level pattern fabrication, the bi-level resist must meet many requirements such as good film quality, high sensitivity, high resolution, and resistance to oxygen-reactive ion etching (O_2-RIE). In particular, low swelling, high contrast, and high thermal stability are essential for a high resolution negative bi-level resist.

A number of polysiloxane resists (2–8) have been proposed as a top imaging resist layer for bi-level resists, because their high silicon content makes them resistant to O_2-RIE. Ladder-type polysiloxane resists exhibit a higher softening

0097–6156/94/0537–0194$06.00/0

temperature and lower degree of swelling than conventional linear polysiloxane resists. We found (7) that ladder-type polysiloxanes containing aromatic rings had a higher softening temperature and lower degree of swelling than even the non-aromatic ladder-siloxanes. This is because bulky aromatic rings suppress molecular mobility, making the ladder polymers more rigid.

To obtain the most rigid polysiloxane resist, we developed a negative organosilicon resist, three-dimensional structure polysilphenylenesiloxane (TSPS), which consists of a three-dimensional mesh silphenylenesiloxane core surrounded by functional groups. TSPS is more rigid than ladder-siloxane resists, because bulky aromatic rings are introduced into the three-dimensional mesh core to correct the Si-O bond angle. The advantages of this rigidity are improved resist contrast, suppressed swelling, and improved thermal stability.

EXPERIMENT

Synthesis

Figure 1 shows the synthesis flow of TSPS. To synthesize the TSPS core, we hydrolyzed 1,4-bis(trialkoxysilyl)benzene, followed by dehydration condensation polymerization using an acid catalyst. After core polymerization, residual silanol groups around the core were terminated with triorganochlorosilane.

The mixture was washed several times with water, then poured into acetonitrile to isolate the solid polymer. The original TSPS polymer was fractionated to obtain a narrow molecular weight distribution.

Bi-level resist process

Figure 2 shows bi-level resist process. In a typical bi-level resist process, a thick bottom layer coats the wafer to provide a uniform surface. Next, a thin resist layer is spun over the bottom layer. After exposure, the upper resist pattern is developed and the bottom layer is etched by O_2-RIE.

We dissolved the TSPS resist in methylisobutylketone (MIBK), then spin coated a 0.1–0.2 μm thick TSPS top layer on a 0.5–1 μm thick hard-baked novolak resist (Shipley MP-1300) bottom layer, and pre-baked the structure at 80 °C.

The TSPS resist was exposed using an electron beam lithography system (Elionix ELS-3300, 30 kV) and a KrF excimer laser stepper (248 nm, NA = 0.37). The exposed resists were spin-developed in alcohol mixture solvents.

The TSPS resist pattern was transferred to the bottom layer by O_2-RIE (Anelva DEM-451) at an oxygen pressure of 2.6 Pa, a gas flow of 10 sccm (standard cm^3 min^{-1}), and an applied power density of 0.16 W/cm^2.

Comparison of resist properties

We compared the TSPS characteristics with ladder-type polysiloxanes. The ladder-type polysiloxanes were synthesized according to known methods (5, 7, 8). The structure of non-aromatic ladder-siloxane {poly-methylsilsesquioxane (PMSS

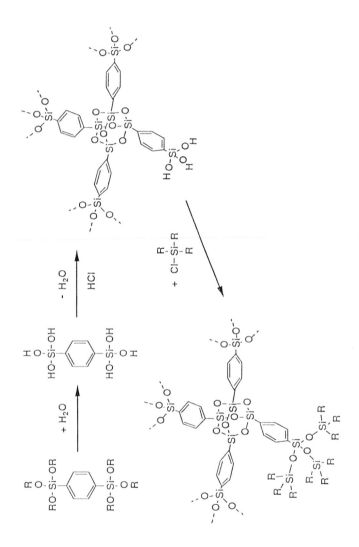

Figure 1. Synthesis flow of TSPS.

(*5*))}, aromatic ladder-polysiloxanes {poly-(vinyl/phenyl)silsesquioxane (PVSS-P (*7*)), and polymethylsilphenylene-siloxane (SPS) (*8*)} are shown in Fig. 3.

RESULTS AND DISCUSSION

Molecular structure

We studied the TSPS structure by ^{29}Si-NMR (Jeol GX500) and GPC-LALLS {gel-permeation chromatography equipped with a low-angle laser light scattering photometer (*9*)} (Tosoh HLC-8020, LS-8000). Figure 4 shows ^{29}Si-NMR spectra of TSPS. In the intermediate synthesis state, the spectrum showed two broad peaks centered at -60 ppm and -70 ppm, corresponding to the presence of residual SiOH and Si(OH)$_2$ groups. As the synthesis advanced, silanol groups were dehydrated. In the final state of synthesis, peaks due to the silanol groups were almost completely eliminated.

The TSPS and ladder-type polysiloxane (SPS) weight average molecular weights (M_w), determined by conventional GPC calibrated with polystyrene and GPC-LALLS, are shown in Table 1. The functional groups of both TSPS and ladder-type are methyl. The M_w of TSPS determined by GPC-LALLS is higher than that of ladder-type, whereas the M_w of TSPS, determined by conventional GPC, is lower than that of ladder-type. This suggests that TSPS's M_w is higher than that of ladder-type, although its molecular size is smaller. It may therefore be concluded that the TSPS molecule has a closely packed three-dimensional mesh structure (see Fig. 5).

Electron Beam Contrast and Sensitivity

Figure 6 compares the electron beam sensitivity curves for the TSPS and ladder-type polysiloxane (SPS) where both samples have a dispersion of 1.6 and methyl functional groups. The film thickness was measured by the alpha-step (Tencor). The TSPS gamma value, obtained from the sensitivity curve, is about 2.8. The gamma of ladder-type systems is about 1.6.

The electron dose giving a remaining 50% resist thickness remaining is 28 μC/cm^2. Introducing more sensitive functional groups, such as chloromethyl, produces even higher sensitivity.

O$_2$-RIE Resistance and Thermal Resistance

Figure 7 shows the oxygen plasma etching rates for TSPS films and the resist bottom layer of hard-baked MP-1300. The TSPS resist was etched at about 10 Å/min (1 nm/min) and the MP-1300 at 1000 Å/min (100 nm/min) on the etching condition, making the rate for TSPS resist one one-hundredth that of MP-1300. TSPS's excellent O$_2$-RIE resistance is due to its high silicon content and good film quality.

Table 2 shows the softening temperature of TSPS and ladder-type polysiloxane (PMSS). The functional groups of both TSPS and ladder-type are methyl. TSPS has no softening temperature under 400 °C, but the ladder-type's softening

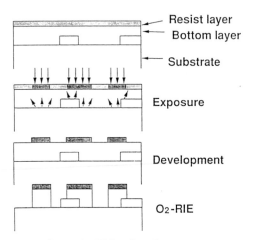

Figure 2. Bi-level resist process.

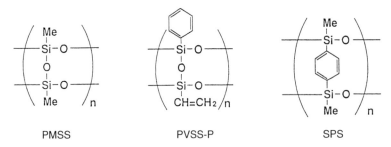

PMSS   PVSS-P   SPS

Figure 3. Structure of ladder-type polysiloxanes.

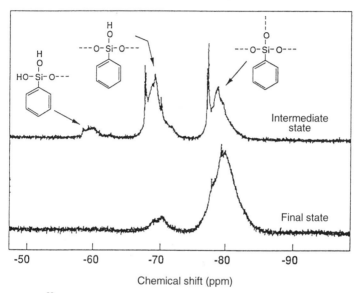

Figure 4. ^{29}Si-NMR spectra: intermediate state and   final state.

Table 1. M_w of TSPS Determined by Conventional GPC and GPC-LALLS

Materials	Conventional GPC	GPC-LALLS
TSPS	4.5×10^4	1.7×10^5
Ladder	5.8×10^4	1.2×10^5

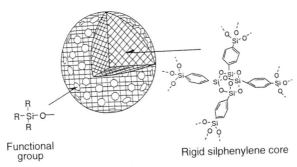

Functional group

Rigid silphenylene core

Figure 5. Structure model of TSPS.

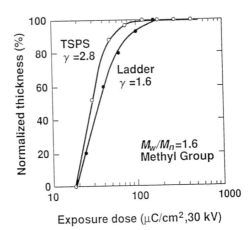

Exposure dose (μC/cm^2,30 kV)

Figure 6. Sensitivity curves for electron beam exposure.

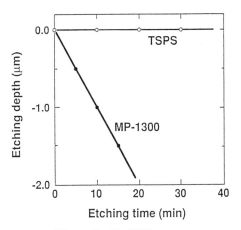

Figure 7. O_2-RIE rate.

Table 2. Softening Temperature

Molecular Weight (M_w)	Softening Temperature (°C)	
	TSPS	Ladder
5 000	>400	150
10 000	>400	210
50 000	>400	360

temperature ranges from 150 to 360 °C at these molecular weight. This suggests TSPS exhibits high thermal stability due to its rigid structure.

Resolution

Table 3 is the comparison of the molecular particle size of TSPS and ladder-type polysiloxane (PMSS) when either is dissolved into a poor solvent, isopropylalcohol (IPA), or a good solvent, methylisobutylketone (MIBK). The particle size is determined by dynamic light scattering system (Otsuka Electronics, DLS-700). Due to swelling of the ladder-type's random coil structure, the molecular size of ladder-type is larger in good solvent than that in poor solvent. In contrast, TSPS shows little variation in particle size indicating that swelling is suppressed due to its rigid structure.

Figure 8 compares the electron-beam exposed imaging-layer resist patterns of TSPS and ladder-type polysiloxanes (PMSS and PVSS-P). Ladder-type polysiloxane containing aromatic rings (PVSS-P) has higher resolution than non-aromatic ladder-polysiloxane (PMSS). TSPS exhibits a highest resolution with its well-defined 0.2 μm line-and-space patterns. The increased resolution is a direct result of the suppressed swelling realized by polymer rigidity.

Figure 9 shows bi-level resist patterns obtained with TSPS/MP-1300 bi-level resist systems using electron-beam exposure. The TSPS functional groups are methyl and chloromethyl, and the exposure dose was 16 μC/cm^2. Using O$_2$-RIE, 0.1 μm line-and-space patterns and 0.2 μm space patterns can be transferred accurately to the bottom layer, without thermal deformation. Figure 10 shows 0.075 μm line-and-space patterns, obtained with TSPS/MP-1300 bi-level resist system, which are possible using the small particle size of TSPS.

Application as a KrF-excimer laser resist

TSPS can also be used as a negative bi-level deep-UV resist when suitable functional groups are introduced, e.g. vinyl and phenyl. These functional groups make the TSPS sensitive to KrF-excimer laser. Figure 11 shows the KrF-excimer laser sensitivity curves for TSPS with a molecular weight (M_w) of 1.3 × 10^4. To increase TSPS sensitivity, a free radical photoinitiator was incorporated into the polymer. We selected 2,2-dimethoxy-2-phenyl acetophenone (DMPA) as a photoinitiator because it is compatible with the polymer, and solvent systems formed uniform film upon spinning. Figure 12 shows the absorption spectra of TSPS. The TSPS by itself only absorbs in the deep-UV region attributed to the phenyl and phenylene groups. The addition of 10% DMPA by weight to TSPS increases the absorption of a 0.2 μm-thick film from 7% to 25% at 248 nm. The addition of DMPA to the TSPS polymer dramatically changes the exposure response of the resist. The sensitivity of TSPS with 10% DMPA by weight is about 20 times that of TSPS alone when either is exposed to N$_2$ atmosphere (see Fig. 11).

Figure 13 shows the ^{29}Si CP-MAS NMR (Jeol EX-270) spectra of the exposed TSPS resist with and without DMPA as compared to the unexposed resist. After exposure, we found that methylene groups are produced and a number of vinyl groups are reduced. These changes are more dramatic when

Table 3. Particle sizes in solvents

Solvent	Molecular Size (nm)	
	Ladder	TSPS
Poor (IPA)	16.2	16.5
Good (MIBK)	54.2	17.3

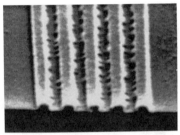

a) Non-aromatic ladder

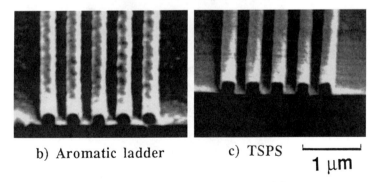

b) Aromatic ladder   c) TSPS

1 μm

Figure 8. Resist patterns after electron beam exposure: (a) non-aromatic ladder, (b) aromatic ladder, and (c) TSPS.

(a) 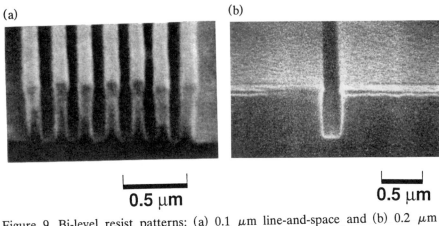 (b)

0.5 μm   0.5 μm

Figure 9. Bi-level resist patterns: (a) 0.1 μm line-and-space and (b) 0.2 μm space.

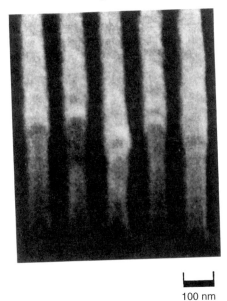

100 nm

Figure 10. 0.075 μm line-and-space patterns.

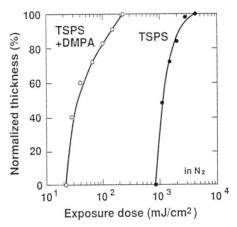

Figure 11. Deep-UV sensitivity of TSPS.

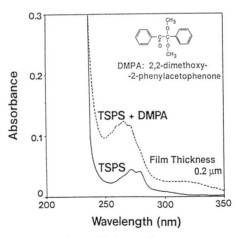

Figure 12. UV absorption spectra.

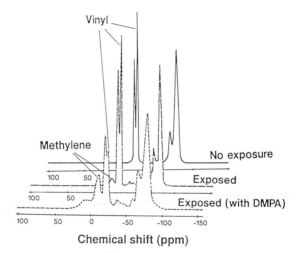

Figure 13. ^{29}Si CP-MAS NMR spectra.

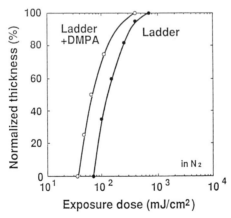

Figure 14. Deep-UV sensitivity of ladder-siloxane.

DMPA is added. This suggests that the addition of DMPA acts as a catalyst for the cross-linking of vinyl groups.

DMPA efficiently generates a benzoyl and substituted benzylic radical with deep-UV irradiation (*10*). The benzylic radical rearranges to give a highly reactive methyl radical, which diffuses readily and initiates crosslinking of the polymer. But the crosslinking efficiency is thus limited by the polymer structure. Adding DMPA to ladder-siloxane having vinyl and phenyl functional groups (PVSS-P, $M_w = 5.3 \times 10^4$) had little effect on sensitivity when either was exposed to N_2 atmosphere (see Fig. 14). We studied the difference of the crosslinking efficiency using spin probe analysis (*11*) with electron spin resonance (ESR) spectrometry (Jeol, JES-FE3XG). Figure 15 shows the ESR spectra of a TEMPO molecular probe dispersed into the TSPS and ladder-type polysiloxane (PVSS-P). The spectrum of the probe dispersed into TSPS is narrower and sharper. Figure 16 shows the relationship between the correlation times of the TEMPO probes dispersed into the polymers and the temperature. The correlation times were determined by Kivelson's method. That the correlation times of the probe dispersed into the TSPS are shorter at 0–80 °C, indicated that the probe in the TSPS is more mobile than that dispersed into ladder-type. This suggests that radicals generated by exposing the DMPA in TSPS can readily diffuse around the vinyl functional groups of the polymer. Thus, crosslinking can be initiated easily.

Figure 17 shows bi-level resist patterns using the TSPS. A 0.25 μm line-and-space pattern and a 0.35 μm hole pattern can be delineated with TSPS/MP-1300 bi-level resist systems using a KrF-excimer laser stepper.

CONCLUSION

We developed the three-dimensional structure polysilphenylene-siloxane (TSPS) organosilicon resist as a negative bi-level electron-beam or KrF-excimer laser resist. The TSPS molecule is structured as a rigid three-dimensional mesh consisting of a silphenylenesiloxane core surrounded by functional groups. The

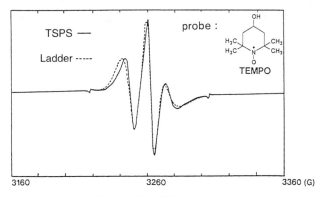

Figure 15. ESR spectra.

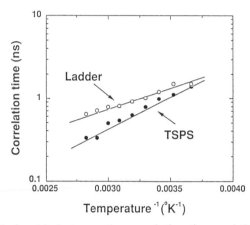

Figure 16. Relationship between the correlation time and the temperature.

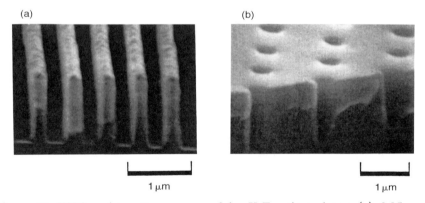

(a)   (b)

1 μm   1 μm

Figure 17. TSPS resist patterns exposed by KrF-excimer laser: (a) 0.25 μm line-and-space and (b) 0.35 μm hole.

structure's advantages are high contrast, low swelling, and high thermal stability. A 0.075 μm rectangular pattern is easily delineated with TSPS/MP-1300 bi-level resist systems using electron-beam exposure. TSPS can also be used as a negative bi-level deep-UV resist when suitable functional groups are introduced, e.g. vinyl and phenyl. Adding 2,2-dimethoxy-2-phenyl acetophenone (DMPA) increases TSPS's KrF-excimer laser sensitivity by almost 20 times. A 0.25 μm pattern can be delineated using a KrF-excimer laser stepper. Bi-level resist systems with TSPS show high sensitivity and high resolution, demonstrating its great potential for application in quarter micron ULSI chip production.

REFERENCES

1. Lin, B.J.: *Solid State Technol.* 1983, 26, 5, 105-112
2. Morita, M., Tanaka, A., Imamura, S., and Kogure, O.: *Jpn. J. Appl. Phys.* 1983, 22, L659-L660
3. Sakata, M., Ito, T., and Yamashita, Y.: *J. Photopolym. Sci. Technol.* 1990, 3 2, 173-178
4. Sugito, S., Ishida, S., and Iida, Y.: *NEC Res. Develop* 1989, 92, 1, 18-24
5. Fukuyama, S., Yoneda, Y., Miyagawa, M., and Nishii, K : U.S. Patent 4657843, 1987
6. Saito, K., Shiba, S., Kawasaki, Y., Watanabe, K., and Yoneda, Y.: *Proc. SPIE* 1988, 920, Vol. (Advances in Resist in Technology and Processing V), 198-202
7. Watanabe, K., Shiba, S., Saito, K., and Yoneda, Y.: *J. Photopolym. Sci. Technol.* 1989, 2, 103-108
8. Oikawa, A., Fukuyama, S., Yoneda, Y., Harada, H., and Takada, T.,: *J. Electrochem. Soc.* 1990, 137, 12, 3923-3925
9. T. Takagi: *Progress in HPLC*, 1, Parrez, eds. VNU Sci Press, 1985, 27-41
10. Tokumaru, K., and Ohkawara, M.,: "4. polymerization Reaction". (in Japanese), Sensitizer, 1st ed., Tokyo, Kodansha, 1987, pp 64-99
11. Kanbara, S.: "4.ESR". (in Japanese), Magnetic Resonance of Polymer, 1st ed., Kyoritsu, 1975, pp 371-555.

RECEIVED December 30, 1992

Chapter 14

Top-Surface Imaging Using Selective Electroless Metallization of Patterned Monolayer Films

J. M. Calvert[1], W. J. Dressick[1], C. S. Dulcey[1], M. S. Chen[2],
J. H. Georger[2], D. A. Stenger[1], T. S. Koloski[1], and G. S. Calabrese[3]

[1]Code 6900, Naval Research Laboratory, Washington, DC 20375–5320
[2]Geo-Centers, Inc., Fort Washington, MD 20744
[3]Shipley Company, Marlborough, MA 01752

A top surface imaging microlithographic process that involves selective electroless (EL) metallization of surfaces modified with ligating organosilane
ultrathin films (UTFs) is described. Fabrication of metal features with 0.4 μm
linewidths using 193 nm exposure is shown. Metal-ligand complexation chemistry is used for covalent attachment of a Pd(II) catalyst to the UTF-treated
surface. The molecular nature of the UTF layer is shown to control the
adhesive strength of the EL metal deposit; values of > 500 psi on single
crystal Si wafers have been obtained. The ligand-based UTF process is a
promising approach for a range of microelectronic applications where high
resolution, adherent, selective metallization is required.

The great majority of useful resist systems for microlithographic applications
involve irradiation of a thick (~ 0.5–2 μm) polymer film (*1, 2*). Exposure takes
place throughout the entire bulk of the film and changes the characteristics of
the polymer in the irradiated regions by a variety of mechanisms (chain scission,
crosslinking, destruction of a dissolution inhibitor, etc.) (*1–3*). Surface imaging
approaches, in which some property of only the outer layer of the resist is
modified, were developed (*4, 5*) to alleviate some of the problems in the thick
film approach. For optical lithography, especially with deep UV (< 250 nm)
sources, imaging in an ultrathin layer is attractive because depth of focus
limitations, excessive absorbance, and standing waves are eliminated (*6, 7*).
Similar considerations pertain to resists for soft x-ray (~ 6-40 nm) projection
lithography (*8, 9*). For beam lithography, ultrathin films (UTFs) are also desirable to minimize the effects of scattered electrons or ions that ultimately limit
feature resolution (*10*).

A number of lithographic processes that use surface imaging techniques
have been described (*6–8, 11–15*). The predominant approach involves post-

0097–6156/94/0537–0210$06.00/0

exposure silylation of a thick organic polymer film (*11, 12*). The polymer film is typically exposed and silylated to a depth of at least 1000 Å to provide sufficient differential plasma etch resistance for pattern transfer.

In contrast to the silylation processes in which the *near-surface* region is modified, far fewer approaches exist for direct manipulation of the *top surface* of a substrate to provide high resolution patterns of etch resistant materials. The initial report of top-surface imaging (TSI) utilized selective deposition of TiO_2 on the photo-oxidized surface of hydrophobic polymers (*4, 5*).

Recently, we described an alternative TSI process that involves patterning and selective electroless (EL) metallization of organosilane ultrathin films chemisorbed onto various substrates (*6, 7, 16–19*). Deep UV irradiation of the UTFs was shown to cleave organofunctional groups from the film and produce hydrophilic surface residues to which a colloidal Pd/Sn catalyst did not adhere. Subsequent electroless (EL) deposition occurred only on those surface regions that were not irradiated. Patterns of EL Ni, Co, and Cu were produced on a variety of substrates including silicon, quartz, alumina, metals, and polymers. The metal patterns served as efficient plasma etch barriers for pattern transfer into an underlying poly-Si substrate. Fabrication of features to 0.3 μm line width has been demonstrated (*6, 7*), and working transistor test structures (*18*) have been produced with this process.

Catalysis of EL plating using the colloidal Pd/Sn system has been successfully used for many years in printed circuit/printed wiring board (PC/PWB) manufacture. However, the detailed nature of the particles, as well as their mechanism of interaction with surfaces, is quite complex and not well understood. For applications in high resolution IC manufacture, process simplicity, control, and latitude are of great importance. We therefore sought to develop a less complicated, more selective EL deposition process using metal-ligand complex formation chemistry.

RESULTS AND DISCUSSION

UTF Surface Modification and Characterization

Nitrogen-containing ligands such as amines and pyridines are known to interact strongly with many transition metal ions and complexes. The nature and strength of the interaction is determined by the number and type of ligating sites and the particular metal involved. This high degree of specificity and control makes metal-ligand binding an attractive approach for use in a selective EL deposition process.

Attachment of various ligands to surfaces of interest was achieved by chemisorption of organosilanes (*20, 21*). Substrates that intrinsically possess (or are treated to possess) polar surface −OH groups were typically immersed in a solution of a ligand-bearing organosilane for several minutes (*22*) (although vapor deposition and spin coating techniques can also be used) to form an ultrathin film that contains a ligating group covalently linked to the substrate. Film formation occurs by condensation of the silane with the substrate, forming a siloxane (Si-O-substrate) bond to the surface (Eq. 1) (23).

$$RSiX_3 + \text{HO-(substrate)} \rightarrow \text{"R-Si-O-(substrate)"} + 3HX \qquad [1]$$

$$X = \text{-Cl, -OCH}_3, \text{-OC}_2\text{H}_5$$

Films containing pyridyl (PYR), ethylenediamine (EDA), and diethylenetri-amine (DETA), and ethylenediaminetriacetic acid (EDTA) ligands have been prepared in this manner. Surface-bound bipyridyl (BPY) and quinolinyl (QUIN) ligands were prepared using chemical coupling reactions, in which amide or sulfonamide bond formation served to graft the ligand onto an aminosilane treated substrate. Chemical structures of various ligand silanes are shown in Figure 1.

Ligand UTFs were characterized by a variety of techniques, including UV spectroscopy, contact angle goniometry, x-ray photoelectron spectroscopy (XPS), and ellipsometry (20, 21, 25). UV spectra of ligand films such as PYR, BPY, PEDA, and QUIN on fused silica exhibit absorptions characteristic of the aromatic chromophores in the molecules; UV spectra of EDA, DETA, and EDTA films are too weak to be useful as diagnostics of film quality. Water contact angles of the films were 10 ° for freshly made films of DETA and EDTA, 20 ° for BPY, 35 ° for EDA, ~45-50 ° for PYR, 60-65 ° for PEDA and ~55-60 ° for QUIN. XPS spectra of ligand films on non-Si substrates (e.g., Pt) showed the presence of both N and Si. Ellipsometry of PYR, EDA, and PEDA films on Si native oxide substrates gave typical thicknesses of ~5–10 Å, indicative of essentially monolayer coverage.

UTF Patterning and Selective Metallization

In our original process, metallization was obtained by the adsorption of a Pd/Sn catalyst (26) to a UTF coated surface. The interaction between the catalyst and the film was relatively non-specific. Polar, hydrophilic surfaces (e.g., treated with aliphatic or aromatic amines) could be metallized as well as non-polar hydrophobic surfaces (e.g., treated with aliphatic or aromatic hydro-carbons). However, the Pd/Sn catalyst was found not to adsorb to clean, smooth silica or silicon surfaces, which have a high density of silanol (Si-OH) groups (6, 16, 17). Patterned metallization could therefore be obtained on UTF treated surfaces only by photochemically removing the *entire* organic functionality of the film, leaving a pattern of silanol groups to which the Pd/Sn catalyst did not adsorb. As a consequence, the molecular design of UTF materials for practical lithographic applications was constrained by the necessity to have high ab-sorbance at the patterning wavelengths *and* efficient photocleavage to generate a surface silanol.

Metallization of substrates using any of the ligand UTFs shown in Fig. 1 was obtained by treating the surface with an aqueous catalyst solution based on $PdCl_4^{2-}$ (20). Binding of Pd(II) by the surface ligands was confirmed by UV and surface spectroscopic techniques. Films of PYR treated with Pd(II)-containing

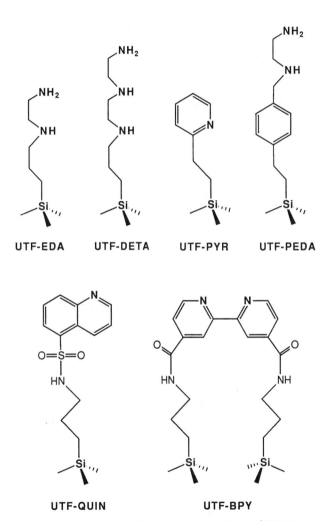

UTF-EDA UTF-DETA UTF-PYR UTF-PEDA

UTF-QUIN UTF-BPY

Figure 1. Chemical structures of ligating organosilanes (EDTA not shown).

solutions have been shown to exhibit increased UV absorption at ~235 nm, consistent with formation of a Pd(II)-pyridine complex (21). Rutherford backscattering spectrometry of ligand UTFs treated with the Pd(II) catalyst gives Pd surface concentrations of ~1 × 10^{15} atoms/cm^2. Immersion of a Pd(II)-catalyzed UTF surface *directly* into an EL plating bath results in homogeneous metallization of the substrate. The ligand-based process is highly selective. For example, surfaces that are treated with non-ligating UTFs (e.g., phenethyl-trichlorosilane, which is structurally analogous to PYR without the N atom) cannot be metallized using this approach (27). The ligand process is also simpler than the Pd/Sn catalyst system because no intermediate "acceleration" step is required to prepare the catalyst for metal deposition.

Pattern formation with the ligand UTF approach has been achieved by exposing the films to masked deep UV radiation from ArF (193 nm) and KrF (248 nm) excimer lasers, and from Hg/Xe lamp-based sources (28). Irradiation destroys the ligating ability of the film, creating regions of intact and modified ligand binding sites. Upon treating the surface with the catalyst solution, Pd(II) is covalently bound only in the regions that have intact ligating sites. After rinsing, the surface is immersed in an EL plating bath, and metal is deposited selectively at the catalytic sites. The ligand-based selective metallization process is shown schematically in Figure 2.

A variety of techniques have been used to characterize the ligating film following photochemical modification. PYR films on fused silica substrates were exposed to either ~1.5 J/cm^2 of 193 nm radiation or ~4.5 J/cm^2 of 248 nm radiation. The UV absorptions between 190–260 nm, characteristic of the pyridyl chromophore, were greatly reduced at these doses (21). A concomitant decrease in hydrophobicity of the PYR surface (water contact angle changes from 45 ° to < 10 °) was observed upon exposure. Laser desorption Fourier transform mass spectrometry showed that the ethylpyridyl group was removed from the surface upon irradiation at 193 nm. XPS analysis of Pt substrates treated with PYR showed that Si remained on the surface after exposure of the ligand film. These observations suggest a photochemical mechanism in which the PYR molecule is cleaved at the Si-C bond, eliminating both the pyridine ligand and ethyl spacer, and leaving Si-OH residues at the surface.

Selective binding of the Pd(II) catalyst to patterned PYR films was demonstrated by scanning Auger microscopy, which indicated the presence of Pd only in the unexposed regions of the surface. Immersion of the catalyzed surface in EL Co or Ni plating baths resulted in patterned metal deposition, yielding an overall positive tone image. Figure 3 is a combination scanning electron micrograph (SEM) and Auger analysis, which shows ~20 μm wide EL Ni lines and superimposed line scans for Si and Ni (29). A SEM showing ~0.4 μm Ni lines produced by 193 nm contact printing with the PYR film is shown in Figure 4.

The photochemistry and selective metallization of other ligand UTFs has also been investigated. The EDA silane, consistent with its weak absorption at 193 nm, has been reported to have a photochemical dose (elimination of the N signal by XPS analysis) of ~13-15 J/cm^2 (23, 30). However, the PEDA silane which has a large absorption at 195 nm due to the phenyl chromophore, exhibits a dose for selective metallization of ~300 mJ/cm^2 at 193 nm on a Si thermal oxide substrate.

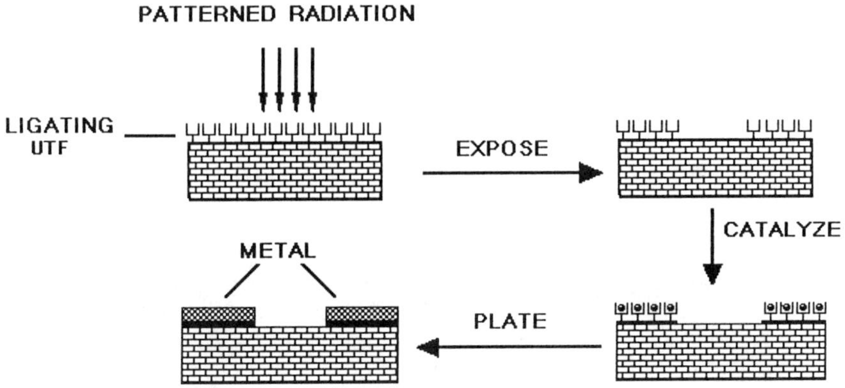

PATTERNED RADIATION

LIGATING
UTF

EXPOSE

CATALYZE

METAL

PLATE

Figure 2. Schematic of ligand-based selective metallization process. The goalpost structures represent ligating groups of the UTF layer. Removal of the goalpost indicates loss of ligating ability, and not necessarily complete removal of the ligand or organofunctional groups in the film.

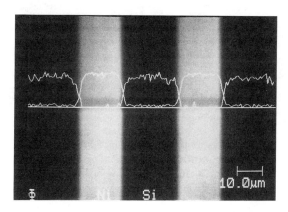

Figure 3. Overlay of SEM photo and Auger line scans of patterned EL Ni plated wafer. A Si native oxide wafer was coated with the PYR film, exposed to ~4 J/cm² of 193 nm radiation (29) (Questek Model 2000 ArF laser; ~4 mJ/cm²/pulse) through a low resolution fused silica mask in hard contact with the wafer. The wafer was treated with the Pd(II) solution for 30 min, rinsed with DI water, and metallized with Shipley NIPOSIT® 468 electroless Ni bath (10% strength) for 20 min. A PHI Model 660 Auger spectrometer was used to analyze the sample. The trace that is low in the dark regions and high in the bright regions is the Ni element scan; the other trace is the Si scan.

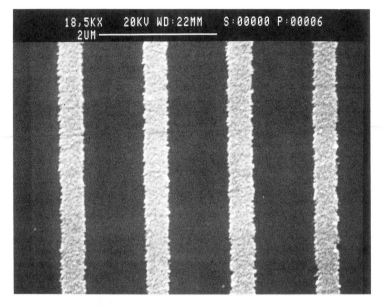

Figure 4. SEM photo of EL Ni patterns produced using a high resolution grating mask. The experimental conditions for producing the patterned wafer were as described in the caption of Fig. 3. The micrograph was obtained using a Cambridge Model S200 electron microscope.

Controlled Adhesion of EL Metal Deposits

In addition to selectivity, another attribute of the ligand-based metallization process is that the strength of the catalyst binding to the surface should be "tunable" by virtue of the formation constant, K_f, of the Pd-ligand complexation reaction (*20*). K_f is known to be affected by factors such as the type of ligand donor atoms (N-, S-, P-, O-, etc.) and the number of binding sites (denticity) (*31*). We have observed that monodentate ligands, such as films of PYR, exhibit lower adhesion than that observed with films of corresponding bi- or tridentate ligands (*21*). For example, ~ 400 Å thick EL Co films can be deposited without flaking on a silicon substrate using the ligand-based metallization process with films of PYR. However, > 70% of the metal film is removed during a tape peel test. In contrast, Co films in excess of 2500 Å can be deposited without flaking onto the same Si surface using the BPY silane as the ligating film. Peel tests show that no Co metal is removed from BPY-treated surfaces. A similar trend has been observed with EL Ni plated Si wafer surfaces: the PYR film exhibits only partial metal adhesion during a Scotch tape peel test, whereas all of the metallized bi- and tridentate ligand films in Fig. 1 pass the test completely. Average stud pull values of ~ 100 psi and ~ 300 psi have been measured for the adhesive strength of Ni plated PYR and BPY surfaces, respectively; however, individual metallized BPY samples have yielded pull strengths in excess of 500 psi (*21*).

CONCLUSIONS

The UTF ligand-based chemistry offers several unique features compared to conventional photoimaging processes for IC fabrication. It is one of the few true top surface, as opposed to near surface, imaging processes currently under development. As such, it is particularly well suited for exposure systems such as 193 nm, projection x-ray, and low voltage electron and ion beams that can best take advantage of a surface imaging resist (*28*). The use of Ni instead of a silicon oxide as the etch mask considerably diminishes the required thickness of the layer and should provide increased latitude in pattern transfer processing. The ultimate resolution of the ligand-based UTF process will likely be limited by the graininess or lateral spread of the catalyst/EL metal, rather than by the latent image created by exposure of the UTF. Further research along these lines should also lead to fundamental insights about the detailed mechanism of the initiation of EL deposition at surfaces. The use of patterned ligand surfaces as a reactivity template for the selective attachment of a variety of species other than Pd catalysts and EL metals is currently under investigation.

The ligand-based chemistry also has several attractive features for use as an advanced metallization process where high throughput, adhesive, selective metallization on a variety of substrates is crucial (*33*). The versatility of the process with regard to surface attachment of UTFs to most key technological materials such as polymers, plastics, and diamond has already been demonstrated (*20, 21, 34*). The elimination of an "acceleration" step simplifies and reduces the overall cost of the EL metallization process; also, the selective metal deposition is additive in nature, so that metal is deposited only where it is required. The ability to manipulate the macroscopic adhesive strength of the EL metal deposit by "tuning" the molecular metal-ligand binding strength parameters is potentially of considerable significance in the fabrication of high density interconnects for PC/PWB and multi-chip module applications, especially where surface roughening is undesirable.

ACKNOWLEDGMENTS

Funding for this work was provided by the Manufacturing Technology Office of the Assistant Secretary of the Navy, Shipley Co., and the Office of Naval Research. Experimental assistance from H. Stever and M. Thomas (both of National Semiconductor Co.) for adhesion measurements and C. Gossett and C. Cotell for RBS measurements is gratefully acknowleged, as well as helpful discussions with M. Peckerar. TSK acknowleges the Office of Naval Technology for a postdoctoral fellowship at the Naval Research Laboratory.

REFERENCES

1. W. Moreau, *Semiconductor Lithography*, Plenum Press, NY, 1988.
2. *Polymers in Microlithography*, E. Reichmanis, S.A. MacDonald and T. Iwayanagi, eds., ACS Symposium Series No. 412, ACS Press, Washington, DC, 1989.
3. R. Dammel *Proc. PMSE* **1992**, *66*, 1841.
4. G.N. Taylor, L. Stillwagon, and T. Venkatesan *J. Electrochem Soc.*, **1984**, *131*, 1658.
5. O. Nalamasu, F.A. Baiocchi and G.N. Taylor, in *Polymers in Microlithography*, E. Reichmanis, S.A. MacDonald and T. Iwayanagi, eds., ACS Symposium Series No. 412, ACS Press, Washington, DC, 1989, p.189-209.
6. J.M. Calvert, et. al. *Solid State Technology* **1991**, *34(10)*, 77.
7. J.M. Calvert, et. al. *J. Electrochem. Soc.* **1992**, *139*, 1677.
8. G.N. Taylor, R.S. Hutton, and D.L. Windt *Proc. SPIE* **1990**, *1343*, 258.
9. H.I. Smith and M.L. Schattenburg in *Semiconductor Materials and Processing Technologies*, J.M. Poate, ed., Kluwer Academic Publishers, Dordrecht, The Netherlands, 1992, p. 1-14.
10. S.W. Kuan, et. al. *J. Vac. Sci. Tech.* **1989**, *B7*, 1745.
11. F. Coopmans and B. Roland *Proc. SPIE* **1986**, *631*, 34.
12. J.W. Thackeray, et. al. *Proc. SPIE* **1990**, *1185*, 2.
13. S.A. MacDonald *Proc. PMSE* **1992**, *66*, 97.
14. G.S. Calabrese, et. al. *Proc. SPIE* **1991**, *1466*, 528.
15. K. Radigan and S. Liddicoat *Proc. SPIE*, **1992**, *1672*, 394.
16. J.M. Schnur, et. al. *U.S. Patent* 5,077,085.
17. J.M. Calvert, et. al. *Thin Solid Films* **1992**, *210 / 211*, 359.
18) J.M. Calvert, et. al. *J. Vac. Sci. Technol. B* **1991**, *9(6)*, 3447.
19. C.S. Dulcey, et. al. *Science* **1991**, *252*, 551.
20. J.M. Calvert, W.J. Dressick, G.S. Calabrese, and M. Gulla *U.S. Patent Application* 07/691,565, pending.
21. W.J. Dressick, et. al. *Proc. MRS*, **1992**, *260*, 659.
22. 2-(trimethoxysilyl)ethyl-2-pyridine (PYR) (*21*), N-(2-aminoethyl)-3-

aminopropyltrimethoxysilane (EDA) (*23*), trimethoxysilylpropyldiethylenetriamine (DETA), and N-[(3-trimethoxysilyl)propyl]ethylenediamine triacetic acid (EDTA) UTFs were prepared by treating a clean substrate with a 1% solution of the appropriate organosilane in acidic aqueous methanol for 20 min. Film treated surfaces were then baked at 120 °C for 5 min. BPY and QUIN UTFs were prepared by treating a propylamine silane-modified substrate with an acetonitrile solution of 2,2'-bipyridine-4,4'-carbonyl chloride or 8-quinolinylsulfonyl chloride in the presence of triethylamine (*20*).

23. The reaction of trimethoxyorganosilanes with a hydroxylated surface typically involves formation of at least one siloxane-surface bond per molecule of coupling agent. The other coordination sites of Si may be involved in intermolecular siloxane bonding, or exist as free Si-OH groups (*24*).
24. B. Arkles in *Huls Silicon Compounds Register and Review, 5th Edition*, R. Anderson, G.L. Larson, and C. Smith, eds., Huls America, Piscataway, NJ, 1991; p. 59.
25. J.H. Georger, et. al. *Thin Solid Films* **1992**, *210/211*, 716.
26. The Pd/Sn catalyst is these studies was obtained from Shipley Co. It should be noted that Pd/Sn catalysts from other manufacturers may have different surface chemical characteristics, and results from these experiments may not be generalizable to all Pd/Sn formulations.
27. Ligand-modified surfaces are also amenable to EL metallization using the conventional Pd/Sn catalyst.
28. With the ligand-based metallization process, pattern formation should be achievable by *any* photochemical or beam process that is capable of spatially modifying the ability of the film to perform its usual function, *e.g.*, binding metal ions. Patterning of UTF ligand films using alternative radiation sources, including focussed ion beams, projection x-ray, and STM is currently under investigation.
29. For experiments in which patterned metal surfaces were to be analyzed by electron beam techniques, Si native oxide was used as the substrate to minimize charging effects. However, a higher UV dose was required for native oxide in comparison to a fused silica or Si thermal oxide substrates (~ 1.5 J/cm^2 for the latter). The necessity of increased UV dose on the less insulating substrate has been observed previously for non-ligating UTFs (*19*) and may be related to more efficient excitation energy relaxation into the substrate.
30. D.A. Stenger, et. al *J. Am. Chem. Soc* **1992**, *114*, 8435.
31. A.E. Martell and R.M. Smith *Critical Stability Constants*, Plenum Press, NY, 1975.
32. Other UTF materials, which utilize the ligand-based metallization chemistry but incorporate a different photochemical reaction mechanism, have recently exhibited selective metallization at doses of <50 mJ/cm^2 at 193 nm.
33. *Metallization of Polymers*, E. Sacher, J.-J. Pireaux, and S.P. Kowalczyk, eds., ACS Symposium Series 400, ACS Press, Washington, DC, 1990, p. 416-514.
34. J.M. Calvert, P.E. Pehrsson, C.S. Dulcey, and M.C. Peckerar *Proc. MRS* **1992**, *260*, 905.

RECEIVED December 30, 1992

Chapter 15

Langmuir–Blodgett Deposition To Evaluate Dissolution Behavior of Multicomponent Resists

V. Rao[1], W. D. Hinsberg[2], C. W. Frank[1], and R. F. W. Pease[3]

[1]Department of Chemical Engineering, Stanford University,
Stanford, CA 94305–5025
[2]IBM Almaden Research Center, 650 Harry Road, San Jose, CA 95120
[3]Department of Electrical Engineering, Stanford University,
Stanford, CA 94305–4055

The development behavior of diazonaphthoquinone resists has been studied extensively. Our work focuses on the influence exerted by photoactive compound (PAC) on the dissolution of an unexposed resist. In particular, we have evaluated the effect of the sensitzer's spatial distribution in the film. Stratified resist films have been fabricated through the combined use of spin casting and Langmuir-Blodgett deposition. With these resist structures, not only can the effect of locally high PAC concentrations on dissolution be evaluated, but the mobility of the sensitizer in the matrix can also be studied. We have found that embedding as little as three monolayers (30 Å) of PAC between two thicker layers of polymer (1000 Å each) causes a significant induction period when the solvent front reaches the PAC layer. In addition, the dissolution rate of the pure spin cast polymer layer underlying the PAC is reduced significantly. Thermal treatment of these resist structures has demonstrated that the PAC appears to be more mobile in Langmuir-Blodgett polymer films than in spin cast polymer films.

As integrated circuit features shrink below 0.5 μm, the need for optimization of all lithographic steps is crucial. In order to define such features with high precision, the processing behavior of the photoresist materials used for patterning must be well-characterized. Hence, we need a fundamental understanding of their development behavior. Although much of the current semiconductor industry uses diazonaphthoquinone-based positive resists, many details of the dissolution mechanism of these materials are still unknown.

Thin film dissolution behavior has been the subject of study for many applications. In the case of positive photoresists, a number of techniques have been used to characterize the kinetics of dissolution (1–12). The earliest such experiments were performed by exposing the resist to a solvent for a fixed time and then measuring the thickness of the remaining film (1). From repeated measurements of this type, a bulk development rate could be determined.

Although inexpensive and straightforward, this method is inaccurate since it is difficult to ensure that development has completely stopped when the film thickness is measured. For this reason, *in situ* measurements of film thickness are currently preferred for characterizing resist dissolution.

Numerous techniques for *in situ* dissolution measurements of positive resists have been developed (*2–13*). Oldham describes a method for measuring capacitance changes between the developer and substrate (*2*). As film thickness decreases during development, the measured capacitance increases proportionally. Although this is a fairly simple and inexpensive technique, results are irreproducible since there is often nonuniform development across the wafer surface, leading to spurious capacitance measurements. Another drawback to this technique is that a conducting solvent is required; therefore it cannot be used for studying many classes of resist materials. Dill and co-workers pioneered the use of an optical system for *in situ* dissolution measurements (*1, 3*). A computer-controlled spectrophotometer is used to scan reflectivity as a function of wavelength. Film thickness is then determined using the absolute reflectivity and optical properties of the resist. Other similar optical-based techniques include the psi-meter (*4*) and the laser interferometer (*5–7*). The psi-meter is essentially an ellipsometer, modified to make continuous measurements in a solvent medium. Although this technique offers the advantage of precise thickness determination, an *a priori* knowledge of the refractive index of the material is necessary (*4*). For the interferometric method, both *in situ* film thickness and resist refractive index can be measured. Reflected light from the substrate/film and developer/film interfaces produces a pattern of constructive and destructive interference. The periodicity of the interference can then be related to the absolute film thickness at any given time. One advantage of interferometry is the ability to delineate the distinct regions in a dissolving film. The formation of a gel layer at the surface of a dissolving resist can be monitored because of reflectivity changes from the interfaces; the behavior of such gel layers can provide some mechanistic understanding.

Unfortunately, there are several difficulties associated with the *in situ* optical techniques discussed here. First, only the dissolution of transparent films can be measured. In addition, some of the techniques require *a priori* knowledge of optical constants. Often, these values can only be obtained through complicated dispersion measurements. Finally, a very serious problem arises from nonuniform film development. Since most resist films demonstrate surface roughening during dissolution, incident light is scattered, resulting in noisy data that are often difficult to interpret (*8*).

An alternative to optical techniques is the measurement of film mass during dissolution, accomplished through the use of piezoelectric quartz crystals (*8–12*). These crystals, fabricated to possess a known oscillation frequency at room temperature, are used as substrates for resist films. Deposition of a mass on the crystal causes it to oscillate at a lower frequency than the uncoated substrate. As the film mass changes during dissolution, the corresponding frequency shift of the crystal can be monitored precisely and correlated to the film thickness through instrument parameters and film density. A rigorous theoretical analysis of the equations relating oscillation frequency to mass is surveyed in Reference 9.

The use of piezoelectric quartz crystals for monitoring thin film deposition processes has been well documented for vapor phase applications (9). Recently a quartz crystal oscillation scheme has been applied in a liquid environment to monitor dissolution of resist films by measuring mass changes (8, 10, 11). The main advantage of this system is that highly accurate frequency measurements can be made. With this quartz crystal microbalance (QCM) technique, frequency can be measured accurately within 20 hz, corresponding to a thickness of 30 Å over a thirty minute experiment (10). In addition, if nonuniform development, such as surface roughening, occurs across the wafer surface, the quality of the rate data obtained is not compromised as would be the case for an optical technique.

One issue associated with the use of a QCM is that polymeric resists may exhibit variable viscoelastic properties during the course of dissolution. This in turn could cause a distortion in the measured oscillation frequency and an inaccurate thickness determination. It has been demonstrated that for films as thick as 7.35 μm, the resist behaved as a rigid film and hence, the mass and film thickness could be correlated to frequency by a simple theoretical analysis (8, 10, 11).

In our study of diazonaphthoquinone dissolution, the quartz crystal microbalance was determined to be the most suitable measurement technique because of its high precision. We are interested in isolating the role of photoactive compound (PAC) in inhibiting dissolution of the matrix resin. There have been numerous theories proposed which address the issue of the role of PAC during development (12–16). Our goal is to understand the physical range of influence exerted on the polymer matrix by individual PAC molecules as well as the effect of PAC spatial distribution on overall development behavior.

Our approach to understanding effects of spatial distribution is to use Langmuir-Blodgett (LB) deposition for fabricating resist films (17). LB films are single monolayer structures which can be formed at an air-water interface and subsequently transferred to a solid substrate one monolayer at a time. In this manner, resist films can be fabricated with stratified sensitizer distributions, and the effects of locally high concentrations of PAC can be directly studied. Previous work has demonstrated the significant influence of sensitizer distribution on development behavior (18–20). In these studies, LB deposition was extremely effective for fabricating structured, spatially segregated resist films; something that would not have been possible with traditional film deposition methods like spin casting.

EXPERIMENTAL

Materials

The resin used for these experiments was poly (3-methyl-4-hydroxystyrene) of weight average molecular weight 12,400 g/mole and polydispersity 1.63 (21). The photoactive compound was a structurally simple 4-substituted diazonaphthoquinone. Both materials were used as received and their chemical structures are shown in Figures 1a and 1b.

Spin Coating

An initial dissolution study was conducted with spin cast films containing different concentrations of PAC dispersed uniformly throughout the polymer matrix. Solutions containing 5 wt% solids in 2-ethoxyethanol were spun at 2000 rpm for 40 seconds yielding films that were approximately 1000 Å. Thicknesses were determined by ellipsometry after the films were baked at 90°C for 30 minutes on an aluminum block in a convection oven to remove residual casting solvent. Samples with PAC concentrations varying from 0-50% by weight of polymer were prepared for study.

Langmuir-Blodgett Deposition

The Langmuir and Langmuir-Blodgett films were fabricated using a single-area Joyce Loebl Langmuir Trough IV equipped with a Wilhelmy plate microbalance (*17*). The subphase was deionized water. The perimeter of the trough is surrounded by a Teflon-coated barrier which is used to alter the area of the subphase and maintain a constant film surface pressure. The sensitizer was spread onto the water surface from a chloroform solution and the polymer from an isopropyl acetate solution, both at concentrations of 1 mg/ml. Following spreading onto the surface of the trough, the solvent was allowed to evaporate for 20 minutes. Pressure/area isotherm measurements were performed for each material as a means to characterize the stability of the films. Isotherms were taken at room temperature. Deposition of monolayers onto the piezoelectric quartz substrates was then performed at half the collapse pressure of the material, as determined from the isotherms. The dipping speed was 5 mm/minute and all layers were deposited at room temperature.

Stratified Resist Film Preparation

The stratified resist films were prepared using a combination of spin casting and LB deposition. First, a spin cast polymer film (1000 Å) was deposited on the quartz substrate as described in the previous section. Next, three monolayers of PAC were deposited by LB deposition (30 Å) on top of the spin cast polymer base. Finally, the stratified structure was completed with an overlayer of 75 monolayers of LB polymer (2000 Å). Figure 2 shows the structure of this embedded sample. The utility of LB deposition is that the deposition of both the PAC layer and polymer overlayer does not cause any underlying layers to redissolve as would be the case if spin casting were used.

We are also interested in understanding the effect of thermal treatment on dissolution behavior of embedded films as a means to study the mobility of the PAC. Therefore, one set of dissolution experiments was performed on films that were unbaked and one set on embedded films that were baked at 90°C for 60 minutes in a convection oven.

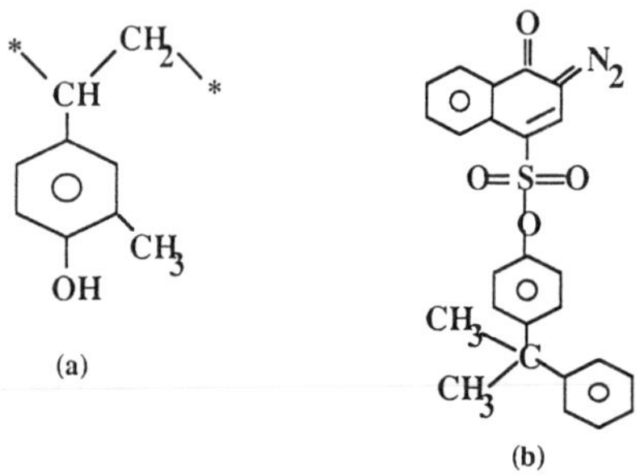

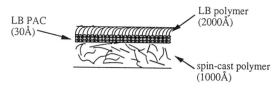

Figure 1. (a) poly (3-methyl-4-hydroxystyrene) polymeric resin 21. The ∗ represent the continuation of the repeat unit to form the full polymer; (b) diazonaphthoquinone PAC.

Figure 2. Embedded film with highly localized PAC concentration distribution for spatial dissolution studies.

Dissolution Rate Measurements

The dissolution measurements were carried out using the QCM apparatus depicted in Figures 3a and 3b (*10, 11*). Measurements were performed at room temperature using Shipley's 2401 potassium hydroxide-based developer diluted 1:5 with deionized water. The solvent flow rate through the sample chamber was 16 ml/minute.

RESULTS AND DISCUSSION

Dissolution Behavior of Spin Cast Films

In order to characterize the effect of PAC distribution on dissolution, we began with a study of spin cast films containing PAC uniformly dispersed through the sample. PAC concentrations in the films were varied from 0-50 wt% (by weight of polymer) and for each PAC loading level, between five and eight samples were tested. Figure 4a is a plot of thickness vs. time for a representative sample of each concentration, normalized to an initial thickness of 1000 Å. Figure 4b is a compilation of the averaged dissolution rates as a function of concentration.

The thickness vs. time plots obtained from QCM data all show a region at the beginning with a rapid linear thickness rise. At the beginning of an experiment, as the flow cell fills up with developer, the frequency of the quartz substrate changes since there is nonuniform contact with the liquid. The frequency change is then seen as a thickness rise in the data, although it does not represent a true thickness increase. This region of the plots is eliminated during data analysis. Typically, at this flow rate, the rise time is only about 10-20 seconds, much less than the time for a typical experiment and does not cause a significant loss of dissolution data.

The thickness vs. time curves in Figure 4a can be divided into two basic groups. The first, consisting of curves 1 (0% PAC) and 2 (15% PAC), show a linear decrease in thickness as a function of time. However, with concentrations of sensitizer greater than 15 wt%, there is an initial region of steeper slope than what is seen in the bulk of the film. The faster region dissolves at about twice the rate as the bulk of the film for all samples. In the case of the 20 wt% curve (plot 3), this initial rapid dissolution region occurs for the first 200 Å of film thickness lost. Curves 4-6, containing 30 wt% or more PAC, also exhibit this faster dissolution region at the surface of the film. However, there is also an apparent initial thickness increase associated with these films. Since the film thickness values are determined from mass measurements, an apparent thickness increase could correspond to an increase in mass due to swelling or solvent absorption by the resist film. There may PAC aggregation at the surface of the film or the formation of polymer/PAC complexes which cause the apparent swelling of the films. At this time, it is not clear what causes this behavior and further studies are currently being conducted. For this study, we focused only on the thickness loss behavior of these materials. The significant observation of two distinct regions which dissolve at considerably different rates indicates that there may be a separate dissolution mechanism at surface of the resist films than in the bulk.

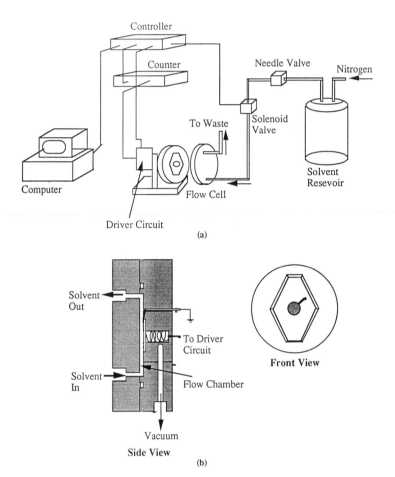

Figure 3. (a) Schematic of the quartz crystal microbalance dissolution rate monitor 11; (b) Detailed view of flow cell containing sample.

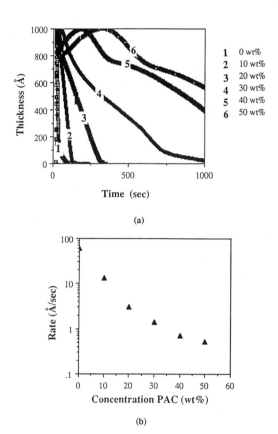

Figure 4. (a) Thickness vs. time plots for spin cast resist films with varying concentrations of PAC; (b) Dissolution rate as a function of increasing PAC concentration.

One possible explanation for these observations is that the polymer present at the surface of the film is preferentially leached out (15). Since the polymer is inherently soluble in the developer and the PAC is not, the matrix material can be removed from the surface of the film while the sensitizer remains behind. A protective PAC layer is then formed of the dissolving film. In this case, a faster initial dissolution rate would be expected at the surface as the polymer at the surface of the film is removed. Once a protective barrier of PAC is formed, however, the film would dissolve at a lower rate since the developer has more difficulty wetting the matrix. Our observations with spin cast films are consistent with a leaching dissolution mechanism and further study is being conducted for a more in depth characterization.

Finally, another notable feature from Figure 4a is that curves corresponding to lower PAC concentrations (curves 1-3) show a constant etch rate in the bulk of the film. By contrast, for films with larger sensitizer concentrations (curves 4-6), the etch rate decreases as development proceeds in the bulk of the film (excluding the initial regions which may show some apparent thickness increase and faster dissolution region). This nonuniformity becomes very pronounced in the films containing 50 wt% PAC. Although the cause of this behavior is not yet understood, there may be significant aggregation or phase separation which prevents the PAC from effectively inhibiting the dissolution of the polymeric resin. Also, large local concentrations of PAC may facilitate the formation of highly insoluble PAC/polymer complexes.

Figure 4b is a summary of the dissolution rates obtained for spin cast films as a function of PAC concentration. Dissolution rates were measured in the bulk of the films (essentially linear thickness loss with time) from a thickness of 500 Å to 100 Å to avoid surface effects of the film. There is a decrease of almost two orders of magnitude in the dissolution rate of the resist as the PAC concentration is varied from 0-50 wt%. All the data were averaged over five to eight samples and were repeatable to within an experimental error of 5-10%.

Langmuir-Blodgett Film Behavior

The majority of studies conducted on Langmuir films have, until recently, been performed on small amphiphilic molecules (17). These materials form highly ordered and oriented films at the air-water interface and upon transfer to a solid substrate. In the case of more complex materials with a variety of chemical functionalities, such as the PAC and polymer, the morphology of monolayers formed at an air-water interface and after transfer is not well-characterized. Although both materials form stable Langmuir films that can be transferred to the quartz substrates, the details of the orientation and morphology of these films are not known and may in fact, be highly disordered.

The main method of characterizing a Langmuir film is through the pressure/area isotherm. The isotherm provides information on the behavior of the film at the air/subphase interface and can be used to determine the point at which the film ceases to be a monolayer. In some cases, films on the water surface may undergo a phase transition or form a multilayer structure (collapse).

The pressure/area isotherms for the PAC and polymer are shown in Figures 5a and 5b respectively.

The sensitizer isotherm, Figure 5a, is typical for many materials that form stable Langmuir films. When a critical surface area is reached (80 Å^2/ molecule), there is a rapid, essentially linear pressure rise with decreasing area. At 5.5 dyn/cm, there is an abrupt transition to a regime with a much smaller slope. This type of transition is either caused by film collapse and the formation of a multilayer structure, or a phase transition at the water surface from a liquid-like two dimensional morphology to a more solid-like state. The limiting area for the PAC, determined from extrapolating the linear region of larger slope, was found to be 55 Å^2/molecule, comparable with previous Langmuir studies of diazonaphthoquinones (22). In the formation of the Langmuir-Blodgett films of the PAC, deposition only occurred on the upstroke of the substrate at a transfer pressure of 2.75 dyn/cm, giving a Z-type film structure. Z-type film structures are formed when monolayers are deposited on the substrate only on the upstroke (as the substrate is raised out of the water).

Figure 5b is the pressure/area isotherm for the polymer. Although this material does form a film on the subphase surface which can then be transferred, the isotherm behavior is indicative of a less stable film. First, there is no distinct phase transition or region of collapse, but rather a continuous rise in pressure until the smallest area on the trough is reached. The limiting area for this material is 12 Å^2/monomer, a value much lower than what would otherwise be expected for a fairly large molecule. The very low limiting area would suggest that the film has partially collapsed on the water surface. There may be looped structures that come out of the plane of the air-water interface or multilayer domains that give rise to the low limiting area. The actual morphology of the film is not known but may be elucidated by further isotherm studies at different subphase temperatures and film equilibration times. The polymer films were transferred at 15 dyn/cm, in the linear region of the isotherm and formed Z-type structures on the quartz crystal substrates.

Dissolution Behavior of Stratified Resists

The major advantage of using the embedded film structure, shown in Figure 2, is that the vertical extent of influence exerted by the PAC on the polymer can be measured. Figure 6a is a plot of thickness vs. time for an unbaked resist structure containing three embedded monolayers of PAC between two layers of polymer. There are three distinct regions in this plot. The first, from 0-25 seconds, indicates a very rapid thickness loss corresponding to the removal of the Langmuir-Blodgett polymer overlayer. This region dissolves at a rate of about 414 Å/sec as compared to the 70 Å/sec for a pure polymer spin cast film, shown in Figure 4a. The very high solubility may indicate that the overlayer is a highly disordered film possessing large pathways for solvent to penetrate. There may also be some modification in the hydrogen bonding properties of the LB polymer because of orientation during film formation at the air-water interface. In the case of a spin cast film, the material can only hydrogen bond with itself. When instead, the film is formed at the air-water interface, interactions with the water

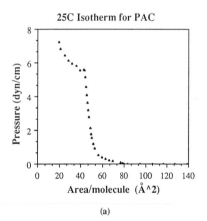

(a)

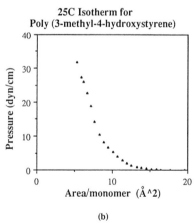

(b)

Figure 5. (a) Room temperature pressure/area isotherm for the photoactive compound; (b) Room temperature pressure/area isotherm for poly (3-methyl-4-hydroxystyrene).

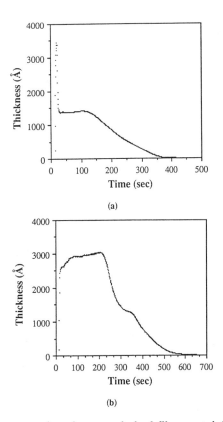

(a)

(b)

Figure 6. (a) Thickness vs. time for an unbaked film containing three embedded monolayers of PAC; (b) Dissolution behavior of embedded layer film after thermal treatment at 90°C for 60 minutes.

surface can conceivably alter the local hydrogen bonding capability and motion of the polymer such that it no longer forms the dense network expected for a spin cast film. The dissolution rate would be much faster in these films because there are now open channels that allow solvent to easily travel throughout the film and are not blocked by hydrogen bonded networks. It may be the case that a densification of the LB film would take place if the film was baked and these open channels would close. Kuan and coworkers have shown that for LB poly(methyl methacrylate) films, after 5000 minutes of baking at 170°C, there was a reduction of film thickness by 11%, presumably from densification of the film (23). We are planning to conduct a series of control experiments studying the effect of baking on dissolution rates of LB polymer films compared to spin cast films.

Between 25 and 125 seconds into the dissolution process, there is no apparent thickness loss (Figure 6a). The point in the film where this occurs, at a thickness of 1200 Å, is where the PAC monolayers were embedded. The presence of only three monolayers of PAC (30Å) inhibited dissolution for 100 seconds. The remainder of the film, consisting of spin cast polymer, dissolves at a rate of 7.7 Å/sec, but this rate is substantially lower than a spin cast film of pure polymer (Figure 5). The most dramatic aspect of these results is that the presence of a segregated photoactive compound layer has not only caused a substantial induction time after LB polymer was removed, but it has also substantially lowered the rate of the underlying spin cast layer. The influence of the PAC is obviously substantial even in the small, isolated quantities used for this experiment.

To study the mobility of PAC in the matrix, embedded films were prepared as before and then subsequently baked them in an oven for one hour at 90°C. There are now less defined regimes present (Figure 6b). As in the case of the heavily loaded spin cast films, there is an apparent thickness increase in the first 100 seconds, which could be due to some type of solvent absorption or swelling of the matrix. Between 75 and 225 seconds into dissolution, there is little or no loss of film thickness. At 1200 Å, the original depth of the three embedded PAC monolayers, there is still some slowing of development. One explanation for the significant difference in the development time of the top layer in Figures 6a and 6b is that the PAC is highly mobile in the LB polymer and may have segregated to the surface of the film. The apparent preferential segregation of the PAC into the LB layer is consistent with the observation from the unbaked films that the polymer layer may be a more loosely networked structure than the spin cast films.

CONCLUSIONS AND FUTURE WORK

The physical distribution of the PAC in a resist film can have a significant effect on the dissolution behavior of these materials. As little as three monolayers of embedded PAC can cause a significant induction time. It also appears that the PAC is more mobile in LB polymer films than in spin cast films. The quartz crystal microbalance measurement technique is a highly accurate method to probe dissolution rates that change throughout the thickness of a single sample.

It can be used to study stratified resist films in and provide information on the spatial extent of the influence that individual monolayers of PAC have on a matrix polymer.

Our future work will be focused on understanding the morphology of the individual layers used to create the stratified resist structures. We will be conducting more experiments with the Langmuir-Blodgett films as well as spectroscopic experiments to obtain a better understanding of the chemical interactions which play a significant role in the dissolution characteristics of these types of films.

ACKNOWLEDGMENTS

We would like to thank Andrew Grenville in the Applied Physics department at Stanford University for his enormous assistance in setting up the quartz crystal microbalance instrument in our laboratory. Support for this work comes from Sematech and SRC through the California Center of Excellence on Lithography and Pattern Transfer Research under contract SRC 91MC515.

REFERENCES

1. F.H. Dill, *IEEE Trans. Electron Dev.*, **ED-22(7)**, 445 (1975).
2. W.G. Oldham, *Optical Eng.*, **18(1)**, 59 (1979).
3. K.L. Konnerth and F.H. Dill, *IEEE Trans. Electron Dev.*, **ED-22(7)**, 452 (1975).
4. W.W. Flack, J.S. Papanu, D.W. Hess, D.S. Soong, and A.T. Bell, *J. Electrochem. Soc.*, **131**, 2200 (1984).
5. F. Rodriguez, P.D. Krasicky, and R.J. Groele, *Solid State Tech.*, **28(5)**, 125 (1985).
6. P.D. Krasicky, R.J. Groele, and F. Rodriguez, *Chem. Eng. Comm.*, **54**, 279 (1987).
7. C.G. Willson, *Introduction to Microlithography: Chapter 3-Organic Resist Materials: Theory and Chemistry,* ACS Symposium Series, **219**, Am. Chem. Soc., Washington D.C., (1983).
8. W.D. Hinsberg, C.G. Willson, and K.K. Kanazawa, *J. Electrochem. Soc.*, **133(7)**, 1448 (1986).
9. C. Lu and Z. Czanderna, eds., *Applications of Piezoelectric Quartz Crystal Microbalances*, Elsevier, New York (1984).
10. W.D. Hinsberg, C.G. Willson, and K.K. Kanazawa, *Proc. SPIE*, **539**, 6 (1985).
11. a: W.D. Hinsberg and K.K. Kanazawa, *Rev. Sci. Instrum.*, **60(3)**, 489 (1989).
    b: W.D. Hinsberg and K.K. Kanazawa, *Rev. Sci. Instrum.*, **60(12)**, 3835 (1989).
12. M. Koshiba, M. Murata, M. Matsui, and Y. Harita, *Proc. SPIE*, **920**, 364 (1988).
13. M. Koshiba, M. Murata, Y. Harita, and Y. Yamaoka, *Polym. Eng. and Sci.*, **29(14)**, 916 (1989).
14. M.K. Templeton, C.R. Szmanda, and A. Zampini, *Proc. SPIE*, **771**, 136 (1987).

15. K. Honda, B.T. Beauchemin, Jr., R.J. Hurditch, A.J. Blakeney, Y. Kawabe, and T. Kokubo, *Proc. SPIE*, **1262**, 493 (1990).
16. M. Hanabata, Y. Uetani, and A. Furuta, *Proc. SPIE*, **920**, 349 (1988).
17. G. Roberts ed., *Langmuir-Blodgett Films*, Plenum Press, New York (1990).
18. L.L. Kosbar, C.W. Frank, and R.F.W. Pease, *Proc. SPIE*, **1262**, 585 (1990).
19. V. Rao, L.L. Kosbar, C.W. Frank, and R.F.W. Pease, *Proc. SPIE*, **1466**, 309 (1991).
20. V. Rao, W.D. Hinsberg, C.W. Frank, and R.F.W. Pease, *Proc. SPIE*, in press.
21. G. Pawlowski, T. Sauer, R. Dammel, D.J. Gordon, W. D. Hinsberg, D. McKean, C.R. Lindley, H.-J. Merrem, H. Roschert, R. Vicari, and C.G. Willson, *Proc. SPIE*, **1262**, 391 (1990).
22. L.L. Kosbar, *Ph.D. Thesis: Department of Chemistry: Chapter 6: Single Component Langmuir Films-manuscript in preparation*, Stanford University, Stanford, CA 1990.
23. S.W.-J. Kuan, Ph.D. Thesis: Department of Chemical Engineering: *Chapter 5: Structural Characterization of Polymer Multilayer Films-manuscript in preparation*, Stanford University, Stanford, CA 1990.

RECEIVED December 30, 1992

Chapter 16

Photochemical Control of a Morphology and Solubility Transformation in Poly(vinyl alcohol) Films

Induced by Interfacial Contact with Siloxanes and Phenol—Formaldehyde Polymeric Photoresists

James R. Sheats

Hewlett Packard Laboratories, 3500 Deer Creek Road, Palo Alto, CA 94304

Poly(vinyl alcohol) (PVA) is found to undergo a dramatic change in morphology when prepared as a thin film (~1000 Å) between a siloxane polymer and an acid-generating novolak photoresist, and heated to 115 °C or more for a few minutes. When the siloxane is stripped off in chlorobenzene, the exposed PVA film is wrinkled and insoluble in water (although soluble in acetone). This effect is entirely prevented if the resist is exposed (before heating the films) to a lithographic dose of deep-ultraviolet (DUV) radiation; patterns can be produced in this way from optical images as small as 0.3 μm. The effect is seen with at least two siloxanes and two resists: poly(phenylsilsesquioxane) (PPSQ), poly(diphenylsiloxane) (PDPS), Shipley SAL-601, and AZ5214E. No effect is seen if either the top or bottom film is omitted, and no effect is seen when the top film is poly(methyl methacrylate) (PMMA), nor when the bottom film is AZ1350. Auger spectroscopy shows trace amounts of Si at the surface of the insoluble film, as well as some N. Atomic force microscopy (AFM) images of the morphology are presented. The transformation appears to be a consequence of the response of polymer conformation to poor solvents, delicately directed by small interfacial concentration gradients.

Resists based on the photochemical generation of acid, which then catalyzes a suitable change in polymer properties, have come into widespread use in recent years; their advantages include much greater sensitivity and considerable flexibility in resist design, as evidenced by the wide variety of systems that have been designed (*1–3*). Although there is an *a priori* concern that the acid might diffuse over sufficient distances to cause loss of resolution, it has been shown in at least some cases that the diffusion range is no more than a few nanometers (*4*).

A second major development in photolithographic chemistry has been that of image enhancement by photobleaching. A number of different variations on this theme exist; the primary ones are contrast enhancement (CEL) (*5*), photochemical image enhancement (PIE) (*6–9*), and built-on mask (BOM) (*10*). CEL was the first of these to appear; in this process a polymer containing a bleachable dye is spun on top of the resist, and both are exposed together. We originated the PIE system shortly thereafter; it likewise uses a top layer of

0097–6156/94/0537–0235$06.00/0

bleachable dye, but only the dye is exposed in the imaging camera, and the resist is exposed by flood exposure during which the dye image is "fixed" so that it cannot further degrade. The result is better performance at the expense of a flood exposure.

In both cases, as well as with the two-layer portable conformable mask (11, 12) and other two-layer processes (1–3) there is concern about intermixing of the layers: it is not easy to devise solvent systems that are adequate for the respective polymers that also do not result in some interpentration of the two layers, with resultant scum or other deleterious effects on resist development. To circumvent this, a barrier layer of a water-soluble polymer such as PVA is often used in between. We used this approach in our early development work on PIE (7), and discovered as a consequence some unexpected effects on polymer morphology that may reveal fundamentally interesting features of interfacial layers.

When PVA is placed on top of certain photoresists, and a siloxane polymer on top of that, and the system baked under conditions similar to the typical post-exposure bakes associated with the resist, the PVA undergoes (after stripping the siloxane in chlorobenzene) a dramatic change, becoming wrinkled and insoluble in water (but soluble in acetone). Most remarkably, this effect can be completely prevented if the system is irradiated before the bake, with a DUV dose suitable for exposing the resist. Since neither PVA nor the siloxane have any photoreactivity at all at those wavelengths and doses, it follows that the photochemical change in the resist is felt throughout the ~ 1000 Å PVA layer thickness.

In this paper we attempt to elucidate this remarkable phenomenon by systematically varying the composition of the bottom and top layers, by Auger analysis of the surface, and by AFM images of the topography. The results have a bearing on interfacial polymer chemistry and suggest the extent to which small variations in composition may dramatically affect solubility and morphology.

EXPERIMENTAL DETAILS

PPSQ (M_w 1200–1600, with "high ladder content") and PDPS (M_w 1000–1400, silanol terminated) (Petrarch Chemicals) were spun from chlorobenzene solution to thicknesses of ~ 5000–8000 Å. For most experiments the PPSQ was mixed with ~ 22–25 wt.% diphenylanthracene (DPA; Aldrich) for historical reasons (this is the formulation used in PIE); however the effect is observed without DPA. The source of PVA was General Electric's barrier coat for use with CEL (5) (these materials are now sold by MicroSci, Inc. (13)); spun at 5000 rpm the thickness was 950 Å (assuming $n_r = 1.51$) (14). AZ5214E was purchased from AZ Photoresist Products. The other acid-generating resist was obtained from Shipley; for the initial experiments it was referred to as ECX-1023 (15, 16); later this became a standard commercial product SAL-601, which was used in some experiments. Despite possible changes in formulation during the intervening 4–5 years, the results were at least qualitatively the same. We will make clear which was used in each experiment; however the discussion will not distinguish between them. Resist thickness was typically ~ 5000 Å for SAL and 1.2 μm for

AZ. MF-312 (Shipley) is a typical full strength alkaline developer for novolak resists.

All experiments used standard Si wafers as substrates (in one case with a ~1000 Å thick thermal oxide). In some cases, a layer of hardbaked (~200 °C, 2 hrs, oven) Hunt HPR-204 was applied first. (A list of each set of films used is given in the Appendix, along with a summary of the results from the associated experiments.) For the experiments using ECX-1023, the resist and PVA were both softbaked at 70 °C for 30 min. in a convection oven; the PPSQ layer was not softbaked. Postbakes (i.e., the bake that produces the effect) were in all cases done on hotplates controlled to ±1 °C. The experiments with SAL-601 and AZ5214E used a hotplate also for the softbakes. AZ5214E was softbaked at 105C for 60 sec, and SAL-601 at 115C for 60 sec. DUV exposure (Lumonics excimer laser, KrF) energy was monitored by a Laser Precision pyroelectric joulemeter (Rj7200). Thickness was measured with a Nanometrics Nanospec. Actual thicknesses in the multilayer stacks are not exact since the refractive indices of the materials are not identical, while a single index is used by the Nanospec (1.64); where possible we provide corrections based on known indices. A Digital Instruments Nanoscope III was used for atomic force microscopy.

RESULTS

Smooth, optically homogeneous films are formed when the three layers are prepared as described, and they remain homogeneous through postbakes of up to 125 °C for several minutes. However, when the top layer (PPSQ or PDPS; most experiments were done with PPSQ but the result with PDPS was visually the same) is removed (after the bake) by rinsing briefly with chlorobenzene, the surface appears gray or a dull silver color. ("Brief rinsing" means, for example, several seconds under a squeeze bottle.) By microscope one sees the wrinkled or convoluted morphology shown in the photos in Figure 1. At the same time, this film is found to be insoluble in water, but soluble in acetone: brief rinsing washes it away instantly. (The untreated PVA does not change thickness after 30 sec. continuous rinsing in acetone, although the surface develops some crack- or ridge-line features visible in the microscope). That the film consists mostly, if not entirely, of PVA, is evidenced by two factors. First, the thickness is close to the same as the original PVA thickness (the Nanospec thickness, which averages over a 20 μm spot, for the film on ECX in Figure 1 is 1300 Å). Second, Auger spectroscopy analysis shows that there is only a small amount of Si at the surface of the film. (This data is discussed in more detail below.)

If the wafer is DUV irradiated with at least 10 mJ/cm^2 (incident on the resist) before the bake, the wrinkled film does not appear: the appearance after chlorobenzene rinse is clear and homogeneous, a thickness measurement corresponds to approximately the resist + PVA, and a brief (~10 sec) water rinse leaves just the resist thickness. If the irradiation is from a pattern, patterns can be produced in the wrinkled film, as shown in Figure 2. These photos were obtained via the PIE process (7). The PPSQ contained DPA and a ketocoumarin sensitizer for 436 nm, and the trilayer was exposed (in air) in a Nikon 0.42 N.A. stepper with a resolution mask containing equal lines and spaces and isolated

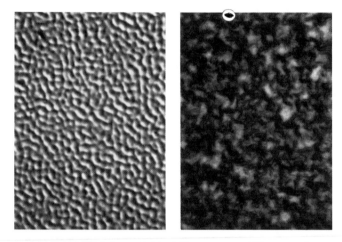

Figure 1. Photos (1000x) of the wrinkled film on two different substrates: (right) ECX; (left) AZ 5214E.

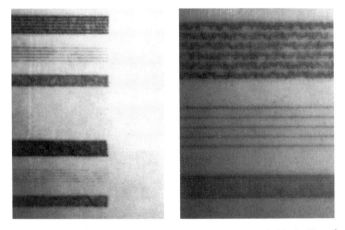

Figure 2. Photos (500x) of patterns printed in the wrinkled film (on ECX): (right) 0.4 μm (top) and 0.3 μm (bottom) features (isolated spaces, isolated lines, and equal lines/spaces, from top to bottom in each set); (left) 0.5 μm features. Since this is a negative process, a line appears in the absence of exposure.

lines and spaces down to 0.3 μm. It was then exposed under N_2 to KrF laser light to give a dose of about 10 mJ/cm^2 to the resist, postbaked at 120C for 3 min, and rinsed in chlorobenzene. It can be seen that even at 0.3 μm, distinct lines of PVA show up. At 0.4 μm, the isolated spaces are resolved, and at 0.5 μm even the equal lines and spaces give a detectable pattern. In all of these, the wrinkles often run at non-zero angles with respect to the trend of the lines as well as parallel, so the spaces in between lines are not completely empty, and the lines are not continuous; this is much better displayed in the AFM images (*vide infra*). The important point is that optical exposure of the resist (which is known to have resolution in this system down to 0.4 μm, and to 0.3 μm for isolated spaces) produces an effect that is transmitted into the PVA and throughout its thickness, so that distinguishable patterns are produced corresponding to this resolution. These patterns are quite stable in water (immersion for 30 min. has no observable effect), but are immediately removed by acetone.

In an attempt to determine the origin of these phenomena, various layers were omitted or altered in composition. The effect was observed either with or without DPA in the PPSQ. PDPS gives a film with the same visual appearance as PPSQ (it was not further studied, e.g., with respect to DUV imaging). PMMA as a top layer has no effect (by "no effect" we mean that the films remain clear, and the PVA water soluble as usual, after a postbake of 125C for 3 or 4 min.). If there is no top layer (i.e., just PVA on top of resist), there is no effect. If PVA is put down on hardbaked HPR-204, and PPSQ spun over it, there is no effect. AZ5214E behaves qualitatively similarly to SAL-601 (Figure 1) (although it may be significant that the lateral scale of the wrinkles is larger), while AZ1350 as a bottom layer leads to no effect. Casting PPSQ on PVA and removing it without baking does not alter the PVA appearance or solubility. PPSQ as both bottom and top layers has no effect.

Thus, it is established that *i*) *both* top and bottom polymer layers are needed, yet the effect that is observed is in the middle layer; *ii*) the top layer must be a siloxane polymer; *iii*) the bottom layer must be an acid-generating photoresist that has not been hardbaked; and *iv*) DUV radiation of a lithographic dose eliminates the effect, with resolution essentially equal to the aerial image resolution. There are some caveats to items (*ii*) and (*iii*): only two siloxane polymers (PPSQ and PDPS, both of low molecular weight) and one hydrocarbon polymer (PMMA) have been tried for the top layer, and only four resists (SAL-601, AZ5214E, AZ1350, and hardbaked HPR-204) for the bottom. AZ5214E was tried for a top layer, but this fails because it becomes insoluble during the postbake (the same would of course happen to SAL). It would be useful to include some other polymers as bottom layers, but it seems clear that the acid is crucial.

An Auger scan of the wrinkled film is shown in Figure 3. The atomic concentrations from peak-to-peak derivative measurements are 94.8% C, 2.7% O, 1.3% Si, and 1.3% N. The corresponding data for a PVA film on Si are 98.2% C, 1.5% O, and 0.3% Si; the latter figure is in the noise (no trace of a peak can actually be seen in the spectrum). Thus the remaining film has a surface layer that contains a small amount of Si; a short sputter etch (0.9 min.) eliminates this signal completely. It is interesting that the N signal remains (at about the same

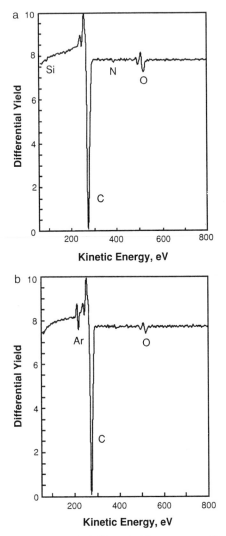

Figure 3. a, Auger scan of wrinkled film; 3–10 μm spot, 10 A/cm^2. Auger scan as in b, after 0.9 min. argon ion sputtering.

intensity) after the sputter etch. (The absence of the expected amount of O is most likely due to degradation by the 10 keV electron beam, which was operating at about ~1 μA current in a spot of ~5 μm diameter, with an acquisition time of 5 min. The results for the unprocessed PVA film on Si, which also show little O, support this explanation.) The N signal suggests that some of the bottom layer (which contains N in the melamine crosslinking agent in SAL, and in the diazonaphthoquinone in AZ) has mixed with the PVA. The possibility that the bottom layer is completely exposed to the electron beam in the low areas between the wrinkles is ruled out because immersion in developer (full strength MF312) for 2 min causes no change. Since SAL is a negative resist, the unirradiated regions would certainly have been developed if they were accessible to developer.

We do not know exactly how close to the surface the Si is confined. Comparison to results obtained on the PVA film (nominally the same thickness) suggest that 0.9 min. etching removes a considerable portion of the film, so it is possible that Si penetrates a significant distance into it.

Some AFM images are presented in Figures 4 and 5. Some of the features that can be observed are: *i*) the roughness is substantial: the peak to valley distance is as much as ~1000 Å in some cases; *ii*) the height of the features varies, with some being only about 500 Å high; *iii*) the wrinkles are smoothly sloped and not very steep; *iv*) the patterned "lines" have wrinkled areas of much smaller height in between (this is probably indicative of incomplete DUV exposure, since regions with clear-field exposure of substantially greater than 10 mJ/cm^2 are optically quite smooth); *v*) an isolated line is in fact a pair of lines, with a deep (greater than the original PVA thickness) narrow trench in between; *vi*) the lines in the equal line and space patterns similarly are not individual, but come in pairs, with a substantially greater gap between pairs than between the members of a pair. This is a very regular and surprising effect: if the remaining film were due to a reaction such as crosslinking, a line would have a typical solid cross-section.

DISCUSSION

The morphological transition described here is remarkable in itself, and an explanation of its origin is not obvious. It clearly involves phenomena at polymer interfaces, since it requires for its appearance the presence of specific polymers adjacent to the film that is actually transformed, yet the Auger data demonstrate that the interpenetration is not great (at least for the top layer). It is clearly not a chemical cross-linking process, since the film is fully soluble in acetone.

Even more remarkable, however, is the photochemical effect. There seems to be no plausible candidate for this effect (at least for its initiation) other than the well-known photoresist reactions. SAL functions by photochemically generating a protonic acid that then catalyzes crosslinking reactions during postbake. AZ5214 operates by changing the resist solubility, but the basis of this change is the conversion of a naphthoquinonediazide into an indenecarboxylic acid of relatively low pK, so a protonic acid is still produced (*17*). AZ1350, on the other hand, produces only a much weaker acid. The strong acid somehow leads to a

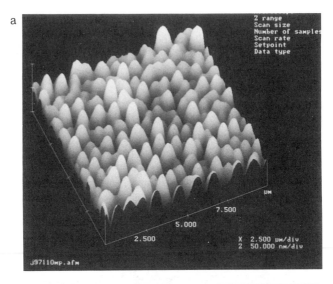

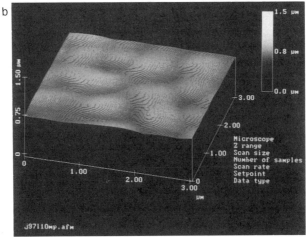

Figure 4. (a) AFM image of the wrinkled film (taken in an area of no DUV exposure). Note the exaggerated vertical scale (50 nm/div.; horizontal 2.5 μm/div.). A cross-section shows profiles as in Fig. 4c, with heights varying from ~ 400 to 1000 Å. (b) Part of the same image as in Fig. 4a, but with vertical scale in the same proportion as the horizontal.

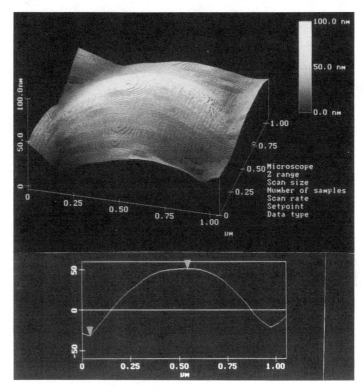

Figure 4c. Magnified view (and cross-section) of one of the "wrinkles" in Figure 4a. Vertical scale on the cross-section is in nm (50 nm/div.).

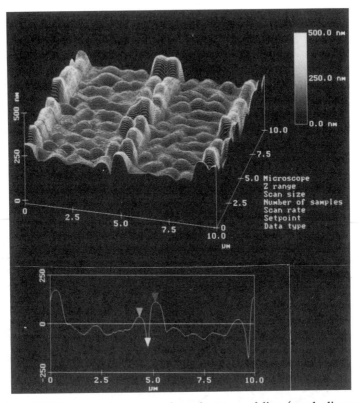

Figure 5a. AFM image and cross-section of patterned line (mask dimension 0.7 μm). The area away from the line, nominally exposed (and therefore free of wrinkled film) still has a considerable amount of it present. Because of the nature of the imaging, it is expected that this region should receive less than a full exposure (the amount received by a large open field), and so there is some film present. Note the trough in the center of the line; this feature is quite straight over the entire image (cf. also Fig. 5b). Its peak-to-valley height is close to 2500 Å. This is a larger variation than seen anywhere in the unpatterned region shown in Figure 4. (Vertical scale of cross-section is 250 nm/div.).

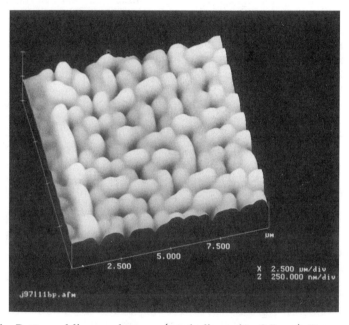

Figure 5b. Patterned lines and spaces (mask dimension 0.7 μm). By comparison with Figure 5a, it appears that the deep valleys, which naively would seem to be the spaces, are actually the centers of the lines, the sides of which touch each other in an irregular fashion.

prevention of the morphology transformation in the PVA film. *A priori*, one might have expected that acid could catalyze the crosslinking of PVA, leading to reduced solubility. However, exactly the opposite trend is seen.

The nature of the transformation is not well understood at this stage. Certainly the phenomenon is worth understanding, in view of its unusual characteristics; a full explication may provide useful insight into polymer interfaces, diffusion, thermodynamics and kinetics. The AFM images show that smoothly rounded domains have formed, and both the solubility and Auger results are in agreement with the assumption that the wrinkled film has a composition that is no longer pure PVA. A plausible hypothesis, then, is that during the bake there is significant interpenetration of PVA and siloxane at the interface, as well as some migration of the molecular components of the resist into the PVA. (Note that although the Auger data indicate the presence of an N-containing compound throughout the entire PVA film, it is possible that this is due to the nonhomogeneous morphology and the Auger spot size. According to the AFM images, the film has both high and low spots on the micron scale; the low spots must be fairly thin ($<<1000$ Å). The Auger beam covers several microns, and therefore sees both thick and thin regions; the N signal may be entirely from the thin regions.)

The diffusion coefficient required to account for penetration of about 500 Å is of the order of 10^{-13} cm^2/sec. (for a 200 s. bake). For small molecules (such as the monomeric resist components) this is a quite reasonable number, either above or below T_g (cf. measurements on camphorquinone in polycarbonate (*18*), which are in the range of $10^{-11} - 10^{-12}$ above T_g and $10^{-12} - 10^{-13}$ below). T_g of PVA is 85C (*19*). The situation is less clear with polymer diffusion. PPSQ apparently does not have a distinct T_g (*20*), but remains glassy up to its thermal decomposition temperature (this report is for high M_w material). Wang, et al., (*21*) obtained diffusion coefficients of $\sim 10^{-15}$ cm^2/sec for high M_w poly(butyl methacrylate) in latex particles, at temperatures well above T_g. It is possible that the low M_w siloxanes used here diffuse at substantially higher rates. Thus the postulated diffusion is plausible albeit not assured.

The interfacial regions thus formed might then be just soluble enough in chlorobenzene for it to penetrate into the nominally PVA layer. As it does so, domains would form with PVA in the interior and a surface region that contains a small amount of siloxane and nitrogen-containing compounds from the resist. This surface would have enough non-PVA composition to be insoluble in water but soluble in acetone (and the mixture form acetone-soluble aggregates). The AFM images are consistent with this picture of a phase separation process.

The simplest mechanism to account for the protection afforded by DUV exposure is to assume that the bake-induced crosslinking of the exposed resist inhibits molecular mobility so as to prevent migration into the PVA. This would then not require that acid diffuse over large distances, and also avoids the problem of what type of reaction might give these results. It would be necessary for the rate of mobility inhibition to be rapid compared to the rate at which interpenetration occurs.

The trough in the center of the patterned lines can be understood in this context. The concentration of penetrant from below will be highest in the center of the line, and the rate of penetration of chlorobenzene will then be highest there; thus one expects that the edges of the phase separated domains should be along that line (they begin forming there).

Although the details remain to be filled in, this picture of interfacial diffusion and the resulting thermodynamic instability leading to phase separation accomodates the data, and we know of no other general hypothesis that is similarly consistent. One major question is why *both* top and bottom films are required, yet the two together have such a dramatic effect. To account for the formation of domains that are impervious to water from all sides, we have assumed that chlorobenzene is initially able to dissolve (or at least permeate and interact with) the entire film, which implies that no pure PVA is left at the end of the bake. The data suggest that penetration from both directions is necessary to accomplish this condition. If this is the case, it would be interesting to determine why one cannot get the same effect by baking longer with a single layer: perhaps the diffusion is self-limiting. Another issue is how the relatively small amount of non-PVA material that is apparently left in the wrinkles is able to solubilize the film in acetone. X-ray photoelectron spectroscopy (XPS), which is not available to us, would be very useful in determining surface compositions in more detail and would shed a great deal of light on these aspects.

SUMMARY

A morphology and solubility transformation in thin films of PVA, induced by the proximity of films of polysiloxane on one side and certain photoresists on the other, coupled with heat treatment and immersion in chlorobenzene, has been described. The effect is eliminated by exposure of the resist with a typical lithographic radiation dose. Patterns (albeit irregular and distorted) are produced from an image with features of only ~0.3 μm. Although we cannot ascertain the mechanism with substantial confidence, and many perplexing questions remain, the process appears to involve diffusional intermixing of polymers (and monomers) from the adjacent layers, followed by phase separation upon contact with chlorobenzene. The photoeffect is suggested to be the result of resist crosslinking preventing the intermixing on the resist side, although migration of photoproducts into the PVA has by no means been ruled out. To provide further illumination, more detailed surface studies, especially using XPS, will be required. Photoluminescence (*21*) might also be useful if the system will bear the appropriate reporter molecules.

ACKNOWLEDGMENT

I thank John Turner and Paul Soo for obtaining the Auger data and help in interpretation, and Hua-Yu Liu for samples of SAL-601.

APPENDIX

We summarize here the specific film stacks and experiments. The symbols $\ominus$ and $\oplus$ are used to indicate no effect and a positive effect (appearance of wrinkled film), respectively. The order of films is from top to bottom.

1) (PPSQ + DPA)/PVA/ECX/HPR204/Si; used in imaging experiments (125C, 3 min. bake; 2 min. development in MF-312). 110C, 3 min. bake: ⊖. 120C, 3 min.; 125C, 3 min.; 120C, 8 min.: ⊕. 112C, 4 min.: ⊕, but film slowly (after many minutes) came off in MF312; not affected (after ~30 min.?) by warm water. 115C, 3 min.: ⊕, but film came off slowly in MF312 and was affected by warm water (flakes came off; film did not dissolve or come off completely). Both of these films were judged "semi-opaque"; they were not examined by microscope, but may have been less wrinkled.

2) (PPSQ + DPA)/PVA/ECX/Si; 125C, 3min. bake: ⊕.

3) PVA/ECX/Si; 125C, 3min. bake: ⊖.

4) PVA/Si; 125C, 3 min. bake: ⊖. MF312 removes PVA completely in ~10-15 sec.

5) PPSQ/Si; 125C, 3 min. bake: ⊖.

6) (PPSQ + DPA)/Si; 125C, 3 min. bake: ⊖.

7) (PPSQ + DPA)/PVA/HPR204/Si; 125C, 4 min. bake: ⊖.

8) (PPSQ + DPA)/PVA/SAL/Si; 125C, 4 min. bake: ⊕. Thickness of PVA/SAL: 6553 ± 6 Å (7 meas.), using $n_r = 1.64$. SAL alone: 5703 ± 15 Å (7 meas.). This implies a PVA thickness of 850A, but the correct refractive index for PVA is 1.51. (14) Thickness of PVA alone (using $n_r = 1.51$) 950 ± 17 (9 meas.); 939 ± 2 for 6 of those measurements. $850 \times (1.64/1.51) = 923$, in close agreement with 939 or 950. Thickness of this sample after chlorobenzene rinse (using $n_r = 1.64$) 6470 ± 86Å (7 meas.). Thickness of SAL after bake is 5446 ± 18 Å (7 meas.) (there is some loss of thickness due to the crosslinking reaction), so this implies 1024Å for the wrinkled film thickness, or 1112Å if corrected to $n_r = 1.51$ (it is not obvious that this value applies to the altered film).

9) PDPS/PVA/SAL/Si; 125C, 4 min. bake: ⊕.

10) PMMA/PVA/SAL/Si; 125C, 4 min. bake: ⊖.

11) (PPSQ + DPA)/PVA/AZ5214E/SiO$_2$; 125C, 4 min. bake: ⊕. After the chlorobenzene rinse, the following thicknesses (in Å) were measured ($n_r = 1.64$) after various times of immersion in water (9 measurements for each): 0 sec.: 12909 ± 284; 30 sec.: 12653 ± 268; 60 sec.: 12510 ± 298; 0.5 hr. more: 12442 ± 223; 24 hrs. more: 12072 ± 223. However, the visual appearance under the microscope is the same as before water treatment, so the meaning of these changes is unclear; the Nanospec measurement of such a nonuniform film is somewhat problematic. Clearly the solubility is small at most.

12) PPSQ/PVA/AZ1350/SiO$_2$: ⊖.

13) PPSQ/PVA/PPSQ/SiO$_2$: ⊖. There is an interesting sidelight to this experiment: there were a large number of particles in the PVA film in this case, and wrinkled, water-insoluble material was formed around the particles after the chlorobenzene rinse. This observation provides strong support for the hypothesis in the text, since one expects that chlorobenzene may easily penetrate along the edge of a particle (i.e., the particle serves the same purpose as the interfacial mixture).

REFERENCES

1. Iwayanagi, T.; Ueno, T.; Nonogaki, S.; Ito, H.; Willson, C.G.; in "Electronic and Photonic Applications of Polymers", ed. M.J. Bowden and S.R. Turner, *ACS Adv. Chem. Ser.* **1988**, *218*.

2. Reichmanis, E.; Thompson, L.F. *Ann. Rev. Mat. Sci.*, **1987**, *17*, 235.
3. Sheats, J.R. *Solid State Technology* **1989**, *32 (6)*, 79-86.
4. Umbach, C.P.; Broers, A.N.; Willson, C.G.; Koch, R.; Laibowitz, R.B. *J. Vac. Sci. Technol.* **1988**, *B6*, 319-322.
5. West, P.R.; Griffing, B.F. *Proc. Soc. Phot. Instr. Eng.* **1983**, *394*, 39-44.
6. Sheats, J.R.; O'Toole, M.M.; Hargreaves, *J.S. Proc. Soc. Phot. Instr. Eng.* **1986**, *631*, 171-177.
7. Sheats, J.R. *Polym. Eng. Sci.* **1989**, *29*, 965-971.
8. Sheats, J.R. Proc. *SPIE* **1989**, *1086*, 406-415.
9. Sheats, J.R. *Amer. Chem. Soc. Symp. Ser.* **1989**, *412*, 332-348.
10. Vollenbroek, FA.; Nijssen, W.P.M.; Kroon, H.J.J.; Yilmaz, B.; *Microcircuit Eng.* **1985**, *3*, 245.
11. Lin, B.J. *Proc. Soc. Phot. Instr. Eng.* **1979**, *174*, 114.
12. Bartlett, K. Hillis, G.; Chen, M.; Trutna, R.; Watts, M.; *Proc. Soc. Phot. Instr. Eng.* **1983**, *394*, 49-56.
13. MicroSci, Inc., 10028 S. 51st St., Phoenix AZ, 85044.
14. Bohn, L. in *Polymer Handbook* (2nd ed.), ed. J. Brandrup and E.H. Immergut (Wiley, New York, 1975), p. III-242.
15. Feely, W.E. *Polym. Eng. Sci.* **1986**, *16*, 1101.
16. Liu, H.-Y., DeGrandpre, M.P.; Feely, W.E. *J. Vac. Sci. Technol.* **1988**, *B6*, 379.
17. Balch, E.W.; Weaver, S.E.; Saia, R.J. *Proc. Soc. Phot. Instr. Eng.* **1988**, *922*, 387-394.
18. Wang, C.H.; Xia, J.L. *Macromolecules* **1988**, *21*, 3519-3523.
19. Lee, W.A.; Rutherford, R.A. in *Polymer Handbook* (2nd ed.), ed. J. Brandrup and E.H. Immergut (Wiley, New York, 1975), p. III-150.
20. Mi, Y.; Stern, S.A. *J. Polym. Sci. B, Polym. Phys.* **1991**, *29*, 389-393.
21. Wang, Y.; Zhao, C.-L; Winnik, M.A. *J. Chem. Phys.* **1991**, *95*, 2143-2153.

RECEIVED January 21, 1993

ELECTRON-BEAM, X-RAY, AND PHOTORESISTS

Chapter 17

Advances in the Chemistry of Resists for Ionizing Radiation

Ralph Dammel

Hoechst Celanese Corporation, 500 Washington Street, Coventry, RI 02816

The lithographic techniques using high-energy radiation may be distinguished according to their use of electromagnetic (x-ray) or particle radiation (e-beam or ion beam). In x-ray lithography, a further distinction may be made depending on whether 1:1 proximity printing or reducing optics are employed; one may further differentiate between systems using synchrotron radiation, and those using e.g. laser focus sources or cathode ray tubes (Figure 1). Similarly, in e-beam lithography, one may distinguish mask-using (projection) or serial writing techniques; the latter are most conveniently further subdivided according to the market segment they serve. Each of these subfields of high-energy lithography has inherent constraints, arising from its physics, that define what resist properties and processes will be required for a successful practical implementation (1–3).

Most organic materials have first ionization energies ranging from 8.5 to 10 eV. Disregarding multi-photon ionization processes, ionizing radiation could therefore be defined as electron beams with > 10 eV kinetic energy, or electromagnetic radiation of wavelength shorter than 124 nm. However, this formalistic definition is somewhat misleading since the energies used in lithographic techniques are typically several orders of magnitude higher: electron beam energies range from 10 to 100 keV, and the radiation used in x-ray lithography lies in the wavelength range from 1.4 nm to 0.4 nm. At these energies, the molecular and electronic structure of the resist material has no influence on the absorption behavior; instead, absorption events occur primarily in the inner non-valence shells of atoms (4). Consequently, the energy deposition within a resist depends solely on its constituent atoms, and may be calculated easily from the atomic composition using atomic absorption cross sections, irrespective of the molecular structure of its components.

Cascades of secondary and tertiary electrons, initiated by the primary event, dissipate the energy of the incoming photon or electron over a more extended volume. While minor deviations caused by different cross sections towards

0097–6156/94/0537–0252$08.50/0

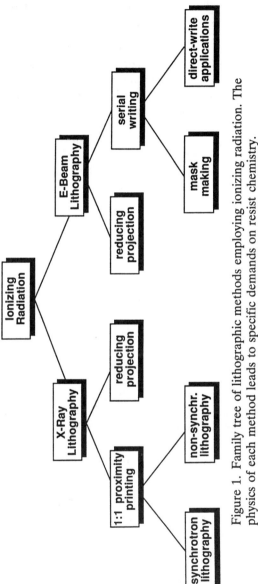

Figure 1. Family tree of lithographic methods employing ionizing radiation. The physics of each method leads to specific demands on resist chemistry.

electrons and x-rays have been reported (5), x-ray and e-beam sensitivities show an approximately linear correlation (see Figure 2). This has traditionally been interpreted to mean that the basic mechanism of energy transfer proceeds via the secondary electrons and is thus very similar for the x-ray and e-beam experiments.

The absorbed energy is thus distributed unspecifically over all atoms in the entire irradiated photoresist volume according to a stochastic physical process. In contrast, in near-UV materials, the photon energy is directly deposited in a specific absorption band of a molecule, leading to an excited state from which a definite chemical reaction occurs, such as e.g. N_2 extrusion from diazonaphthoquinones (DNQs): in other words, the energy is directly delivered to the chemical bond that is to be broken. It is therefore not surprising that the sensitivity of DNQ resists in x-ray and e-beam resists (e.g. AZ1450J in Figure 2) is much lower than for near-UV irradiation. The challenge facing photoresist chemists has therefore been to design a resist material for high energy radiation that provides a chemical mechanism for focusing the indiscriminately deposited energy into specific chemical reaction channels.

TAXONOMY OF PHOTORESIST MATERIALS FOR IONIZING RADIATION

The multitude of materials described for high-energy applications may be classified in a "taxonomy" of resists for ionizing radiation: the two main sub-kingdoms of resists, positive and negative tone, each may be subdivided into three phyla according to the mechanism of the photoinduced solubility change (Figures 3 and 4). Each phylum may be further classified according to whether the solubility change is effected directly by the incoming radiation, or whether it proceeds by the intermediacy of a photogenerated cationic agent. The distinction of chemically amplified and non-amplified resists (6) yields a subdivision of these classes into different orders; for each order, one specific resist genus (e.g. PBS) is given as an example. Actually, to keep the analogy to zoological taxonomy, an additional taxonomic level, the family, will have to be recognized; e.g. tertiary polyalkenesulfones (such as PMPS) and polyaldehyde resists both belong to the family of low-T_c unzipping polymers. In this paper, resist taxonomy has not been carried to this level.

The high sensitivity required of resists for ionizing radiation is usually achieved by some kind of chemical amplification scheme (7). In the case of polybutenesulfone (PBS), the bonds between monomer units are so weak that the chain scission process, with its concomitant decrease in molecular weight and increase in dissolution rate, occurs with great efficiency. In PBS, the polymer is, however, not so unstable that a single chain break event would lead to a large number of consecutive monomer split-offs. In the closely related poly(methylpentenesulfone) (PMPS), the ceiling temperature of the polymer is so low that one chain break can lead to an "unzipping reaction" in which the polymer thread unravels outward in both directions from the original scission site, reverting to monomers and leading to substantial volatilization of the resist.

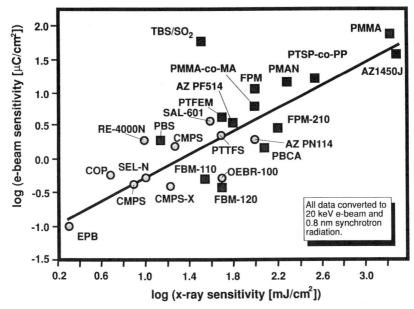

Figure 2. Double logarithmic plot of e-beam vs. x-ray sensitivity. Circles denote negative, squares positive-tone resists. For the off-diagonal point (TBS/SO$_2$) see section 4, for further references see [1]. AZ PF514: acetal-based 3-component system, AZ PN114: 3-component system based on electrophilic novolak crosslinking, AZ1450J: diazonaphthoquinone/novolak resist, CMPS, CMPS-X: chlorinated polymethylstyrene, COP: poly(glycidylmethacrylate-co-ethylmethacrylate), EPB: epoxidized polybutadiene, FBM types: poly(hexafluorobutylmethacrylate), FPM types: poly(tetrafluoroisopropylmethacrylate), OEBR-100: poly(glycidylmethacrylate), PBCA: polybutyl(cyanoacrylate), PBS: polybutenesulfone, PMAN: poly(methacrylonitrile), PMMA-co-MA: poly(methylmethacrylate-co-methacrylic anhydride), PMMA: polymethylmethacrylate, PTFEM: poly(tetrathiafulvenylstyrene), PTSP-co-PP: poly(trimethylsilylpropanal-co-3-phenylpropanal), RE-4000N: polyiodostyrene, SAL-601: 3-component system based on electrophilic novolak crosslinking, SEL-N: poly(glycidylmethacrylate-co-methylmaleate), TBS/SO$_2$: poly(t-butoxycarbonylstyrene-co-sulfur dioxide).

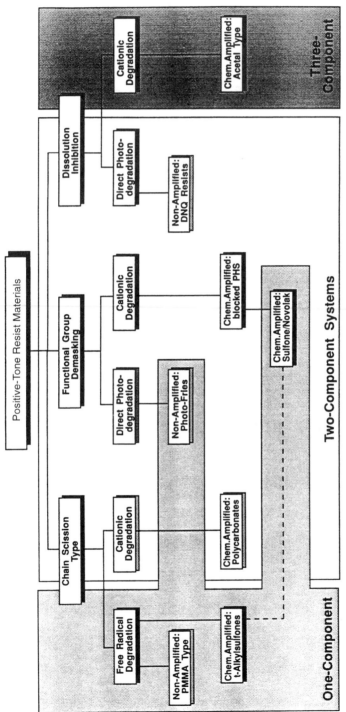

Figure 3. Taxonomy of positive-tone resists for ionizing radiation. Superimposed on the taxonomic scheme is a commonly used classification of resist materials according to the number of components. Black backdrops denote taxa which are particularly interesting for high-performance radiation resists.

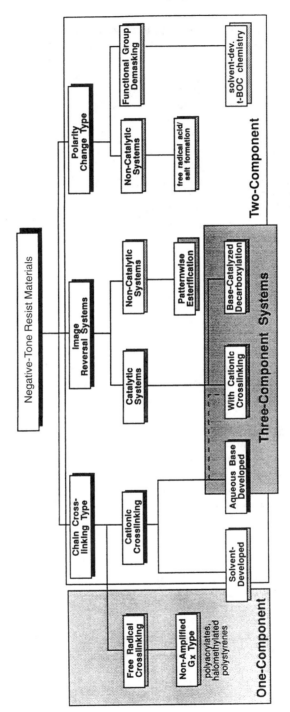

Figure 4. Taxonomy of negative-tone resists for ionizing radiation.

More complicated chemical amplification schemes are required for resists which combine etch stability and high sensitivity. In Figure 3, those classes of resist chemistries best suited to the particular demands of x-ray and e-beam lithography have been marked by a black backdrop. The rationale for this selection will be discussed in more detail in the following sections.

Traditional negative-tone radiation resists have been based on free-radical crosslinking of such polymers in which the number of chain crosslinking events (G_x) per unit of incoming energy is larger than the number of chain scission events (G_s) (8). While very high sensitivities may be achieved with such an approach (cf. Figure 2), the extensive swelling observed during solvent development in these materials has limited their use to lower resolution applications in which speed was paramount. More recently, non-swelling, aqueous base developable materials based on cationic crosslinking have been described (9–12) which exhibit such an advantageous combination of properties that they have become the materials of choice whenever a negative-tone image is desired.

X-RAY RESIST MATERIALS

As already outlined in the Introduction, every subfield of high-energy lithography requires a different set of resist properties for optimum performance. In x-ray lithography, this set of requirements has changed considerably since the first projections in the early eighties, at which time x-ray lithography was considered to be necessary for submicron lithography, e.g., the 1 Mb DRAM. This prediction was quickly invalidated partly because of improvements in conventional lithographic technology, and partly because the elements of x-ray lithography could not be brought into place rapidly enough; a projection around 1986 saw x-ray lithography coming into its own with the 16 or 64 Mbit DRAM generation. The advent of high-NA optical steppers, further refinements in DNQ photoresist chemistry and, most recently, the development of phase shift mask technology and annular illumination have further postponed the manufacturing debut of x-ray lithography. The present, "realistic" prediction sees x-ray lithography as a candidate for 256 Mbit or 1 Gbit DRAM technology in the next decade (Figure 5). The similarity of the market growth curves, with their ever lengthening flat section followed by the predicted quick upturn, with a sporting implement has lent this phenomenon its name (hockey stick effect); the hockey stick effect is not particular to x-ray technology but is quite commonly observed for advanced technologies (13). As an aside, there are signs that DUV lithography, one of the main competitors of x-ray lithography, may be beginning to suffer from similar "hockey sticking".

The shift in the production implementation of synchrotron x-ray lithography caused by the hockey stick effect has had a profound effect on the performance required of a production-ready x-ray resist: while initially, resolution requirements were set at 0.5 μm and later at 0.3 μm, one must now conclude that the resolution demonstrated under laboratory conditions must be of the order of 0.1 μm. Sensitivity and processing requirements have changed to a lesser degree; whereas initially (14), a sensitivity of <100 mJ/cm^2 was considered adequate, most users would nowadays probably prefer <50 mJ/cm^2. Only chemically

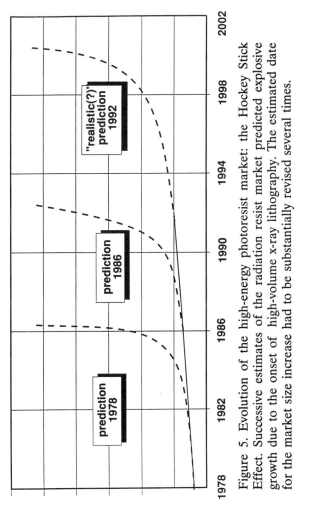

Figure 5. Evolution of the high-energy photoresist market: the Hockey Stick Effect. Successive estimates of the radiation resist market predicted explosive growth due to the onset of high-volume x-ray lithography. The estimated date for the market size increase had to be substantially revised several times.

amplified resists can currently provide the combination of such speeds with dry etch stability. In the positive-tone mode, the t-butoxycarbonyl (t-BOC)-blocked polymer (15–18) (cf. Eq. 1) and acetal resist types (19–22) (cf. Eq. 2) have been extensively studied, and may be assumed to be capable of delivering the required resolution (Figure 6).

The general concept of the t-BOC and acetal deprotection schemes may be implemented in a large variety of ways using quite different monomeric or polymeric components. A collection of such approaches is given in Scheme I for t-BOC deprotection, and in Scheme II for acetal cleavage systems. While some of the above systems have been developed for DUV applications, it is evident from their chemistry that they will be just as well suited for, e.g., X-ray applications.

As with all chemically amplified resists, a major concern is, however, the latent image stability and the susceptibility to environmental conditions. With t-BOC deprotection systems, the influence of airborne nucleophilic contaminants has been recently demonstrated (23); the observation of surface residues in a number of such materials (23, 24) may be traced back to the presence of ppb amounts of volatile bases. In the case of the acetal systems (19–21), the influence of trace bases is less pronounced, as even amine hydrochlorides are still sufficiently acidic to have some catalytic activity. Linewidth changes with the interval between exposure and post exposure bake have been observed for both the t-BOC and the acetal systems. In the case of the t-BOC systems, long intervals (several hours) between exposure and post-exposure bake will lead to a decrease of apparent sensitivity, which manifests itself as a linewidth increase, or, in extreme cases, as failure to open the imaged areas. These effects are normally due to contamination by base traces, or, in cases where the presence of even ppb amounts of bases can be excluded, may be assumed to be the result of the same, unspecified chain termination (acid annihilation) mechanism which is responsible for the containment of the calalytic reaction to the immediate vicinity of the imaged resist.

For the acetal systems, the activation energy of the catalyzed hydrolysis (Eq. 2) is sufficiently low for the reaction to slowly occur at room temperature, so that a marked dependence of dose-to-clear on the interval between exposure and development ensues. Fortunately, a simple post exposure bake is sufficient to complete the reaction in the irradiated areas, and to obliterate the post-exposure thermal history of the resist (12). The situation is somewhat different in high vacuum exposures, where the absence of an external water source may result in acid-annihilating side reactions. A mechanism for acid traps in acetal systems has been proposed (25).

An engineering solution to the delay time problem may lie in the use of automated, single-wafer processing exposure equipment, where the wafer is immediately postbaked, and possibly developed, after exposure. While such an approach seems feasible, it would represent a major paradigm shift in the industry, which typically has expected the resist chemistry to adjust to the existing equipment.

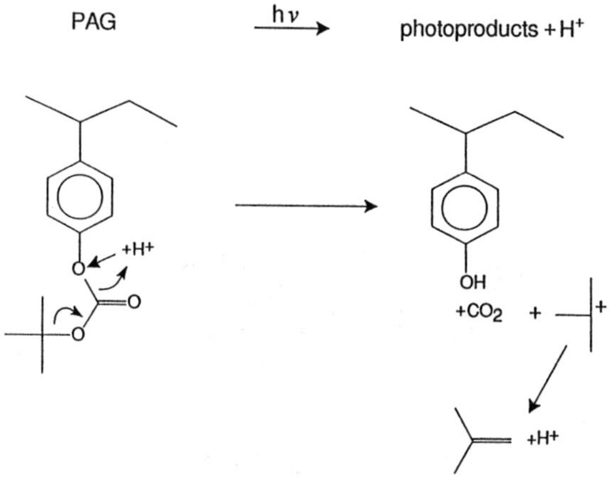

$$\text{PAG} \xrightarrow{h\nu} \text{photoproducts} + H^+$$

(1)

$+H_2O$

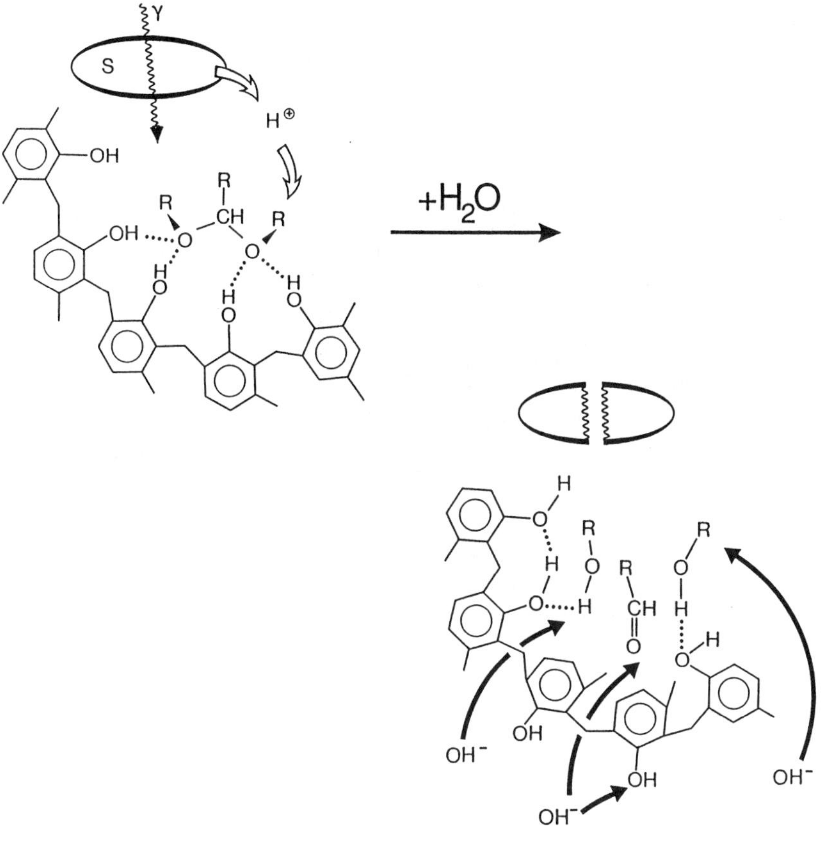

(2)

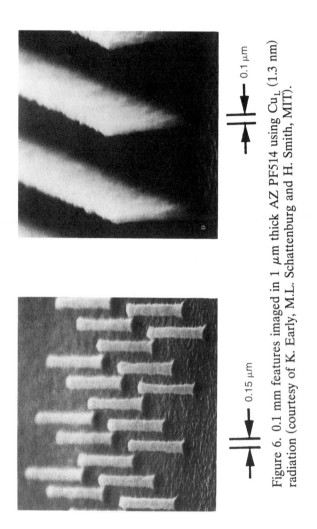

Figure 6. 0.1 mm features imaged in 1 μm thick AZ PF514 using Cu_L (1.3 nm) radiation (courtesy of K. Early, M.L. Schattenburg and H. Smith, MIT).

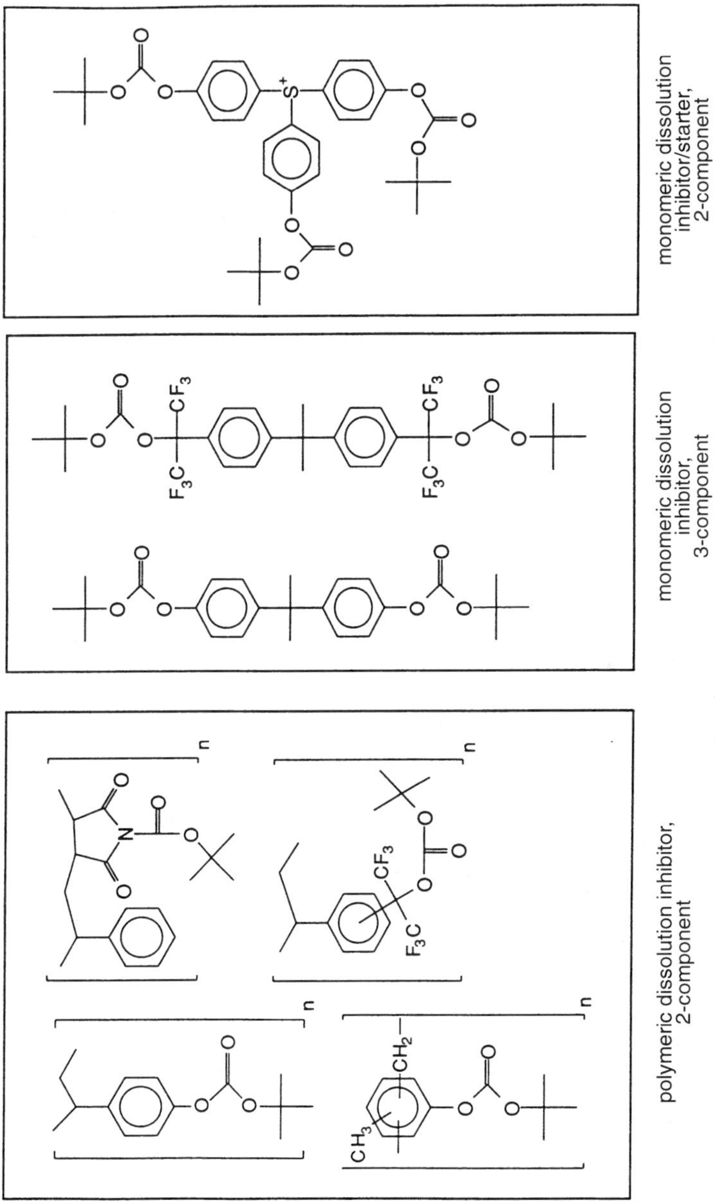

monomeric dissolution inhibitor/starter, 2-component

monomeric dissolution inhibitor, 3-component

polymeric dissolution inhibitor, 2-component

Scheme I. A non-comprehensive collection of t-BOC deprotection systems reported in the literature.

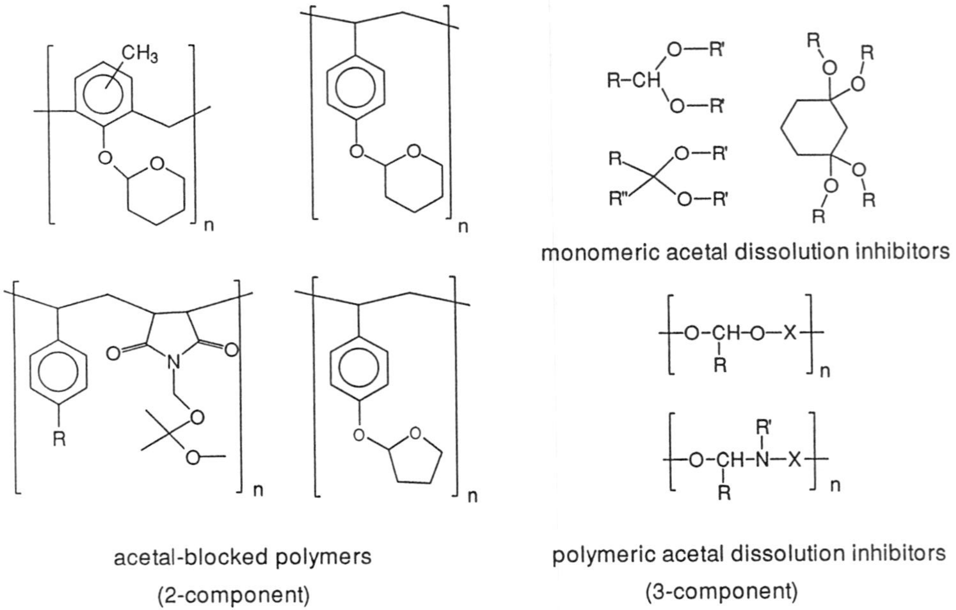

Scheme II. A non-comprehensive collection of acetal cleavage systems reported in the literature.

Another very sensitive x-ray resist based on a combination of functional group deprotection and chain scission chemistry has recently been described by workers from AT & T (*26, 27*). In a new twist on the old PBS chemistry, SO_2 may be reacted with 4-t-butoxycarbonylstyrene (TBS) to yield a copolymer the composition of which may be determined by the feed ratio and reaction temperature. The chemical reactions under x-ray or e-beam irradiation are fairly complex: in the primary step, the copolymer undergoes main chain scission (G_s 3.6 for the 2.1:1 TBS/SO_2 copolymer); the sulfinyl radicals generated at the chain ends may abstract hydrogen from a suitable donor, e.g the polymer itself, and yielding sulfinic and, by oxidation or disproportionation reactions, sulfonic acids. Alternatively, SO_2 may be eliminated from the sulfinyl radical terminals, and may react further with ubiquitous moisture to yield sulfurous acid, or possibly sulfuric acid in the presence of oxidants. The resulting benzyl radical may stabilize by loss of a hydrogen radical to yield a terminal double bond, or by hydrogen abstraction with formation of a terminal methyl group. The total G-factor for acid generation was found to be $G_{acid} \approx 1.6$, which corresponds to nearly 1 acid molecule being formed in two chain scission events. If an "unzipping mechanism" were operative to a substantial degree, one would expect a higher G_s, and, through the SO_2 formed, also G_{acid} factor.

In a second step, the acid generated catalytically deprotects the OH functionality, rendering the exposed sections base-soluble. The combination of main chain scission and catalytic deprotection leads to a high sensitivity of ca. 10 mJ/cm^2 (laser focus source at 1.4 nm). While no etch rate data were communicated in the original publication, the etch rate was found to be comparable or even superior to novolak after an initial phase of film thickness loss due to t-BOC decomposition (*27*), a finding that is corroborated by the reported plasma stability exhibited by the related aryl sulfones, despite their high G_s value (*see* Eq. 3).

Etch stability typically is not an issue with negative-tone x-ray resists. Even for one component systems, very satisfactory etch resistance may be obtained; however, even for the chloromethylated polystyrenes (*28*), arguably the most advanced such systems, resolution is severely limited due to resist swelling during solvent development. Solvent-developed negative-tone systems based on a polarity change, e.g. the t-BOC-polyhydroxystyrene resists developed in negative-tone mode (*15-18*), do not exhibit swelling but are still subject to environmental contamination and process interval constraints. The new class of aqueous base-developed negative-tone resists based on crosslinking by electrophilic substitution, which has emerged in the past 5 years (*9-12*), shows swelling-free development as well as a much lower influence of environmental contaminants and process intervals; with these systems, a delay of 24 h or more between exposure and development during which the wafer is exposed to vacuum or normal cleanroom air does not result in appreciable linewidth changes (*29*). Novolak-based systems are now commercially available (*12*) which combine high sensitivity with superior resolution and process stability (cf. Figure 7). The chemistry of these materials (cf. Eq. 4) contains a built-in acid trap from which the acid catalyst may be thermally regenerated (Eq. 5), a process which is thought to

(3)

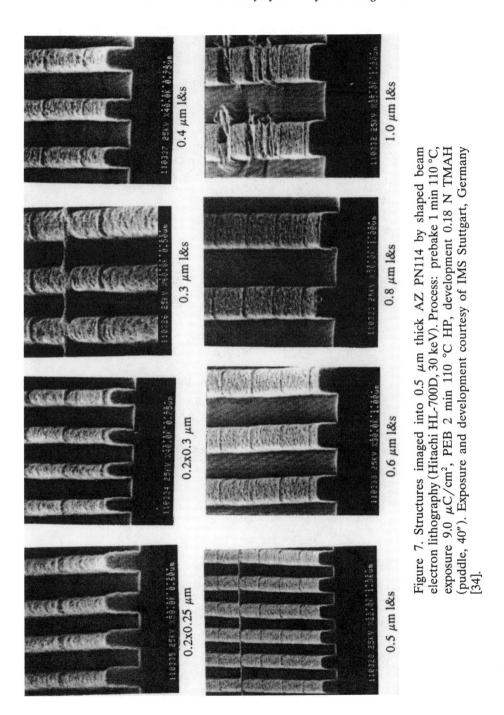

Figure 7. Structures imaged into 0.5 μm thick AZ PN114 by shaped beam electron lithography (Hitachi HL-700D, 30 keV). Process: prebake 1 min 110 °C, exposure 9.0 μC/cm², PEB 2 min 110 °C HP, development 0.18 N TMAH (puddle, 40″). Exposure and development courtesy of IMS Stuttgart, Germany [34].

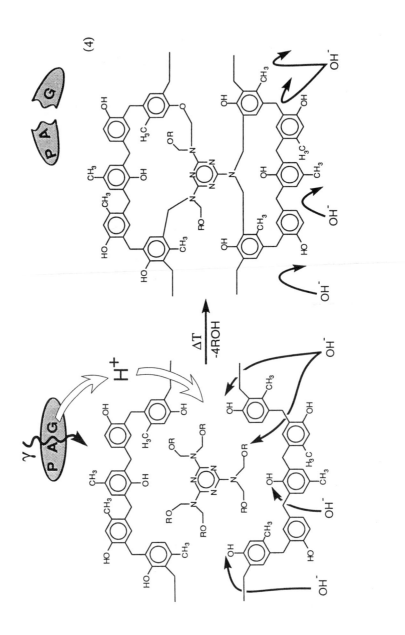

(4)

(5)

contribute substantially to the resist's latent image stability and lack of environmental susceptibility. These resist materials have come a long way towards removing a prevailing prejudice against negative-tone resists, and their superior performance has in many cases resulted in a re-orientation of mask tone for such levels in which use of negative-tone is feasible.

Requirements for resists for non-synchrotron x-ray lithography are basically the same as for synchrotron sources, with the exception that their lower luminance places even greater importance on resist sensitivity. For laser focus sources, a resist speed of ca. 10 mJ/cm^2 is close to optimum; the required sensitivity increase is largely obtained automatically due to increased absorption at the longer wavelength of the laser focus source (typically centered at 1.4 nm vs. 0.8 nm for synchrotron radiation): e.g. AZ PF514 resist which shows a sensitivity for synchrotron radiation of 35-45 mJ/cm^2 may be imaged at 10 mJ/cm^2 in a laser focus experiment (*30*). The increased absorptivity at this wavelength is still below the optimum value of 0.4 (or $\alpha = 0.92$ μm^{-1}) (*31*) even for resist materials containing heavy elements, so that no deleterious effects due to x-ray attenuation in the resist are observed.

At still longer wavelengths, however, the x-ray cross section of all organic materials rises so quickly that a single layer resist material with the approx. 1 μm film thickness required for semiconductor device patterning is no longer feasible (cf. e.g. carbon at 0.8 nm: $\approx$ 1, 1.4 nm: 3.32, 14.56 nm: 3,460 [cm^2/g]). Since projection x-ray lithography (*32*) must employ very soft x-rays in the range of 13-15 nm due to the limitations of multilayer (diffracted Bragg) x-ray mirrors, device manufacturing with this technique must therefore resort to surface imaging techniques. The nearly complete energy absorption in the top tenths of a micrometer may have the beneficial side effect of increasing resist sensitivity.

E-BEAM RESISTS

While the mechanism of secondary electron-mediated energy transfer is similar for x-ray and e-beam lithography, the different laws governing propagation of particles lead to a major limitation of e-beam lithography: while (synchrotron-generated) x-rays exhibit little scattering and attenuation in organic materials, electron beams show substantial forward scattering which causes the beam to progressively expand horizontally with its vertical passage through the resist. For low energy electrons, the penetration depth into the resist is limited; if forward scattering is minimized by using sufficiently high electron energies to move the scattering "bulb" below the resist layer, backscattering from the substrate becomes increasingly important.

For one of the two major market segments of e-beam lithography, these phenomena constitute no major limitation: in photomask making, wet etching of the mask substrates allows of the use of thin films which minimize the impact of these effects. The standard mask making resist used nowadays is PBS with a sensitivity of about 0.88 C/cm^2. PBS processing is not trivial (e.g. the moisture in the development chamber has to be tightly controlled) but good reproducibility has been obtained through extensive automation. Due to the complexity of advanced photomasks, write times may still exceed 10 h per mask even with PBS.

The advent of phase-shift technology has added dry etch stability to the requirements for e-beam resists for maskmaking in order to allow dry etching of the phase shifter materials. Wet etching cannot be very well controlled for submicron geometries, so that submicron mask technology has to resort to dry etch processes. However, the etch stability of PBS is insufficient even for the structuring of a few nanometers of chrome; while other, more etch-stable single-component (main chain scission type) resists such as EBR-9 are available, their use incurs a substantial photospeed penalty (typical EBR-9 range: 10 - 20 mC/cm^2). The recently reported cyanoacrylate resists (*33*) may offer a better compromise between photospeed and dry etch resistance. Phaseshift mask making may require the use of thicker films, compounding the problem and adding the additional difficulties associated with e-beam scattering.

Chemically amplified resists would be able to provide the required combination of high photospeed and dry etch stability. However, the susceptibility of positive-tone resist materials towards processing delays and environmental conditions has up to now severely limited the use in production: while with the single component systems, it is irrelevant whether an exposed mask blank irradiated on, say, a Friday is developed immediately thereafter, or left for the next shift on Monday, it is hard to imagine such tolerance to process delays for e.g. a t-BOC system. Other materials, such as the acetal-based resists, require water for the solubility-changing reaction (cf. Eq. 2), which is not present in a high-vacuum environment; moreover, volatile hydrolysis products of the dissolution inhibitor which act as dissolution promoters may escape from the resist layer, leading to a sensitivity reduction with long residence times in a vacuum. Recently, it has been found that a simple DUV flood exposure will reverse the sensitivity change, presumably by removing the top few nanometers which are particularly susceptible to the vacuum effects (*34*). Tests are presently underway to evaluate the feasibility of such a scheme for mask making.

A similar phenomenon occurs in the TBS copolymer system described above ((*26, 27*), cf. Eq. 3) in which a dramatic e-beam sensitivity decrease is observed: whereas the TBS system is a very sensitive x-ray resist, its e-beam speed of 90 mC/cm^2 @30 keV is at best moderate. The ratio of e-beam to x-ray sensitivity is thus very different for this resist (cf. Figure 2). One may speculate that lack of water and evaporation of SO_2 result in reduced acid formation, and hence lower sensitivity.

A clever variation of the monomeric dissolution inhibitor concept (3-component system) makes use of the base-catalyzed opening of the lactone ring in cresolphthalein which becomes possible after catalytic deprotection (*17b*). In the t-BOC protected inhibitor, the quinomethane system cannot form, and the lacton ring is not hydrolysed. The additional acidic functionality improves the dissolution rate in the exposed resist (*see* Eq. 6). The e-beam sensitivity of a 3-component system based on this inhibitor in combination with a novolak matrix and a triphenylsulfonium triflate photoacid generator was reported to be 2-3 $\mu C/cm^2$ (*17b*).

The negative-tone radiation resists based on electrophilic substitution (described above) exhibit sufficient tolerance to process delays and environmental conditions, and their use for such an application has been demonstrated (*35*).

(6)

However, their application is limited to masks with little transparent area, so that a positive-tone resist will still be required for, e.g., contact hole layer masks. Photomask makers have up to now not shown much inclination to introduce two separate resist systems into their processing lines.

E-beam direct write applications do in general not involve long residence times in a vacuum, and they require both positive- and negative-tone resists. Excellent dry etch resistance is one of the most important requirements for this application; resist speed is primarily important because of charging effects and resist heating observed with low sensitivity resists; throughput reasons while still important usually take second place. While water-soluble charge dissipation topcoats based on amine hydrochlorides have been described (*36*) which only moderately add to process complexity, they may not be compatible with chemically amplified systems. Resist heating which may become a limiting factor in high-throughput shaped-beam lithography is essentially beyond the control of chemistry, since the thermal conductivity of all resist materials is very similar, and must be addressed by changes in writing strategy.

For positive-tone direct-write work, DNQ-based resists have traditionally been used to take advantage of standard device processing, although their low e-beam sensitivity does not make them ideally suited for this application. Recently, commercially available acetal-based resist chemistry has made some inroads into this market segment; the high resist sensitivity which may be increased to nearly PBS level allows users to largely avoid charging and heating problems observed in particular with the high throughput shaped-beam systems.

In the negative-tone market segment, the commercially available aqueous-base developed resists based on nucleophilic substitution have had a tremendous impact, to the extent of virtually displacing all other negative-tone systems. However, while their performance is not limited by resist resolution, they cannot ultimately evade the constraints of e-beam physics: forward and back-scattering in the thick films (up to 2.5 m) required for ASIC manufacture limit the practical resolution in single-layer systems to about 0.7 to 1 m (at 20 keV), depending on film thickness and substrate (*36*). While some improvements may be expected with the development of more efficient (and more widely available) proximity correction algorithms, the 0.5 m design rules that ASICs are headed for in the early nineties probably cannot be achieved in all cases with shaped-beam lithography using a classical single layer resist.

A recently presented processing scheme proposes to redress the depth-dependence of e-beam energy deposition by subjecting the resist to a blanket DUV exposure after e-beam irradiation (*36, 37, 38*). While the energy deposited by the electron beam is lower at the top than at the bottom, optical exposure will yield an inverse profile, particularly at wavelengths at which the resist is little transparent. Combination of the two exposures may be carried out in such a way (e.g. by judicious choice of the blanket exposure wavelength according to resist parameters and electron energy) that a nearly depth-independent energy profile is obtained (Figure 8).

An attractive strategy which minimizes the effects of electron scattering and proximity while avoiding the complexity and particle contamination problems of multilayer systems is top-layer imaging, in which only the lithographic information in the uppermost few hundred nanometers of a resist is actually used to form a mask of a refractory oxide in a dry etch step. Top-layer resist schemes described in the literature include DESIRE (*39*) and Top-CARL (*40*) (negative-tone) as well as PRIME (*41*) and SUPER (*42*) (positive-tone) The first three of these systems use DNQ-based chemistry, and therefore show only moderate to low e-beam sensitivity (e.g. 190 C/cm^2@50 keV in the case of PRIME). The SUPER (SUbmicron Positive-tone E-beam Resist) scheme is based on crosslinking of the irradiated resist areas in a chemically amplified reaction, which causes diffusivity differences of a silylating agent (cf. Figure 9), and results in high resist sensitivity (ca. 4-10 $\mu C/cm^2$ @20 keV). The chemistry of SUPER resists is basically the same as that of the single-layer, aqueous-base developed negative-tone resists described earlier (cf. Eq. 4); actually, such commercially available resists have been used successfully in top-layer e-beam lithography (*43*), in which resolution down to 0.5 m was obtained for 1.0 m thick resist layers. t-BOC resists may be very successfully used in silylation-type dry development, yielding a negative-tone image. Although very similar in mechanism, t-BOC-PHS strictly speaking cannot be addressed as a top-layer imaging resist since the silylating agent permeates the entire resist depth in the exposed areas (*44*). An analogous phenomenon which was observed for SUPER-type resists at high silylation temperatures has been linked to the viscoelastic properties of phenolic polymers which allow self-diffusion of the silylated polymer chains (*43*).

Further improvements in the performance of SUPER systems will require optimization of all resist components for a top-layer imaging application. Since no aqueous development is involved, it may even be feasible to make the resist conductive by suitable additives to avoid charging effects. While such optimization work does not seem prohibitively difficult, it appears that no photoresist vendor has until now undertaken this task, presumably because of the small present market size.

INFLUENCE OF RESIST PERFORMANCE ON LITHOGRAPHIC EVOLUTION

The above observation is not an uncommon one: photoresist vendors find it increasingly difficult to justify assigning resources to low-volume high-technology areas which for them may essentially be niche markets. While profit margins may be good in a niche, the return on investment (ROI) for any substantial research investment will tend to be low due to the small market size. Conventional business sense therefore requires that limited resources will be directed to the larger mainstream business. This may lead to a lopsided situation in which large investments in an installed tool base and high tool development expenses are not properly balanced by an appropriate investment into the photoresist chemistry. To some extent, such a situation is already in effect today for high-energy photoresists: take e.g. the investment in x-ray lithographic "hardware", with worldwide over 20 dedicated storage rings, which assuming a very conservative cost of ca. $30 million per storage ring may well be of the order of over half a billion dollars for the sources alone; adding the development cost and

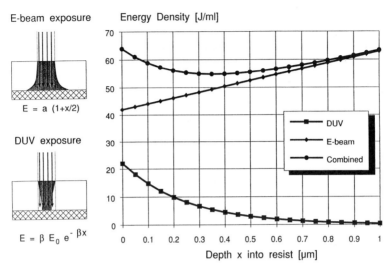

Figure 8. Schematic representation of the depth dependence of energy deposition of the double exposure process. Neglecting bleaching, the energy density deposited is described by $\varepsilon = \beta E_0\, 10^{-\beta x} + a(1 + x/2)$, where β is the optical absorption coefficient, E_0 the incident optical energy per surface unit, x the depth into the resist, and **a** a coefficient proportional to the incident electron dose Q.

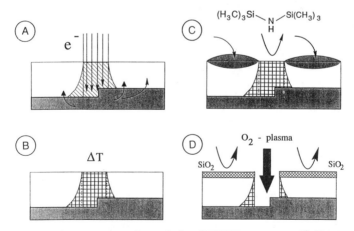

Figure 9. Schematic representation of the SUPER process: A) Exposure and acid catalyst generation. Forward and backscattering during exposure lead to degradation of the image with increasing resist depth; B) acid-catalyzed thermal crosslinking step; C) gas-phase silylation. Si is only incorporated into unexposed parts of the pattern; D) O_2 plasma dry etching. Si in top layer acts as etch barrier.

capital of e-beam tools, laser focus sources and dedicated x-ray steppers, as well as of their lithographic environment, easily brings the total to over $2 billion. The corresponding worldwide investment in chemical research for high-energy resists may be estimated at less than 1% of that amount—if these were all production facilities operating at capacity, a better photoresist which results in a modest 4% yield improvement would pay back all research costs in less than a year (45). For the introduction of new lithographic processes and technologies, the economic benefits derived from improved photoresist chemistry are harder to estimate but may very well be even larger.

If photoresist research is so cost-effective, why is it that the ratio of tooling costs to chemistry research is so lopsided? One of the answers may be that in the corporate culture of most semiconductor manufacturers, it is easier to sell a project calling for $4 million for a new lithography tool than one asking for $4 million for resist chemistry; perhaps in part because chemical research does not lend itself as easily to planning as do engineering projects, or because its results are less visible and tangible. The main reason is, however, that in the case of semiconductor manufacturers and resist vendors, benefits and costs occur on opposite sides of the business relationship, and customers would be unwilling to pay the outrageous prices photoresist vendors would have to charge to recoup high research investments into accelerated resist improvement, and doubly so for low-volume high-tech areas. A partial answer is provided by the formation of alliances between photoresist and semiconductor manufacturers; such alliances occur naturally in the Japanese 'keiretsu' (families of companies) but have been unknown until recently in Western economies.

Borrowing a leaf from business management techniques, it is possible to evaluate lithographic hardware and 'wetware' (resist) performance in a sort of "portfolio analysis" of lithographic technologies (Figure 10). Rating resist and tool (stepper, source, mask etc.) performance on a scale from "inadequate" to "fully adequate", four different regions may be distinguished: on one end the mature technologies, exemplified by i-line and g-line lithography, in which both tools and resists are fully in place for manufacturing; on the other end the experimental technologies in which neither can measure up to the requirements of a production environment.

Moving up along the tool readiness axis, one enters the region of "resist-limited" technologies. A typical example is DUV lithography, for which light source, mask and other hardware components of the lithographic process are more or less in place but which is hampered by the unavailability of a suitable, commercially available positive-tone resist material. Due to the large, worldwide effort aimed at producing a viable DUV resist, it may be expected that such materials will soon be available. IBM, where an in-house, t-BOC based resist has been moved out of the experimental stage and into a production environment, may be considered the forerunner of this development.

The opposite end of the spectrum is completed by the "tool-limited" methods, for which adequate resist materials are available but where one or more hardware components still have not been developed to the level required for their use in a production environment. X-ray lithography may be considered a tool-limited method: whereas both synchrotron and non-synchrotron x-ray

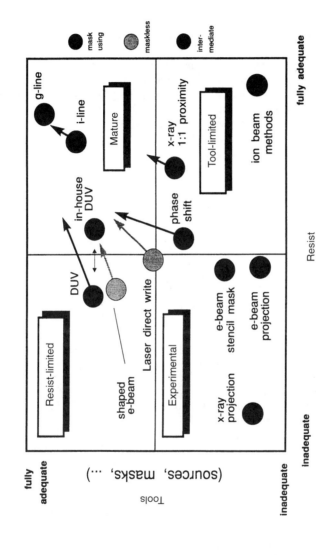

Figure 10. Portfolio analysis of lithographic technologies based on commercial availability. The arrows denote approximate direction and speed of evolution for the various methods.

sources are available on the market, and the resist problem has been considered essentially solved (*46*), the extraordinary difficulties involved in manufacturing error-free thin-membrane 1:1 x-ray masks at the quarter micron level and below must at present be accounted the Achilles heel of this technology.

Where do we go from here? It is probably safe to say that while e-beam technology both for ASIC and mask making will see a stable evolution, the future of mass-production methods is less certain. Will the prevailing theme of the last decade, the extension of conventional near-UV technology to ever higher performance, continue until the end of the millenium? Even as the hockey sticks in the corporate planning departments grow longer, the crystal balls remain cloudy: it looks like all three competing lithographic technologies (phase shift i-line/annular illumination, DUV, and x-ray) will be able to produce 256 Mbit devices - any decision will not be based on feasibility but on cost-effectiveness. Can one of the contenders win the race outright, or shall we be entering a phase in which several technologies coexist, until one (x-ray lithography?) ultimately prevails? The answer to these questions will in part be decided by photoresist chemistry.

REFERENCES

1. For previous reviews of resists for ionizing radiation, cf. J. Lingnau, R. Dammel, J. Theis, Solid State Technology Sept. **1989**, 105-112: Oct. **1989**, 107-111, as well as [2].
2. A.A. Lamola, C.R. Szmanda, and J.W. Thackeray, Solid State Technol. August **1991**, 53-60; O. Nalamasu, M. Cheng, A.G. Timko, V. Pol, E. Reichmanis, and L.F. Thompson, J. Photopolym. Sci. Technol. **4**, 299 (1991).
3. E. Reichmanis and J.H. O'Donnel (eds.), *The Effects of Radiation on High-Technology Polymers*, ACS Symp. Ser. 381, ACS, Washington, D.C., 1989
4. Cf. e.g. Farhataziz, M.A.J. Roberts (eds.), *Radiation Chemistry*, VCH Publ., New York 1987.
5. Cf. e.g. J.O. Choi et al., J. Vac. Sci. Technol. **B6**, 2286 (1988).
6. Chemical amplification, a term coined by C.G. Willson, J.M.J. Frechet, and H. Ito denotes the multiplication of a primary radiation event by a series of chemical reactions. Typically, an acidic catalyst is generated in the photostep which in a subsequent dark reaction changes the solubility characteristics of a large number of acid-reactive sites, thus amplifying the original photoevent trough the catalytic process. The terms seems to have been used for the first time in ref. [7.
7. C.G. Willson, H. Ito, J.M.J. Frechet, F Houlihan, Proc. 28th IUPAC Macromol. Symp., Oxford, UK.
8. G_s and G_x are usually defined as the number of chain scission events per 100 eV of incident energy.
9. R. Dammel, K.F. Dössel, J. Lingnau, J. Theis, H.L. Huber, H. Oertel, and J. Trube, Microelectronic Engineering *9*, 575-578 (1988): cf. also [12].
10. Cf. e.g. J.W. Thackeray, G. Orsula, D. Canistro, and A.K. Berry, J. Photopolym. Sci. Tech. **2**(3), 129 (1989); L. Blum, M.E. Perkins, and H.Y. Liu, J. Vac. Sci. Technol. **B6**, 2280 (1988), and literature quoted therein; cf. also [2].

11. For other electrophilic substitution negative-tone chemistry, cf. J.M.J. Frechet, S. Matuszcak, S.M. Lee, J. Fahey, and C.G. Willson, Proc. SPE Conf. Photopolym. (Ellenville) 1991, 31-39; H.D.H. Stöver, S. Matuszcak, C.G. Willson, and J.M.J. Frechet, Macromolecules 24, 1741-45, 1746-54 (1991); R. Sooriyakumaran, H. Ito, and E.A. Mash, Proc. SPIE 1466, 419-428.
12. Commercial negative-tone high-energy resist materials based on electrophilic crosslinking (in alphabetical order): Hoechst AZ PN100 series, Shipley SAL 600 series.
13. One might formulate the hockeystick rule as follows: "If a market size prediction for an advanced technological field is made, the result will be that while the market size at evaluation time is close to zero, it is predicted to increase exponentially shortly after the evaluation date. This effect is independent of the evaluation date and the number of previous erroneous evaluations."
14. A. Wilson, Proc. SPIE 537, 85-101 (1985).
15. H. Ito and C.G. Willson, Polym. Eng. Sci. 23, 1012-1018 (1983); H. Ito, C.G. Willson and J.M. Frechet, US Patent 4,491,628 (1985); J. G. Maltabes, S.J. Holmes, J.R. Morrow, R.L. Barr, M. Hakey, G. Reynolds, W.R. Brunsvold, C.G. Willson, N.J. Clecak, S.A. MacDonald, and H. Ito, Proc. SPIE 1262 (1990), and references quoted therein; cf. also R. Schwalm, Polym. Mat. Eng. Sci. 61, 278 (1989). For the effect of polymer end groups, cf. H. Ito, W. P. England, and S.B. Lundmark, Proc. SPIE 1672, 2 (1992).
16. C.E. Osuch, K. Brahim, F.R. Hopf, M.J. MacFarland, A. Mooring, and C.J. Wu, Proc. SPIE 631, 68-75 (1986): S. Chatterjee, S. Jain, P.H. Lu, R.E. Potvin, and D.N. Khanna, Proc. SPE Conf. Photopolym. (Ellenville) 1991, 239-253.
17. a) K.J. Przybilla, R. Dammel, G. Pawlowski, H. Röschert, and W. Spiess, Proc. SPE Conf. Photopolym. (Ellenville) 1991, 131-144; .K.J. Przybilla, H. Röschert, W. Spiess, C. Eckes, S. Chatterjee, D.N. Khanna, G. Pawlowski, and R. Dammel, Proc. SPIE 1466, 161-173 (1991); K.J. Przybilla, H. Röschert, and G. Pawlowski, Proc. SPIE 1672, 500 (1992). b) H. Koyanagi, S. Umeda, S. Funukaga, T. Kitaori, K. Nagasawa, Proc. SPIE 1672, 125 (1992).
18. I.S. Daraktchiev, D. Goossens, P. Matthijs, M. Thirsk, A. Blakeney, O. Nalamasu and M. Cheng, Proc. SPIE 1672, 553 (1992); O. Nalamasu, E. Reichmanis, J.E. Hanson, R.S. Kanga, L.A. Hainbrook, A.B. Emerson, F.A. Baiocchi, and S. Vaidya, Polym. Eng. Sci. 32, 1571 (1992); O. Nalamasu, E. Reichmanis, M. Cheng, V. Pol, J.M. Kometani, F.M. Houlihan, T Neenan, M.P. Bohrer, D.A. Mixon, L.F. Thompson, and C. Takemoto, Proc. SPIE 1466, 12 (1991); O. Nalamasu, M.Cheng, J.M. Kometani, S. Vaidya, E. Reichmanis, and L.F. Thompson, Proc. SPIE 1262, 32 (1990); T.X. Neenan, F.M. Houlihan, E. Reichmanis, J.M. Kometani, B.J. Bachman, and L.F. Thompson, Proc. Spie 1086, 1 (1989); R.G. Tarascon, E. Reichmanis, F.M. Houlihan, A. Shugard, and L.F. Thompson, Polym. Eng. Sci. 29, 850 (1989); F.M. Houlihan, E. Recihmanis, R.G. Tarascon, G. Taylor, and L.F. Thompson, Macromolecules 22, 2999 (1989).
19. a) K.F. Dössel, H. Huber, and H. Oertel, Microelectronic Engng. 5, 97 (1986); K.F .Dössel, EPA0,312,751 (1986); b) L. Schlegel, T. Ueno, H.

Shirashi, N. Hayashi, and T. Iwayanagi, Chem. Mater. **2**, 299 (1990); J. Vac. Sci. Technol. **B9** (2), 278 (1991).

20. R.-U. Ballhorn, R. Dammel, H.H. David, Ch. Eckes, A. Fricke-Damm, K. Kreuer, G. Pawlowski, and K. Przybilla, Microelectronics Eng. *13*, 73-78 (1991), and references quoted.

21. R. Dammel, C.R. Lindley, G. Pawlowski, U. Scheunemann, and J. Theis, Proc. SPIE **1262**, 378-390 (1990).

22. The concept of acetal protective groups for imaging applications seems to have originated with a group of researchers at 3M Corp.: cf., e.g., G.H. Smith, J.A. Bonham,, US 3,779,778; US 224,818 (1972).

23. S.A. MacDonald, N.J. Clecak, H.R. Wendt, C.G. Willson, C.D. Snyder, C.J. Knors, N. B. Deyoe, J.G. Maltabes, J.R. Morrow, A.E. MacGuire, and S.J. Holmes, Proc. SPIE **1466**, 2-7 (1991); W.D. Hinsberg, S.A. MacDonald, N.J. Clecak, and C.D. Snyder, Proc. SPIE **1672**, 24 (1992).

24. O. Nalamasu, E. Reichmanis, M. Cheng, V. Pol, J.M. Kometani, F.M. Houlihan, T.X. Neenan, M.P. Bohrer, D.A. Mixon, L.F. Thompson, and C. Takemoto, Proc. SPIE **1466**, 13-25, and references quoted.

25. J. Lingnau, R. Dammel, C.R. Lindley, G. Pawlowski, U. Scheunemann, and J. Theis, in: Y. Tabata, I. Mita, S. Nonogaki, K. Horie, and S. Tagawa (eds.), in: *Polymers for Microelectronics—Science and Technology*, pp.445-462, Verlag Chemie, Weinheim/Bergstr., 1990.

26. A.E. Novembre, W.W. Tai, J.M. Kometani, J.E. Hanson, O. Nalamasu, G.N. Taylor, E. Reichmanis, and L.F. Thompson, Proc. SPIE **1466**, 89-99 (1991).

27. O. Nalamasu, M. Cheng, J.M. Kometani, S. Vaidya, E. Reichmanis, and L.F. Thompson, Proc. SPIE **1262**, 32-49 (1990).

28. N. Yoshioka, Y. Suzuki, and T. Yamazaki, Proc. SPIE **537**, 51 (1987); N. Yashioka, Y. Suzuki, N. Ishio, and T. Yamasaki, J. Vac. Sci. Technol. **B5**, 546 (1967)..

29. M. Padmanaban, H. Endo, Y. Inoguchi, Y. Kinoshita, T. Kudo, S. Masuda, and Y. Nakajima, Proc. SPIE **1672**, 141 (1992).

30. This value was obtained on a Hampshire Instruments laser focus stepper.

31. A. Neureuther and C.G. Willson, J. Vac. Sci. Technol. **B6**, 167 (1988).

32. D.M Tennant, J.E. Bjorkholm, R.M. D'Souza, L. Eichner, R.R. Freeman, T.E. Jewell, A.A. MacDowell, W.M. Mansfield, J.Z. Pastalan, L.H. Szeto, W.K. Waskiewicz, D.L. White, D.L. Windt, and O.R Wood,II, J. Vac. Sci. Technol. **B9** (1991), in press; cf. also Solid State Technol., July **1991**, 37-42; Proc. SPE Conf. Photopolym. (Ellenville) **1991**, 255-257, and references quoted therein.

33. A. Tamura and M. Sato, Proc. SPIE **1465**, 35 (1991).

34. Ch. Eckes, G. Pawlowski, K. Przybilla, W. Meier, M. Madore, and R. Dammel, Proc. SPIE **1466**, 394-407 (1991).

35. Ch. Ehrlich, R. Demmeler, U. Goepel, S. Pongratz, K. Reimer, R. Dammel, J. Lignau and J. Theis, Proc. Microprocess Conf. 1989 (Kobe); cf. also ref. [21] and S. Pongratz, R. Demmeler, C. Ehrlich, K. Kohlmann, K. Reimer, R. Dammel, W. Hessemer, J. Lingnau, U. Scheunemann, and J. Theis, Proc. SPIE **1089**, 303 (1989).

36. This 'water soluble conductive layer' (WSCL) technology has been pioneered by Hitachi.

37. O. Suga, et al., J. Vac. Sci. Technol. **B6** (1), 1988.
38. F. Lalanne, A. Weill, G. Amblard, and J.P. Panabiere, Proc. Microcircuit Engng. Conference, Leuven 1990; Proc. SPIE **1465** - 34 (1991).
39. F. Coopmans and B. Roland, Proc. SPIE **631**, 34 (1986); F. Coopmans and B. Roland, Solid State Technology, 93 (June 1987); R.J. Visser, J.D.W. Schellekens, M.E. Reuhman-Huiskens and L.J. Ijzendoorn, Proc. SPIE **771**, 110 (1987); D. Nichols, A.M. Goethals, P.DeGeyter and L. Van den Hove, Microelectronic Engng. **11**, 515 (1990); M. Tipton, C. Garza and T. Seha, Proc. SPIE **1086**, 416 (1989); C.M. Garza, Proc. SPIE **920**, 223 (1987); B. Roland, R. Lombaerts, C. Jacus, and F. Coopmans, Proc. SPIE **771**, 69 (1987); F. Vinet, M. Chevalier, J. Guibert, and C. Pierrat, Proc. SPIE 1086, 433 (1989); C.M. Garza, G. Misium, R. Doering, B. Roland and R. Lombaerts, Proc. SPIE **1086**, 229 (1989).
40. R. Sezi, R. Leuschner, M. Sebald, H. Ahne, S. Birkle, and H. Börndorfer, Microelectronic Engng. **11**, 535 (1990); R. Sezi, M. Sebald, R. Leuschner, H. Ahne, S. Birkle, and H. Börndorfer, Proc. SPIE **1262**, 84 (1990).
41. C. Pierrat, S. Tedesco, F. Vinet, T. Mourier, M. Lerme, B. Dal'Zotto, and J.C. Guibert, Microelectronic Engng. **11**, 507 (1990)
42. J.P.W. Schellekens and R.J. Visser, Proc. SPIE **1086**, 220 (1989), and literature quoted.
43. T.G. Vachette, P.J. Paniez, and M. Madore, Proc. SPIE **1262**, 258-272 (1990)
44. S.A. MacDonald, H. Schlosser, H. Ito, N.J. Clecak, and C.G. Willson, Chem. Mater. 3, 435-442 (1991); S.A. MacDonald, H. Schlosser, R.D. Allen, R.J. Twieg, N.J. Clecak, and C.G. Willson, Proc. SPE Conf. Photopolym. (Ellenville) **1991**, 235.
45. A corresponding treatment of the presently used g- and i-line technology comes up with comparable numbers.
46. A. Heuberger, Proc. SPIE **1089**, 140 (1989).

RECEIVED May 17, 1993

Chapter 18

Out-of-Plane Expansion Measurements in Polyimide Films

Michael T. Pottiger and John C. Coburn

Experimental Station, DuPont Electronics, P.O. Box 80336, Wilmington, DE 19880–0336

Out-of-plane linear CTEs (α_z) were calculated from the difference between the volumetric CTE and the sum of the in-plane linear CTEs (α_x and α_y). Volumetric CTEs were obtained from a pressure-volume-temperature (PVT) technique based on Bridgeman bellows. Although the linear CTEs vary significantly with processing, the volumetric CTE is essentially constant, independent of molecular orientation. For all of the polyimide films studied, the out-of-plane linear CTEs (α_z) were higher than the in-plane linear CTEs (α_x and α_y).

Polyimide films are used in a variety of interconnect and packaging applications including passivation layers and stress buffers on integrated circuits and inter-layer dielectrics in high density thin film interconnects on multi-chip modules and in flexible printed circuit boards. Performance differences between polyimides are often discussed solely in terms of differences in chemistry, without reference to the anisotropic nature of these films. Many of the polyimide properties important to the microelectronics industry are influenced not only by the polymer chemistry but also by the orientation and structure. Properties such as the linear coefficient of thermal expansion (CTE), dielectric constant, modulus, strength, elongation, stress and thermal conductivity are affected by molecular orientation. To a lesser extent, these properties as well as properties such as density and volumetric CTE are also influenced by crystallinity (molecular ordering).

A typical microelectronics device construction consists of multiple layers of different materials, e.g., metals, ceramics and polymers, in contact with one another. The materials are exposed to repeated thermal cycling during device fabrication leading to the development of thermally induced stresses. The thermally induced stresses result from the mismatch in linear CTEs between the various materials. Polyimides with low in-plane linear CTEs, such as those based on BPDA/PPD, were developed to address this problem. The low in-plane linear CTE of this polymer arises from the high degree of planar molecular orientation.

0097–6156/94/0537–0282$06.00/0

While considerable attention has been focused on the relationship of stress to in-plane properties such as the linear CTE and Young's modulus, there is little data on the relationship of stress to out-of-plane properties, in part due to the difficulty of measuring these properties. A high out-of-plane linear CTE of the polymer can result in a high CTE mismatch leading to interlayer delamination and/or cracking. For example, cracking of copper plated-through holes during thermal cycling of printed circuit boards is related to the mismatch between the out-of-plane linear CTEs of the dielectric and the copper (*1–2*).

Processing plays a critical role in affecting molecular orientation and structure and in turn the properties of the resulting film. Molecular orientation in spin coated polyimides develops from a competition between the planar conformation induced by the volume collapse during processing and the tendency of the polymer molecules to assume a random equilibrium conformation (*3–16*). During processing, the evaporation of solvent and/or reaction by-products leads to shrinkage forces. A spin coated polyimide film is constrained in the plane by the substrate, therefore, the bulk of the shrinkage occurs in the thickness direction. The residual stresses that develop during this volume collapse induce anisotropy in the film resulting in in-plane molecular orientation.

The development of in-plane orientation is complicated by the conversion of the relatively flexible and soluble poly(amic acid) precursor into the relatively inflexible and insoluble polyimide. During the conversion, solvent decomplexes from the poly(amic acid) and subsequently evaporates, and water is released as a reaction by-product (*17–20*). In addition, the molecular weight of the poly(amic acid) is believed to initially decrease, and then slowly build during the later stages of conversion (*21*). The loss of solvent, the conversion of the relatively flexible and soluble poly(amic acid) precursor into the relatively inflexible and insoluble polyimide, and the increase in molecular weight during cure severely restricts the ability of the molecule to relax to its equilibrium conformation.

The formation of the relatively inflexible polyimide and the molecular ordering that develops during processing, lock in the orientation induced by the volume collapse during solvent evaporation. The degree of molecular ordering developed is affected by the heating rate during cure and the final cure temperature, relative to the glass transition temperature of the film (*11*). The effect of heating rate during cure on the in-plane orientation and molecular ordering can be explained in terms of an effective glass transition of the film. For slow heating rates, the effective glass transition temperature increases faster than the film temperature. Molecular mobility is restricted and significant relaxation does not occur. In contrast, during rapid heating, the film temperature increases faster than the effective glass transition temperature. Above the effective glass transition, significant molecular relaxation can occur. Increasing the heating rate may also increase the depolymerization reaction (*21*) leading to lower molecular weight and increased molecular mobility. The increased mobility as a result of rapid heating leads to a loss of molecular orientation and an increase in crystalline order as evidenced by both x-ray and dynamic mechanical data. Lower in-plane orientation results in a higher in-plane linear CTE, leading to an increase in the in-plane residual stress (*11, 14, 22–23*).

The degree of molecular orientation in many polyimides has been shown to be a function of film thickness (*10–16*). The decrease in orientation with increasing film thickness observed in PMDA-ODA and BPDA-PPD polyimides has been attributed to the rate of solvent loss during the curing process (*3–4, 16*). A skin is believed to form at the free surface of the film during cure (*5–6*). The skin slows the evaporation rate of the solvent from the film increasing the time available for relaxation, which leads to a loss in orientation. With increasing film thickness, increasing time, due to longer solvent resident time in the curing film, is available for relaxation to occur.

The increase in birefringence with increasing final cure temperature observed in PMDA-ODA and BPDA-PPD films cured using a slow heating rate (2 °C/min) is attributed to increased in-plane orientation and subsequently to ordering at final cure temperatures above the glass transition of the fully cured film. The slow heating rate during cure assures that the curing process takes place predominantly in the glassy (vitreous) state. Molecular mobility is limited in the glassy state and the molecular orientation present in the film cannot relax. Curing above the glass transition provides sufficient mobility to allow for molecular ordering.

The influence of processing, molecular orientation and molecular ordering on the linear CTEs (α_x, α_y and α_z) and volumetric CTE (β) will be discussed. The determination of the out-of-plane linear CTEs (α_z) of PMDA-ODA and BPDA-PPD films prepared by two different processes (spin coating and casting) will be described. A comparison of our results with the findings of others will also be presented.

EXPERIMENTAL

Materials

Free films of poly(4,4'-oxydianiline-pyromellitic dianhydride) (PMDA-ODA) and poly(p-phenylenediamine-3,3',4,4'-biphenyltetracarboxylic dianhydride) (BPDA-PPD) were prepared from Pyralin PI-2540 and PI-2611 poly(amic acid) in NMP solutions, respectively, by spin coating the precursor solutions onto 5 inch silicon wafers containing a 1000 Å thermally grown oxide layer. The films were dried in a VWR Clean Room convection oven at 135 °C in air for 30 minutes. The dried films were cured in a Blue M AGC-160F programmable oven under nitrogen purge by heating the wafer at 2 °C/minute to 200 °C, holding at 200 °C for 30 minutes, heating at 2 °C/minute to the final cure temperature and holding for one hour. The fully cured films were removed from the wafers by dissolving the oxide layer in a 6:1 buffered HF solution. 50 μm Kapton HN (PMDA-ODA) and 25 μm Upilex S (BPDA-PPD) films prepared by commercial casting processes were obtained from DuPont and Ube, respectively.

Measurement Techniques

The in-plane linear CTEs (α_x and α_y) were measured on fully cured films using a Perkin-Elmer TMA-7 thermomechanical analyzer. A constant force of

0.030 N was applied to film samples 15 mm long and 2 mm wide. The samples were initially heated at 20 °C/min to 220 °C in the TMA and held for 60 minutes to erase previous thermal history. The samples were then cooled to 20 °C at a rate of 20 °C/min and held for 10 minutes. The samples were reheated at 5 °C/min to 220 °C. An average linear CTE was determined between 50 and 200 °C from the second heat cycle using the following equation:

$$\bar{\alpha} = \frac{1}{L_0}\left(\frac{\Delta L}{\Delta T}\right) \tag{1}$$

where L_0 is the initial length of the sample between the grips and ΔL is the change in the length of the specimen over the temperature interval ΔT. Spin coated polyimide films are isotropic in the plane of the film, therefore the in-plane linear CTEs are identical, i.e. $\alpha_x = \alpha_y = \alpha_{\text{in-plane}}$. In contrast, the commercial films, prepared by a casting process, exhibit some in-plane anisotropy, i.e. $\alpha_x \neq \alpha_y$. Therefore, two in-plane CTE measurements are needed to describe the in-plane CTE behavior. For Kapton HN and Upilex S, α_x and α_y were measured parallel and normal to the optical axis, respectively.

The pressure-volume-temperature (PVT) data were acquired on predried (to remove moisture) film samples using a Gnomix Research PVT Apparatus in isothermal mode (*24–26*). The PVT apparatus consists of a sample cell with a flexible bellows on one end containing approximately 1 gram of film and mercury as a confining fluid. The cell is placed inside a pressure vessel. The deflection of the bellows as a result of temperature and/or pressure changes is measured by a linear variable differential transducer (LVDT) located outside the pressure vessel. The deflections are converted to volume changes of the sample using the known PVT properties of mercury. The accuracy of the PVT apparatus is ± 0.002 cm^3/g up to 250 °C and ± 0.004 cm^3/g at higher temperatures, with a sensitivity of better than 0.0005 cm^3/g. The data are acquired at constant temperature and at pressures ranging from 10 MPa up to 200 MPa in 10 MPa increments. Once a measurement has been made at a pressure of 200 MPa, the pressure is reduced back to 10 MPa and the temperature is increased by 10 °C prior to the next series of measurements. This cycle is repeated until a final temperature of approximately 400 °C is reached. The data at 0 MPa is found by extrapolation using the Tait equation. The film density at room temperature and atmospheric pressure was obtained using a Micromeritics autopycnometer.

DATA ANALYSIS

The PVT data is fitted to the Tait equation, which describes the volume dependence along isotherms as follows (*25*):

$$V(P,T) = V(0,T)\left(1 - 0.0894\ln\left[1 + \frac{P}{B(T)}\right]\right) \tag{2}$$

where B(T) is the temperature dependent Tait parameter, often given by

$$B(T) = B_1 \exp(-B_2 T) \tag{3}$$

and V(0,T) is the temperature dependent volume at zero pressure. The expression for the compressibility $\kappa(P,T)$ is

$$\kappa(P,T) \equiv -\frac{1}{V}\left(\frac{dV}{dP}\right)_T$$

$$= \left([P + B(T)]\left\{\frac{1}{0.0894} - \ln\left[1 + \frac{P}{B(T)}\right]\right\}\right)^{-1} \tag{4}$$

and the expression for the volumetric coefficient of thermal expansion $\beta(P,T)$ is

$$\beta(P,T) \equiv \frac{1}{V}\left(\frac{dV}{dT}\right)_P$$

$$= \beta_0 - PB_2\kappa(P,T) \tag{5}$$

where β_0 is the zero pressure thermal expansivity, taken from the expression for V(0,T). An average volumetric CTE was determined between 50 and 200 °C from the expression for V(0,T) as follows:

$$\bar{\beta} = \frac{1}{V_0}\left(\frac{\Delta V}{\Delta T}\right) \tag{6}$$

where V_0 is the volume of the sample at 25 °C and ΔV is the change in the volume of the specimen over the temperature interval ΔT.

The relationship between the linear CTEs and the volumetric CTE is given by

$$\beta = \frac{1}{V}\left(\frac{dV}{dT}\right)_P = \alpha_x + \alpha_y + \alpha_z \tag{7}$$

The out-of-plane linear CTE (α_z) is calculated by subtracting the sum of the in-plane linear CTEs from the volumetric CTE.

RESULTS

Isobaric specific volume versus temperature curves are shown in Figure 1 for Kapton HN, PI-2540, Upilex S and PI-2611. The temperature dependent volume at zero pressure V(0,T) was obtained from an extrapolation of the Tait equation fit to the 10 to 50 MPa data. The V(0,T) data were fit with a second order polynomial. The coefficients for the zero pressure volume, V(0,T), and the Tait parameter, B(T), for each film are listed in Table 1.

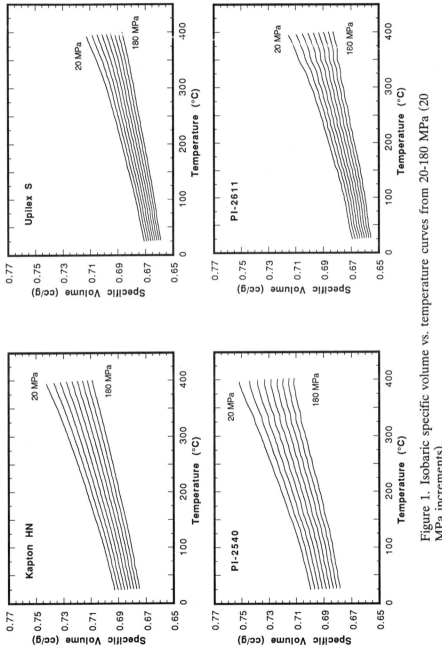

Figure 1. Isobaric specific volume vs. temperature curves from 20-180 MPa (20 MPa increments).

Table 1. Coefficients for Tait Equation

	Kapton HN	PI-2540	Upilex S	PI-2611
V_0	0.6934	0.7004	0.6720	0.6710
V1	9.157×10^{-5}	9.851×10^{-5}	6.230×10^{-5}	5.197×10^{-5}
V2	1.337×10^{-7}	1.686×10^{-7}	1.470×10^{-7}	2.250×10^{-7}
B1	504	425	778	704
B2	2.092×10^{-3}	2.442×10^{-3}	2.554×10^{-3}	3.116×10^{-3}

The average volumetric and linear CTEs from 50–200 °C, and the compressibility (κ) and density (ρ) data at 25 °C are reported in Table 2. The average out-of-plane linear CTE ($\bar{\alpha}_z$) is calculated from the difference between the average volumetric CTE ($\bar{\beta}$) and the sum of the two average in-plane linear CTEs ($\bar{\alpha}_x$ and $\bar{\alpha}_y$).

PI-2540 processed in the manner described in this paper has relatively little crystallinity (11). In contrast, Kapton HN has appreciable crystalline order (27). The higher volumetric CTE for PI-2540 is due to the lower crystallinity in this sample. This is consistent with a lower density and higher compressibility compared with Kapton HN. The lower average in-plane CTEs ($\bar{\alpha}_x$ and $\bar{\alpha}_y$) of Kapton HN are attributed to greater in-plane molecular orientation and more crystalline order.

The volumetric CTEs and densities of the two BPDA-PPD films are comparable although the average in-plane CTEs ($\bar{\alpha}_x$ and $\bar{\alpha}_y$) of Upilex S are approximately four times larger than those of PI-2611. This illustrates that although molecular orientation has a dramatic effect on the linear CTEs, it has relatively little or no effect on the volumetric CTE. The lower compressibility of Upilex S compared with PI-2611 may be due to a slightly different crystal structure in Upilex S.

The volumetric and in-plane linear CTEs of the BPDA-PPD films are lower than those for PMDA-ODA films. The BPDA-PPD films are highly anisotropic and the lower in-plane linear CTEs are due to a larger degree of in-plane

Table 2. Average Volumetric and Linear CTEs from 50-200 °C, and Compressibility and Density at 25 °C

	Kapton® HN	PI-2540	Upilex® S	PI-2611
$\bar{\beta}$ [$\mu m/m$ °C]	176 ± 5	195 ± 5	143 ± 5	154 ± 5
$\bar{\alpha}_x$ [$\mu m/m$ °C]	23 ± 2	33 ± 1	11 ± 2	3 ± 1
$\bar{\alpha}_y$ [$\mu m/m$ °C]	31 ± 2	33 ± 1	15 ± 2	3 ± 1
$\bar{\alpha}_z$ [$\mu m/m$ °C]	122	129	117	148
κ [$\times 10^4$ MPa^{-1}]	1.869	2.236	1.225	1.373
ρ [g/cm^3]	1.437	1.423	1.485	1.487

molecular orientation in BPDA-PPD compared with PMDA-ODA (*11*). The lower volumetric CTEs in the BPDA-PPD films are a result of a more rigid polymer backbone. The out-of-plane linear CTEs for both PMDA-ODA and BPDA-PPD films are much larger than the in-plane linear CTEs. This behavior is due to the in-plane chain axis orientation.

Values for the out-of-plane linear CTE reported in the literature vary, reflecting the difficulty in accurately measuring the out-of-plane linear CTE of thin films. Tong et al. (*28–29*) have used both a capacitance charge technique and a Fabry-Perot laser interferometric technique to measure the out-of-plane CTEs of Kapton HN films. A comparison of our results with their data is listed in Table 3. Both the capacitance and interferometric techniques involve determining the spacing between two parallel plates separated by a polymer spacer. Tong et al. have commented that their elastic thermal stress model for the parallel plate capacitor does not account for errors arising from plate bending and/or tilting. In addition, the measured change in the spacing between the plates does not correspond to the average thickness of the film, but the thickness of the film's high points (*29*). The largest sources of error in parallel plate capacitance measurements are the determination of the sample thickness and the introduction of air gaps between the sample and the electrodes (*30*). An advantage of the PVT technique is that it is a direct measure of the volumetric CTE.

DISCUSSION

The effect of molecular orientation on the linear CTEs has been modeled (*31–32*) in terms of intrinsic linear CTEs of individual chain segments using an approach similar to that used to account for the effect of orientation on polarizabilities (*33*). Assuming cylindrical symmetry about the chain axis, only two intrinsic linear CTEs, α_1 parallel and α_2 normal to the chain axis, are needed to account for the thermal expansion. The volumetric CTE, β, is related to the linear CTEs in the three principle directions, α_x, α_y and α_z, and the two intrinsic linear CTEs by the expression

$$\beta = 3\alpha_0 = \alpha_x + \alpha_y + \alpha_z = \alpha_1 + 2\alpha_2 \tag{8}$$

where α_0 is the linear CTE of an isotropic material. In the case of spin coated (planar) films or uniaxially drawn or extruded polymers, $\alpha_x = \alpha_y = \alpha_\perp$ and

Table 3. Comparison of Out-of-Plane CTE Results for Kapton HN

	Pottiger & Coburn PVT	Tong et al. Capacitance	Tong et al. Interferometric
Film Thickness [μm]	50	125	125
$\bar{\alpha}_x$ [μm/m °C]	23 ± 2	34 ± 2	34 ± 2
$\bar{\alpha}_y$ [μm/m °C]	31 ± 2	34 ± 2	34 ± 2
$\bar{\alpha}_z$ [μm/m °C]	122 ± 15	81	134 ± 13

$\alpha_z = \alpha_{\parallel}$, where $\alpha_{\perp}$ and $\alpha_{\parallel}$ are perpendicular and parallel to the axis of symmetry. For spin coated films, the axis of symmetry is normal to the plane of the film. For uniaxially drawn or extruded polymers, the axis of symmetry is along the draw or extrusion direction.

The model mentioned above has been shown to account for the behavior of several polymer systems. Retting (31) and Wang et al. (34–35) have shown that for both polystyrene (PS) and poly(methymethacrylate) (PMMA), the linear CTEs, $\alpha_{\parallel}$ and $\alpha_{\perp}$, varied as a function of the draw ratio, but the isotropic CTE, α_0, was constant. Retting reached similar conclusions for a series of rubber modified polystyrenes (36). Hellwege et al. (37) studied polycarbonate (PC), poly(vinyl chloride) (PVC), polystyrene (PS) and poly(methymethacrylate) (PMMA) and concluded that α_0 for each of these polymers was constant, independent of orientation.

The independence of β on orientation can also be explained in terms of thermodynamic principles. The volume (V) is an extensive thermodynamic state variable, defined for a material in thermodynamic equilibrium under specific conditions of temperature (T) and pressure (P). Variations of molecular orientation do not dramatically perturb the density of the film. Therefore, the volumetric coefficient of thermal expansion, which is based on the temperature dependence of the volume, is unaffected by changes in molecular orientation. Morphological changes, such as changes in the crystallinity (molecular ordering) or cross-linking, that affect the density of the film, can be expected to affect the volumetric CTE. For polymers well below their glass transition, however, the effect is not expected to be substantial because the difference in the degrees of freedom between the amorphous and crystalline states is small, leading to only minor differences in the volumetric CTE for the two phases.

In contrast, the linear CTEs are affected by changes in molecular orientation. This behavior can be explained primarily in terms of intermolecular forces. Along the chain axis, strong intermolecular forces (i.e. covalent bonds) result in low thermal dependence of the interatomic spacing. Normal to the chain axis, weaker intermolecular forces (in the absence of hydrogen bonding) lead to a large increase in intermolecular spacing with temperature. As a result, the linear CTE along the chain axis (α_1) is substantially lower than the CTE normal to the chain axis (α_2). Therefore, an increase in molecular orientation leads to a decrease in the in-plane linear CTE. A less important effect in these polymers is the contribution from entropic effects (32). The practical implication is that in spin coated (planar) films, a decrease in the in-plane linear CTE occurs at the expense of the out-of-plane linear CTE.

CONCLUSIONS

A technique for determining the volumetric CTE (β) from pressure-volume-temperature measurements was demonstrated. Out-of-plane linear CTEs (α_z) were calculated from the difference between the volumetric CTE and the sum of the in-plane linear CTEs (α_x and α_y). Although the linear CTEs vary significantly with processing, the volumetric CTE is essentially constant. Therefore, in spin coated (planar) films, a decrease in the in-plane linear CTE ($\alpha_x = \alpha_y$)

occurs at the expense of an increase in the out-of-plane linear CTE. This must be taken into account when building devices where both in-plane and out-of-plane linear CTEs are considered important.

REFERENCES

1. Nankey, R. W. Presented at the 1976 Institute of Printed Circuits Fall Meeting, San Francisco, CA, September 1976; IPC-TP-121.
2. Torres, L. A. Presented at the 1984 Institute of Printed Circuits Fall Meeting, San Francisco, CA, September 1984; IPC-TP-510.
3. Croll, S. G. *J. Coat. Tech.*, **50(638)**, 33-38 (1978).
4. Croll, S.G. *J. Appl. Poly. Sci.*, 23, 847-858 (1979).
5. Prest, W. M., Jr.; Luca, D. J. *J. Appl. Phys.*, **50(10)**, 6067 (1979)
6. Prest, W. M., Jr.; Luca, D. J. *J. Appl. Phys.*, **51**, 5170 (1980)
7. Cherkasov, A. N.; Vitovskaya, M. G.; Bushin, S. V. *V. Ysokomol. Soyed.*, **A18(7)**, 1628 (1979).
8. Russell, T. P. ; Gugger, H.; Swalen, J. H. D. *J. Poly. Sci., Polym. Phys.*, 21, 1745 (1983).
9. Machell, J. S.; Greener, J.; Contestable, B. A. *Macromolecules*, 23, 186 (1990).
10. Pottiger, M. T.; Coburn, J. C. "Internal Stress Development in Polyimide Films"; In **Materials Science of High Temperature Polymers for Microelectronics**, Grubb, D. T.; Mita, I.; Yoon, D. Y., Eds.; MRS Symposium Proceedings 227; Materials Research Society: Pittsburgh, 1991; p 187.
11. Coburn, J. C.; Pottiger, M. T. In **Proceedings of the Fourth International Conference on Polyimides**; Feger, C., Ed.; Elsevier: New York, 1991 (to be published).
12. Noe, S.C.; Pan, J. Y.; Senturia, S. D. *Proceedings*, 49th Annual Technical Conference, Montreal, Canada; Society of Plastics Engineers, 1991; Vol. 32; p1598.
13. Noe, S.C; Senturia, S.D. In **Proceedings of the Fourth International Conference on Polyimides**; Feger, C., Ed.; Elsevier: New York, 1991 (to be published).
14. Coburn, J. C.; Pottiger, M.T. *Proc.eedings*, Spring Meeting of the ACS Division of Polymeric Materials: Science and Technology, San Francisco, CA; American Chemical Society: Washington, DC, 1992; Vol. 66, p 194.
15. L. Lin and S. A. Bidstrup, *Proc.eedings*, Spring Meeting of the ACS Division of Polymeric Materials: Science and Technology, San Francisco, CA; American Chemical Society: Washington, DC, 1992; Vol. 66, p 265.
16. Coburn, J. C.; Pottiger, M. T.; Noe, S. C.; Senturia, S. D. *J. Poly. Sci., Poly. Phys.* (to be published).
17. Brekner, M.-J.; Feger, C. *J. Poly. Sci., Poly. Chem.*, 25, 2005 (1987).
18. Brekner, M.-J.; Feger, C. *J. Poly. Sci., Poly. Chem.*, 25, 2479 (1987).
19. Han, B.; Gryte, C.; Tong, H.; Feger, C. *Proccedings*; 46th Annual Technical Conference; Society of Plastics Engineers, 1988; p 994.

20. Feger, C.; Tong, H. M.; Han, B. J.; Gryte, C. C. *Proccedings*; 49th Annual Technical Conference, Montreal, Canada; Society of Plastics Engineers, 1991; p 1742.
21. Dine-Hart, R. A.; Wright, W. W. *J. Appl. Poly. Sci.*, **11**, 609-627 (1967).
22. Nomura, H.; Eguchi, M.; Asano, M. *J. Appl. Phys.*, **70**(11), 7085 (1991).
23. Jou, J.-H.; Huang, P.-T.; Chen, H.-C.; Liao, C.-N. *Polymer*, **33**, 967 (1992).
24. Zoller, P.; Bolli, P.; Pahud, V.; Ackermann, H. *Rev. Sci. Instrum.*, **47**(8), 948 (1976).
25. Fakhreddine, Y. A.; Zoller, P. *Proceedings*; 49th Annual Technical Conference, Montreal, Canada; Society of Plastics Engineers, 1991; p 1642.
26. Gnomix Research, 3809 Birchwood Drive, Boulder, CO 80302.
27. Gardner, K. H.; Edman, J. R.; Freida, J. E.; Freilich, S. C.; Manring, L. E. *Abstracts of Papers*, Recent Advances in Polyimides and Other High Perfomance Polymers, San Diego, CA; American Chemical Society: Washington, D.C., 1990.
28. Tong H. M.; Hsuen, H. K. D.; Saenger, K. L.; Su, G. W. *Rev. Sci. Instr.*, **62**(2), 422-430 (1991).
29. Tong H. M.; Saenger, K. L.; Su, G. W. *Proccedings*; 49th Annual Technical Conference, Montreal, Canada; Society of Plastics Engineers, 1991; p 1727.
30. ASTM D 150-87; **Annual Book of ASTM Standards**; American Society for Testing and Materials: Philadelphia, 1987.
31. Retting, W. *Colloid & Poly. Sci.*, 259, 52-72 (1981).
32. Struik, L. C. E. **Internal Stresses, Dimensional Instabilities and Molecular Orientations in Plastics**; John Wiley: New York, 1990; Part IV.
33. Samuels, R. J. **Structured Polymer Properties**; Wiley: New York, 1974.
34. Wang, L.-H.; Choy, C. L.; Porter, R. S. *J. Poly. Sci., Poly. Phys.*, **20**, 633-640 (1982).
35. Wang, L.-H.; Choy, C. L.; Porter, R. S. *J. Poly. Sci., Poly. Phys.*, **21**, 657-665 (1983).
36. Retting, W. *Pure Appl. Chem.*, **50**, 1725 (1978).
37. Hellwege K. H.; Hennig, J.; Knaooe, W. *Kolloid-Z.*, **188**, 121 (1963).

RECEIVED January 21, 1993

Chapter 19

Radiation-Induced Modifications of Allylamino-Substituted Polyphosphazenes

M. F. Welker[1], H. R. Allcock[1], G. L. Grune[2,3], R. T. Chern[2], and
V. T. Stannett[2]

[1]Department of Chemistry, Pennsylvania State University, University
Park, PA 16802
[2]Department of Chemical Engineering, North Carolina State University,
Raleigh, NC 27695

Efforts to synthesize allylamino-substituted polyphosphazenes and characterize their sensitivity to radiation and determine the value of using such
polymers as new and better resist materials for microlithographic applications
continues. An initial attempt to synthesize an amino-substituted polyphosphazene specifically tailored toward providing better resist properties was
unsuccessful in that the polymer was quite radiation insensitive. The elastomeric models of an earlier study led us to the realization that the addition of
allylic substituents might enhance the radiation sensitivity of such polymers
significantly. Also, these amino-substituted polyphosphazenes were found to
exhibit the glassy, thermal, and film forming properties required for conventional resists. Other potentially favorable resist characteristics of these
polyphosphazenes includes relatively high RIE and high temperature resistance and good adhesion to SiO_2 substrates. This work indicates our progress
involving the use of these polymer systems including results for G(X) values, (a
measure of the efficiency of crosslinking) and grafting attempts to induce
complete solubility changes. Silicon wafers coated with thin films of the
allylamino-substituted polyphosphazenes were patterned with E-beam lithographic techniques and illustrated excellent sensitivity ($<6\mu C/cm^2$) and resolution (to 0.1 μm).

A collaborative effort to synthesize amino-substituted polyphosphazenes and
examine their sensitivity to radiation has been recently undertaken. The objective was to determine the value of using such polymers as new and better resist
materials for microlithographic applications. An initial study was carried out
using elastomeric phenoxy substituted polyphosphazenes as models for radiation
and grafting (1). However, it became necessary to synthesize several new
amino-substituted polyphosphazenes for several reasons. First, the usefulness of
elastomeric polymers for resist applications is severely limited by their inability
to remain dimensionally stable at normal temperatures. In contrast, glassy
polymers have been shown by many (2-5) to provide the necessary properties
required for resist film formation on silicon wafers. Second, the amino-substituted polyphosphazenes are excellent film forming polymers with high molecu-

[3] Corresponding author

0097–6156/94/0537–0293$06.00/0
© 1994 American Chemical Society

lar weights. Finally, it seemed possible that these polymers would exhibit similar RIE resistance to those of a phenoxy-substituted counterpart (6, 7), and experimental work was performed to determine RIE values for the most radiation sensitive of the new polymer candidates.

EXPERIMENTAL

Synthesis of several allyl amino-substituted polyphosphazenes were accomplished via nucleophilic substitution techniques described previously (8–10) and are represented in Figure 1.

Following the synthesis of these polymers, molecular and materials characterization was accomplished by means of ^{31}P and ^{1}H NMR, and elemental analysis, followed by GPC for molecular weight determination and thermal analysis by DSC for Tg measurements.

Films of .0045–.0065" thickness, prepared from solutions of the polymer in THF, were cast on a clean glass plate using a precision blade to spread the 20

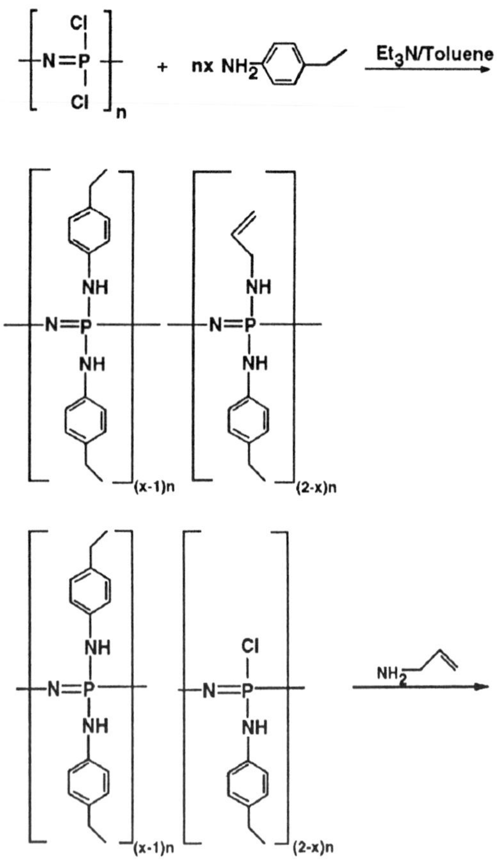

Figure 1. Nucleophilic aminolysis reaction for synthesis of allyl-amino substituted polyphosphazenes.

wt.% solution. Deionized water was used to remove the dried film from the glass substrate.

To determine their sensitivity to radiation, the polymer films were irradiated in a Gammacell 220 ^{60}Co γ-ray source manufactured by Atomic Energy of Canada Limited. Films weighing approximately .0100 g were placed in sealed evacuated glass vials (1 x 10^{-6} torr) before exposure. The dose rate was 0.52 kGy/hour. Extraction of the soluble portion was performed by placing irradiated films in fritted glass vials (20–50 μm pore size) and refluxing with THF to constant weight.

RESULTS AND DISCUSSION

Initially, model polymers were investigated for their sensitivity to both E-beam and gamma radiation. Experimental determination of the G(X) (the number of crosslinks/100 eV) values for chemically different poly(organophosphazenes) was used as an initial indicator of their behavior as resist materials. In that study (*1*), it was found that the presence of an allylic substituent (8.5 mol%) could greatly enhance not only gelation, but also the grafting of reactive monomers to the elastomers. The first attempt to synthesize an amino-substituted polyphosphazene specifically tailored to provide better resist properties was unsuccessful because the polymer was quite insensitive to radiation. Incorporation of the allylic groups into the amino-substituted polymers, resulted in sufficient gelation after irradiation to suggest possible negative resist applications.

Experience with the irradiation of polyphosphazenes in any form (rubbery elastomer, fibrous glassy, or film), has been limited to that of only two or three groups of researchers, including those from both of these laboratories (*11–16*). Stannett et.al. (*15, 16*), investigated eight different polyorganophosphazenes and obtained Dg (the dose where the gel is first formed), G(X) and G(S) values for each. Depending on the nature of the substituent, the G(X) values found ranged from 0.050 to 2.49. The smallest G(X) value was found for the amino-substituted phosphazene, possibly due to the stabilizing influence of the "hindered amines" within its structure (*17*). Studies by Beggiato (*18*), Hiraoka (*19*), and Lora (*20*), have reviewed the different aspects of irradiating polyphosphazenes, but neglected the use of irradiation grafting techniques. Recently, a study (*21*) was conducted where dimethylaminoethyl methacrylate monomer was grafted to various poly(organophosphazenes) to increase biocompatibility.

Grafting experiments with purified acrylic acid have also been performed in our laboratories. We speculate that the stabilizing influence of the amine substituent has hindered the graftability of the allyl-amino substituted polyphosphazenes. It is possible that the amine acts as an internal radical scavenger, thus trapping radicals formed during gamma or electron-beam irradiation. Attempted grafting with a 50/50 mixture of acrylic acid and water has resulted in little or no success. Pure acrylic acid was found to dissolve these polyphosphazene polymers.

Synthesis and characterization of allyl amino-substituted polyphosphazenes were carried out using previously described techniques (*8–10*). Initially, the polymer shown in Figure 2, containing only 4-ethyl anilino substituents, was synthesized, characterized and exposed to varying doses of gamma radiation.

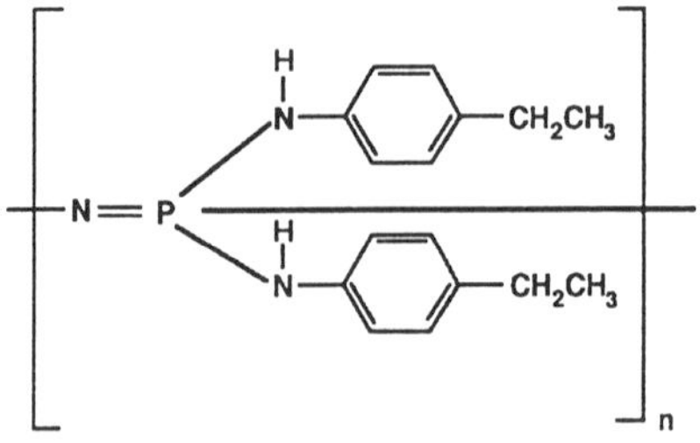

$$\left[\begin{array}{c} \overset{H}{N}-\!\!\!\left\langle\!\!\!\bigcirc\!\!\!\right\rangle\!\!\!-CH_2CH_3 \\ -N\!=\!P \\ \overset{H}{N}-\!\!\!\left\langle\!\!\!\bigcirc\!\!\!\right\rangle\!\!\!-CH_2CH_3 \end{array}\right]_n$$

Tg = 86 C

Figure 2. Structure of 4-ethyl-anilinophosphazene homopolymer.

Unfortunately, this polymer was found to be very radiation insensitive when exposed to gamma rays under vacuum (10^{-6} torr) even at temperatures of 95°C–well above the Tg. This stabilizing influence has recently been confirmed by adding a small amounts (1.25–2.5 wt.%) of the polymer to styrene monomer. This resulted in small, but significantly retarded, polymerization yields and lower molecular weights of the polystyrene. Subsequently, polymers with the three structures shown in Figures 3 (a), (b), and (c) were obtained, and G(X) values for two of the polymers were determined.

This involves the use of the Charlesby-Pinner (CP) (22–24) treatment which describes the determination of the gel fraction of the polymer as a function of radiation dose. Figures 4(a) and 4(b) illustrate that polymers with structures as those shown for Polymers **1** and **2** crosslink when exposed to ^{60}Co radiation. Polymer **1** has been synthesized with approximately 7 % of the allylic substituent, while the synthesis of polymer **2** allowed for 3-4% of the same allylic group. Glass transition temperatures for these polymers are not quite that of conventional resists such as polymethyl methacrylate (> 100°C), and it is speculated that as allyl content is increased, Tgs will decrease slightly.

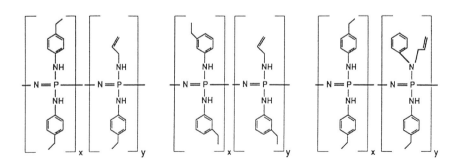

Tg = 87 C   **Tg = 52 C**   **Tg = ?**

Figure 3. Structures of polymers 1, 2, and 3.

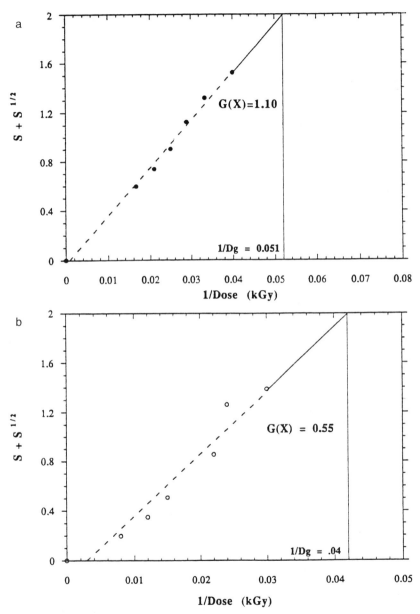

Figure 4. Charlesby-Pinner Plots for a, Polymer 1 and b, Polymer 2.

The Charlesby-Pinner equation used to determine the G(X) values for these polymers is given as:

$$s + s^{1/2} = p_0/q_0 + 2/q_0 MwD \tag{1}$$

where
p_0 = density of main chain fractures per unit dose
D = radiation dose
q_0 = density of crosslinks per unit dose
s = soluble fraction of the polymer

and reduces to allow the simple determination of the G(X) value when p_0 is close to zero:

$$G(X) = \frac{4.52 \times 10^6}{Dg \times Mw} \tag{2}$$

where   Dg = dose at which the gel first appears in kGy
        Mw = weight average molecular weight

This equation is very dependent on proper determination of Dg, which can be facilitated by using a method made popular by Lyons (25)—that of a log-log plot of $s + s^{1/2}$ vs. dose. Other restrictions in using this technique include:

a) crosslinks are distributed randomly along molecular chains during irradiation

b) crosslinks are assumed independent of the absorbed dose

This was confirmed by the data in Figures 4(a) and (b), which indicate the best straight-line fit of the data goes through or close to the origin.

The best polymer film candidate based on sol/gel analysis performed as described, was dissolved (5 wt.%) in methyl-isobutyl ketone (MIBK) and spin coated onto a 4″ silicon wafer at 2000 rpm using hexamethyldisilazane (HMDS) as a primer. Small sections of the more homogenoeus wafers were sectioned and exposed to a 15 keV electron beam using an electron beam/scanning electron microscope lithography tool. The E-beam exposure of the wafer was performed under vacuum at 2×10^{-6} torr. Subsequent development was achieved in 10 seconds, again using THF as the solvent.

A cross-section of a spin-coated silicon wafer and SEM analysis indicated a 1780 A^0 film thickness of polymer 1 on its surface. Figure 5 is an E-beam lithographic pattern obtained using a 15 keV exposure at a dose of 6 μC/cm^2. Submicron resolution (to 0.1 μm) was achieved using THF as a developing solvent and a development time of 10 seconds. No pre- or postbaking of the resist was performed, yet adhesion of the polyphosphazene film was excellent. Solvents other than THF were used for this system, however it was THF that was found not to cause the swelling found with conventional crosslinked resists.

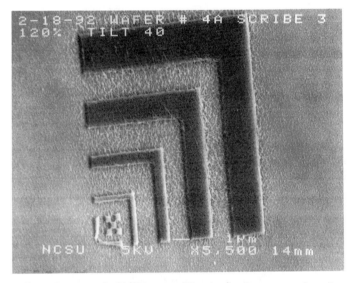

Figure 5. Microphotograph (5,500 magnification) of patterned wafer using 15 keV electron beam/SEM at 6 μC/cm^2 illustrating submicron (0.1 μm) resolution for Polymer **1**. Development time in THF was 10 seconds.

Reactive ion etching rates as low as 585 A^0/min. have been determined (after E-beam irradiation and subsequent crosslinking) for this film. These rates are given in Tables 1 and 2 from a previous study and from the results of our work. The effect of the differences in vacuum and two types of equipment used for plasma etching is reflected in the etch rate differences between the diphenoxyphosphazene and the allyl-amino substituted derivatives. A more common description of the process performed at 900 mTorr would perhaps be "high pressure oxygen plasma etching", and is given simply for comparison purposes based on the only other work in this area performed by Hiraoka and Chiong (6).

A problem with the E-beam patterning of polymer 1 on the silicon substrate (SiO_2 over silicon) is the "growth" of unexposed regions or other regions near the unexposed areas, which were purposely "written" with the computer driven E-beam unit. These effects are universal when using E-beam writing tools, and are commonly referred to as "proximity effects". These effects are the major problem associated with establishing E-beam lithography for industrial use in the large scale production of microelectronic devices.

One type of proximity effect—the interproximity effect—involves the exposure of features near the ones that are intentionally exposed, referred to as near neighbors. Unexposed spaces between lines are exposed by backscattered electrons. This effect is seen in the work performed for this study. The interproximity effect causes unexposed spaces in positive resists to become thinner and narrower. For negative resists, such as ours, the unexposed region is filled with "scumming" resist.

Reduction of interproximity effects involves a series of trade-offs depending on the geometry of the desired patterned features. One of many methods to substantially reduce proximity effects for positive and negative resists, is by using a membrane 5-10 fold thinner than the dimension of the feature to be used (26, 27) over the actual resist layer. Because the proximity effect is almost directly proportional to resist thickness, multilayer resists are similarly practical solutions to this problem as well. Top surface imaging, thus again plays an important role in the future use of E-beam lithographic imaging. The use of a 6-Hema monomer at the very top surface of an already highly sensitive RIE resistant polyphosphazene, may be one method of overcoming the interproximity effects found in this study and is discussed in a future publication.

SUMMARY

It has been shown that allyl-amino substituted polyphosphazenes, specifically tailored to form glassy films from suitable solvents can be used for negative resist/mask applications for microlithography. As suggested by model studies, the addition of allylic double bonds to amino-substituted polyphosphazenes increases the sensitivity to radiation-induced crosslinking. However, it is probable that the stabilizing influence of the hindered amine substituent has reduced the graftability of these polymers. In addition, the inherent RIE resistance of the phosphazene family has been demonstrated, which underscores the possible usefulness of these systems for future resist work. Currently, other related

Table 1. Reactive Ion Etching Rates for Various Negative Resist Materials Based on Work by Hiraoka (6) (−250 V Bias, 40 SCCM, 60 mTorr, 0.35 W/cm^2)

Resist Type	Etch Rate
Poly(diphenoxyphosphazene)[a]	30 A^0/min.
AZ-1350J	1300 A^0/min.
Silyated AZ-1350J	30 A^0/min.
Poly(chloromethylstyrene)	1400 A^0/min.

[a] UV-hardened films

Table 2. Reactive Ion Etching Rates for Allyl-Amino Substituted Polyphosphazenes and Commercially Available Resist Materials (−250 V Bias, 40 SCCM, 900 mTorr, 0.08 W/cm^2)

Resist Type	Etch Rate
Shipley 1400-31 Novolak Resin	4,224 A^0/min.
Polymer 1 - Exposed w/E-beam	585 A^0/min.
Polymer 1 - Unexposed	830 A^0/min.

polymer structures with higher allylic loadings are being investigated for enhanced radiation sensitivity.

ACKNOWLEDGMENTS

The authors would like to thank Professor J.A. Moore of Rensselaer Polytechnic
Institute, for helpful advice and unpublished data involving the grafting enhancements which can be accomplished for potential resist applications. In addition,
R. Greer, an undergraduate in Chemical Engineering at North Carolina State
University was 'extremely helpful in preparing polymer films for irradiation and
grafting studies. Professor Phillip Russell and Ph.D. candidate Terry Stark of
North Carolina State University were responsible for E-beam writing of the
silicon wafers. Dr. Sam Nablo of Energy Sciences Inc., Wilmington, Massachusetts has been helpful with technical information and enlightening discussions.

The work at North Carolina State University has been supported with an
SUR (Shared University Research) grant from the IBM Corporation. Drs. R.C.
Sanwald and J.R.Kirby of IBM-RTP, North Carolina and Jane Shaw of IBM-
Yorktown, New York, have been extremely gracious with financial and technical
support for this effort.

The polymer synthesis studies at the Pennsylvania State University were
conducted with support from the Office of Naval Research.

REFERENCES

1. Stannett, V.T., Chern, R.T., Grune, G.L., and Harada, J., in *Polymer
   Preprints*, **1991**, *32*(2), 34-36.
2. Atoda, N., Komuro, M., and Kawakatsu, H., *J. Appl. Phys.*, **1979**, *50*, 3707.
   (1979).
3. Moreau, W.M., *Semiconductor Lithography;* Plenum Press, New York, N.Y.,
   1988; pp. 330-333.
4. Thompson, L.F., Feit, E.D., and Heidenreich, R.D., *Poly. Eng. Sci.*, **1974**,
   14(7), 529.
5. Bowden, M.J., and Novembre, A.E., *Poly. Eng. Sci.*, **1983**, *23*(17), 975.
6. Hiraoka, H., and Chiong, K.N., *J. Vac. Sci. Technol. B*, **1987**, *1*(5), 386-388.
7. Welker, M.F., Allcock, H.R., Grune, G.L., Stannett, V.T., and Chern, R.T.,
   in *PMSE Preprints*, **1992**, *66*, 259-260.
8. Allcock, H.R., Cook, W.J., and Mack, D.P., *Inorg. Chem.*, **1972**, *11*(11),
   2584-2590.
9. Allcock, H.R., and Kugel, R.L., *Inorg. Chem.*, **1966**, *5*(10), 1716-1718.
10. White, J.E., Singler, R.E., and Leone, S.A., *J. Poly. Sci., Chem. Ed.*, **1975**,
    13, 2531-2543.
11. Allcock, H.R., Kwon, S., Riding, G.H., Fitzpatrick, R.J., and Bennett, J.L.,
    Biomaterials, **1988**, *19*, 509-513.
12. Allcock, H.R., Gebura, M., Kwon, S., and Neenan, T.X., *Biomaterials*, **1988**,
    19, 500-508.
13. Bennett, J.L., Dembek, A.A., Allcock, H.R., Heyen, B.J., and Shriver, D.F.,
    Chemistry of Materials, **1989**, *1*, 14-16.

14. Bennett, J.L., Dembek, A.A., Allcock, H.R., Heyen, B.J., and Shriver, D.F., *Polym. Prepr.* (ACS Polym. Div.), **1989**, 437-438.
15. Stannett, V.T., Yanai, S., and Squire, D.R., *Radiat. Phys. Chem.*, **1984**, 23(4), 489-490.
16. Stannett, V.T., Babic, D., Souverain, D.M., Squire, D.R., Hagnauer, G.L., and Singler, R.E., *Radiat. Phys. Chem.*, **1986**, 28(2), 169-172.
17. Hodgeson, D.K.C., *Developments in Polymer Degradation;* Grassie, N. Ed., Applied Science Publishers: London, 1982; pp. 189-234.
18. Beggiato, G., Bordin, P., Minto, F., and Busulini, L., *Eur. Poly. J.*, **1979**, *15*, 403.
19. Hiraoka, H., *Macromolecules*, **1979**, *12*(4), 753-757.
20. Lora, S., Minto, F., Carenza, M., Parma, G., and Faucitano, A., *Radiat. Phys. Chem.*,**1988**, *31*(4-6), 629-638.
21. Lora, S., Carenza, M., Palma, G., Pezzin, G., Caliceti, P., Battaglia, P., and Lora, A., *Biomaterials*, **1991**, *12*, 280.
22. Charlesby, A., *J. Polym. Sci.*, **1953**, *11*, 513, 521; *Proc. R. Soc. London*, **1954**, *(A222)*,60, 542, **1954**, *(A224)*, 120, **1955**, *(A231)*, 521.
23. Charlesby, A., and Pinner, S.H., *Proc. R. Soc. London*, **1959**, *(A249)*, 367.
24. Charlesby, A., *Atomic Radiation of Polymers*; Permagon Press: Oxford, 1960; pp.142-148.
25. Lyons, B.J., *Radiat. Phys. Chem.*, **1983**, *22*, 136.
26. Moore, R., Caccoma, G., Pfeiffer, H., Weber, E., and O. Woodard, *J. Vac. Sci. Technol.*, **1981**, *19*, 950.
27. Adesida, I., and Everhart, T., *J. Appl. Phys.*, **1980**, *51*, 5994.

RECEIVED May 7, 1993

Chapter 20

Synthesis of Perfluorinated Polyimides for Optical Applications

Shinji Ando, Tohru Matsuura, and Shigekuni Sasaki

NTT Interdisciplinary Research Laboratories, Midori-cho 3—9—11, Musashino-shi, Tokyo 180, Japan

This study reports the first synthesis of perfluorinated polyimides that have high Tgs over 270 °C and high optical transparency over the entire wavelengths of optical communications. Their high thermal stability and optical transparency are due to the fully aromatic molecular structure and the absence of hydrogen atoms. The use of diamine, which has relatively high reactivity, and the new per-fluorinated dianhydrides, which has a flexible structure, makes it possible to obtain strong, flexible perfluorinated polyimide film. In addition, these polymers have low dielectric constants, low refractive indices, and low birefringence. Perfluorinated polyimides are promising materials for optical communication applications.

Polymers are expected to be used as media for transmitting near-infrared light in optical communication applications such as waveguides in opto-electronic integrated circuits (OEIC) and in multichip interconnections (1, 2). The current manufacturing process for ICs and multichip modules includes soldering at 260 °C and short-term processes at temperature of up to 400 °C. Waveguide polymeric materials should therefore have high thermal stability—that is, a high glass transition temperature (Tg) and a high polymer decomposition temperature —as well as high transparency at the wavelengths of optical communications (WOC), 1.0–1.7 μm.

Conventional waveguide polymeric materials, such as poly(methyl methacrylate) (PMMA), polystyrene (PS), or polycarbonates (PC), do not have such thermal stability. In addition, their optical losses at the WOC are much higher than in the visible region (0.4–0.8 μm), because carbon–hydrogen (C–H) bonds harmonically absorb infrared radiation. Figure 1 shows the visible-near-infrared absorption spectrum of the PMMA dissolved in chloroform with a concentration of 10 wt%. The same amount of chloroform was used as a reference. Two types of C–H bonds—those in methyl and methylene groups - give broad and strong absorption peaks in the infrared region. Although the wavelengths currently used for long distance optical communication, 1.3 and 1.55 μm, are located in

0097–6156/94/0537–0304$06.00/0

what are called windows, absorption peaks originating from C-H bonds increase the optical losses at these wavelengths.

Polyimides, on the other hand, have been investigated as optical waveguide materials because they have excellent thermal, chemical and mechanical stability (*3–5*). We have recently reported new fluorinated polyimides using 2,2'-bis(trifluoromethyl)-4,4'-diaminobiphenyl (TFDB) as a diamine (*6–8*). These compounds show high transparency in the visible region as well as low dielectric constants, low refractive indices, and low water absorption. It has been reported that optimally cured partially fluorinated polyimides can be used to decrease optical losses below 0.1 dB/cm in the visible region (at 0.63 μm), and that these losses are stable at temperatures up to 200 °C (*5*). As described below, however, they also have some absorption peaks in the near-infrared region that originate from the C-H bonds in their phenyl groups.

Figure 2 shows a schematic representation of the fundamental stretching bands and their harmonic absorption wavelengths for the carbon-hydrogen (C-H), carbon-deuterium (C-D), and carbon-fluorine (C-F) bonds. The wavelengths were measured for benzene, hexadeuterobenzene, and hexafluorobenzene with a near-infrared spectrophotometer. For simplicity, we have not shown the absorptions that originate from the fourth and fifth harmonics of the stretching vibration, and from the combinations of the harmonics and the deformation vibration. The harmonics of C-D and C-F bonds are displaced to longer wavelengths than the C-H bond because the wavelengths for the fundamental stretching vibrations of C-D and C-F bonds are about 1.4 and 2.8 times longer than that of C-H bond (*9, 10*). Since the absorption band strength decreases about one order of magnitude with increase in the order of harmonics (i.e., the vibrational quantum number) (*11, 12*), the losses in the visible and near-infrared region can be appreciably reduced by substituting deuterium or fluorine for hydrogen atoms. Kaino et al. (*13, 14*), have produced low-loss optical fibers from deuterated PMMAs and fluorodeuterated PSs. Most recently, Imamura et al. (*15*) fabricated low loss waveguides of less than 0.1 dB/cm at 1.3 μm using deuterated and fluorodeuterated PMMAs. Although this substitution effect must be greater in the near-infrared region, perdeuteration would nonetheless seem inadequate for decreasing optical losses over the entire WOC. The strength of absorption due to the harmonics of C-D bonds is smaller than that due to C-H bonds, but the third harmonics of C-D bond stretching appearing at 1.55 μm (*8*) are not negligible. On the other hand, perfluorinated amorphous polymers, such as Cytop (Asahi Glass Co.), have been reported to have no absorption peaks between 1.0 and 2.5 μm (*16*). For the reduction of optical losses in the WOC, perfluorination is, in principle, superior to perdeuteration.

The combination of low optical losses over the entire WOC and high thermal, chemical, and mechanical stability must therefore be attained by the perfluorination of polyimides. In addition, perfluorination should decrease the dielectric constant, refractive index, and water absorption. These characteristics are desirable for optical and opto-electronic applications. This study reports the first synthesis and the properties of perfluorinated polyimides.

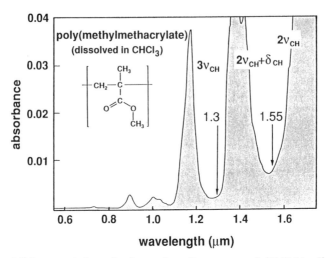

Figure 1. Visible-near-infrared absorption Spectrum of PMMA dissolved in chloroform.

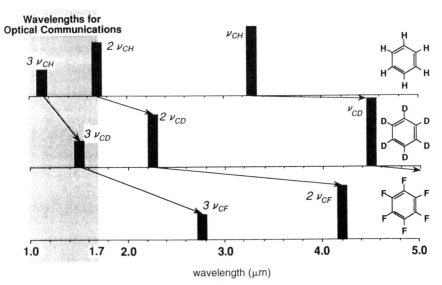

Figure 2. Schematic representation of fundamental stretching bands and their harmonic absorption wavelengths for C–H, C–D, and C–F Bonds.

RESULTS AND DISCUSSION

Figure 3 lists the already known or commercially available perfluorinated dian-hydrides and diamines that can be used for synthesizing perfluorinated poly-imides. Because of the high electronegativity of fluorine, the substitution of fluorines for hydrogens considerably decreases acylation reactivity of diamine monomers and increases the reactivity of dianhydride monomers. To generate high molecular weight perfluorinated polyimides, it is first necessary to know how fluorine affects the reactivity of the monomers. In particular, the fluorine substituting effect to the diamine reactivity is important because kinetic studies of the acylation of conventional monomers have revealed that acylation rate constants can differ by a factor of 100 between different dianhydrides, and by a factor of 10^5 between different diamines (*17*).

To estimate the acylation reactivity of perfluorinated diamines, we prepared poly(amic acid)s from the five diamines listed in Figure 3 using 2,2-bis(3,4-di-carboxyphenyl)hexafluoropropane dianhydride (6FDA) as a dianhydride. Table I shows the end-group contents of the poly(amic acid)s determined from ^{19}F NMR. 4FMPD shows the lowest end-group content that means the highest reactivity, and 4FPPD shows the next. However, for all the diamines, the acylations were not complete, and end-group contents were still high even after 6-days reaction. This considerable decrease in reactivity should be induced by the fluorination of diamines. The synthesis of polyimides from perfluorinated diamines and conventional dianhydrides seems fairly difficult even when the most reactive diamine, 4FMPD, is used. When 8FBZ diamine, the least reactive one, was reacted with 6FDA, no NMR signal for the poly(amic acid) could be detected.

In a previous paper (*18*), we discussed the relationships between the NMR chemical shifts and the rate constants of acylation (k) as well as such electronic-property-related parameters as ionization potential (IP), electronic affinity (EA), and molecular orbital energy for a series of aromatic diamines and aromatic tetracarboxylic dianhydrides. The usefulness of NMR chemical shifts for estimat-ing the reactivity of polyimide monomers was first reported by Okude et al. (*19*). We have revealed that the ^{15}N chemical shifts of the amino group of diamines (δ_N) depend monotonically on the logarithm of k (log k) and on IP.

In this study, we attempted to estimate the reactivity of the five perfluori-nated diamines from ^{15}N and 1H NMR chemical shifts of amino groups (δ_N and δ_H) and calculated IPs. Figure 4 plots δ_N against δ_H, where the upfield displacement of chemical shifts (δ_N and δ_H are decreased) corresponds to the higher reactivity for acylation (*18*). The IPs calculated using MNDO-PM3 semi-empirical molecular orbital theory (*20*) are also incorporated in the figure. From the δ_N, δ_H, and IPs of the diamines, 4FPPD is suggested to have the highest reactivity among the five and 4FMPD is the next. However, this does not coincide with the end-group contents of poly(amic acid)s derived from the experiments described above. As shown in Figure 5, the acylation starts with a nucleophilic substitution in which diamine donates an electron to dianhydride (*17*). This reaction is called 'first acylation' and it affords a monoacyl derivative (MAD). Poly(amic acid)s are generated by the succeeding 'second acylation' , in which MAD reacts with dianhydride, diamine or MAD. Therefore, the reactivity

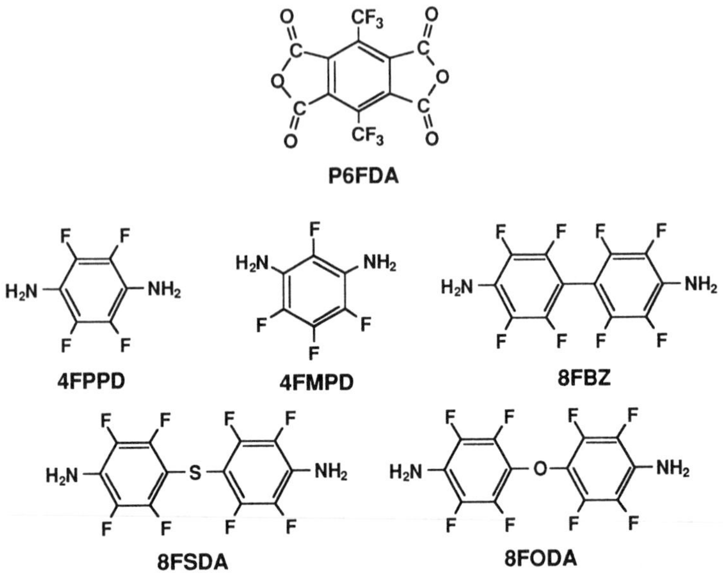

P6FDA

4FPPD   4FMPD   8FBZ

8FSDA   8FODA

Figure 3. Structures of dianhydride and diamines.

Table I. End-Group Contents of Poly(amic acid)s Synthesized from
Perfluorinated Diamines and 6FDA

Diamine	Endgroup Content (%)
4FPPD	42%
4FMPD	15%
8FODA	75%
8FSDA	91%
8FBZ	>99%

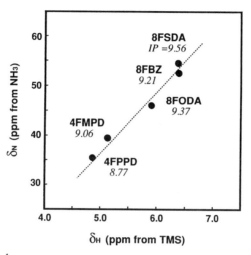

Figure 4. ¹⁵N and ¹H NMR chemical shifts and calculated ionization potentials (eV) of perfluorinated diamines.

1st. Acylation

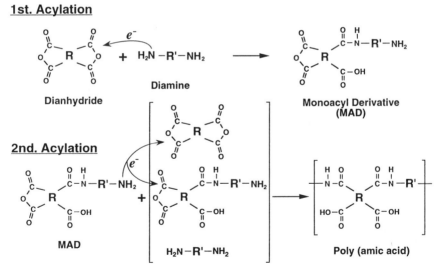

Figure 5. Two-step acylation reactions that generate Poly(amic acid) from diamine and dianhydride.

of MADs rather than that of diamines should be examined for synthesizing high molecular weight poly(amic acid)s.

The five perfluorinated diamines were reacted with equimolar amounts of phthalic anhydride to estimate their MAD reactivity. The molecular structures, δ_N, and δ_H of the diamines and the MADs are shown in Figure 6. The diamine of 8FBZ is not shown because no MAD could be obtained. The δ_N of 4FPPD was displaced downfield by 12.5 ppm in changing to MAD that corresponds to a more than 10^3 decrease of acylation rate constant. On the other hand, the displacements of δ_N and δ_H for the other diamines are much smaller. This means that the reactivity of the residual amino group is little affected by the first acylation, unless two amino groups are located at *para*-position in the same benzene ring. As a result, 4FMPD-MAD shows the highest reactivity that coincides with the result of the end-group content of poly(amic acid)s. Despite the fact that the δ_N and δ_H of 4FPPD-MAD and 8FODA-MAD resonate at near positions, the end group content of the poly(amic acid) derived from 4FPPD and 6FDA was lower than the case of 8FODA and 6FDA (Table I). Some difference may exist in the steric effects during the generation of poly(amic acid)s between one and two benzene ring diamines.

The introduction of fluorine or fluorinated groups into dianhydrides, on the other hand, increases the reactivity. The ^{13}C NMR chemical shift of carbonyl carbons (δ_C) of P6FDA (*13*), the only existing perfluorinated dianhydride, resonated at about 4 ppm upfield from those of conventional unfluorinated dianhydrides (*18*). Although δ_C are not as closely correlated with rate constants (*18*), this upfield shift suggests considerable increase of reactivity. P6FDA is therefore expected to compensate for the low reactivity of perfluorinated diamines.

P6FDA was then used to synthesize perfluorinated polyimides with the five diamines by a conventional method (described in the experimental section). Although P6FDA was suggested to be more reactive than 6FDA, the end-group content of the poly(amic acid) synthesized from 4FMPD and P6FDA was 36% which is higher than that of poly(amic acid) prepared from 4FMPD and 6FDA. The resultant perfluorinated polyimide (P6FDA/4FMPD, Structure 1) was cracked, brittle, and did not form a continuous film. For the other four diamines, the polyimides prepared using P6FDA are coarse powder or films that have many cracks. The primary reason for the non-continuous film is probably that the high reactivity of the perfluorinated dianhydride could not compensate for the very low reactivity of perfluorinated diamines. However, the effect of the rigidity of the polymer chain cannot be neglected for the cases of P6FDA and one-benzene-ring diamines. In this situation, bond rotation is permitted only at the imide linkage (nitrogen-aromatic carbon bonds). However this rotation is restricted by steric hindrance between the fluorine atoms and carbonyl oxygens. The rigidity of the polymer chain has to be improved by introducing flexible linkage groups into the dianhydride component. Accordingly, continuous and flexible films of perfluorinated polyimides are expected to be obtained by combining diamines, which have high reactivities, with dianhydrides, which have flexible molecular structures.

A novel perfluorinated dianhydride, 1,4-bis(3,4-dicarboxytrifluorophenoxy)-tetrafluorobenzene dianhydride (10FEDA), was synthesized according to Scheme

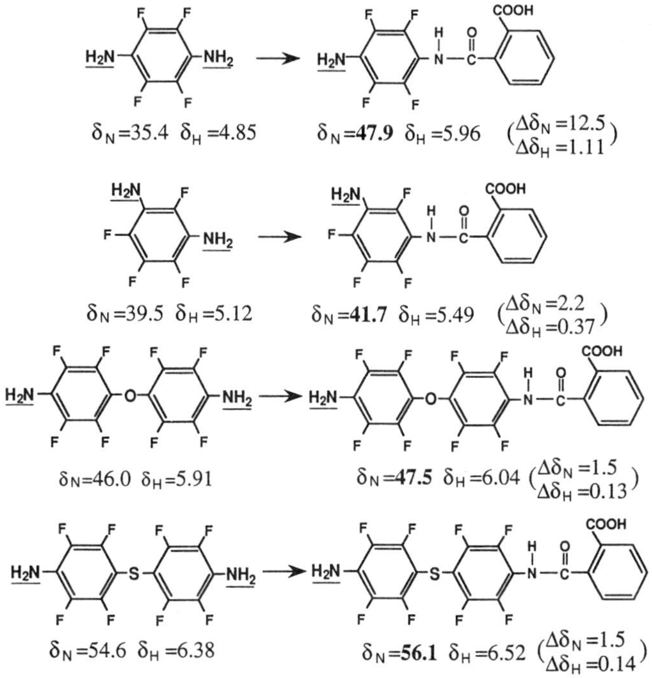

Figure 6. ^{15}N and ^{1}H chemical shift changes of perfluorinated diamines caused by the first acylation.

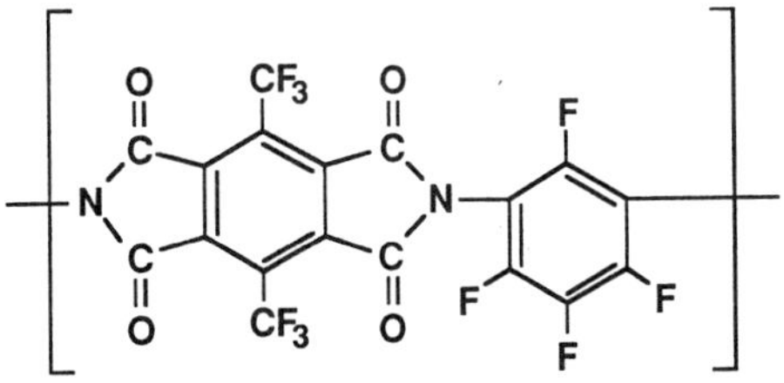

Structure 1. P6FDA/4FMPD.

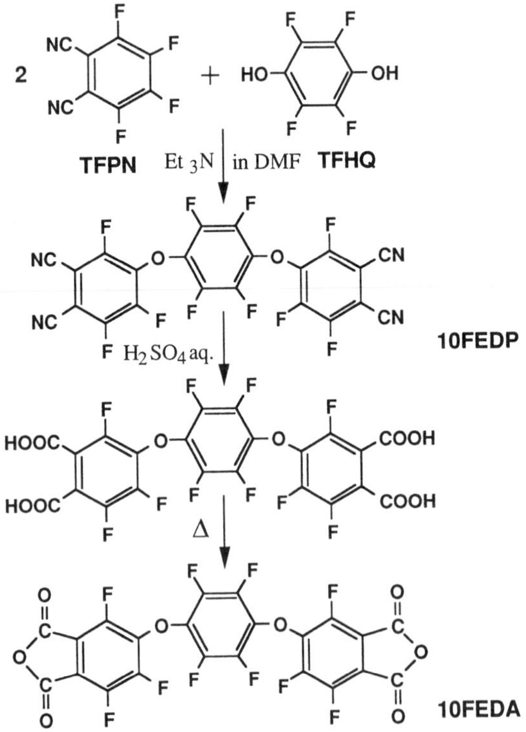

Scheme I.

I. It should be noted that this molecule has two ether-linkages that give flexibility to the molecular structure. In addition, δ_C of 10FEDA (157.5 ppm from TMS) was almost the same as that of P6FDA (157.6 ppm), so this dianhydride should have higher reactivity than unfluorinated and partially fluorinated dianhydrides. The end group content of the poly(amic acid) prepared from 10FEDA and 4FMPD was 6% that is much less than that of poly(amic acid) from P6FDA and 4FMPD (36%). This can be ascribed primarily to the considerable increase of the flexibility of dianhydride structure. The resultant perfluorinated polyimide (10FEDA/4FMPD, Scheme II) was a 9.5-μm thick, strong and flexible film, pale yellow like Dupont's Kapton^R. The film was not soluble in polar organic solvents, such as N-methyl-2-pyrrolidinone (NMP), acetone, N,N'-dimethyformamide (DMF), and N,N'-dimethylacetamide (DMAc). The infrared spectrum of the film (Figure 7) had absorption peaks specific to imide groups (at 1755 cm^{-1} and 1795 cm^{-1}) and no peaks due to C-H bonds (around 3000 cm^{-1}). Thermal mechanical analysis (TMA) showed that the glass transition temperature was 301 °C (Figure 8). This is about 40 degrees higher than the soldering temperature. Thermal gravimetric analysis showed the initial polymer decomposition temperature was 407 °C. At 10 kHz, the dielectric constant of the film was 2.8.

Furthermore, the 10FEDA/4FMPD film cured at 200 °C (stepwise at 70 °C for 2 hours, at 160 °C for 1 hour, and at 200 °C for 2 hours) was soluble in polar organic solvents. The same phenomena have been observed for partially fluorinated polyimides (*22*). The ^{19}F NMR spectrum of 10FEDA/4FMPD dissolved to a concentration of 5 wt% in acetone-d$_6$ (Figure 9) also confirmed the completion of imidization at 200 °C. The signals in the spectrum were assigned by using substituent effects determined from model compounds (*23*). In addition, signals for the end groups of polyimides were under the observation limit. This suggests that polycondensation proceeds during the imidization. The difference in solubility between the polyimides cured at 200 °C and 350 °C may be explained by increased aggregation of the polyimide molecules with curing above Tg, and/or by condensation of unreacted end groups of the polyimide.

The visible-near-infrared absorption spectrum of the 10FEDA/4FMPD, cured at 200 °C and dissolved in acetone-d$_6$ (10 wt%), is shown in Figure 10. The same amount of acetone-d$_6$ was used as a reference. The solid line indicates the absorbance of 10FEDA/4FMPD and the dashed line indicates that of partially fluorinated polyimide (6FDA/TFDB (*6*), Structure 2). Except for a small absorption peak due to moisture (1.4 μm) absorbed in the solvent or adsorbed on the polyimide film, the perfluorinated polyimide has no substantial absorption peak over the entire WOC. Partially fluorinated polyimide, on the other hand, has an absorption peak due to the third harmonic of the stretching vibration of the C-H bond (1.1 μm), a peak due to the combination of the second harmonic of stretching vibration and deformation vibration of the C-H bond (1.4 μm), and a peak due to the second harmonic of the stretching vibration of the C-H bond (1.65 μm). The electronic transition absorption of 10FEDA/4FMPD that appears in the visible region is slightly shifted to longer wavelengths from that of 6FDA/TFDB. This shift causes the yellowish color of 10FEDA/4FMPD but does not affect the transparency in the near infrared region.

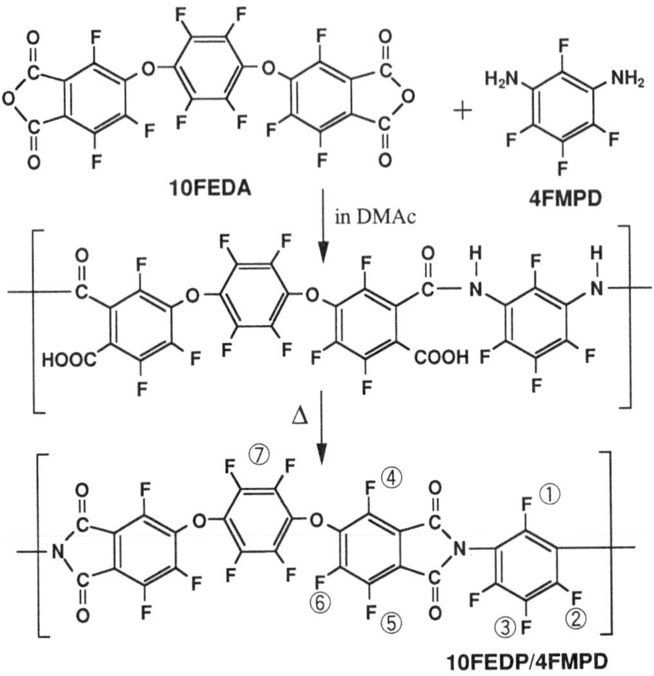

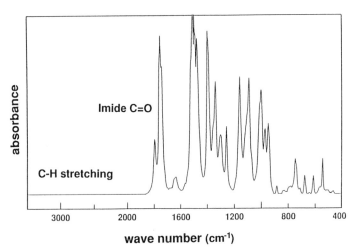

Scheme II.

Figure 7. Infrared spectrum of 10FEDA/4FMPD film.

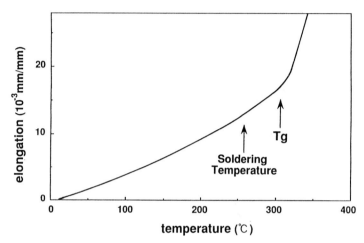

Figure 8.   TMA curve of 10FEDA/4FMPD film.

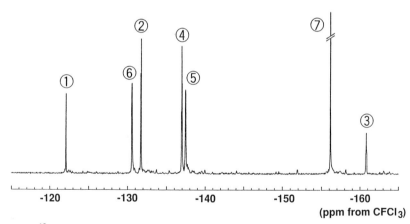

Figure 9. ^{19}F NMR spectrum of 10FEDA/4FMPD dissolved in acetone-d$_6$ (the numbering of peaks corresponds to the fluorines in Scheme II).

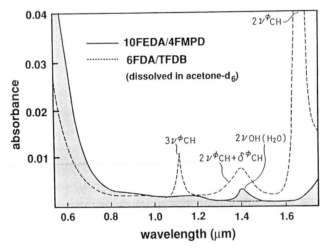

Figure 10. Visible-near-infrared absorption spectrum of 10FEDA/4FMPD and 6FDA/TFDB polyimides dissolved in acetone-d$_6$.

Structure 2.

Table II lists the strength and flexibility of perfluorinated polyimide films synthesized from the two dianhydrides and five diamines. Polymerizing 10FEDA with 8FODA or 8FSDA produced continuous and flexible films, but the films were slightly brittle compared with 10FEDA/4FMPD. As described above, P6FDA did not give any continuous films.

Table III lists the properties of perfluorinated polyimides, along with those of partially fluorinated and unfluorinated polyimides. The structure of PMDA/ODA is the same as that of Dupont's Kapton. Because of the direct introduction of fluorines into the aromatic rings and the flexible structure of the 10FEDA component, the polymer decomposition temperatures and Tg's of perfluorinated polyimides are slightly lower than those of conventional polyimides. The thermal stability of these films is nonetheless high enough to withstand the manufacturing process for IC's and multichip modules.

Their dielectric constants (ϵ) at 1kHz and average refractive indices are as low as those of the partially fluorinated polyimides. This is because the fluorine contents of perfluorinated polyimides are comparable to those of partially fluorinated polyimides. It is worth noting that the birefringence of perfluorinated polyimides is lower. This originates from the steric effect between perfluorinated aromatic rings and from a number of bent structures like an ether, a thioether and a *meta*-phenylene linkage. The low birefringence is convenient in designing the waveguide structures in OEICs and in multichip interconnections.

CONCLUSIONS

A novel polymeric material, perfluorinated polyimides, was synthesized. This material is resistant to soldering (260 °C) and highly transparent at the wavelengths of optical communications (1.0-1.7 μm). To generate high molecular weight perfluorinated poly(amic acid)s, the reactivity of five kinds of diamines and their monoacyl derivatives was estimated from the end group content of poly(amic acid), ^{15}N and ^{1}H NMR chemical shifts, and calculated ionization potential. A new perfluorinated dianhydride, 10FEDA, which has two ether linkages was synthesized in order to provide flexibility to the polymer chain. The perfluorinated polyimides prepared from 10FEDA dianhydride gave flexible films that have Tgs over 270 °C and high optical transparency over the entire optical communication wavelengths. In addition, their dielectric constants and refractive indices are as low as those of conventional fluorinated polyimides, and their birefringence is lower. These characteristics indicate that perfluorinated polyimides are promising materials for optical communication applications.

EXPERIMENTAL SECTION

Materials

We synthesized 1,4-bis(trifluoromethyl)-2,3,5,6-benzenetetracarboxylic dianhydride (P6FDA) (7), and prepared bis(2,3,5,6-tetrafluoro-4-aminophenyl)- ether (8FODA) and bis(2,3,5,6-tetrafluoro-4-aminophenyl)sulfide (8FSDA) according to the schemes reported by Kobrina (24) and Furin (25). Tetrafluoro-*p*-phenyl-

Table II. Strength and Flexibility of Perfluorinated Polyimide Films

Diamine \ Dianhydride	10FEDA	P6FDA
4FPPD	—	no film
4FMPD	strong and flexible	brittle and cracked
8FODA	flexible	no film
8FSDA	flexible	no film
8FBZ	no film	no film

Table III. Fluorine Contents, Polymer Decomposition Temperatures,
Glass Transition Temperatures, Dielectric Constants, Average
Refractive Indices, and Birefringences of Perfluorinated,
Partially Fluorinated, and Unfluorinated Polyimides

	Fluorine content (%)	Decomp. temp. (°C)	T_g (°C)	ε	$\overline{n}$	$n_{TE} - n_{TM}$
10FEDA/4FMPD	**36.6**	**501**	**309**	**2.8**	**1.562**	**0.004**
10FEDA/8FODA	**38.4**	**485**	**300**	**2.6**	**1.552**	**0.004**
10FEDA/8FSDA	**37.7**	**488**	**278**	**2.6**	**1.560**	**<0.01**
10FEDA/TFDB	35.1	543	312	2.8	1.569	0.009
6FDA/TFDB	31.3	553	327	2.8	1.548	0.006
PMDA/TFDB	22.7	613	>400	3.2	1.608	0.136
PMDA/ODA	0	608	>400	3.5	1.714	0.088

enediamine (4FPPD) and 4,4'-diaminoocttafluorobiphenyl (8FBZ) were obtained from Tokyo Kasei Co. Ltd., and tetrafluoro-m-phenylenediamine (4FMPD) was obtained from Fuji Chemical Industries Ltd. These materials were purified by sublimation under reduced pressure.

Preparation of Monoacyl Derivatives (MAD)

The five kinds of perfluorinated diamines listed in Figure 3 were added to tetrahydrofuran (THF) with equimolar amounts of phthalic anhydride to a concentration of 10 wt% and stirred at room temperature for 7 days. The monoacyl derivatives were slowly precipitated and washed by n-hexane. From the ^{13}C, ^{1}H, and ^{19}F NMR, the powder samples thus obtained were mixtures of MAD, diamine, phthalic anhydride, and di(amic acid). However, ^{15}N and ^{1}H NMR signals of amino groups of MADs were easily assigned. For 8FBZ diamine, no signals of MAD and di(amic acid) was observed.

Synthesis of 10FEDA dianhydride

Tetrafluorophthalonitrile (TFPN) and it's half-molar amount of tetrafluoro-hydroquinone (TFHQ) were stirred at room temperature, in the presence of triethylamine, in dimethylformamide (DMF) as a solvent. This reaction mixture was then poured into water, and the oily lower layer was extracted and washed again with water. This substance was recrystallized from methanol to form 1,4-bis(3,4-dicianotrifluorophenoxy)-tetrafluorobenzene (10FEDP). The 10FEDP was then stirred in 80% sulfuric acid at 200 °C for 2 hours. After cooling the acid to room temperature, the precipitated white solid was filtered and quickly washed with water, and dried. The chemical shifts and signal ratios observed using ^{19}F NMR spectroscopy were consistent with the assigned structure. A dianhydride of 10FEDA thus obtained was purified by sublimation under reduced pressure.

Preparation of Poly(amic acid)s and Polyimides

Equimolar amounts of a dianhydride and a diamine were added to DMAc to a concentration of 15 wt% and stirred at room temperature for 7 days under nitrogen. The solution of poly(amic acid) was spin-coated onto a silicon wafer and heated first at 70 °C for 2 hours, then at 160 °C for 1 hour, at 250 °C for 30 minutes, and finally at 350 °C for 1 hour.

Measurements

High resolution ^{15}N, ^{1}H, ^{13}C, and ^{19}F NMR spectra were measured at 40.56, 400.13, 100.61, and 376.49 MHz with a Bruker MSL-400 spectrometer at room temperature (22 °C $\pm$ 2 °C). Samples were dissolved to a concentration of about 6 wt% in dimethylsulfoxide-d$_6$ (DMSO-d$_6$). The ^{1}H and ^{13}C chemical shifts and ^{19}F chemical shift were read directly from internal tetramethylsilane (TMS) and tri-chlorofluoromethane (CFCl$_3$), respectively. The ^{15}N chemical shifts were calibrated indirectly through the nitromethane signal (380.4 ppm from NH_3) (*26*).

The infrared (IR) spectra were measured with a Hitachi 270-30 spectrometer using a silicon wafer as substrate. The visible-near-infrared absorption spectra were measured with a Hitachi U-3400 spectrophotometer using 10 mm quartz cells. The same amount of solvent was used as a reference. The in-plane and out-of-plane refractive indices (n_{TE} and n_{TM}) of the polyimide films were measured with an Atago 4T-Type refractometer at 23 °C. Light from the sodium D-line with a wavelength of 589.3 nm was used, and a polarizer was inserted in the light-path. Average refractive index, $\bar{n}$, and birefringence were calculated as $(2n_{TE} + n_{TM})/3$ and $n_{TE} - n_{TM}$, respectively.

The glass transition temperatures (Tg) were measured by thermomechanical analysis (TMA) with a Sinku Riko TMA-7000 analyzer. Specimen dimensions were 5 mm wide, 15 mm long, and 9-15 μm thick. The measurements were carried out during elongation with a heating rate of 5 °C/min under nitrogen at a load of 3 g. The polymer decomposition temperatures were measured by thermogravimetric analysis (TGA) with a Shimadzu TGA-50 analyzer. The measurements were conducted with a heating rate of 10 °C/min under nitrogen.

The dielectric constants were measured with a YHP 4278 capacitance meter at a frequency of 1 kHz and a temperature of 23 °C. Samples were preconditioned at 1 Torr and 120 °C for 2 h to eliminate adsorbed water.

Calculation

Ionization potentials of perfluorinated diamines were calculated as energies of the highest occupied molecular orbitals using the MNDO-PM3 semiempirical molecular orbital approximation (*20*). Calculations were performed with MOPAC Ver.6 (*27*) program with a Sony News-830 work station. Bond lengths, bond angles, and dihedral angles were fully optimized within the MNDO-PM3 framework.

REFERENCES

1. Schriever, R.; Franke, H.; Festl, H.G.; Kratzig, E. *Polymer* **1985**, 26, 1426
2. Kurokawa, T.; Takato, N.; Katayama, T. *Appl. Opt.* **1980**, 19, 3124
3. Franke, H; Crow, J.D. *SPIE* **1986**, 651, 102
4. Sullivan, C.T. *SPIE* **1988**, 994, 92
5. Reuter, R.; Franke, H.; Feger, C. *Appl. Opt.* **1988**, 27, 4565
6. Matsuura, T.; Hasuda, Y.; Nishi, S.; Yamada, Y. *Macromolecules* **1991**, 24, 5001
7. Matsuura, T.; Ishizawa, M.; Hasuda, Y.; Nishi, S. *Macromolecules*, **1992**, 25, 3540
8. Matsuura, T.; Yamada, N.; Nishi, S.; Hasuda, N. *Macromolecules* **1993**, 26, 419
9. Weeler, O.H. *Chem. Rev.* **1959**, 59, 629
10. Schleinitz, H.M. *Wire Cable Symp.* **1977**, 25, 352
11. Kaino, T.; Fujiki, M.; Nara, S. *J.Appl.Phys.* 1981, 52, 7061
12. Groh, W. *Makromol. Chem.* **1988**, 189, 2861

13. Kaino, T.; Jinguji, K.; Nara, S. *Appl. Phys. Lett.* **1983**, 42, 567
14. Kaino, T. *Appl. Phys. Lett.* **1986**, 48, 757
15. Imamura, S.; Yoshimura, R.; Izawa, T. *Electron. Lett.* **1991**, 27, 1342
16. Aosaki, K. *Plastics (Japan)* **1991**, 42, 51
17. Bessonov, M.I.; Koton, M.M.; Kudryavtsev, V.V.; Laius, L.A. *"Polyimides, Thermally Stable Polymers"* Consultants Bureau, **1987**, New York Chapter 2
18. Ando, S.; Matsuura, T.; Sasaki, S. *J.Polym.Sci. Part A, Polym Chem.*, **1992**, 30, 2285
19. Okude, K.; Miwa, T.; Tochigi, K.; Shimanoki, H. *Polym. Prep.*, **1991**, 32, 61
20. Stewart, J.J.P. *J. Comput. Chem.*, **1989**, 10, 209
21. Koton, M.M.; Kudryavtsev, V.V. *Vysokomol. Soedin. Ser. A*, **1977**, 16, 2081
22. Ando, S.; Matsuura, T.; Nishi, S. *Polymer*, **1992**, 33, 2934
23. Ando, S.; Matsuura, T.; Sasaki, S. *to be published*
24. Kobrina, L.S.; Furin, G.G.; Yakobsen, G.G. *Zh. Obshch. Khim.* **1968**, 38, 514
25. Furin, G.G.; Kurupoder, S.A.; Yakobson, G.G. *Izv. Sib. Akad. Nauk SSSR Ser. Khim. Nauk* **1976**, 5, 146
26. Jameson, C.J.; Jameson, A.K.; Oppusunggu, D; Wille, S.; Burrell, P.M.; Mason, J. *J. Chem. Phys.*, **1981**, 74, 81
27. Stewart, J.J.P. *QCPE Bull.*, **1989**, 9, 10

Received December 30, 1992

Chapter 21

Charged Species in σ-Conjugated Polysilanes as Studied by Absorption Spectroscopy with Low-Temperature Matrices

K. Ushida[1,3], A. Kira[1], S. Tagawa[2], Y. Yoshida[2,3], and H. Shibata[2]

[1]Institute of Physical and Chemical Research, 2−1 Hirosawa, Wako-shi, Saitama 351−01, Japan
[2]University of Tokyo, 2−22 Shirakata-Shirane, Tokai-mura, Ibaraki 319−11, Japan

Optical absorption spectra of the radical cation and anion of polysilanes were measured by a γ-irradiated matrix isolation technique. For both radical cation and anion, a sharp UV band and a broad IR band were observed. Although the UV bands are compatible with those reported in previous steady-state and pulse radiolysis studies, the broad band in the IR was observed even for polymethylpropylsilane (PMPrS) without any aryl pendant groups. The negative ions also exhibit similar bands in the IR which have *not* been reported in the previous matrix studies. In addition to the UV bands, which is due to the well-known charge delocalization along the σ-conjugated main chain, the IR bands indicate the existence of the charge-resonance (CR) interaction among the σ-conjugated segments. (Inter-segment charge resonance: ISCR) In the present radical ions, the charges are delocalized over more than one σ-conjugated segment in the polymer chain through ISCR.

Recently, considerable attention has been paid to polysilanes not only because of their *potential usefulness* as resists but also because of their unusual *photochemical* and *photophysical* properties (*1*). For example, the measurements of photoconductivity indicate the existence of very high mobile positive carriers (*2*). However, the electronic states of charged carriers have not been revealed yet.

From this point of view, it is useful to know the electronic properties of the charged species (the radical cations and anions) of polysilanes. In these species, a single electron or hole is trapped on the polymer and useful information about the charge distribution can be obtained by, for example, optical absorption spectroscopy (*3*).

As for the experiments with the radiation chemical technique, one of the present authors (Tagawa) presented the optical absorption spectra of both radical cation and anion of polysilanes and their pairing properties of both radical ions in his earliest work on pulse radiolysis (*4*). Tagawa and his co-workers reported that strong UV absorption bands due to the delocalized charge in the

---

[3]Corresponding authors

0097−6156/94/0537−0323$06.00/0

σ-conjugated main chain were observed for both polysilane radical cation (5) and anion (5, 6).

Irie et al also published their results of pulse and steady-state radiolyses reporting that absorption bands in the IR region were observed *only for* the radical cations of polysilanes with aryl pendant groups (7, 8). They concluded that this IR band should be attributed to the charge resonance (CR) between two aryl pendant groups. This is incompatible with our earliest conclusion because their assignment for the IR band indicates that the positive charges are not delocalized along the main chain but localized on the pendant aryl groups. Moreover, their results for the radical anion, which showed no IR band, are very different from ours and inconsistent with the pairing properties which should be observed for the radical cations and anions.

The present article describes optical absorption spectra of positive and negative ions of polysilane observed for γ-irradiated low temperature matrices. Our results agree with the previous conclusion that the charge is delocalized along the σ-conjugated main-chain. In addition, we proposed that the IR absorption should be attributed to the inter-segment charge resonance (ISCR) interaction.

EXPERIMENTAL

In the present study, we use freon mixture (FM: mixture of $CFCl_3$ and $CFBr_2CFBr_2$ in a 1:1 volume ratio) (3, 10) and *sec*-butyl chloride (sBuCl) for radical cation and MTHF for radical anion as matrices. Purification methods are the same as those described elsewhere (3, 11, 12). Polymethylpropylsilane (PMPrS) and *Polymethylphenylsilane* (PMPS) were synthesized according to the conventional method (1, 13).

Absorption spectra measured in 1.5 mm quartz cells at 77K were recorded on a Cary 14RI spectrophotometer. All the samples were sealed in the cell and irradiated with γ-ray from a ^{60}Co source at The Institute of Physical and Chemical Research. Photo-bleaching was performed with a tungsten lamp incorporated in the spectrophotometer through Toshiba glass filters (3, 11, 12, 14).

All the polymers were dissolved in a matrix for 0.1-20mM (monomer units). Only PMPS, however, could not be successfully dissolved in FM or sBuCl at sufficient concentration to produce PMPS radical cation successfully.

RESULTS

Charge Transfer Reaction From The Matrix To Polysilanes

The method of preparation of radical ions used in this work is widely known in the field of radiation chemistry (3, 11, 12). The radical ions of the solute are produced by a charge- (hole- or electron-) transfer process from γ-irradiated low temperature matrices. This method is applicable for the formation of radical ions of polymers including polysilanes (7, 8), and a number of reports about polymer ion radicals have been published so far (12). However, some problems exist in formation of polymer radical ions because of their low solubility and the tendency to be aggregated in low-temperature matrices. Under such conditions, the solvent and solute systems are sometimes mixed inhomogeneouly and some difficulties occur: (i) Since part of the polymers cannot be mixed homogeneously with solvent molecules, the yield of the charge scavenging process is lower than expected from the solute concentration characterized by the monomer units. (ii) Since the solubility of polymers are quite low in general, the effective concentration of the polymer is sometimes insufficient for the charge-scavenging process from the solvent to the polymer. (iii) If the charge can travel through the inside of polymers by some sequential charge-transfer processes or charge delocalization, additional charge recombination process will be feasible and the yield of the ions will decrease: Therefore, assuring the occurrence of charge-scavenging process with no side reactions, careful examination should be indispensable for assignment of absorption spectra obtained for polymer systems.

Actually, residual holes or electrons were sometimes observed and the dose dependence check and the stoichiometric consideration were performed in each case. The irradiation dose was carefully restricted within a magnitude where all the spectral lineshape was not changed and the intensity growth was linearly dependent on the irradiation dose.

Positive Charge Trapped on PMPrS

Figure 1 indicates absorption spectra taken for γ-irradiated FM glassy solution of PMPrS which should be attributed to the radical cation of PMPrS. Immediately after irradiation, there exist several peaks in the spectrum marked as (a). After photobleaching with IR light (> 900 nm), the spectral lineshape changed gradually to become spectrum (b), which consists of two band peaks at 350 nm and 1600 nm. Similar two bands, which decayed with the same lifetime, were observed for the cationic species of the same compounds in the pulse radiolysis study (*4–6, 9*). Photobleaching with monochromatic light of 350 nm slowly reduced the intensity of both bands. This indicates that both bands should be attributed to the same cationic species which decomposes with a slow photochemical reaction. Since the spectral change on the IR photobleaching progresses irreversibly, the last lineshape (b) should be attributed to the fully relaxed radical cation.

The strong peak at 350nm in the spectrum (b) is narrow and was unchanged during photobleaching; therefore, we assigned these UV peaks to a σ-σ * transition within a single σ-conjugated segment (*1, 15*). In contrast to the ISCR transition discussed later, here we term this transition *intra-segment excitation* (ISE). This assignment is essentially the same as those reported in previous reports (*4–6*).

The broad band observed in visible-IR region (400-2000nm) in the initial spectrum (a) is similar to the spectrum for the residual hole trapped in the FM matrix (*10*) but the position of the spectral peak is different from those obtained for γ-irradiated neat FM glass. On the spectral change from (a) to (b), no radical cation seems to be newly produced by photobleaching because the intensity of the UV peak was constant. Therefore we concluded that the whole absorption bands from the visible to the IR region should be attributed to the charge resonance (CR) interaction between two σ-conjugated segments each of which acts as a trap for positive holes. (see Appendix)In this article, we term this CR interaction *Inter-Segment Charge Resonance* (ISCR). The whole band in seems to consist of a number of CR bands having various transition energy gaps depending on the distribution of the σ-conjugated segments. A detailed explanation will be presented in the Discussion section.

Essentially the same result was obtained for the sBuCl matrix.

Negative Charge Trapped on PMPS

The absorption spectrum for the radical anion of PMPS was obtained in MTHF matrix as is indicated in Figure 2, although no valid data was obtained for positive charges of PMPS. Initially, the strong absorption of the trapped electron was observed (a) but disappeared on photobleaching. (see insertion) The residual spectrum also has two main peaks, an ISE peak at 365 nm and a ISCR peak (> 2000 nm). Structural relaxation could not be distinguished because the unrelaxed anion may be bleached together with trapped electrons. However, spectral conversion of the ISCR band seems fairly small or non-existent because the spectrum displayed in the insertion is almost the same as the spectrum for trapped electrons reported previously (*3*). Accordingly, the final spectrum should be assigned to the fully-relaxed radical anion. The IR peak position seems identical with those for PMPrS(-), which indicate that only a small amount of negative charge would be distributed on the side-chain phenyl groups in PMPS.

Negative Charge Trapped on PMPrS

For PMPrS/MTHF system, the strong absorption of the trapped electrons was observed immediately after irradiation and disappeared on photobleaching, which is compatible with the results for PMPS in MTHF. (These results are not shown.) Finally a weak absorption band remained in the IR region, of which the peak seemed to exist at > 2000 nm. Considering the results for PMPS in MTHF described below, the band should also be assigned to the ISCR band of the PMPrS radical anion. The low yield of the radi cal anion should reflect upon the small electron affinity of PMPrS. However, the UV (ISE) peak could not be measured because the self-absorption of PMPrS obscures it in such a highly concentrate solutions. For diluted solution, the scavenging process was not completed and no absorption spectra for the radical anion could be observed.

Comparative Analysis with Pulse Radiolysis (*4–6, 9*)

In the pulse radiolysis study for the liquid phase, similar UV bands have been reported previously (*4–8*). Our recent pulse radiolysis study also indicates that all the radical ions studied here exhibit the ISCR band in the IR region (*9*). Moreover, the slow (several microseconds) spectral relaxation of the IR band was also observed in the pulse radiolysis study (*9*). This result is compatible with the present matrix study especially for PMPrS(+). However, several differences were observed for peak positions and intensities. In the liquid phase, the

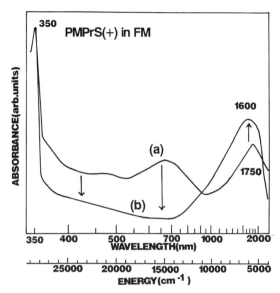

Figure 1. Optical Absorption Spectra Observed for γ-irradiated PMPrS/FM(20mM). (a) Immediately After Irradiation (b) After Photobleaching With IR Light.(> 900nm)

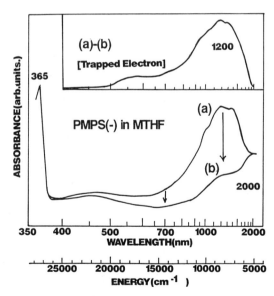

Figure 2. Optical Absorption Spectra Observed for γ-irradiated PMPS/MTHF(20mM). (a) Immediately After Irradiation (b) After Photobleaching With IR Light. (> 900nm): The Differential Spectrum in the Insertion is Attributed to the Trapped Electron.

polymer main chain should cause fast fluctuations, and the structural dependence of absorption spectra should be averaged out.

DISCUSSION

The Intra-Segment Excitation (ISE) Band Observed in UV Region

The sharp UV band was observed for both PMPrS(+) and PMPS(-) in our experiment. Similar bands have also been reported for both radical cations and anions of various polysilane derivatives previously (4-8). This band is extremely sharp and was unchanged by any photobleaching by IR light. This result indicates that the transition energy of this band is insensitive to the geometry of the main chain in contrast to the IR band. Therefore this band should be attributed to the σ- $\sigma *$ excitation within a single Si-Si conjugated segment (ISE).

Although we could not observe this band for PMPrS(-) and PMPS(+) in the present study, well-resolved absorption spectra were obtained in the previous pulse radiolysis experiment for liquid phase (4-9). The radical cation and anion of the same polysilane have the UV band at almost the same wavelength. This result indicates that *the Pairing Theorem* is valid for the polysilane radical ions and that the interaction with other σ-orbital is quite small. (see Appendix)

The Inter-Segment Charge Resonance (ISCR) Band Observed in the IR Region

For PMPrS(+), PMPrS(-) and PMPS(-), strong, broad bands were observed in the visible-IR region. We concluded that all these bands should be attributed to the charge resonance among delocalized positive or negative charges in main chain conjugated segments, (ISCR) for the following reason.

Irie et al (7, 8) have reported that the radical cation of phenyl-substituted polysilanes exhibit a similar band in the IR region and assigned it to the CR band between two phenyl groups. They also reported that no IR absorption was observed for the radical anion. In the present study, however, we observed this IR band also for PMPrS(+) and for two polysilane radical anions. Since PMPrS has no aromatic groups, the optical transition should be related with the delocalized positive charge on *the σ-conjugated polymer main chain.*

We consider that the whole spectrum observed in the visible-IR region (400-2000 nm) at the first stage (Figure 1a) should consist of a number of overlapping bands which were finally relaxed into a single band centered at 1300 nm. (Figure 1b). Since photobleaching is thought to realign the geometry of the Si-Si main chain, the experimental results indicate that these optical transitions in the IR should depend delicately on the geometry of the polymer main chain. Based on these results, we finally assigned these strong, broad IR bands to the ISCR bands.

Another choice of the interpretation of this IR band could be a phonon-side band of the trapped charge based on a polaron model (16, 17). However, we suppose that the ISCR model is more appropriate to explain our present results in which the spectrum indicates a variety of transition energies including relatively large values such as 2 eV at the first stage before photobleaching. (Figure 1a) The magnitude of the electron-phonon coupling in this system is known to be very small. Moreover, no spectral shifts were observed for the UV band which correspond to a wide spectral shift of the IR band.

Essentially the same interpretation is applicable to the IR bands observed for PMPrS(-) and PMPS(-).

The Dynamic Behavior on Photo-Bleaching Observed for PMPrS(+)

It is well-known that polysilanes have large σ-conjugated system on the main chain. Under the present experimental conditions the polymers are frozen in randomly folded structures and some gaps are formed in the σ-conjugated system where Si-Si bonds are fixed in some irregular conformation. Accordingly, the main chain is separated into a number of σ-conjugated segments. Each of these segments should act as the traps for either positive holes and electrons, which have various sizes and depths (*18*).

Immediately after irradiation, positive charges are trapped randomly at one of the nearest-neighbor traps (which include shallow and unstable ones), and accordingly the complex spectral lineshape as in Figure 1a reflects the existence of various types of positive hole traps.

On photobleaching of the next stage, these holes are excited again and obtain enough energy to migrate towards a more stable trap. Moreover, an amount of thermal energy sufficient to change the folded structure (maybe on local annealing of the surrounding matrices) should be ejected on the non-radiative decay of the excited radical cation. Although a similar spectral shift on annealing the solution has previously been reported by Irie et al for polysilastyrene, they did not find any shift with photobleaching (*7*). Finally the bulk system should attain the most stable hole distribution on the cation polysilane main chain.

As is referred above, the spectrum indicates a variety of ISCR transition energies including large values up to 2 eV. Accordingly, if there are various segment pairs in various geometries, there will exist a variety of CR-transition energies. This explanation based on ISCR is displayed in schematic diagram of Figure 3c.

After photobleaching, the IR band was converged into a single peak band (Figure 1b) and the average value of the CR transition energy seemed to be lowered. This may indicate the enhancement of the CR area. The irreversible relaxation process on photobleaching should be interpreted as a charge transfer from one short σ-conjugated segment to another longer one. Since each segments are thought to be sufficiently long, the splitting of ISE energy gap is not sensitive to the length of the segment. However, the magnitude of ISCR interaction will be smaller when the charges are more delocalized in one segment. (see Appendix) In addition, this relaxation may include the reorientation of the Si-Si main chain itself on annealing due to local heating of the matrix. The converged value of the ISCR energy should reflect the most stable geometry of CR interacting σ-conjugated segments which corresponds to the most preferred conformation of the bridging part of the main chain.

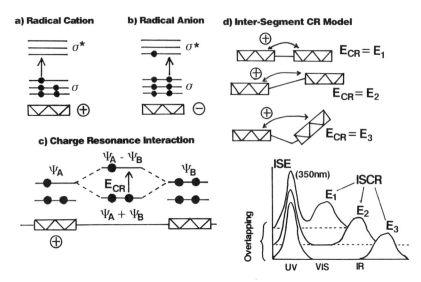

Figure 3. Electronic Energy Diagram of Polysilane Radical Ions (a) Radical Cation (b) Radical Anion (c) Energy Diagram of Charge Resonance Absorption Band in The Case of A Pair of Cationic and Neutral Segments. (d) Two Centered Charge Resonance (CR) Model Applied For Two Segments Taking Part In the Interaction. The CR Band Appears Extremely Broad Because of Number of Overlapping CR Bands Corresponding to a Variety of CR Transition Energies.

For radical anions, almost no photochemical relaxation was observed and fully relaxed anions seemed to have existed immediately after irradiation. It is well-known for the present matrix method that the stabilization energy of negative charge is always smaller than for the positive charge. Therefore the negative charge can be delocalized along the polysilane main chain from the beginning to make a most stable distribution of the negative charges.

CONCLUSION

Based on our measurements of optical absorption spectroscopy, we obtained several conclusions as follows.

(1) Both positive and negative radical ions of the two types of polysilanes exhibit two prominent absorption bands at the UV (350-365nm) and IR (1300-2000nm) regions. These results are compatible with those obtained by recent pulse radiolysis studies (9).

(2) The UV band is the σ-σ^* transition band within a single conjugated segment. (ISE band)

(3) The IR band is attributed to the charge-resonance interaction among a number of conjugated segments of the polysilane main chain.(ISCR band)

(4) The structural relaxation of the main chain was observed on photobleaching.

(5) Our results are incompatible with those reported by Irie et al (7, 8) which includes following remarkable difference from ours:

a) The IR absorption bands were observed even for the radical ions of polymethylpropylsilane (PMPrS), which has no aromatic groups on the sidechain. b) We also observed the same IR bands for radical anions of polysilane. c) Our results support that the charges of radical ions are delocalized on polymer main chain, however Irie et al. assumed that they are localized on the aromatic side chain.

APPENDIX

Background and Definition of the Terms Used for the Assignments in This Article

Once radical cations are produced, characteristic absorption often emerges in visible-IR region because new transitions involving the SOMO (singly occupied molecular orbital) become possible; this is illustrated schematically in Figure 3a and 3b for the case of radical cation and anion of polysilanes, respectively. The energy diagram is based on the case of the Sandorfy model C (*1a, 15*).

When the parent neutral molecule has well-conjugated electronic orbitals without any interactions with orbitals in other symmetries, we note that the observed absorption transition for the radical cation and anion should be identical. This nature is widely known as "*the Pairing Theorem*". Concerning the σ-conjugation of the polysilane main chain, this theorem should be closely related to the Sandorfy C and H orbital models (*1a, 15, 19*). In the case that this theorem is approximately valid, the electronic property of the main chain can be explained by the model C, which is a more simple approximation than the other.

When a molecule has equivalent or near equivalent groups strong and broad *charge resonance* (CR) bands are frequently observed (*20*), for which a diagram is also indicated in Figure 3c. The CR band has a transition energy of E_{CR}, as is indicated in Figure 3c. In the text we applied this interpretation to the present case, regarding each segment of the polysilane as an interacting groups. When the segments are sufficiently long, each electronic states should converge into one level and should give almost no dependence on the length of the segments. Consequently, we can regard each segments approximately equivalent.

In the CR interaction, we can include both static (time-independent) and dynamical (time-dependent) perturbation such as direct overlap of the orbitals (static perturbation), intramolecular vibrational coupling and dynamical exchange interaction induced by surrounding systems (dynamic perturbation). This optical transition is always symmetrically allowed and has a fairly large transition dipole moment which lies along the line connecting the two centers. Since ECR should be corresponding to the magnitude of the interaction energy between two chromophores, it depends delicately on the geometry (e.g. distance and orientation) and dynamic property of the two chromophores.

The CR band is well-known as an evidence for formation of dimer radical ions (*11, 12, 21*).

REFERENCES

1. a) Miller, R. D.; Michl, J. Chem. Rev. 1989, 89, 1359 b) Wallraff, G. M.;

Miller, R. D.; Baier, M.; Ginsburg, E. J.; Kurz, R. R. J. Photopolym. Sci. Technol. 1992, 5, 111.

2. a) Kepler, R. G.; Zeigler, J. M.; Harrah, L. A.; Kurtz, S. R. Phys. Rev. 1987, B35, 2818 b) Fujino, M. Chem. Phys. Lett. 1987, 138, 451 c) Abkowitz, M. A.; Stolka, M.; Weagley, R. J.; McGrane, K. M.; Knier, F. E., 1990 p467 "Silicon-based Polymer Science" ACS Adv. Chem. Ser. 224, edited by Zeigler, J. M.; Fearonin, F. W. G.

3. Shida, T., Electronic Absorption Spectra of Radical Ions; Elsevier 1988

4. Tagawa, S. Abstract of IBM Polymer Colloquium '87, May, 1987, Fuji, Japan and Abstract of Intern. Topical Workshop, "Advances in Silicon-based Polymer Science", 1987, Hawaii, 20.

5. a) Tagawa, S.; Washio, M.; Tabata, Y.; Ban, H.; Imamura, S. J. Photopolymer Sci. Technol. 1988, 1, 323 b) Tagawa, S. Polymer Preprints, ACS, Washington, 1990, 31, 242.

6. a) Ban, H.; Sukegawa, K.; Tagawa, S. Macromolecules, 1987, 20, 1775, ibid, 1988, 21, 45 b) Ban, H.; Tanaka, A.; Hayashi, N.; Tagawa, S.; Tabata, Y. Radiat. Phys. Chem., 1989, 34, 587.

7. Irie, S.; Oka, K.; Irie, M. Macromolecules, 1988, 21, 110

8. Irie, S.; Irie, M. Macromolecules, 1992, 25, 1766

9. Tagawa, S; Yoshida, Y; Ushida, K.; Kira, A. in preparation.

10) Grimson, A.; Simpson, G.A. J. Phys. Chem., 1968, 72, 1776

11. Shida, T.; Haselbach, E.; Bally, T. Acc. Chem. Res. 1984, 17, 180

12. Kira, A. Chapt, VIIIA in Handbook of Radiation Chemistry edited by Tabata, Y.; Ito, Y.; Tagawa, S. CRC Press 1991

13. Trefonas, P.; Djurovich, P. I.; Zhang, X.; West, R.; Miller, D.; Hofer, D. J. Polym. Sci., Polym. Lett. 1983, 21, 819

14. Ushida, K.; Shida, T.; Shimokoshi, K. J. Phys. Chem. 1989, 93, 5388

15. Sandorfy, C. Can. J. Chem. 1955, 33, 1337

16. Rice, M. J.; Phillpot, S. R. Phys. Rev. Lett. 1987, 58, 937

17. Tagawa, S. J. Photopolym. Sci. Technol. 1991, 4, 231

18. Klingensmith, K. A.; Downing, J. W.; Miller, R. D.; Michl, J. J. Am. Chem. Soc. 1986, 108, 7438

19. Herman, A.; Dreczewski, B.; Wojnowski, W Chem. Phys. 1985, 98, 475

20. Ishitani, A.; Nagakura, S. Mol. Phys. 1967, 12, 1

21. Ushida, K.; Shida, T. Chem. Phys. Lett. 1984, 108, 200

RECEIVED January 28, 1993

Chapter 22

Acid-Sensitive Phenol–Formaldehyde Polymeric Resists

W. Brunsvold, W. Conley, W. Montgomery, and W. Moreau

IBM Technology Products Division, Route 52, B 300–40E, Hopewell Junction, NY 12533

Novolak (cresol-formaldehyde) resins were converted into positive working resists by partial esterification of the phenolic groups with di-t-butylcarbonate. In the presence of an acid source, fast resists were formulated for 248 nm (8 mj/cm^2), I line, electron beam, and X- ray exposure. Lithographic performance issues of poor adhesion and film shrinkage ($> 25\%$) encountered with resists based on 100% TBOC esters of novolak were eliminated by a minimum bock content (< 20 mole%). For top surface imaging, negative type images were produced by the gas phase and by the liquid phase silylation of the novolak groups formed by the acid catalyzed removal of TBOC groups.

Novolak resins in conjunction with diazoquinone dissolution inhibitors have been the dominant photoresist for the last two decades (*1*). The novolak resin of cresol formaldehyde has served as the base soluble resin for many posiitve formulations operating in the 300-500 nm region. In the last five years, new acid catalyzed resists based on the dissolution inhibition of polyhydroxystyrene(PHOST) have been formulated for deep UV (240-260 nm) lithography (*1*). For example, the dissolution inhibition of novolak or PHOST can be accomplished by addition of small molecules such as diazoquinones or t-butylcarbonate (TBOC) esters of bis-phenol A (TBOCA) (*2, 19*).

In the presence of a photoacid, the TBOC group of TBOCA is converted to the free phenol group to induce the dissolution of the exposed region in the alkaline developer. Another example of a TBOC based resist is the TBOC ester of PHOST which undergoes an acid catalyzed removal of TBOC sites (*3*) to form an alkaline soluble PHOST.

0097–6156/94/0537–0333$06.00/0

PTBOC $\longrightarrow$ PHOST

The TBOC based resists are examples of resists which can be aqueous alkaline developed. The PHOST resin used in acid sensitive resists are more transparent in the deep UV region than their novolak counterparts and thus have been the focus of more investigations. Previous attempts to use diazquinone novolak resist in the deep UV region are hampered by the higher absorbance of the novolak resin and the diazoquinone photoproducts (4). Only thin films less than 500 nm thick can be used for the deep UV region (4).

In the deep UV region, exposure tools which use the conventional Hg lamps require fast resists (< 15 mj/cm^2) in order to provide short exposure times and sufficient throughput. To meet the photospeed requirements, resists based on the incorporation of TBOC groups onto the PHOST backbone pioneered the era of acid catalyzed amplified systems. Both negative, positive, and top surface imaging (TSI) resists based on TBOC have been formulated for the deep UV region (3, 5).

Although PHOST resins based systems are more transparent than their novolak counterparts in the deep UV region, the lower costs of novolak have prompted this investigation of novolak based deep UV resist for the deep UV region. For deep UV and for electron beam lithography, three component resists consisting of a novolak or PHOST resin, an acid source(AG), and an acid labile dissolution inhibitor have been investigated. The dissolution inhibiitors mixed with novolak were of the molecular type such as TBOCA (2) or more recently of the polymeric type such as the TBOC ester of PHOST (6) or the TBOC ester of novolak (7). In the deep UV, the TBOCA-novolak formulation exhibited a low contrast of < 2 presumably due the high absorbance (B > 0.5/um) of the novolak (2). In the other cases for electron beam exposure, the TBOC-PHOST was immiscible with the novolak (6) or the TBOC-NOVOLAK formed negative images with a triphenylsulfonium triflate (7) acid generator. Finally a recent investigation of TBOC esters of novolak as electron beam resist have found negative working images are formed with 100% TBOC esters of novolak (8).

Although TBOC based resist are fast positive resist system some shrinkage issues due to loss of TBOC groups have been noted such as sidewall image distortion and the loss of the film ($> 25\%$) during conditions such as in reactive ion etching (RIE).

The purpose of this work was to formulate TBOC esters of novolak (NOVOBOC) for deep UV, near UV, E-beam, and X-ray and TSI lithography. Lithographic performance issues were addressed by examining novolaks with varying amounts of TBOC groups.

Table 1. NOVOBOC (Novolak) Resin Precursors

Novolak	M_w	M_n	M_w/M_n	$T_g,°C$	248 nm A/um	254 nm A/um
meta cresol	19.5K	3.4K	5.8	119	0.53	0.36
meta cresol	8.3K	1.7K	4.9	110	0.52	0.38
meta/para	7.5K	1.2K	6.3	94	0.43	0.34
bis-phenol A	3.7K	0.9K	3.8	79	0.41	0.28
cresylic acid	17.9K	2.3K	7.6	105	2.15	1.92
para cresol	3.7K	0.7K	5.6	81	0.33	0.33

EXPERIMENTAL

Various novolak resins, Table 1, were obtained from commercial sources(Schenectady Chemical, Reichold, and Rhone-Poulec) and characterized for their absorbance(A), molecular weight by gel permeation chromatography, and glass transition (T_g). Various amounts of TBOC groups were introduced on the phenolic site of the novolaks by reaction with di-t-butylcarbonate using a trace of amine as a catalyst (9). Photoacid generators (PAG) of the triflic acid type were selected for deep UV(PDT) and I line (DPMT) from derivatives of N-hydroxymaleimide-triflates (10).

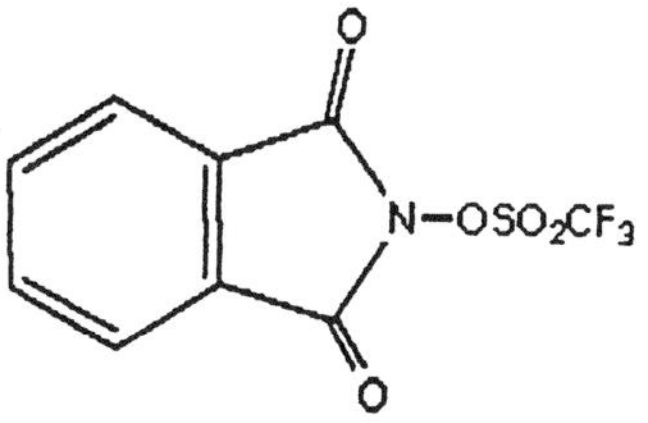

PDT
PAG DEEP UV

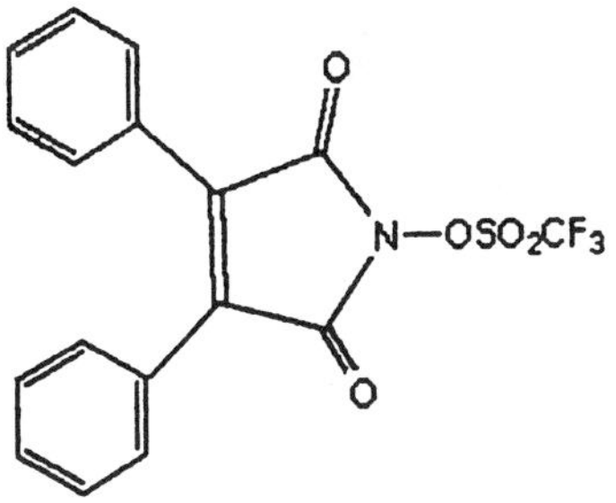

DPMT
NEAR UV PAG

For top surface imaging, after deep UV exposure, silylation of the image was performed in the gas phase using N,N-diethylaminotrimethylsilane (DEATS) or in the liquid phase (HMCTS) using hexamethylcyclotrisilazane (11). Photospeed, resolution, and contrast were determined using 50 KeV IBM EL-3 electron beam, Brookhaven synchrotron X-ray radiation, GCA I line stepper, Perkin Elmer 500(deep UV mode) and ASM excimer laser deep UV stepper.

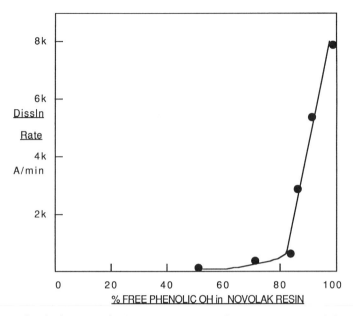

Figure 1. Dissolution rate in 0.263 N TMAH of meta/para novolak esterified with varying amounts of TBOC.

RESULTS AND DISCUSSION

The acid catalyzed removal of TBOC groups from PHOST is a highly efficient reaction for forming positive or negative images. For the case of positive working resists, the presence of many TBOC groups left on remaining resist, can significantly lower or alter the physical, thermal and reactive ion etch properties of the resist. In addition, the masking of phenolic sites by TBOC may reduce the adhesion to silicon type surfaces. Image distortion can occur due to the loss of TBOC groups during post expose bake. A reduction in the number of TBOC groups could enhance the overall lithographic performance of the resist. Thus we sought to determine the minimum of TBOC esters on the novolak for dissolution inhibition. Using a fixed developer concentration of 0.263 N tetramethylammonium hydroxide(TMAH), the minimum TBOC of the novolak to suppress the dissolution of the resin was in the range of 15-20 mole%. For a 55/45 meta/para cresol novolak of 7K M_w, the minimum TBOC content to provide a unexposed film thinning rate of $< 500A^0$ nm/min., Figure 1, was 17 mole%.

After exposure to generate an acid and subsequent heating, the TBOC ester of the phenol group of PHOST or novolak is removed to reform the phenol structure. The exposed region rapidly dissolves in the alkaline developer. In the case of a novolak resin with 100% TBOC content, Figure 1, the TBOC level has to be lowered to < 10 mole% in order for the exposed region to dissolve faster than $6000A^0$/min. The net loss of TBOC group is of the order 90 mole%. If an initial lower TBOC content of 20 mole% is used, a net smaller conversion (10 mole%) of TBOC groups to free phenol is required before onset of dissolution (> 6000A^0/ min). A novolak with a 90 mole% TBOC content required a dose of 30 mj/cm^2 while a TBOC with 17 mole% content required 3X lower dose. Thus, fewer chemical events are required and hence faster resists.

To evaluate the imaging capability of various TBOC contents of a meta/para NOVOBOC, a Meyerhofer type plot (*1, 18*), Figure 2, reveals a projected R/R$_0$ (dissolution rate ratio of exposed to unexposed thinning rate) of > 10 for a TBOC content of 15 mole%.

When compared to novolaks of high TBOC contents, other lithographic benefits can be derived from novolaks of reduced TBOC contents. The benefits are related to the presence of more phenolic groups. The adhesion, for example, to oxide surfaces was improved (no lifting of submicron islands). However, in processing, care must be exercised in the post apply bake step as to not exceed 110^{0}C since the phenolic group is sufficiently acidic (*12, 13*) to catalyze the removal of TBOC groups, Figure 3. After exposure, the TBOC based resists are baked to complete the acid catalyzed conversion to an aryl-OH group. However, with high TBOC contents, considerable film shrinkage (> 25%) of the exposed region occurs. Film cracking at the edges of the image into the unexposed region was observed for high TBOC contents (> 50 mole%). Later, in processing, the remaining resist mask can be subject to elevated temperatures such as in ion implantation or in RIE. In this case considerable film loss (> 25%) can occur. The shrinkage and loss of the resist film can be significantly reduced to < 5% for NOVOBOC with a TBOC content of < 20 mole%. In the deep UV region, novolak resins, in general, absorb a significant amount of radiation. For an effective transparency, (< 0.4/um) the novolak has to be carefully chosen or its structure modified (*14*). At 254 nm, when compared to PHOST, some novolaks exhibit a low absorbance window, Table 1, Figure 4. However,at 248 nm, a sharp increase is noted. Novolaks of para cresol structure are the closest match of PHOST transparency, but unfortunately because of strong intramolecular H-bonds are not very soluble in alkaline developers. Novolaks based on bis-phenol A,

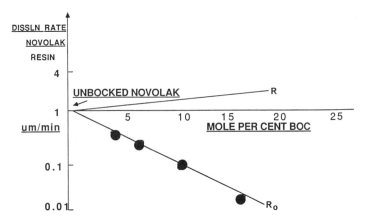

Figure 2. Effect of TBOC content of meta/para NOVOBOC on dissolution rate of exposed resist (R) and unexposed resist rate (R_0) in 0.263 N TMAH.

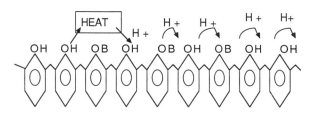

Figure 3. Schematic of thermolysis of NOVOBOC activated by acidic phenolic group.

possess an additional phenolic group which impart base solubility. NOVOBOC based on bis-phenol A and meta-para novolak were further investigated as deep UV, X-ray and electron beam resists.

In the deep UV region, the photoacid generator of pthalimide-triflate (PDT) has negligible absorbance at the concentration used (5 wt%). The absorbance of the NOVOBOC of meta para novolak, Figure 5, is dominated by the phenolic group. Previous studies of electron transfer reactions of model compounds of phenol type (15, 16) and of novolak resin in the deep UV (19) suggest that the energetics, Figure 6, are favorable for photoacid generation. In the deep UV the highly absorbing phenolic group of TBOC photosensitizes the generation of acid from the weakly absorbing PDT.

In device lithography with resists, reflective notching and interference effects caused by substrate reflectivity, can be suppressed by using thin antire-

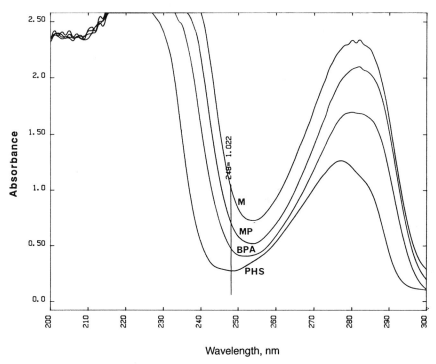

Figure 4. UV absorbance spectra of 1000 nm thick films of novolaks, M-cresylic acid type, MP-meta/para, BPA-bis phenol A, and PHS-polyhydroxystyrene.

flective layers or by using multilayer resists consisting of a silicon containing resist on top of an absorbing planarizing layer (*17*). TSI combines the separate layers into one film and incorporates Si by silylation of the exposed resist (*5*).

For TSI, the highest absorbing novolak of a meta cresol type (Table 1 and Figure 7) was fully esterified with TBOC, and used at a film thickness of 1500 nm. Two silylation and development schemes were used for the gas phase or for the liquid phase silylation, Figure 8. In the gas phase process, after a dose of 35 mj/cm^2 at 248 nm, DEATS was used to incorporate Si into the novolak image structure. A negative image was formed by oxygen reactive ion etch (RIE) development, Figure 9.

Liquid phase silylation process can use silylation agents such as HMCTS (*11*) which can also crosslink the silylated region and prevent flow of the image

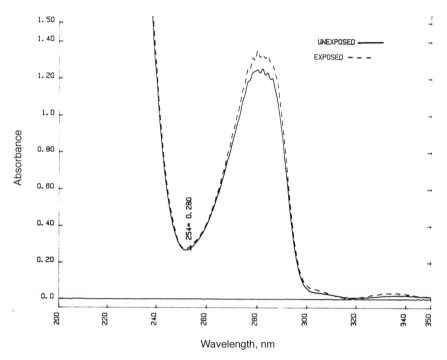

Figure 5. Deep UV absorbance spectrum of meta/para NOVOBOC before and
after post expose bake.

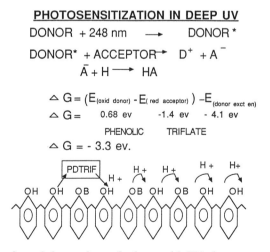

Figure 6. Energetics of formation of photoacid HA by energy transfer from
donor phenolic group of NOVOBOC to PDTRIF (acid generator).

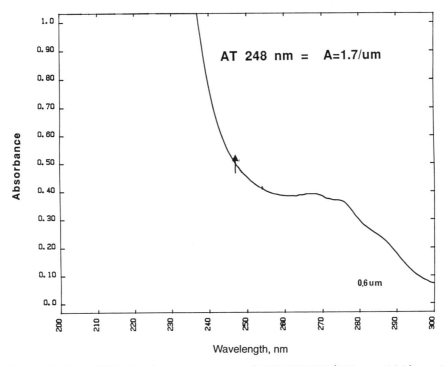

Figure 7. Deep UV absorbance spectrum of NOVOBOC (900 nm thick) used for TSI.

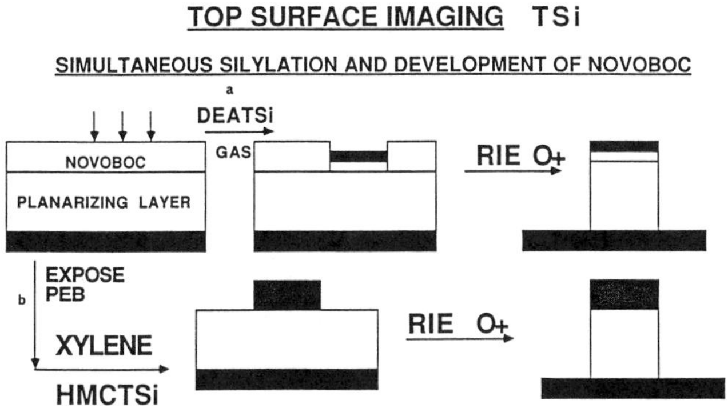

Figure 8. Schematic of TSI imaging processes for NOVOBOC, (a) gas phase silylation with DEATS followed by RIE in oxygen, (b) simultaneous liquid phase development and silylation with HMCTS followed by RIE in oxygen.

for high temperature lift off. In this application, a NOVOBOC (500 nm thickness) was coated on top of a polyimide layer (1000 nm thickness). The exposed NOVOBOC was baked and then developed in a mixture of xylene and 10% by wt. of (HMCTS) which simultaneously developed the unexposed region and silylated the exposed region. The final image was transferred through the polyimide layer by oxygen RIE, Figure 10.

For normal deep UV type imaging, the most transparent NOVOBOC resist of bis phenol A was used and the contrast was determined to be 3.9, Figure 11. Submicron images, Figure 12, were formed at doses of < 10 mj/cm^2. For electron beam exposure (Figure 13) or for X-ray lithography, the opacity of the novolak is not of concern. For I line applications, the NOVOBOC resin is transparent and the acid generator of DPMT absorbs all of the radiation. For electron beam, X-ray and I line, a NOVOBOC meta-para, Table 1, was used. The contrast of each type of resist were determined to be in the range of 3-4, Table 2.

Although this study of the NOVOBOC resist system has concentrated on two component formulations consisting of a NOVOBOC and acid generator (AG),the NOVOBOC resin can also be used as a dissolution inhibitor of PHOST or novolak. PHOST or novolak completely esterified with TBOC are noted to be immiscible with PHOST or novolak resins (6, 7). Partially esterified NOVOBOCS of this study are miscible with PHOST and novolak (single T_g and no phase separation). The good intermixing has been ascribed to efficient hydrogen bonding between the phenolic and TBOC groups (12). As an example, a meta cresol of 20 mole% TBOC content was mixed at various wt.% with a meta-para novolak. A Meyerhofer plot, Figure 14, reveals that at 8% loading, produces a $R/R_0 > 100$.

Lastly, for applications of the resist such as in ion implantation or as a RIE mask for metal etching, post image hardening by deep UV of diazoquinone-novolak is often employed (20). The high absorbance of the novolak or diazoquinone photoproducts limits the penetration of the radiation and also requires high doses of > 1 J/cm^2. The NOVOBOC resist is more transparent in the deep UV and lower doses (3X) are required. Acid induced crosslinking of novolaks has been reported (21). In this case, the additional acid generated from the AG may faciliate the novolak crosslinking.

SUMMARY AND CONCLUSIONS

The introduction of less than 100% of TBOC leaving groups onto a low cost novolak resin improves the adhesion, photospeed, and film retention during lithographic processes. NOVOBOCS were formulated from novolaks at 15-20

Table 2. Contrast of NOVOBOC Type Positive Resists

	I LINE	DEEP UV	E BEAM (50 KeV)	X-Ray	TSi
CONTRAST	3.5	3.9	4.1	4.5	
DOSE/cm^2	27 mj	8 mj	3 uC	12 mj	35 mj

Figure 9. TSI image of (500 nm) in 1200 nm thick NOVOBOC.

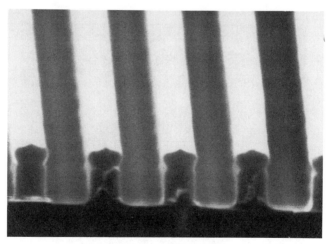

Figure 10. Bilayer image of 500 nm in NOVOBOC/polyimide formed by liquid phase silylation/development followed by RIE oxygen transfer.

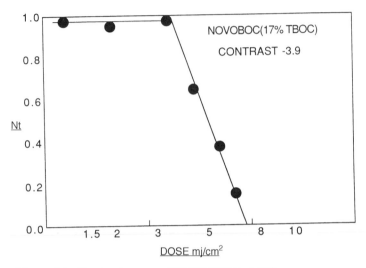

Figure 11. Contrast curve of NOVOBOC at 248 nm exposure.

Figure 12. Submicron images (400 nm) formed in NOVOBOC with dose of 8.5 mj/cm^2 at 248 nm.

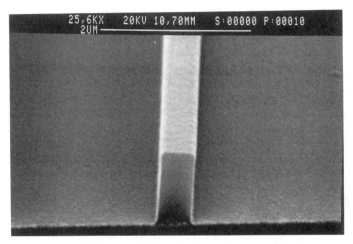

Figure 13. Submicron images of 500 nm formed by electron beam dose of 3 microcoul/cm².

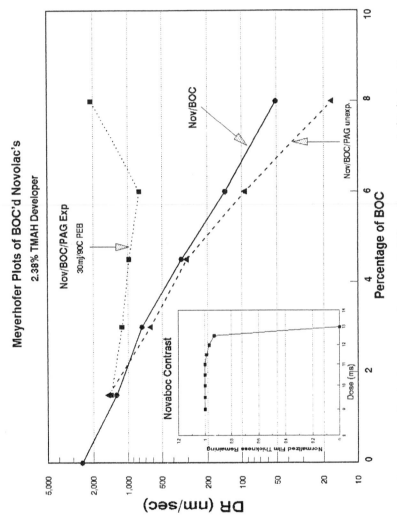

Figure 14. Dissolution rate of exposed (R) and unexposed (R_0) for three component resist of novolak (nov), % NOVOBOC (BOC), and photoacid generator (PAG).

mole% TBOC content and demonstrated to be fast I line, electron beam, X-ray and TSI resist. The low amount of TBOC groups greatly reduced the film shrinkage in processing to less than 5% loss.

For deep UV lithography,novolaks were chosen to be of highest transparency and for TSI of the highest opacity. In the deep UV region, evidence suggests that the phenolic group of NOVOBOC can act as a photosensitizer for acid generation. In a three component formulation, the NOVOBOC can act as a dissolution inhibitor and a fast resist of submicron resolution can be produced.

For further extension, a TSI NOVOBOC resist system was demonstrated with gas phase silylation followed by RIE transfer. A bilayer Si system was also formed and processed by simultaneous liquid phase development, liquid phase silylation and RIE pattern transfer.

REFERENCES

1. W. Moreau, SEMICONDUCTOR LITHOGRAPHY, Plenum Press, 1989, Chapter 2.
2. D. McKean, SPIE, *920*, 61(1988).
3. H. Ito and C. Willson., Pol. Eng., and Sci., *23*, 2012(1983).
4. T. Wolf, R. Hartless, A. Shugard, and G. Taylor, J. Vac. Sci., and Tech., *B5*, 396(1987).
5. S. MacDonald, H. Ito, H. Hiraoka, and C. Willson, SPE RETEC on Photopolymers, 1985, p.177.
6. H.Shiraishi, H. Takumi, T. Ueno, T. Sakamizu, and F. Murai, J. Vac. Sci. and Tech., *B9*, 3743(1991).
7. H.Ban, J. Nakamura, K. Deguchi, and A. Tanaka, J. Vac. Sci., and Tech., *B9*, 3387(1991).
8. A. Gozdz and J. Shelburne, SPIE Proc., *1672*, 184(1992).
9. J. Frechet, E. Eichler, H. Ito, and C. Willson, Polymer, *24*, 995(1983)
10. W.Brunsvold, W.Montgomery, and B. Hwang, SPIE Proc., *1466*, 368(1991).
11. J. Shaw, M. Hatzakis, E. Babich, J. Paraszczak, D. Witman, and K. Stewart, J. Vac. Sci. and Tech., *B7*, 1709, (1989).
12. H. Ito, J. Pol. Sci, *25*, 2971(1986).
13. P. Paniez, D. Demattei, and M. Abadie, Microlectronic Eng., *17*, 279(1992).
14. H.Bogan and K. Graziano, SPIE Proc.,*1262*, 180(1990).
15. N. Hacker and K. Welsh., SPIE Proc., *1466*, 384(1991).
16. W. Brunsvold, R. Kwong, W. Montgomery, W. Moreau, H. Sachdev, and K. Welsh, SPIE Proc., *1262*, 162(1990).
17. M. OToole, E. Liu, and M. Chang, IEEE Trans. Elec. Dev., *ED-28*, 1405(1981).
18. D.Meyerhofer, IEEE Trans. Elec. Dev., ED-27, 921(1980).
19. L. Schlegel, T. Ueno, H. Shiraishi, N. Hayshi, and T. Iwayanagi, Chem Mater., *2*, 299 (1990).
20. J. Pacansky and H. Hiroaka, J. VAC. Sci. and Tech., *19*, 1132(1981).
21. A. Knop and W. Scheib, CHEMISTRY and APPLICATIONS of PHENOLIC RESINS, Springer-Verlag, 1979.

RECEIVED March 8, 1993

Chapter 23

Superiority of Bis(perfluorophenyl) Azides over Nonfluorinated Analogues as Cross-Linkers in Polystyrene-Based Deep-UV Resists

Sui Xiong Cai[1], M. N. Wybourne[2], and John F. W. Keana[1]

Departments of [1]Chemistry and [2]Physics, University of Oregon, Eugene, OR 97403

A comparative study of polystyrene (PS) with bis(perfluorophenyl) azides 1-2 and the corresponding non-fluorinated bisazides 3-4 as deep-UV resists is reported. Inclusion of as low as 1.2 wt-% of 1 in PS led to 70% retention of film thickness after photolysis and development. PS containing 2.4 wt-% of 1 is > 100 times more sensitive as a deep-UV negative resist than PS itself. The presence of 1 in PS also increased the contrast of the resist. On a molar basis, 1 was about 10 times as effective as non-fluorinated bisazide 3 in cross-linking PS while 2 was about 6 times as effective as 4. PS containing 2.4 wt-% of 1 was found to have a deep-UV sensitivity of 5-10 mJ cm^{-2} and resolution of about 0.5 μm.

The miniaturization of integrated circuits in the microelectronics industry (1) has stimulated research in the development of technological alternatives to conventional photolithography (2, 3). Deep-UV lithography (4), electron beam lithography (5) and X-ray lithography (6, 7) all have been demonstrated to give submicron resolution and are undergoing vigorous development.

A negative deep-UV resist composed of poly(p-vinylphenol) and a bisazide has been developed (8). A limitation has been that a high percentage of bisazide (20 wt%) to resin is required such that a resist film of 1 μm thickness is virtually opaque in the 200-300 nm region. Consequently, undercut profiles are typically observed after development and the processing conditions have to be carefully controlled to maintain line-width and reproducibility (2). Other negative resists such as novolac resin with a bisazide (15-20 wt%) (9), an acidic resin with a bisazide (30 wt%) (10) and poly(methyl methacrylate) with a bisazide (20-25 wt%) (11) also suffer because of the poor cross-linking efficiency of the bisazide.

We (12, 13) and others (14) have developed perfluorophenyl azides (PFPAs) as a new class of photolabeling agents (15) with improved CH insertion efficiency in hydrocarbon solvents over nonfluorinated analogues. The nitrene intermediates derived from photolysis of PFPAs do not undergo ring expansion (12), which is probably the main wasteful reaction pathway of nonfluorinated

0097–6156/94/0537–0348$06.00/0

aryl nitrenes (*16*). Since the overall cross-linking efficiency of a bisazide depends on the square of the efficiency of the individual azide groups (*17*), bis-PFPAs are expected to be highly efficient cross-linking agents. We have already reported the development of bis-PFPAs as efficient cross-linking agents for cyclized poly(isoprene) (*18*), polystyrene (*18*) and poly(3-octylthiophene) (*19*) and their application in deep-UV and electron beam lithography. Herein, the deep-UV cross-linking properties of bis-PFPAs **1-2** are compared and found to be superior to those of their respective non-fluorinated analogues **3-4**.

RESULTS AND DISCUSSION

Bis-PFPA **1** (*18*) was prepared from 4-azido-2,3,5,6-tetrafluorobenzoyl chloride[12] and ethylene glycol. Bis-PFPA **2** (*18*) was obtained as a colorless solid from the reaction of decafluorobenzophenone and NaN_3. Bisazide **3** (*20*) was synthesized from 4-azidobenzoyl chloride (*21*) and ethylene glycol. Bisazide **4** (*22*) was prepared via diazotization of 4,4'-diaminobenzophenone followed by treatment with sodium azide.

Photolysis of **1** in cyclohexane gave bis-CH insertion product **5** (45%), mono-CH insertion product **6** (21%) and bis-aniline **7** (*18*). The isolation of bis-amine **5** appears to be the first instance in which a bisazide has been demonstrated photochemically to give a bis-CH insertion product. In contrast, photolysis of **3** under identical conditions gave no isolable product other than red tar.

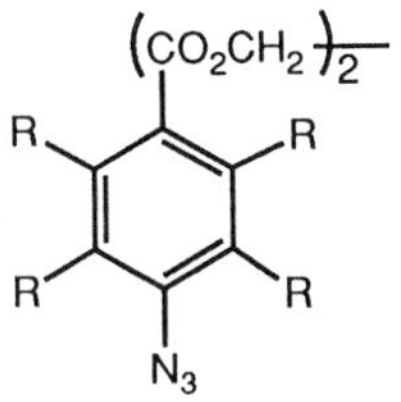

1 R = F
3 R = H

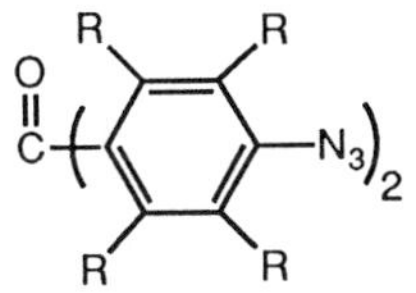

2 R = F
4 R = H

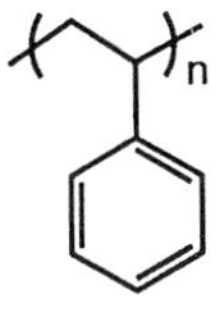

PS

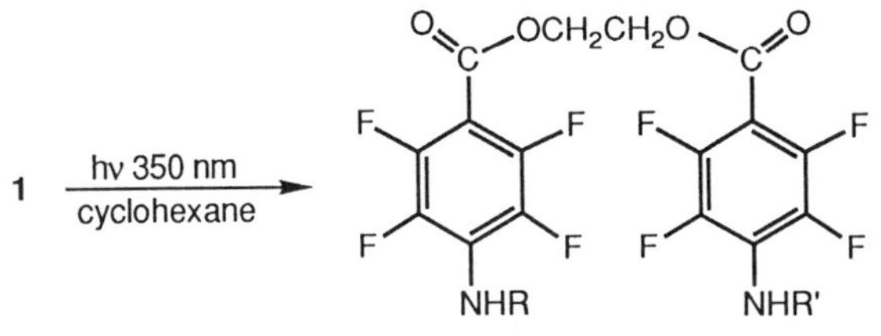

1 $\xrightarrow[\text{cyclohexane}]{\text{hv 350 nm}}$

5 R = R' = cyclohexyl
6 R = cyclohexyl, R' = H
7 R = R' = H

Polystyrene (PS) is a negative deep UV (23) and electron beam resist showing high resolution (24) but low sensitivity (25). Apparently no attempt has been made to improve the sensitivity of PS by addition of a bisazide. We expected that addition of bis-PFPA to PS should result in a deep-UV resist with increased sensitivity. Thus, various amounts of bis-PFPA **1** or **2** and the corresponding nonfluorinated bisazide **3** or **4** were separately added to PS. The resist solutions were then spin-coated on NaCl discs, baked, photolyzed and developed. The intensity of the IR CH stretching absorption at 2924 cm^{-1} before and after development was used to estimate the retention of film thickness.

Table 1 shows that as low as 1.2 wt-% of **1** (#4) in PS is enough to retain 70% of film thickness after complete decomposition of the azido group. In comparison, about 15 wt-% of the corresponding non-fluorinated bisazide **3** (#3) was needed to retain a similar film thickness. Thus on a molar basis, bis-PFPA **1** is about 9-14 times as effective as its non-fluorinated counterpart **3** in cross-linking PS. Also, a longer photolysis time was needed to decompose bisazide **3** (#3) compared to **1** (#3). Bis-PFPA **2** is less effective on a molar basis than bis-PFPA **1** in cross-linking PS (**2** #3 vs **1** #3). On a molar basis, bis-PFPA **2** is about 6-7 times (**2** #2 vs **4** #1) as effective as its corresponding nonfluorinated bisazide **4** in cross-linking PS. As a control, a resist consisting of 5.9 wt-% of methyl 4-azidotetrafluorobenzoate (12) (a mono-PFPA) in PS was found to retain less

Table 1. Bisazides in Polystyrene as Negative Deep-UV Resists

bisazide	#	wt-%[a]	mmol/g[a]	$h\upsilon$(min)[b]	retention[c]
1	1	7.0	0.14	1	0.94
1	2	4.6	0.092	2/3	0.83
1	3	2.3	0.046	1/3	0.74
1	4	1.2	0.024	1/3	0.70
1	5	0.6	0.012	1/3	0.08
1	6	0.0	0.0	1	0.0
3	1	27.5	0.78	3	0.97
3	2	19.3	0.55	3	0.83
3	3	14.7	0.42	2	0.78
3	4	9.7	0.27	2	0.16
3	5	5.0	0.14	1	0.0
2	1	5.7	0.14	1	0.87
2	2	3.8	0.093	1	0.63
2	3	2.0	0.049	1	0.0
4	1	14.6	0.55	4	0.58
4	2	11.4	0.43	2	0.24
4	3	7.6	0.29	2	0.12
4	4	3.7	0.14	1	0.0

[a] In PS. [b] Complete decomposition of the azido group was observed by IR at 2123 cm^{-1} after photolysis. [c] The retention (normalized film thickness) was determined by IR at 2924 cm^{-1} (CH absorption) after development in xylene for 25 sec and rinsing in isopropanol for 10 sec.

than 10% of the film after photolysis and developing. It seems likely that the bis-PFPAs are cross-linking the PS by a bis-CH insertion mechanism similar to that observed in cyclohexane, rather than a polymer radical recombination mechanism (*26*).

PS containing 2.4 wt-% and 4.6 wt-% of bis-PFPA **1** and 14.7 wt-% and 19.3 wt-% of bisazide **3** as well as PS itself were evaluated for their sensitivity as deep-UV resists. Figure 1 shows the exposure characteristics for these resists. PS itself is a negative deep-UV resist with very low sensitivity. The sensitivity of PS is remarkably increased by the addition of small amounts of bis-PFPA **1**. PS with 2.4 wt-% **1** is more than two orders of magnitude more sensitive than PS itself. The sensitivity curve also shows that the presence of **1** enhances the contrast of the resist. The presence of bisazide **3** in PS also increased the sensitivity of PS. However, **1** is much more effective than **3** (about 9 times on a molar basis) in crosslinking PS and resists of PS containing **1** are about 5-10 times more sensitive than resists of PS containing **3** (**1**, 2.4 wt-% vs **3**, 14. 7 wt%). From the data shown in Figure 1, we estimate the resist of PS with **1** (4.6 wt-%) has a contrast of about 2.9.

The UV absorption maximum of **1** in ethanol occurs at 264 nm (log ε = 4.6, log ε_{254} = 4.5). PS itself has only weak absorption at 254 nm. UV absorption spectra for the resist film of PS with 4.6 wt-% of **1** before and after photolysis are shown in Figure 2. The low concentration of **1** in the resist resulted in a relatively low absorption at 254 nm, thus allowing homogeneous deep-UV exposure of the resist. The resist is partially bleached at 254 nm by the exposure. This observed partial bleaching is consistent with the observation that the UV absorption maximum of the main photolysis products **5** (292 nm) and **6** (285 nm) of **1** in cyclohexane appear at wavelengths longer than 264 nm.

Lithographic evaluation of the resists was carried out in a KSM Karl Suss deep-UV contact aligner. We find that the sensitivity of the resists of PS with 2.4 wt-% of **1** is 5-10 mJ cm^{-2}, while a sensitivity of 30-40 mJ cm^{-2} is found for the resists of PS with 14.7 wt-% of **3**. PS alone has a sensitivity of about 3000 mJ cm^{-2}. The smallest feature in the mask (0.5 μm) could be resolved (Figure 3). Swelling of the polymer was observed under these non-optimized conditions, suggesting that further improvement of the resolution of the resists may be possible.

In conclusion, we have demonstrated that bis-PFPAs **1-2** are about 10 times better than the corresponding non-fluorinated bisazides **3-4** in cross-linking polystyrene. Utilization of the highly efficient cross-linking agents **1-2** significantly reduces the required amounts of cross-linkers in PS-based resists and increases the sensitivities of the resists.

EXPERIMENTAL

Resist Preparation and Evaluation

Bis-PFPAs **1** and **2** were prepared as reported (*18, 27*). Bisazides **3** and **4** were prepared following literature procedures (*20, 22*). Resist solutions were prepared by mixing xylene (2 ml) with 100 mg of polystyrene (MW 125,000-250,000, Polyscience Inc.) and various amounts (Table 1) of bis-PFPA **1** or **2**, or

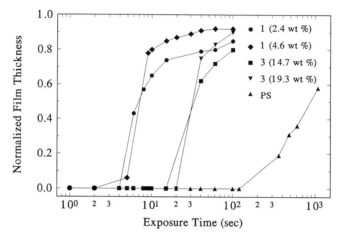

Figure 1. Effect of concentration on exposure characteristics for resists of PS-bis-PFPA **1**, PS-bisazide **3** and PS itself. Resist films were exposed in a Rayonet photoreactor (254 nm) and developed in xylene for 30 sec then rinsed in isopropanol for 10 sec. The normalized film thickness was determined by IR at 2924 cm^{-1} (CH absorption) before photolysis and after development.

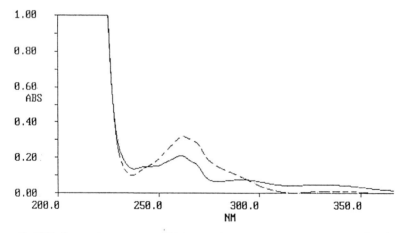

Figure 2. UV absorption spectra of PS containing 4.6 wt-% of **1** before (- - -) and after (----) photolysis for 20 sec. The film thickness was about 0.7 μm.

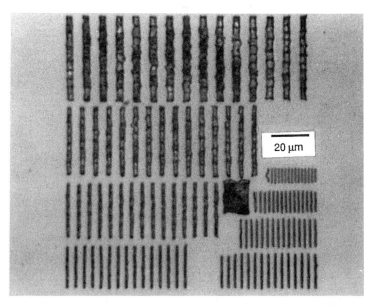

Figure 3. Optical micrograph of images from a resist of PS containing 2.4 wt-% of **1**. The deep-UV exposure dose was 5 mJ cm^{-2}. The exposed film was developed in xylene for 30 sec and rinsed in isopropanol for 20 sec.

bisazides **3** or **4**. The resist solutions were spin-coated on NaCl discs by a spin-coater (Headway Science Inc.) set at 1000 rpm and the films were baked at 60 °C for 30 min. Virgin film thicknesses were measured with an ellipsometer (Rudolph Science) to be about 0.7 μm. The films were photolyzed at ambient temperature in a Rayonet photoreactor (254 nm). The progress of photolysis was followed by a Nicolet 5DXB FTIR spectrometer at 2123 cm^{-1} (azide absorption). The films were developed in xylenes for 25 sec and rinsed in isopropanol for 10 sec and air dried. The retention of film thickness (normalized) was determined by FTIR at 2924 cm^{-1} for the CH absorption. The UV absorption of the films coated on quartz discs was measured with a Perkin-Elmer Lambda 6 UV/vis spectrophotometer. For lithographic evaluation, resist solutions were spin-coated (1000 rpm) on silicon wafers, baked and exposed in a KSM Karl Suss deep-UV contact aligner. The samples were developed in xylene for 30 sec and rinsed in isopropanol for 20 sec and air dried.

ACKNOWLEDGEMENT

This work was supported by grants from the Oregon Resource Technology Development Corporation, the Office of Naval Research and the National Institute of General Medical Sciences GM 27137. We thank Dr. Gary Goncher of Tektronix Inc. for his help in carrying out the deep-UV exposure of the resists.

REFERENCES

1. Fuller, G. E. Solid State Technology. 1987, 114.
2. Reichmanis, E.; Thompson, L. F. Chem. Rev. 1989, 89, 1273.
3. Reichmanis, E.; Houlihan, F. M.; Nalamasu, O.; Neenan, T. X. Chem. Mater. 1991, 3, 394.
4. Willson, C. G.; Bowden, M. J. in Electronic and Photonic Applications of Polymers, Bowden, M. J.; Turner, S. R. Eds. ACS Advances in Chemistry Series 218, American Chemical Society, Washington, DC, 1988, Ch. 2.
5. Hohn, F. J. J. Vac. Sci. Technol. B. 1989, 7, 1405.
6. Heuberger, A. J. Vac. Sci. Technol. B. 1989, 7, 1387.
7. Novembre, A. E.; Tai, W. W.; Kometani, J. M.; Hanson, J. E.; Nalamasu, O.; Taylor, G. N.; Reichmanis, E.; Thompson, L. F. Chem. Mater. 1992, 4, 278.
8. Iwayanagi, T.; Kohashi, T.; Nonogaki, S.; Matsuzawa, T.; Douta, K.; Yanazawa, H. IEEE Trans. Electron Devices, 1981, ED-28, 1306.
9. Yang, J.-M.; Chiong, K.; Yan, H.-J.; Chow, M.-F. Proc. SPIE, Adv. Resist Technol. Proc. I, 1984, 469, 117.
10. Endo, M.; Tani, Y.; Sasago, M.; Nomura, N. J. Electrochem. Soc. 1989, 136, 2615.
11. Han, C. C.; Corelli, J. C. J. Vac. Sci. Technol. B, 1988, 6, 219.
12. Keana, J. F. W.; Cai, S. X. J. Org. Chem. 1990, 55, 3640.
13. Keana, J. F. W.; Cai, S. X. J. Fluorine Chem. 1989, 43, 151.
14. Soundararajan N.; Platz, M. S. J. Org. Chem. 1990, 55, 2034.

15. Aggeler, R.; Chicas-Cruz, K.; Cai, S. X.; Keana, J. F. W.; Capaldi, R. A. Biochemistry. 1992, 31, 2956.
16. Li, Y.-Z.; Kirby, J. P.; George, M. W.; Poliakoff, M.; Schuster, G. B. J. Am, Chem. Soc. 1988, 110, 7209.
17. Reiser, A. Photoactive Polymers. Wiley-Interscience Publication, New York, 1989. Ch 2.
18. Cai, S. X.; Nabity, J. C. ; Wybourne, M. N.; Keana, J. F. W. Chem. Mater. 1990, 2, 631.
19. Cai, S. X.; Keana, J. F. W.; Nabity, J. C.; Wybourne, M. N. J. Mol. Electron. 1991, 7, 63.
20. Mistr, A.; Vavra, M,; Adlerova, H.; Babak, Z. Collect. Czech. Chem. Commun. 1969, 34, 3811.
21. Tsuda, M.; Tanaka, H.; Tagami, H.; Hori, F. Makromol. Chem. 1973, 167, 183.
22. Reiser, A.; Wagner, H. M.; Marley, R.; Bowes, G. Trans. Faraday Soc. 1967, 63, 2403.
23. Imamura, S.; Sugawara, S. Jap. J. App. Phy. 1982, 21, 776.
24. a) Brewer, T. L. 6th Int. Conf. Electron Ion Beam Sci. Technol. Ed. Bakish, R. 1974, p. 71. b) Lai, J. H.; Shepherd, L. T. J. Electrochem. Soc. 1979, 126, 696.
25. Reichmanis, E. Thompson, L. F. Ann. Rev. Mater. Sci. 1987, 17, 235.
26. Nonogaki, S.; Hashimoto, M.; Iwayanagi, T. Shiraishi, H. Proc. SPIE, Adv. Resist Technol. Proc. II, 1985, 539, 189.
27. Cai, S. X. Ph. D. Thesis, University of Oregon, May, 1990.

RECEIVED December 30, 1992

Chapter 24

New Photoresponsive Polymers Bearing Norbornadiene Moiety

Synthesis by Selective Cationic Polymerization of 2-(3-Phenyl-2,5-norbornadiene-2-carbonyloxy)ethyl Vinyl Ether and Photochemical Reaction of the Resulting Polymers

T. Nishikubo, A. Kameyama, K. Kishi, and C. Hijikata

Department of Applied Chemistry, Faculty of Engineering, Kanagawa University, Rokkakubashi, Kanagawa-ku, Yokohama 221, Japan

New functional monomers such as 2-(3-phenyl-2,5-norbornadiene-2-carbonyloxy)ethyl vinyl ether (PNVE), 2-{3-[(phenyl)carbamoyl]-2,5-norbornadiene-2-carbonyloxy}ethyl vinyl ether (PCNVE), 2-[(4-vinylphenyl)oxy]carbonyl-3-phenyl-2,5-norbornadiene (VPNB), and 2-[(4-vinylphenyl)oxy]carbonyl-3-(phenyl)carbamoyl-2,5-norbornadiene (VPCNB) containing both polymerizable vinyl groups and photoresponsive norbornadiene (NBD) moieties were synthesized in high yields by the reaction of potassium 2,5-NBD-2-carboxylates with 2-chloroethyl vinyl ether or p-chloromethylated styrene using a phase transfer catalyst. Although cationic polymerization or copolymerization of PNVE proceeded successfully to give soluble polymers bearing NBD moiety in high yield without any gel products at -10 and -40 °C in dichloromethane solution, the cationic polymerization of PCNVE and VPCNB did not occur under similar conditions. Radical polymerization or copolymerization of VPNB, VPCNB and PCNVE was also tried in order to obtain the corresponding polymers with NBD moieties. However, soluble polymers were not obtained in high yield without gel production. Photochemical valence isomeriza tion of NBD moiety in poly(PNVE) and the resulting quadricyclane group in the polymer were also investigated in the film state.

Although photochemical valence isomerization of polymers bearing pendant norbornadiene (NBD) moieties has been of interest from the viewpoint of solar energy conversion and storage (*1*, *2*), these polymers with pendant NBD moieties also appear to be useful as new photoresponsive polymers in imaging technology, and also as new functional polymers for switch or device material in the opto-electronics field. This is because the quadricyclane (QC) groups produced in the polymer reverted to the corresponding NBD moieties under photo-irradiation of about 250 nm light or by heating.

In previous papers, we reported the synthesis of polymers bearing pendant NBD moieties by the substitution reaction (*3–6*) of poly(chloromethylstyrene) with potassium 2,5-NBD-2-carboxylates, or by the addition reaction (*7, 8*) of poly(glycidyl methacrylate-co-methyl methacrylate)s with 2,5-NBD-2-carboxylic acids or the acid chlorides. We also examined photochemical valence isomeriza-

0097–6156/94/0537–0356$06.00/0

tion of pendant NBD moieties in the obtained polymers, and catalytic reversion of the resulting QC groups using (5,10,15,20-tetraphenyl-21H,23H-porphine)cobalt as a catalyst in solution or in the film state.

This paper describes the synthesis of some new functional monomers bearing both polymerizable vinyl groups and photorespon sive NBD moieties, and the synthesis of polymers containing pen dant 2,5-NBD-carboxylate moieties by cationic polymerization and radical polymerization of these monomers. Also investigated were photochemical valence isomerization of pendant NBD moieties in the polymer films, and photochemical reversion of the produced QC groups in the polymer films.

EXPERIMENTAL

Synthesis of 2-(3-Phenyl-2,5-norbornadiene-2-carbonyloxy)ethyl Vinyl Ether (PNVE)

The reaction of 5.10 g (20 mmol) of potassium 3-phenyl-2,5-NBD-2-carboxylate, which was prepared by Dield-Alder reaction of phenylpropyonyl chloride with cyclopen tadiene according to the method reported previously (*4*, *5*), and 21.31 g (0.2 mol) of 2-chloroethyl vinyl ether (CEVE) was carried out using 0.64 g (2 mmol) of tetrabutylammonium bromide (TBAB) as a phase transfer catalyst at boiling point of CEVE for 5 h. The reaction mixture was diluted in toluene, and the potassium chloride produced was filtered. The filtrate was washed several times with water, and dried with anhydrous $MgSO_4$. Excess CEVE and toluene were evaporated, and then the crude 2-(3-phenyl-2,5-norbornadiene-2-carbonyloxy)ethyl vinyl ether (PNVE) thus ob tained was purified by distillation at 149-152 °C/ 0.43 mmHg, followed by silica gel column using ethyl acetate/hexane (2/8) as eluents. The isolated yield of PNVE was 4.47 g (81%). IR (neat); 1700 (C = O, ester), 1640, 1620 (C = C), 1235 (C-O-C, ester), 1100 cm^{-1} (C-O-C, ether). UV (CH_2Cl_2); λ_{max} = 295 nm. ^{1}H-NMR ($CDCl_3$, TMS); δ = 2.2 (dd, 2H, bridged CH_2 in NBD), 3.8-4.4 (m, 8H, O-CH_2-CH_2-O, CH_2 = C-O, C-H in NBD moiety), 6.4 (dd, 1H, C = CH-O), 6.9 (m, 2H, CH = CH), 7.2-7.7 ppm (m, 5H, aromatic protons). Anal. Calcd for $C_{18}H_{18}O_3$: C, 76.57; H, 6.43. Found: C, 76.25; H, 6.26.

Synthesis of 2-{3-[(Phenyl)carbamoyl]-2,5-norbornadiene-2-carbonyloxy)} ethyl Vinyl Ether (PCNVE)

The reaction of 4.67 g (50 mmol) of potassium 3-(phenyl)carbamoyl-2,5-NBD-2-carboxylate , which was synthesized by the reaction of 2,5-NBD-2-carbo-xylic acid with aniline using DCC according to the method reported pre viously (*6*), and 106.6 g (1 mol) of CEVE was performed in the presence of 1.61 g (5 mmol) of TBAB under the same conditions as the above. The crude PCNVE was purified by silica gel column using ethyl dichloromethane as eluents. The yield of PCNVE (m.p. 80.5-81.4 °C) was 13.34 g (82%). IR (neat); 1689 (C = O, ester), 1656 (C = O, amide) 1623, 1610 (C = C), 1254 (C-O-C, ester), 1107 cm-1

(C-O-C, ether). UV (CH_2CL_2); λ_{max} = 330 nm. ^{1}H-NMR $(CDCl_3,$ TMS); δ = 2.1 (dd, 2H, bridged CH_2 in NBD), 3.8-4.6 (m, 8H, $O-CH_2-CH_2-O$, CH_2 = C-O, C-H in NBD moiety), 6.5 (dd, 1H, C = CH-O), 6.7-7.8 (m, 7H, CH = CH and aromatic protons), 11.1 ppm (s, 1H, NH). Anal. Calcd for $C_{19}H_{19}O_4N$: C, 70.12; H, 5.89; N, 4.33. Found: C, 70.15; H, 5.95; N, 4.42.

Synthesis of 2-[(4-Vinylphenyl)oxy]carbonyl-3-phenyl-2,5-norbornadiene (VPNB)

Potassium 3-(phenyl)carbamoyl-2,5-NBD-2-carboxylate [9.01 g (36 mmol) was reacted with 4.58 g (30 mmol) of p-chloromethylated styrene (CMS) in 25 mL of DMF using 0.97 g (3 mmol) of TBAB at 50 °C for 24 h. The reaction mixture was poured into water, filtered, then washed with water. The crude VPNB was recrystallized twice from hexane. The yield of VPNB (m.p. 87.3-88.2 °C) was 9.67 g (98%). IR (neat); 1680 (C = O), 1620, 1610 (C = C), 1240 cm^{-1} (C-O-C). ^{1}H-NMR $(CDCl_3,$ TMS); δ = 2.2 (dd, 2H, bridged CH_2 in NBD), 4.0 (m, 2H, C-H in NBD moiety), 5.0-5.3 (m, 4H, CH_2 = C and CH_2-O), 6.7 (dd, 1H, C = CH-), 6.8-7.6 ppm (m, 11H, CH = CH and aromatic protons). Anal. Calcd for $C_{23}H_{20}O_2$: C, 84.12; H, 6.14. Found: C, 84.07; H, 6.04.

Synthesis of 2-[(4-Vinylphenyl)oxy]carbonyl-3-(phenyl)-carbamoyl-2,5-norbornadiene (VPCNB)

VPCNB (viscous oil) was synthesized in 79% yield from the reaction of 21.12 g (72 mmol) of potassium 3-(phenyl)carbamoyl-2,5-NBD-2-carboxylate with 10.99 g (60 mol) of CMS in 50 mL of DMF using 1.93 g (6 mmol) of TBAB in the same way as above. IR (neat); 1690 (C = O, ester), 1670 (C = O, amide), 1620, 1610 (C = C), 1260 cm^{-1} (C-O-C). ^{1}H-NMR $(CDCl_3,$ TMS); δ = 2.0 (dd, 2H, bridged CH_2 in NBD), 4.1-4.4 (m, 2H, C-H in NBD moiety), 5.2-5.8 (m, 4H, CH_2 = C and CH_2-O), 6.7 (dd, 1H, C = CH-), 6.8-7.7 (m, 11H, CH = CH and aromatic protons), 11.7 ppm (s, 1H, NH).

Cationic Polymerization of PNVE

PNVE (0.63g, 2.2 mmol) was dissolved in dichloromethane, and the solution was cooled with dry ice/methanol. To the monomer solution, 8.4 μL (0.067 mmol) of $BF_3 \cdot O(Et)_2$ was added with stirring, and then the polymeriza tion was carried out at -40 °C for 3 h under dry nitrogen. The polymerization was terminated by the addition of a small amount of triethylamine, then the solution was poured into methanol to give a precipitate. The obtained polymer was reprecipitated twice from THF into methanol, and dried *in vacuo* at room tempera ture. Yield of soluble poly[2-(3-phenyl-2,5-norbornadiene-2-carbonyloxy)ethyl vinyl ether] [P(PNVE)] was 0.51 g (80%). To evaluate the solubility, the resulting polymer was dissolved in THF, and the insoluble part was filtered off by a 3G type glass filter and dried. Reduced viscosity of the polymer in DMF was 0.10 dL/g, measured at a concentration of 0.5 g/dL at 30 °C. IR (film); 1700 (C = O), 1610 (C = C), 1237 (C-O-C, ether), 1100 cm^{-1} (C-O-C, ether). UV (film); λ_{max} = 295 nm. ^{1}H-NMR (CDCl$_3$, TMS); δ = 0.9-2.5 (m, 4H, CH$_2$ in NBD and main chain in the polymer), 3.2-4.4 (m, 7H, O-CH$_2$-CH$_2$-O, C-H in NBD and main chain in the polymer), 6.9 (m, 2H, CH = CH), 7.0-7.6 ppm (m, 5H, aromatic protons).

Other cationic polymerizations or copolymerizations were also carried out with the same methods.

Radical Polymerization of VPNB

Radical polymerization of 2.62 g (8 mmol) of VPNB was performed in 8 mL of THF using 13 mg (0.08 mmol) of AIBN as a radical initiator at 60 °C for 6 h under nitrogen, and then the solution was poured into methanol to give a precipitate. The obtained polymer was reprecipitated twice from THF into methanol, and dried in vacuo at room temperature. Yield of the polymer bearing pendant 3-phenyl-2,5-NBD-2-carboxylate moiety was 0.73 g (28%). To evaluate the solubility, the resulting polymer was dissolved in THF, and the insoluble part was filtered off by a 3G type glass filter and dried. Reduced viscosity of the soluble part of the polymer in DMF was 0.11 dL/g, measured at a concentration of 0.5 g/dL at 30 °C.

Radical polymerization of other monomers bearing NBD moieties was also performed with the same methods.

Photochemical Valence Isomerization between Pendant NBD moiety and the QC Group in Polymer Film

A solution of P(PNVE) (0.01 g) in THF (3 mL) was cast on the inside wall of a quartz cell and dried. The polymer film (about 3.8 μm) on the quartz was irradiated by a 500-W xenon lamp (Ushio Electric Co., UXL-500D-O) through a monochrometer (JASCO Model CT-10) or a filter (Toshiba UV-31), in which the exposed energy of the irradiating light was monitored by an electric photon-counter (ORC Model UV-M30). Rates of disappearance or appearance of the absorption peak at 295 nm due to the NBD moiety were measured by a UV spectrophotometer (Shimadzu Models UV-200 and UV-2100S).

RESULT AND DISCUSSION

Cationic Polymerization and Copolymerization of Monomers with NBD Moieties

New functional monomers such as PNVE PCNVE, VPNB, and VPCNB containing corresponding NBD moieties were syn thesized in high yields by the substitution reaction of potassium 2,5-NBD-2-carboxylates with CEVE or CMS using TBAB as a phase transfer catalyst (Scheme 1).

The cationic polymerization of PNVE having both vinyl ether group and 3-phenyl-2,5-NBD-2-carboxylate moiety in the molecule was performed in toluene using BF3 · O(Et)2 as a catalysts at -10 or -75 °C. Polymers thus obtained contained 3-5 wt% of insoluble gel products. On the other hand, polymers without any insoluble gel products were obtained in high yields when the cationic

Table I. Conditions and Results of Cationic Polymerization
of Monomers Bearing NBD Moieties[a]

run no.	monomer, mmol	solvent	temp, °C	yield, % sol.	insol.	η_{red},[b] dL/g
1	PNVE (8.0)	toluene	-10	93	3	0.16
2	PNVE (8.0)	toluene	-75	76	5	0.12
3	PNVE (1.8)	dichloromethane	-10	95	0	0.13
4	PNVE (2.2)	dichloromethane	-40	80	0	0.10
5	VPNB (8.0)	dichloromethane	-75	0	—	—
6	VPCNB (8.0)	toluene	-50	trace	—	—

[a]Polymerization was carried out with the concentration of 100 mmol/L using 3 mol% of $BF_3 \cdot O(C_2H_5)_2$ as a catalyst for 3 h. [b]Measured at a concentration of 0.5 g/dL in DMF at 30 °C.

polymerization of PNVE took place in dichloromethane at -10 or -40 °C under the same catalytic conditions (Table I).

The IR spectrum of the soluble P(PNVE) showed absorption peaks at 1710 and 1620 cm^{-1} due to C = O and C = C bonds, respec tively, and showed disappearance of the absorption peak due to C = C bond of the vinyl ether group at 1640 cm^{-1}. The ^{1}H-NMR spectrum of the polymer showed a dramatic change from the spectrum of the PNVE monomer, and showed the disappearance of the signal at δ = 6.4 ppm based on C = CH-O, although a signal at δ = 6.8 ppm due to CH = CH remained, and new signals appeared at δ = 1.6 and 3.5 ppm derived from methylene and methine protons of the polymer backbone.

This result suggests that selective cationic polymerization of vinyl ether groups in the PNVE molecule proceeded successfully to give the corresponding polymer [P(PNVE)] containing pendant NBD moiety (Scheme 2).

The cationic polymerization of PCNVE having both vinyl ether group and 3-(phenyl)carbamoyl-2,5-NBD-2-carboxylate moiety in the molecule was carried out under similar reaction conditions; however, no polymer was obtained. The cationic polymerization of styrene type monomer VPCNB with 3-(phenyl)-carbamoyl-2,5-NBD-2-carboxylate moiety was also attempted; however, polymerization was not observed. It seems that the active proton on the car bamoyl group in those monomers inhibits the cationic polymeriza tion of the vinyl ether and vinyl groups. The cationic polymerization of styrene type monomer VPNB with 3-phenyl-2,5-NBD-2-carboxylate moiety could not be tried, because this monomer precipitates from the toluene or dichloromethane solutions at low temperatures.

The cationic copolymerizations of PNVE with 50 mol% of other vinyl ethers such as (2-phenoxy)ethyl vinyl ether (PEVE), 2-isobutyl vinyl ether (IBVE), and CEVE were performed in dich loromethane using 3 mol% of BF3 · O(Et)2 as the catalyst at -40 °C.

As summarized in Table II, soluble copolymers P(PNVE-PEVE), P(PNVE-IBVE), and P(PNVE-CEVE) were obtained in high yield without any unwanted

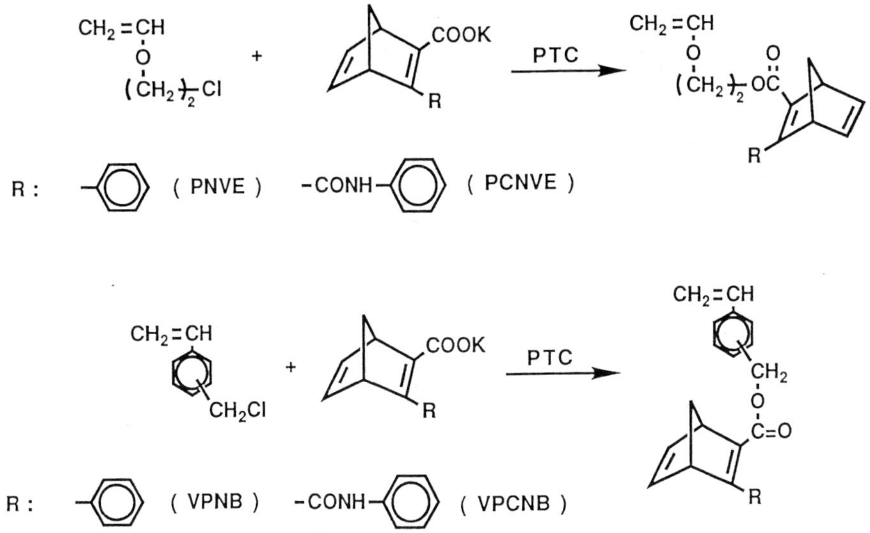

Scheme 1.

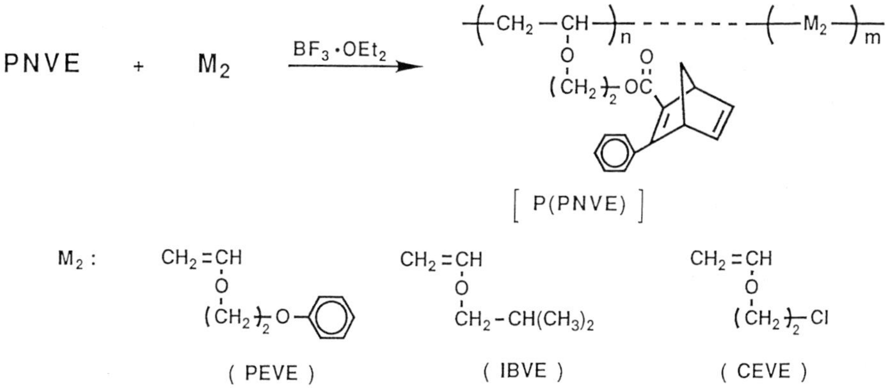

Scheme 2.

Table II. Conditions and Results of Cationic Polymerization and
Copolymerization of Monomers Bearing NBD Moieties[a]

run no.	comonomer,	feed ratio, I : II	yield, % sol.	insol.	composition,[b] I : II	η_{red},[c] dL/g
7	none	100 : 0	80.2	0	100 : 0	0.10
8	PEVE	50 : 50	93.8	0	53 : 47	0.22
9	IBVE	50 : 50	85.0	0	57 : 43	0.16
10	CEVE	50 : 50	73.9	0	49 : 51	0.12

[a] Polymerization was carried out with the concentration at 100 mmol/L using 3 mol% of $BF_3 \cdot O(C_2H_5)_2$ as a catalyst in dichloromethane at -40 °C for 3 h. [b] Composition of copolymer was estimated by a UV spectrum. [c] Measured at a concentration of 0.5 g/dL in DMF at 30 °C.

gel products. In these cationic copolymerizations, compositions of the resulting copolymers, which were determined by UV-spectrum, were directly related to the feed ratio of the monomers.

This result means that selective cationic copolymerization of the vinyl ether group in PNVE with other vinyl ether monomers also gave soluble photoresponsive copolymers in high yields without any insoluble gel productions using BF3 · O(Et)2 in dich loromethane solution.

Radical Polymerization and Copolymerization of Monomers with NBD Moieties

Since polymers bearing pendant NBD moieties were not obtained by the cationic polymerization of PCNVE, VPNB, and VPCNB, the radical polymerization of those monomers was tried. As summarized in Table III, the polymer bearing 3-phenyl-2,5-NBD-2-carboxylate moiety was obtained in 28% yield without any gel products, when radical polymerization of VPNB was carried out in THF at 60 °C for 6 h. The IR and ^{1}H-NMR spectra of the thus-obtained polymer

Table III. Conditions and Results of Radical Polymerization and
Copolymerization of Monomers Bearning NBD Moieties[a]

run no.	monomer, mmol	comonomer, mmol	solvent	temp, °C	time, h	yield, % sol.	insol.	η_{red},[b] dL/g
11	VPNB (8.0)	—	THF	60	6	28	0	0.11
12	VPCNB (8.0)	—	THF	50	72	4	61	—
13	VPCNB (8.0)	—	toluene	60	6	7	27	—
14	VPCNB (8.0)[c]	—	toluene	60	6	17	4	0.33
15	PCNVE (4.0)	PMI (4.0)	dioxane	60	6	trace	66	—
16	PCNVE (4.0)[c]	PMI (4.0)	dioxane	60	6	14	75	0.20

[a] Polymerization was carried out with the concentration at 100 mmol/L using 1 mol% of AIBN as an initiator under nitrogen. [b] Measured at a concentration of 0.5 g/dL in DMF at 30 °C. [c] Solution of monomers bearing NBD moiety in glass flask was irradiated with high-pressure mercury lamp for 60 min.

showed the same spectra as the polymer syn thesized (*4*, *5*) by the reaction of poly[(*p*-chloromethyl)styrene] with potassium 3-phenyl-2,5-NBD-2-carboxylate using the phase transfer catalyst.

The polymers bearing 3-(phenyl)carbamoyl-2,5-NBD-2-carboxylate moiety were produced in 65 and 35 % yields by the radical polymerization of VPCNB at 50 °C for 72 h and 60 °C for 6 h, respectively. However, most of the obtained polymers were gel products. These results suggest that although selective radical polymerization of vinyl group on styrene residue in VPNB and VPCNB pro-ceeded at low conversions of the monomers, some radical reactions such as copolymerization and chain transfer reaction of two C = C bonds on NBD residues in these monomers and polymers also occurred at high conversions of the monomers. This is a meaningful result concerning the radical polymerization of poly-functional vinyl monomers and is contrary to that of Kamogawa et al. (*9*) who reported that the polymers bearing some 3-(phenyl)carbamoyl-2,5-NBD-2-carboxylate moieties were obtained in 29-69% yields without any gel production by the radical polymerization of the corresponding monomers under the speci-fied conditions.

It is of interest that the yield of gel products decreased and the yield of soluble polymer increased when the photo-irradiated VPCNB monomer was used for the radical polymerization. This means that some NBD residues in the VPCNB were changed to QC groups by the photo-irradiation, and the side reactions due to NBD residue decreased during the the polymerization of the monomers.

Because the radical polymerization of vinyl ethers is very slow, the radical copolymerization of PCNVE was carried out with the electron poor monomer N-phenylmaleimide (PMI) in dioxane. As shown in Table III, although the copolymerization proceeded smoothly to give the polymer with 67% yield at 60 °C for 6 h, the resulting polymer was mostly gel products (Scheme 3). It seems that copolymerization of C = C bonds in the NBD moieties may also be occurred as a side reaction during the copolymerization. On the other hand, when the photo-irradiated PCNVE was used, the yield of soluble polymer increased somewhat under the same condi tions. This result suggests that some C = C bonds in the NBD moieties related to the gel production.

From the results of cationic polymerization and radical polymerization of the vinyl monomers bearing NBD moieties, the following conclusions can be drawn.

(1) The cationic polymerization of the corresponding monomer is better than the radical polymerization of the cor responding monomer in giving the soluble polymer bearing NBD moiety in high yield.

(2) PNVE having both vinyl ether group and 3-phenyl-2,5-NBD-2-carboxy-late moiety is a suitable monomer for the cationic polymerization.

Photochemical Isomerization of Pendant NBD moiety in Polymer Film

Photochemical valence isomerization of the pendant NBD moiety in P(PNVE) film was performed on a quartz cell by photo-irradiation using a xenon lamp. An absorption peak at 295 nm due to NBD moiety in P(PNVE) decreased rapidly indicating that the NBD moiety isomerized to the QC group after only 10 min irradia tion with light beyond 310 nm. As shown in Figure 1, two isos bestic points were also observed at 237 and 251 nm. This result indicates that the photochemical reaction of the pendant NBD moiety to the corresponding QC group in the polymer occurred very smoothly and selectively without any side reaction upon irradia tion beyond 310 nm light (Scheme 4).

Accordingly, the photochemical reaction of the pendant NBD moieties to the QC groups in the polymer film was carried out by irradiation with 293, 311, 338, 345, and 365 nm light, respec tively. As shown in Figure 2, the rate of photochemical reaction of the pendant NBD moiety in the polymer was strongly affected by the wavelength of the irradiation. That is, the pendant NBD moiety was isomerized to the corresponding QC group by only 20 mJ/cm2 irradiation with 311 nm light. However, the degree of conversion of NBD moiety in the polymer film was about 40 mol% upon 100 mJ/cm2 irradiation with 365 nm light.

It was also found that the initial rate of the photochemical reaction of the NBD moiety in the polymer film obeyed first-order kinetics with each photo-irradiation, and the observed first-order rate constants of the reaction from NBD to QC are sum marized in Table IV.

Furthermore, the rates of photochemical reaction of pendant NBD moieties in the copolymers also obeyed first-order kinetics upon irradiation with 311 nm light. As summarized in Table IV, the rate of photochemical reaction of the pendant NBD moiety in P(PNVE) film was faster than those in the copolymer films, and the rates of photochemical reaction of pendant NBD moieties in the copolymers decreased as follows: $P(PNVE) > P(PNVE_{49}\text{-}CEVE_{51}) > P(PNVE_{57}\text{-}IBVE_{43}) > P(PNVE_{53}\text{-}PEVE_{47})$.

When the pendant QC group produced in P(PNVE) film on the quartz cell was irradiated with 248 nm light, its UV spectrum reverted to that of the corresponding NBD moiety, with two isos bestic points at 237 and 251 nm (Figure 3). However, the degree of reversion of the QC groups was mostly saturated at 60 mol% conversion.

As shown in Figure 4, the rate of reversion of the QC group to the NBD moiety in the polymer film was strongly affected by the wavelength of the irradiating light. That is, the rate of the reversion upon irradiation with 248 nm light was faster than rates upon irradiation with 240, 244, 250, or 254 nm light.

It was also found that the initial rate of the reversion from QC group to NBD moiety in P(PNVE) film upon irradiation with each wavelength obeyed first-order kinetics, and the observed first-order rate constants are summarized in Table V.

Furthermore, the photochemical reversions of the pendant QC groups in the copolymer films took place upon irradiation with 248 nm light. As summa-rized in Table V, the initial rate of the reversion of pendant QC group in P(PNVE) was faster than those of the pendant QC groups in the copolymers. It was also found that although the degree of reversion of the pendant QC group in $P(PNVE_{53}\text{-}PEVE_{47})$ film upon 5 min irradiation was lower than those of the QC groups in other copolymer films under the same irradia tion conditions, interest-

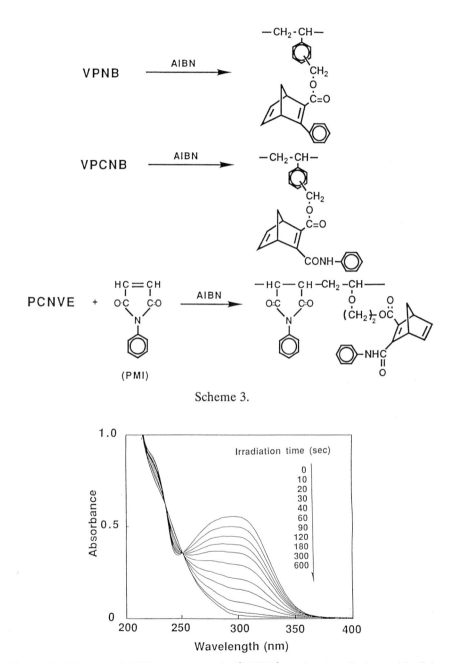

Scheme 3.

Figure 1. Change of UV spectrum of P(PNVE) under irradiation with light beyond 310 nm in the film state.

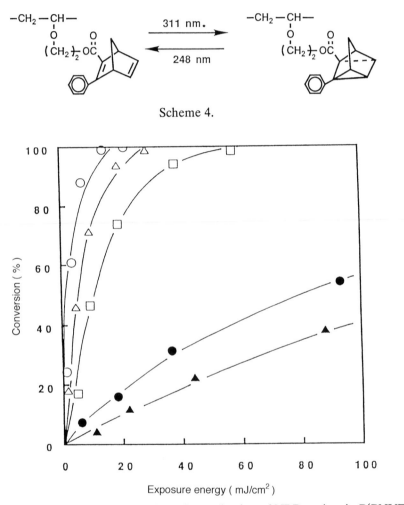

Scheme 4.

Figure 2. Plots of the degree of photo-isomerization of NBD moiety in P(PNVE) film vs. exposure energy by xenon lamp: (△) 293 nm; (○) 311 nm; (□) 338 nm; (●) 345 nm; (▲) 365 nm.

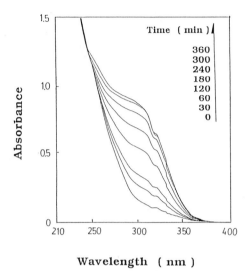

Figure 3. Change of UV spectrum of the produced QC group in P(PNVE) under irradiation with 248 nm light in the film state.

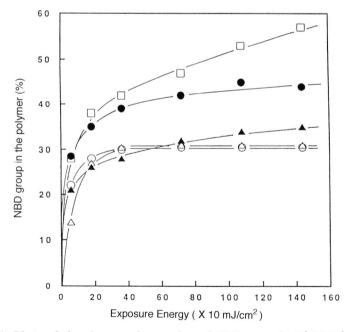

Figure 4. Plots of the degree of reversion of QC group in P(PNVE) film vs. exposure energy by xenon lamp: (△) 240 nm; (○) 244 nm; (□) 248 nm; (●) 250 nm; (▲) 254 nm.

Table IV. First-Order Rate Constant of Valence Isomerization of NBD Moiety
in the Polymers upon Photo-Irradiation[a]

polymer	wavelength of irradiation, nm	$k_{obsd} \times 10^3, s^{-1}$
P(PNVE)	365	4.26
P(PNVE)	345	5.77
P(PNVE)	338	20.91
P(PNVE)	311	30.69
P(PNVE)	293	20.89
P(PNVE$_{49}$-CEVE)	311	24.51
P(PNVE$_{57}$-IBVE)	311	16.23
P(PNVE$_{53}$-PEVE)	311	10.52

[a]The photo-irradiation was carried out by a 500-W xenon lamp through a monochrometer.

ingly enough the initial rate of the re version of the pendant QC group in the copolymers, which obeyed first-order kinetics, decreased as follows: P(PNVE) > P(PNVE$_{53}$-PEVE$_{47}$) > P(PNVE$_{57}$-IBVE$_{43}$) > P(PNVE$_{49}$-CEVE$_{51}$) (Table V).

From the results of photochemical valence isomerization be tween pendant NBD moiety and the QC group in the polymer film, the following conclusions can be drawn.

(1) The photochemical valence isomerization between pendant NBD moiety and the QC group in polymer film is strongly affected by both the wavelength of the photo-irradiation and by the com position of the polymer backbone.

(2) The rate of the photochemical reaction of pendant NBD moieties was much faster than that of pendant QC groups in the polymer films under each irradiation condition.

Table V. First-Order Rate Constant of Reversion of
QC Group in the Polymers upon Photo-Irradiation[a]

polymer	wavelength of irradiation, nm	$k_{obsd} \times 10^3, s^{-1}$
P(PNVE)	254	0.42
P(PNVE)	250	0.53
P(PNVE)	248	0.91
P(PNVE)	244	0.40
P(PNVE)	240	0.17
P(PNVE$_{49}$-CEVE)	248	0.57
P(PNVE$_{57}$-IBVE)	248	0.61
P(PNVE$_{53}$-PEVE)	248	0.68

[a]The photo-irradiation was carried out by a 500-W xenon lamp through a monochrometer.

REFERENCES

1. Hautala, R. R.; Little, J.; Sweet, E. *Solar Energy* **1977**, 19, 503.
2. Hautala, R. R.; King, R. B.; Kutal, C. *Solar Energy*; Hummana: Clifton, NJ, 1979.
3. Nishikubo, T.; Sahara, A.; Shimokawa, T., *Polym. J.*, **1987**, 19, 991.
4. Nishikubo, T.; Shimokawa, T.; Sahara, A. *Macromolecules* **1989**, 22, 8.
5. Nishikubo, T.; Hijikata, C.; Iizawa, T. *J. Polym. Sci., Polym. Chem. Ed.* **1991**, 29, 671.
6. Iizawa, T.; Hijikata, C.; Nishikubo, T. *Macromolecules* **1992**, 25, 21.
7. Nishikubo, T.; Kawashima, T.; Watanabe, S. *Polym. Preprints, Jpn.* **1990**, 39, 512.
8. Kishi, K.; Kameyama, A.; Nishikubo, T. *Polym. Preprints, Jpn.* **1992**, 41, 813.
9. Kamogawa, H.; Yamada, M. *Macromolecules*, **1988**, 21, 918.

RECEIVED December 30, 1992

Chapter 25

Photoinitiated Thermolysis of Poly(5-norbornene 2,3-dicarboxylates)

A Way to Polyconjugated Systems and Photoresists

Ernst Zenkl, Michael Schimetta, and Franz Stelzer

Christian Doppler Laboratorium für katalytische Polymerisation Institut für Chemische Technologie organischer Stoffe, Technische Universität Graz, Stremayrgasse 16, A–8020 Graz, Austria

Linear and aromatic conjugated polymers have gained great interest in last decade for several reasons. The problem of inprocessibility was circumvented either by the use of substituted monomers [e.g. poly(alkylthiophene) (1), poly(alkylphenylene) (2), etc.] or by preparation via precursor polymers as shown for polyacetylene (3), poly-p-phenylene (4) and poly(arylene vinylene) (5). The latter ones are mainly prepared via sulfonium polymers in a condensation reaction. This gives bad control of the average molar mass and of the molar mass distribution.

The ring opening metathesis polymerization (ROMP) of bicycloalkenes leads to polymers where a ring is embedded between two vinylene groups. If this ring is substituted with the proper number of leaving groups, double bonds can be created thus forming a conjugated system. This has already been shown for the formation of poly(cyclopentadienylene vinylene) (6, 7) from poly(2,3-diacetoxy-5-norbornene) 1, equ.1.

$$\text{with} \quad R = \quad \text{-COCH}_3 \ (\underline{1}), \ \text{-COC}_6\text{H}_5 \ (\underline{2}) \\ \text{-COOCH}_3 \ (\underline{3})$$

There are several disadvantages in this method: a) high temperature (> 300 °C for the *exo,exo*-diester) is necessary for the thermal elimination; b) this temperature is even higher if an *endo,endo*-diester is used, because the elimination is a typical *syn*-elimination (8) according to equation 2, which means that the favoured structure for the elimination is that of the polymer from the

0097–6156/94/0537–0370$06.00/0

exo,exo-monomer; c) at so high temperatures different side reactions (crosslinking and degradation) may occur; and d) the acid elimination from the ester does not come to completeness before oxidative degradation begins and therfore is not well controllable.

$$\qquad\qquad\qquad\qquad\qquad\qquad\qquad\qquad\qquad\qquad (2)$$

It is also known, that the temperature of elimination is lower for different leaving groups in the following order: -COCH3 > -COPh > -(CO)OCH3 > -(CS)SR > -(CO)NHPh. Therefore we varied the acid groups of the diesters (see polymers 2 and 3 in equ.1) and the stereochemistry (*exo,exo*-disubstituted = **a**, *endo,endo-* = **b**, see Table 1).

Thermolysis of esters can be catalyzed through H + . In this case the stereochemistry plays a less important role. The proton may be created from several photo acid generators (PAG) through UV-irradiation (e.g. by photochemical decomposition of onium salts) (*9*). So we used triphenylsulfonium salts as PAG to catalyze the thermal extrusion of the acid groups. Furthermore this method offers an easy and quick way to build conducting structures in a nonconducting matrix. In this paper we present our first results of these experiments.

EXPERIMENTAL

Monomers were synthesized in standard esterification procedures from exo,exo-2,3-dihydroxy-5-norbornene (prepared from bicyclo[2.2.1]hepta-2,5-diene by selective oxidation of one double bond with $KMnO_4$ (*10*) or with OsO_4/N-Methylmorpholine-N-oxide) (*11*) or from *endo*-5-norbornenylene-2,3-carbonate (prepared from cyclopentadiene and vinylene carbonate in a Diels-Alder condensation) (*12*).

Table 1. Characterization of Polymers

Polymer	M_w	PDI	T_d (Temp of decomposition) (°C)	Heat of decomposition[**] (J/g)	(J/unit)
1 a1	82000	1.07	330	$-140 \cdot 10^3$	$-3 \cdot 10^7$
a2	34000	1.09	-"-	-"-	-"-
b[***]	1250000	1.24	390	-	-
2	52700	1.09	310	$-0.9 \cdot 10^3$	$-3 \cdot 10^5$
3	82000	1.08	270	$-27 \cdot 10^3$	$-6 \cdot 10^6$

[*]minimum of the DSC-Plot [**]determined by DSC [***]polymerized with K_2RuCl_5

ROMP was carried out in absolute chlorobenzene under N2 in a glove box using Mo(CH-*t*-but)(NAr)(O-*t*-but)$_2$ (*13*) as the catalyst. Than the polymers were precipitated in a > 5 fold excess of methanol, washed and dried. If impurities (monomer, catalyst, etc.) were detected in the polymer (colored polymer, or by means of IR- or NMR-spectroscopy) the polymer was reprecipitated from CH$_2$Cl$_2$/MeOH for further purification. In one case K$_2$RuCl$_5 \cdot$ H$_2$O and 2,3-diacetoxy-7-oxa-5-norbornene (ca. 5 mol% of the 2,3-diacetoxy-5-norbornene) in ethanol/water (1:1) was used as the initiator, but no drastic differences in the polymer properties were observed (*9*) (more details about the initiating system see ref.) (*14*).

Films were spin cast on discs of BaF$_2$, KBr or glass from solutions of the polymer (10 wt%) and the photoinitiator (12 mol% relative to the number of ester functions in the dissolved polymer) in dichloromethane (filtered through a 0.5 μm Teflon filter). Then the films were dried *in vacuo* for 15 minutes, exposed to the light of a low pressure Hg-lamp and developed by heating *in vacuo*.

RESULTS AND DISCUSSION

In Table 1 the experimental data (weight average of molar mass, polydispersity index, decomposition temperature and heat of decomposition) of polymers with varius acid groups — OR (see equ.1) and different stereochemistry are listed. It can be seen, that the decomposition temperature decreases from ca. 390 °C for poly(*endo,endo*-2,3-diacetoxy-5-norbornene), polymer **1 b**, down to ca. 270 °C for poly(*exo,exo*-5-norbornene-2,3-bis(methylcarbonate)), polymer **3 a**. As expected, the polymers from the *exo,exo*-isomers decomposed at lower temperatures as the *endo,endo*-isomers because the thermal elimination reaction is a typical *syn*-elimination:

By means of mass spectroscopy we proved, that only the acids and products from the photolysis of the PAG (such as biphenylsulfide and fragments of the BrFnsted acid HAsF6) were eliminated at temperatures lower than the DSC-maximum. At temperatures above 400 °C scission of the polymer backbone was detectable. Solid state MAS-NMR spectra showed that the elimination of the acid was still incomplete even after thermolysis at this high temperature with theoretical loss of weight.

Amongst the onium salts investigated triphenylsulfonium hexafluoroarsenate was the best PAG to catalyse the extrusion of the acid for polymer **1**. A sample with a content of 12% PAG (referred to the number of ester groups in the polymer) showed the best results in the catalyzed decomposition. As can be seen from the DSC plots in Figure 1, the decomposition maximum shifted down to ca 150 °C with an onset already below 80 °C. In the DSC plot a small peak remained at the original decomposition temperature, even after a strong increase of the initiator concentration. In contrast to the uncatalyzed thermolysis, polymer **1** showed the best effect and no difference was found between the isomers **a** and **b**. With our equipment at least one hour of irradiation was necessary, shorter irradiation times showed lower yields of conversion, the high temperature conversion peak remained with higher intensity.

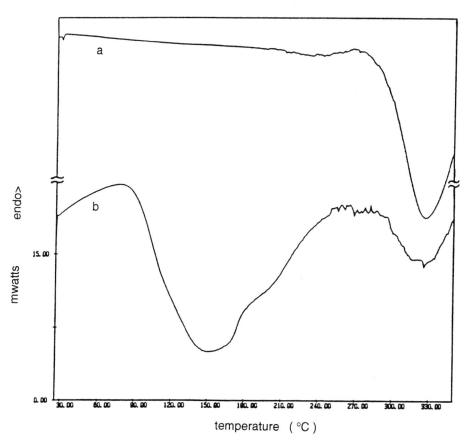

Figure 1. DSC-plots of poly(2,3-diacetoxy-5-norbornene) containing 12 mol% triphenylsulfonium hexafluoroarsenate a) not irradiated b) after UV irradiation (low pressure Hg-lamp, 60 min)

Inspite of the drastic changes in color, the IR spectra of polymers treated at temperatures below 150 °C even for several hours show that the elimination reaction does not come to completion at this temperature, see Figure 2, plot c. The baseline drift of this graph comes from the slope of the broad absorption peak seen in the UV-VIS-IR spectrum with an onset at 2080 cm^{-1} (= 0.26 eV). Only after heating to higher temperatures (> 200 °C) the intensity of the ester modes decreased drastically, plot d. Further heating up to the original Td led to a far going destruction of the polymer structure, plot e. The color of these overheated films was reddish brown.

Upon thermal treatment within the DSC range (50–400 °C) the color of the exposed film changed from colorless to green, metallic dark blue and finally to black. UV-VIS-NIR-spectroscopy can be used to follow this changes in color. Figure 3 shows the UV-VIS-NIR-spectra of a poly(norborn-5-ene-2,3-dioldi-benzoate)/TPS-photoresist. Graph a shows the original film without irradiation and heating. Graph b shows the spectrum of the film after irradiation. Note the appearance of the double peak at 630–680 nm that increases upon heating to 85 °C (Figure 2c–Figure 2d). Upon further heating to higher temperatures broad peaks at 870 and 1150 nm appear and increase which we assign to conjugated systems (Figure 2d–Figure 2e). After heating to 105 °C the two peaks grow together to one broad peak at 950 nm (indicating a bandgap of ca. 1,3 eV at peak maximum), whereas the double peak at 630–680 nm disappears. Judging from the optical absorption spectrum the final product shows the properties typical for a low band gap semiconductor with a gap energy of 0.5 eV, see Figure 3., graph f).

Irradiated and heated films of poly(norborn-5-ene-2,3-dioldiacetate)/PAG were not soluble anymore, whereas films heated to the same temperature without irradiation or without any photoinitiator at all remained soluble, thus functioning as a negative photoresist. In combination with their intensive color, this offers technological applications in the field of photoimaging and as negative deep-UV photoresists. Figure 4 shows a relief image generated by employing the lithographic process described in the experimental section on the poly(2,3-di-acetoxy-5-norbornene)/TPS-system. Due to the extrem sensitivity to thermal treatment (heating rate and maximum temperatureJ) and not appropriate technical equipment for our imaging experiments (simple masque technique combined with an ordinary low pressure Hg lamp without any optical system) it was yet not possible to find out the resolution limits of this system.

Conductivity measurements of the native polymer showed higly insulating properties, which did not change very much upon only thermal or photoacid catalyzed thermal elimination ($\gg 10^9$ W cm). Doping with I_2 brought an increase in conductivity by more than 4 orders of magnitude measured as a resistivity r $\approx$ 8.10^5 W cm showing a semiconducting characteristic. This clearly proofs the possibility of the formation of (semi)conducting structures in a nonconducting matrix following a standard lithographic procedure.

SUMMARY

This paper describes the design of a negative tone resist. The polymer matrix is based upon esters of endo,endo- and exo,exo-norborn-5-ene-2,3-diol that were

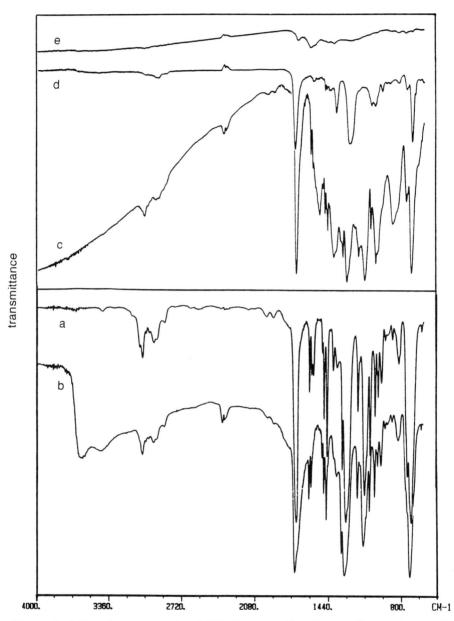

Figure 2. Infrared spectra of poly(2,3-diacetoxy-5-norbornene) containing 12 mol% triphenylsulfonium hexafluoroarsenate a) not irradiated b) after UV irradiation c) heated to 150 °C for 2 min d) heated to 200 °C for 10 min e) heated to 350 °C for 30 min

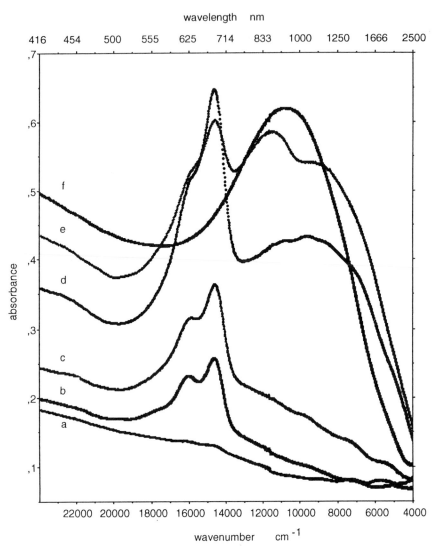

Figure 3. UV-VIS-IR spectra of poly(2,3-diacetoxy-5-norbornene) containing 12 mol% triphenylsulfonium hexafluoroarsenate (TPS) a) after UV irradiation at room temperature b) at 60 °C for 5 min c) at 75 °C for 5 min d) at 95 °C for 5 min e) at 105 °C for 5 min

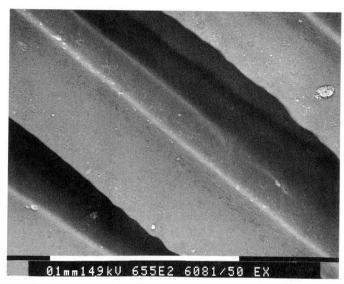

Figure 4. Scanning electron micrograph of negative images delineated in a poly(2,3-diacetoxy-5-norbornene)/TPS resist after development with ethylacetate.

polymerized via ring opening metathesis polymerization (ROMP) using Mo(CH-*t*-Bu) (NAr)(O-t-Bu)$_2$ as catalyst. Elimination of the ester groups lead to poly(cyclopentadienylene vinylene). Differential scanning calorimetry of the various polymers showed a strong dependence of the temperature of decomposition on the acid group. The elimination temperature decreased in the following order : acetate > benzoate > methylcarbonate and exo,exo- > endo,endo- . For acid catalysed elimination of esters we used triphenylsulfonium hexafluoroarsenate (TPS) as PAG. Photolysis of TPS produces the strong Brønsted acid HAsF6 that lowers elimination temperature for all investigated polymers from > 270 °C to ≈ 80 °C (onset of thermal conversion). Upon exposure to UV-irradiation a double peak at 630 - 680 nm appears in the UV-VIS-NIR-spectra of poly(norborn-5-ene-2,3-dicarboxylate)/TPS resists. After postexposure bakes up to 100 °C the resists show a broad absorption in their UV-VIS-NIR-spectra (Jmaximum at E 1.3 eV, onset at ≈ 0.5 eV for poly(norborn-5-ene-2,3-diol-diacetate)/TPS) which we believe are due to charged conjugated sequences. Although IR-spectra indicate that elimination of esters is not complete at low temperatures the properties of the system show drastic changes in solubility (well soluble to insoluble in ethylacetate) and color (colorless to dark blue). These changes after post exposure bakes at low temperatures can be applied for use as a negative tone resist. This system also offers a way to create conjugated structures in an isolating nonconjugated matrix.

ACKNOWLEDGEMENTS

We thank the Christian Doppler Gesellschaft, Vienna, Austria, for financial support. We want to appreciate the practical support by Dr. Peter Pölt (electron microscopy, Forschungsinstitut für Elektronenspektroskopie und Feinstrukturforschung, Technische Universität Graz). Furthermore we thank Wibke Tritthard (conductivity measurements), Christian Heller (UV-VIS measurements) and Prof. Günther Leising (all three at the Institut für Festkörperphysik, Technische Universität Graz) for many constructive discussions.

REFERENCES

 1. Wellinghoff, S.T. In *Polymers for Electronic Applications*; Lai, J.H., Ed.; CRC Press: Boca Raton, FL, **1989**; p. 93.
 2. Rehahn, M.; Schlüter, A.-D.; Wegner, G.; *Makromol. Chem.* **1990**, *191*, 1991.
 3. Edwards, J.H.; Feast, W.J.; *Polymer* **1980**, *21*, 595.
 4. Ballard, D.G.H.; Courtis, A.; Shirley, I.M.; Taylor, S.C.; *J. Chem. Soc., Chem. Commun.* **1983**, 954.
 5. Machudo, J.M.; Masse, M.A.; Karasz, F.E.; *Polymer* **1989**, *30*, 1992.
 6. F. Stelzer, lecture held at the 8th International Symposium on Olefin Metathesis, Bayreuth (Germany), 4–9 Sept. 1989.
 7. Bazan, G.C.; Schrock, R.R.; Cho H.-N.; Gibson, V.C.; *Macromolecules* **1991**, *24*, 4495.
 8. Saunders, H. jr.; Cockeril, A.F.; *Mechanisms of Elimination Reactions*, John Wiley & Sons; New York, **1973**, p. 406 ff.
 9. Reiser, A.; *Photoreactive Polymers*, Wiley & Sons, New York **1989**, Chapters 4 and 7, and literature therein.
10. Fulmer Shealy, Y.; Clayton, J.D.; *J. Am. Chem. Soc*, **1969**, *21*, 3075.
11. VanRheenen, V.; Kelly, R.C.; Cha, D.Y.; *Tetrahedr. Lett.* **1979**, *23*, 1973.
12. Zenkl, E. Ph.D. Thesis, TU Graz, **1992**
13. We thank Prof. R.R. Schrock (M.I.T) for the donation of a sample of this catalyst used for some peliminary investigations at the beginning of this project.
14. Zenkl, E.; Stelzer, F.; *J. Molec. Catal.*, **1992**, *76*, 1-14.

RECEIVED January 19, 1993

POLYIMIDES AND DIELECTRIC
POLYMERS

Chapter 26

Recent Progress of the Application of Polyimides to Microelectronics

Daisuke Makino

Yamazaki Works, Hitachi Chemical Company, Ltd., Hitachi, Ibaraki 317, Japan

Highly purified and high heat resistant PI for microelectronics was first developed in the early 1970s (1). Purpose of the development was the application to the interlayer dielectric of dual metal semiconductor device. By coating liquid PI precursor on a stepped first Al wiring layer, flat PI insulating layer is obtained and step-free second Al layer can be formed as shown in Fig. 1. As a result, reliability of the electrode, especially the second level electrode, is remarkably improved.

Result of high temperature and high humidity life test of the PI interconnect ICs is shown in Fig. 2 (2). Time to reach to 1% cummulative failure is about 10 years under 65 °C-95%RH condition. This life is comparable with the PSG (phosphorous doped silica glass) passivated single layer device.

In this way high reliability of PI interconnect ICs was confirmed in the market, and application of PI to microelectronics rapidly expanded. Fig. 3 summarizes the application of PIs to microelectronics in chronological order (3). They include α-ray shield of memory device, interlayer insulation of thin film facsimile thermal head, thin film magnetic head and so forth. Out of these applications, this paper introduces the recent progress of the application to LSI, Liquid Crystal Display, Multi Chip Module and Optical Wave Guide.

PI BUFFER COAT

Mounting method of semiconductor package to printed wiring board is shifting from pin inserting to surface mounting. In a surface mount, soldering is carried out by dipping the package in a molten solder or exposing to high temperature gas, vapor phase soldering. During this process mold stress generates in a package because of the difference of thermal expansion of mold resin and semiconductor chip. This mold stress gives fatal damage to the package such as passivation crack, Al electrode slide and package crack. Accordingly, it is a very

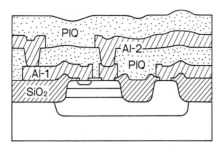

Figure 1. Two layer metal device using PI dielectric.

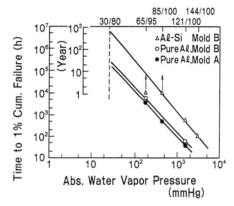

Figure 2. Failure time dependence of PI interconnect devices on absolute water vapor pressure.

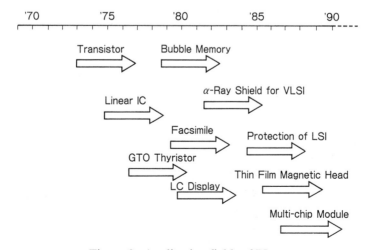

Figure 3. Application fields of PIs.

important issue to reduce the mold stress during soldering process. For this purpose, coating the surface of the chip by PI is currently remarked and utilized in the industry.

Prevention of Passivation Crack

Fig. 4 compares the humidity resistance of the package of PSG passivated chip dipped in solder with non dipped sample (4). Failure rate of solder dipped package is higher. However, when 2 μm thick PI (PIQ:PMDA/BTDA/ODA/ Diaminodiphenylether carbonamide) film was coated on the PSG passivation, failure rate was remarkably improved to the equivalent value to non dipped package as shown in Fig. 5 (4).

Role of PI chip coat film can be explained as follows. In a case of PSG passivation alone, cracks are generated in the PSG film by the mold stress because of the brittleness of PSG film, and Al electrode corrodes by the moisture penetrating through these cracks. On the other hand, for a package with PI chip coat, mold stress is absorbed by this PI film and the passivation crack is prevented.

Fig. 6 compares the passivation crack for various PI chip coat (5). PI-A is low thermal expansion, PI-B is low modulus and PI-C is conventional (Hereafter conventional PI refers to PMDA/ODA PI unless otherwise specified). Among them, low modulus PI-B coating gives the best result. This can be explained by its big stress absorbing ability. On the other hand, relatively good preventing effect is observed for PI-A, which has the highest modulus. In this case stress generation is suppressed by the smaller difference of coefficient of thermal expansion between PI and passivation film. Namely, both PI-A and -B have an ability to prevent passivation crack but their function is different. This can be understood by the fact that stress is proportional to the product of the difference of coefficient of thermal expansion of PI film and substrate and modulus of PI film.

In some cases mold stress affects the active area of the device. Table 1 shows the effect of PI chip coat on the threshold voltage shift of Bipolar Interface Device having piezosensitive circuit (6). By applying 12 μm thick PI coat, threshold failure rate decreases from 10 to 0 %. PI plays an important role to stabilize the electrical property of the circuit by relaxing the mold stress.

Prevention of Electrode Slide

Al electrode in a chip can slide by the mold stress during heat cycle. Effect of PI chip coat on Al slide is shown in Table 2 (5). No Al slide was observed for both low thermal expansion and low stress PIs compared to the large Al slide for non-coated device. Thus, PI coating is also effective to Al slide by absorbing the stress generated by heat cycle.

Prevention of Package Crack

Recently, so called package crack attracts attention. Package crack occurs as a result of the abrupt evaporation of moisture trapped at the interface of mold

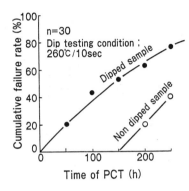

Figure 4. Degradation of humidity resistance by dipping in solder.

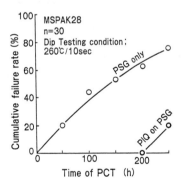

Figure 5. Effect of PIQ formed on PSG for the humidity resistance of solder dipped device.

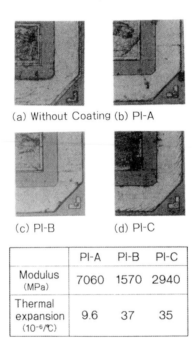

(a) Without Coating (b) PI-A

(c) PI-B   (d) PI-C

	PI-A	PI-B	PI-C
Modulus (MPa)	7060	1570	2940
Thermal expansion $(10^{-6}/°C)$	9.6	37	35

Figure 6. Effect of PI chip coat on passivation crack after thermal cycle test.

Table 1. Electrical Test and Yield Data of Polyimide Coated BPI Devices

Thickness microns	Bake[1] Type	Threshold Failure(%)	Cont. Fail(%)	Yield (%)
None	N/A	10	0.0	90.0
2	A	1	0.0	98.4
2	N	2	0.0	97.3
12	A	0	0.0	98.9
12	N	0	0.6	98.7

1) N : N_2 atmosphere
   A : Air atmosphere

Table 2. Al Slide by Thermal Shock Test

Number of cycles	Al Slide (μm)			
	No Coat	PI-A	PI-B	PI-C
100	1	0	0	0
300	3	0	0	0
500	5	0	0	1
1000	8	0	0	3

Thermal shock test : $-196\,°C/2min \leftrightarrows 150\,°C/2min$

Item	PI-A	PI-B	PI-C
Modulus (MPa)	7060	1570	2940
CTE ($10^{-6}/°C$)	9.6	37	35

resin and chip or lead frame during high temperature soldering. Table 3 summarizes the mode of package crack and the factors affecting the cracks. There are three kinds of package crack. Top crack occurs from the delamination between mold resin and chip surface. Bottom crack from the delamination between mold resin and back side of die pad. Side crack relates to die attaching material.

Fig. 7 shows a mechanism of bottom package crack induced by the vapor pressure (7). Moisture trapped between die pad and mold resin evaporates in the soldering process. From this mechanism, it is thought to be possible to prevent crack by the improvement of adhesion between die pad and mold resin. By forming a thick PI film on the back side of die pad as shown in Fig. 8, occurrence of package crack was investigated. Results are shown in Table 4 (8). By forming PI layer, package crack can be depressed, but it depends on the type of PI. Of these PIs, PI-B which has good adhesion, T_g higher than soldering temperature, 260 °C, and low moisture absorption rate shows the best result.

Next example is the prevention of top crack by coating PI on a chip surface. Table 5 shows the number of cracked package for various PI coated and non coated device after moisture absorption followed by vapor phase soldering and heat cycle (9). Cracking can be drastically suppressed by forming PI chip coat. No remarkable difference was observed among PIs, but big difference was observed between molding compound A and B. Observing the interface of chip and molding compound, delamination was found for the cracked package. Namely, top crack can be prevented by improving the adhesion between molding compound and chip by PI chip coating and selection of molding compound.

Required Properties of PI for Chip Coat Application

As so far described, PI chip coat is effective to prevent damages to the package during soldering or heat cycle. Table 6 summarizes the required properties of PI for chip coat. These values can be deduced from the reliability data mentioned above. Table 7 shows the 5 step ranking of various types of PIs for these properties. Conventional and fluorinated PIs show the balanced properties but the stress is rather high. Low modulus PI has low T_g and high moisture absorption. There is a problem of low adhesive property in low thermal expansion PI. If the adhesion of low thermal expansion PI can be improved by a certain method, it must be the best fit material.

Reliability of chip coated device is strongly affected by the adhesion of mold resin and PI film. Mechanism of the adhesion of mold resin and PI and the method to improve it will be introduced in the following. Fig. 9 is a typical viscoelastic spectrum of PI (10). β dispersion due to the localized motion of the molecular chain is observed at around 130 °C. Relation between adhesive strength and β dispersion temperature is shown in Fig. 10 (10). Good correlation is observed between them. The lower the β dispersion temperature, the higher the adhesion. Especially, when β temperature exceeds the curing temperature of mold resin, 180 °C in this case, the adhesive strength drastically drops. This means that, in order to obtain high adhesion, stress of PI has to be relaxed

Table 3. Factors Affecting the Package Crack

Crack mode	Factors 1	Factors 2
Top Crack	Delamination between mold resin and chip	Poor adhesion between mold resin and chip
Side Crack	Delamination of die attaching layer	Poor mechanical strength, adhesion. and voids of die attaching material
Bottom Crack	Delamination between mold resin and back side of die pad.	Poor adhesion between mold resin and die pad
Common factors to each mode	Large quantity of moisture pick up of package Insufficient mechanical strength of mold resin	

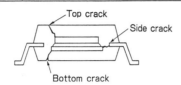

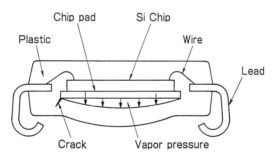

Figure 7. Package cracking mechanism induced by vapor pressure.

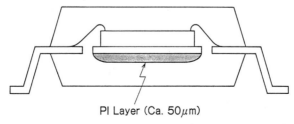

PI Layer (Ca. 50μm)

Figure 8. Prevention of package crack by forming PI layer on the back side of die pad.

Table 4. Dependence of Package Crack on PI Chip Coat

Material	Adhesion	Tg (°C)	Water Absorption (Wt %)	Package Crack Time kept under 85 °C/85 % RH	
				72h	168h
PI − A	Good	245	1.1	Not observed	Observed
PI − B	Good	260	0.9	Not observed	Not observed
PI − C	Bad	290	1.0	Observed	−
No Coat	−	−	−	Observed	−

Soldering Condition : 85 °C/85 % RH for 75 or 168h → Solder dip 280 °C/10s

Table 5. Number of Cracked Package After VPS and Heat Cycle Test

Molding compound	MC − A			MC − B		
No. of heat cycle / PIs[1]	Blank[2]	0[3]	250	Blank[2]	0[3]	250
PI − A	0/10	1/10	2/10	0/10	0/10	0/10
PI − B	0/10	2/10	2/10	0/10	0/10	0/10
PI − C	0/10	0/10	1/10	0/10	0/10	0/10
PI − D	0/10	0/10	1/10	0/10	0/10	0/10
No Coat	0/10	1/10	8/10	0/10	0/10	1/10

Number of cracked package / total package
1) PI- A, PI- B : Conventional thermal Cure PIs
   PI- C : Ion bond type photosensitive PI
   PI- D : Covalent bond type photosensitive PI
2) Before moisture absorption
3) After VPS

Table 6. Required Properties of PI for Chip Coat Application

Item	Required properties
Adhesive strength	PI- Molding Compound = PI- Chip > Molding Compound- Chip (no coat)
Residual stress	< 20 MPa
Moisture absorption	< 0.5 %
Tg	> 260 °C (soldering temp.)

Table 7. Ranking of PIs for Chip Coat Application

Item	Conventional PI	Fluorinated PI	Low modulus PI	Low thermal expansion PI
Adhesive strength	5	3	5	1
Residual stress	3	3	5	5
Moisture absorption	3	5	2	5
Tg	4	4	1	5

1 : Worst .   5 : Best

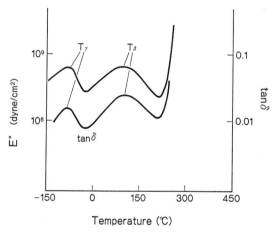

Figure 9. Visco-elastic property of PI.

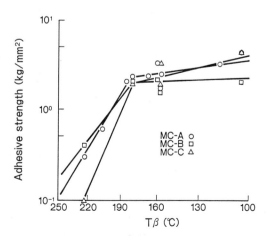

Figure 10. Relation of $T\beta$ of PIs with adhesive strength of PIs and molding compounds.

at the cure temperature of mold resin. In other words, PI having β dispersion temperature lower than curing temperature shows good adhesion to the mold resin.

But some PIs such as low thermal expansion PI has no or very high β dispersion temperature and no stress relaxation is expected. In such a case, it is expected that the PI surface treatment is effective from the analogy of the improvement of PI-PI adhesion. By treating the PI surface by N_2 or O_2 ashing, contact angle decreased and good adhesion was obtained as shown in Table 8 (11). By the plasma treatment PI surface is roughened. Anchoring effect may also be the factor to enhance the adhesive strength. In conclusion, poor adhesion of low thermal expansion PI can be improved by the plasma treatment, and low thermal expansion PI is the most promising material for the chip coat application.

APPLICATION TO MULTI CHIP MODULE (MCM)

PI has began to be applied to multi chip module, which can reduce the machine cycle of computer. Cross section of typical MCM is shown in Fig. 11 (12). 5 layers of metallization insulated by 15 to 20 μm thick PI film is formed on ceramic wiring board. Table 9 is the results of the evaluation of various PIs for MCM (13). Among them, low stress PI, namely low thermal expansion PI, seems to be best balanced except for adhesion and planarity. In the following, recent works to improve these properties of low thermal expansion PI will be introduced.

In MCM there exists three adhesion interfaces, metal on PI, PI on PI and PI on metal as shown in Fig. 12. Among them adhesion of PI on metal is sufficient. Fig. 13 compares the adhesive strength of conventional PI, PIQ, and low thermal expansion PI, L-100, for various metals (14). Adhesion of low thermal expansion PI is extremely low. But, as shown in Fig. 14, by treating the PI surface by oxygen, CF4 or nitrogen gas plasma, high adhesion can be obtained for Cu and Ti, which shows poor adhesion when untreated (14). Among the gases nitrogen is most effective. Reason of the improvement of adhesion by nitrogen plasma was investigated by the surface analysis using ESCA (14). Imide group decomposes to produce more active group such as amide and amino as shown in Fig. 15. These groups are thought to work to improve adhesion.

The second interface, PI-PI adhesion is also able to be improved by either oxygen plasma or Ar sputtering of PI surface as shown in Fig. 16. PI-PI adhesion is influenced by the baking atmosphere. Fig. 17 shows the experimental results of the dependence of the baking atmosphere on the adhesion (15). By baking in nitrogen atmosphere high adhesion is maintained under high humid condition for PI-C, low thermal expansion PI.

Planarity of PI film is affected by the solid content of PI precursor. Good planarity can be obtained for high solid content varnish as shown in Fig. 18 (16). However, usually, high solid content means high viscosity and gives the ununiformity of the resultant film. Most effective method to obtain low viscosity varnish with high solid content is to decrease molecular weight by using capped monomer such as half ester of acid dianhydride. Liquid properties of PI precursor obtained by ester oligomer method is shown in Table 10 (16).

Table 8. Effect of PI Surface Treatment on the Adhesion Between
PI and Mold Resin

Test pieces	PI surface treatment	Contact angle (degree)	Adhesion
A	No ashing	68.1	Poor
B	N₂ ashing	48.2	Good
C	O₂ RIE	41.6	Good
D	O₂ ashing	32.1	Good

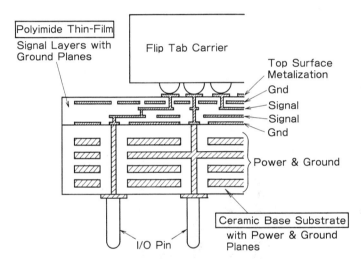

Figure 11. Cross section of PI thin film multi chip module.

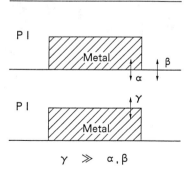

Figure 12. Three adhesion interfaces in PI multi chip module.

Table 9. Evaluation of Various PIs for MCM Application

Property	Standard PI	Fluorinated PI	Silicone PI	Low stress PI	Acetylene-terminated PI	Benzocyclo-butene
Solvent compatibility	1	1~5	1	1	1~5	1
Thermal properties	2	3	3	1	2	3
Moisture absorption	5	3	3	2	4	1
Stress	5	5	5	1	5	5
Adhesion	1	2	2	5	2	2
Planarization	5	5	5	5	1	1
RIE etchability	1	1	4	1	1	4

Relative ranking 1 — Best : 5 — Worst

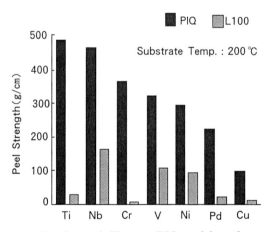

Figure 13. Peel strength of metal films to PIQ and low thermal expansion PI (PIQ-L100).

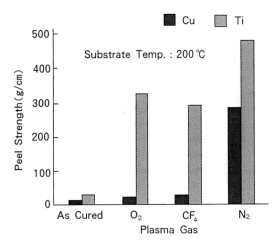

Figure 14. Improvement of adhesion to low thermal expansion PI by plasma surface treatment.

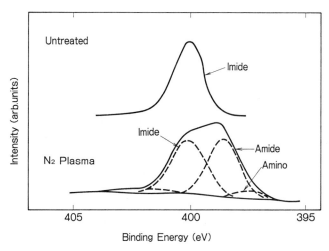

Figure 15. N(1s) spectra of untreated and N_2 plasma treated surfaces.

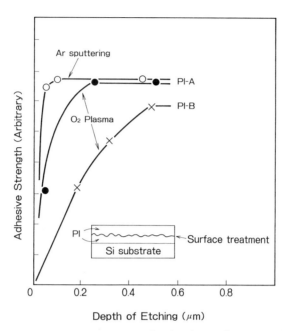

Figure 16. Improvent of PI-PI adhesion by surface treatment.

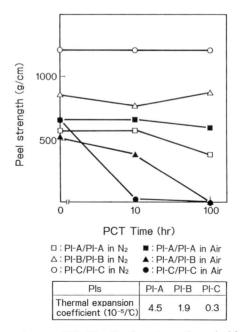

Figure 17. Dependence of PI-PI adhesive strength on baking atmosphere.

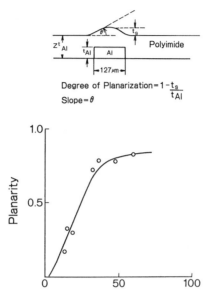

Degree of Planarization $= 1 - \dfrac{t_s}{t_{Al}}$

Slope $= \theta$

Figure 18. Relation between resin content and planarity.

Table 10. Synthesis of High Solid Content and Low Viscosity Polyamic Acid

	Ester Oligomer Method	Polyamic acid Method
Synthesis Method		
Resin Content	40%	14.5%
Viscosity	100cp	1100cp
Molecular Weight	500~1000	$\cong 30,000$

Cu is widely used as the wiring metal of MCM. When PI precursor is baked in contact with Cu, its properties deteriorates as shown in Table 11. This deterioration is explained by the decomposition of imide ring as a result of the reaction of migrated Cu ion and carboxyl group of polyamic acid (*17*). So it is necessary to cover Cu wire by another metal such as Cr or to use preimidized PI instead of polyamic acid.

Table 12 is a summary of selection of PI and processing in MCM application. Low thermal expansion PI is most promising. Future subject is etching. O_2 -RIE is most suitable from the point of fine patterning but the etching time is too long for the production line. Therefore, photosensitive low thermal expansion PI seems to be most promising for MCM application, hopefully preimidized in order to prevent the reaction with Cu, and positive working to get high accurate patterning.

APPLICATION TO LIQUID CRYSTAL DISPLAY (LCD)

In LCD picture image is obtained through the change of the transmission of light by orienting the liquid crystal molecule under the electrical field (Fig. 19). The orientation of liquid crystal is controlled by the alignment surface, and PI is dominantly utilized for this purpose. One of the advantages of PI is to be able to control the pretilt angle. As shown in Fig. 20 pretilt angle is defined as an angle between liquid crystal molecule and alignment surface under the absence of electrical field. This pretilt angle is produced by rubbing the surface of PI. If this pretilt angle is not constant throughout the panel, orientation of liquid crystal is dispersed and gives an ununiform image.

Fig. 21 shows the relation between surface tension of alignment surface and pretilt angle for two kinds of PIs (*18*). Pretilt angle increases with the decrease of surface tension. [Usually pretilt angle increases with the increase of surface tension. This result seems to be the mistyping of the original paper]. However, for PI containing long alkyl chain, pretilt angle increases drastically. This indicates that pretilt angle is also affected by the physical structure of the surface.

PIs having various methylene chain length were synthesized using alkyl chain containing diamine, and the relation between chain length and pretilt angle was investigated (*19*). Results are shown in Fig. 22. Low pretilt angle is obtained for odd carbon number PI, and high pretilt angle for even number PI. This phenomenon can be interpreted from the difference of the conformations of PIs as shown in Fig. 23 (*19*). Assuming that PI is oriented to rubbing direction, PI using odd carbon number diamine has cis conformation. If liquid crystal molecule is adsorbed to alkyl chain no pretilt angle is produced. But for even number PI, since its conformation is trans zigzag, liquid crystal molecule is aligned in one direction with pretilt angle α.

Table 13 shows some of the example of the required properties to the alignment surface of TFT LCD and molecular design of PI. Preimidized varnish reacted from alkyl chain containing diamine and cycloaliphatic dianhydride, which is closely related to the voltage retention rate, seems to be the most appropriate PI for the alignment surface of TFT LCD.

Table 11. Comparison of the Characteristics of PIs Cured on Various Substrates

Underlying Film	Decomposition Temperature (°C)	Tensile Strength (kg/mm²)	Elongation (%)
Cu	380	11.0	8.0
SiO₂	450	13.5	17.0
Cr	450	13.4	16.8
Al	450	13.5	16.7

Cure : N₂. 350°C

Table 12. Summary of the Selection of PI and Processing in MCM Application

Required properties	Countermeasure
Low stress	Low thermal expansion PI
High adhesion	Metal on PI N₂ sputter
	PI on PI Ar sputter, O₂ plasma, N₂ bake
High planarity	High solid type polyamic acid
	Electro / Electroless plating
No reaction with Cu	Cr over coat
Etchability	O₂- RIE but etching time is too long

Photosensitive low thermal expansion PI

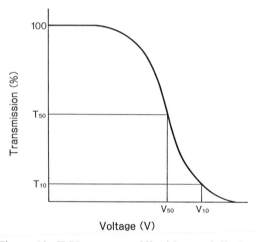

Figure 19. T-V property of liquid crystal display.

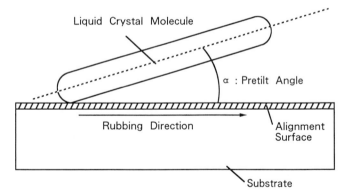

Figure 20. Generation of pretilt angle by rubbing the surface of alignment surface.

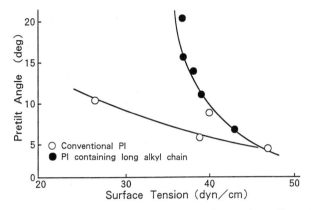

Figure 21. Relation between surface tension and pretilt angle.

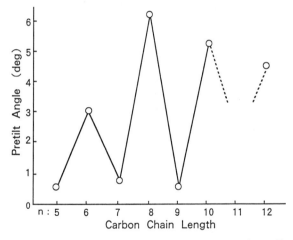

Figure 22. Diamine carbon chain length of PI and pretilt angle.

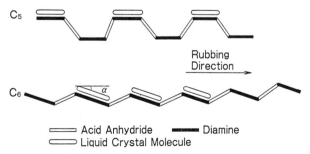

Figure 23. Schematic representation of the conformation of PIs derived from diamine having odd and even numbered carbons and the generation of pretilt angle.

Table 13. Example of Required Properties of Alignment Surface of TFT LCD and Molecular Design of PI

Required properties		Molecular design
Cure temp.	<180℃	Preimidized varnish
Pretilt angle	5°	Introduction of alkyl chain
Voltage retention rate	>98%	Cycloaliphatic monomer

⇩

Preimidized varnish reacted from alkyl chain containing diamine and cycloaliphatic dianhydride

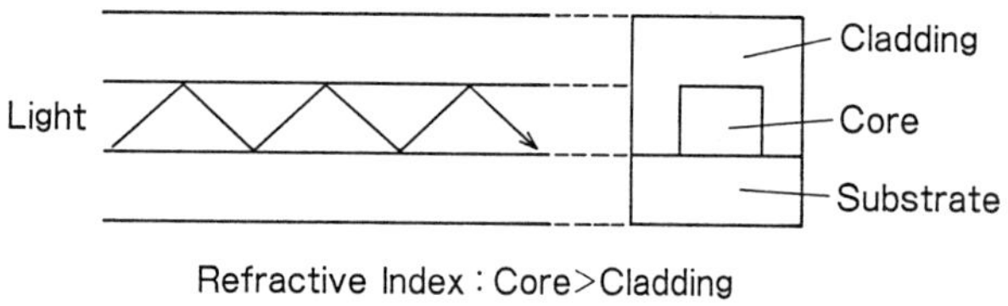

Cladding
Light
Core
Substrate
Refractive Index : Core>Cladding

Figure 24. Structure of passive optical wave guide.

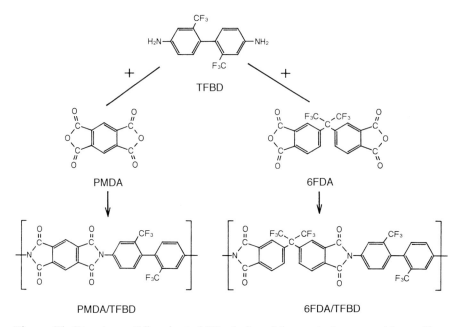

TFBD
PMDA
6FDA
PMDA/TFBD
6FDA/TFBD

Figure 25. Structure of fluorinated PIs designed for optical wave guide application.

APPLICATION TO OPTICAL WAVE GUIDE (OWG)

Optical wave guide which is used in the opto-electronic packaging composes of core and cladding and when refractive index of core is higher than cladding light progresses in the core as shown in Fig. 24. PI optical wave guide has been extensively studied (*20, 21*) since PI stands better for soldering temperature and various circumstances than PMMA or polycarbonate.

PI for optical wave guide must have a high transmittancy to the light. Fluorinated PI is known to be transparent to the visible light. Fig. 25 shows the two types of fluorinated PIs (*22*). Properties of these PIs are shown in Table 14. In order that light progresses in a core without leaking to cladding, refractive index difference should be greater than 0.005 to 0.01. At the same time, since cladding and core are fabricated as an unit, their properties are required to be similar. By changing the mole ratio of 6FDA and PMDA as shown in Fig. 26 (*22*), it is possible to design two kinds of PIs, one for core and one for cladding, with different refractive index and similar properties.

CONCLUSION

So far the recent activities on the application of PI to microelectronics are reviewed and the following conclusion can be obtained.

(1) To add new functions is required to PI with the spread of the application of PI to microelectronics.

(2) PI chip coat is effective for the prevention of passivation crack, Al slide and package crack of semiconductor devices. Low thermal expansion PI can be regarded as the most suitable material for this application.

(3) Low thermal expansion PI is also the most appropriate material for MCM application. But it has disadvantages in adhesion, planarity and reaction with Cu, which can be improved by the chemistry of PI, processing and wire structure.

(4) PI plays an important role as an alignment surface of STN and TFT LCDs. Pretilt angle can be controlled by introducing alkyl chain to PI backbone. At the present stage preimidized varnish reacted from alkyl chain containing diamine and cycloaliphatic dianhydride seems to be most promising.

Table 14. Characteristcs of Fluorinated PIs

Item	6FDA/TFDB	PMDA/TFDB
Fluorine content, %	31.3	23.0
Refractive index (589.6 nm)	1.556	1.647
Dielectric constant	2.8	3.2
Tg, °C	335	> 400
Water absorption, %	0.2	0.7
Thermal expansion coefficient, $10^{-5} \times °C^{-1}$	4.8	0.3
Decomp. temp, °C	569	610

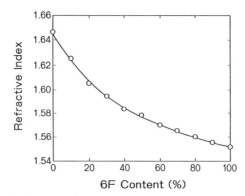

Figure 26. Copolymer ratio of 6FDA and PMDA and refractive index.

(5) PI is also the promising material for the cladding and core of optical wave guide. Fluorinated PI is extensively evaluated because of its high light transmittancy and controlability of refractive index.

REFERENCES

1. K. Sato et al., IEEE Trans. Parts, Hybrid and Packaging, PHP-9, 176 (1973).
2. T. Nishida et al., Proc. Int. Rel. Phys. Symp. (1985).
3. S. Numata et al., "Polymers for Microelectronics", Kodansha, 689 (1989).
4. S. Sasaki et al., Proc. 34th Electr. Comp. Conf., 383 (1984).
5. T. Takeda and A. Tokoh, Proc. 38th Electr. Comp. Conf., 420 (1988).
6. M. M. Khan et al., Proc. 38th Electr. Comp. Conf., 425 (1988).
7. A. Nishimura, Tech. Proc. Semicon Japan, 502 (1991).
8. K. Fujita et al., Industry Technical Seminor, Japan, T2106, 66 (1988).
9. E. Takeuchi et al., Proc. 40th Electr. Comp. Conf., 818 (1990).
10. M. Tomikawa et al., "Polymers for Microelectronics", Kodansha, 655 (1989).
11. N. Shimizu et al., Nikkei Microdevices, Japan, March, 83 (1992).
12. D. Akihiro et al., Proc. 40th Electr. Comp. Conf., 525 (1990).
13. T. G. Tessier et al., Proc. 39th Electr. Comp. Conf., 127 (1989).
14. K. Miyazaki et al., Proc. 176th Meeting of Electro Chemical Soc., No220 (1989).
15. T. Takizawa et al., Proc. 203rd ACS National Meeting, Division of PMSE, 283 (1992).
16. S. Uchimura et al., "Polymers for Microelectronics", Kodansha, 603 (1989).
17. K. Miyazaki et al., Proc. 174th Meeting of Electro Chemical Soc., No208 (1988).
18. T. Abe et al., Organic Synthesis Chemistry, Japan, 49, 506 (1990).
19. H. Yokokura et al., Proc. 16th Liquid Crystal Symp., Japan, 308 (1990).
20. R. Reuter et al., Appl. Opt., 27, No21 (1988).
21. R. Selvaraj et al., J. Lightwave Tech., 6, 1034 (1988).
22. T. Matsuura et al., Macromolecules, 24, 5001 (1991).

RECEIVED May 20, 1993

Chapter 27

Base-Catalyzed Cyclization of *ortho*-Aromatic Amide Alkyl Esters

A Novel Approach to Chemical Imidization

W. Volksen, T. Pascal, J. W. Labadie, and M. I. Sanchez

IBM Research Division, Almaden Research Center, 650 Harry Road, San Jose, CA 95120–6099

It has been discovered that poly(amic alkyl esters) display a pronounced sensitivity toward organic amines, resulting in the partial imidization of these materials in solution at ambient conditions. Model compound studies have revealed that this apparent base-catalyzed reaction is first- or pseudo first-order in nature. In addition to basicity, solvent and temperature, the chemical nature of the parent molecule also plays an important role in the rate of base-catalyzed imidization. Thus, electron withdrawing ester groups as well as amide groups greatly enhance the reactivity. In translating these findings to polymeric systems, it has been possible to significantly lower the temperature required to effect complete imidization.

The conversion of *ortho*-aromatic amide-carboxylic acid units, as found in poly(amic acid) based polyimide precursors, is readily achieved via thermal or chemical means (*1–3*). Polyimide precursors based on poly(amic alkyl esters), however, are only known to cyclize to the corresponding polyimide by thermal conversion at temperatures of 250–300 °C (*4, 5*). In our study of poly(amic alkyl esters) we have made a number of unusual observations, which appeared to be related to the partial imidization of these polyimide precursors by the presence of amines. Characterization of PMDA/ODA based poly(amic ethyl esters), prepared via the low temperature solution polycondensation of p,p-oxydianiline (ODA) and diethyl pyromellitate diacyl chloride (PMDA), by size exclusion chromatography (SEC) in N-methylpyrrolidone (NMP) resulted in cloudy solutions when dilute samples were allowed to stand overnight (*6*). This behavior was attributed to the partial imidization and concomitant precipitation of the highly insoluble imide segments. Distillation of the NMP from P_2O_5 to remove low levels of methylamine, a known impurity in this particular solvent, eliminated this behavior. One should keep in mind that SEC samples tend to be fairly dilute (ca. 0.1% wt/v) and thus yield a significant ratio of methylamine (impurity) to alkyl ester units.

Along similar lines, when investigating the methylation of a poly(amic acid) with 3-methyl-1-p-tolyltriazine, a reaction which liberates a full equivalent of

0097–6156/94/0537–0403$06.00/0

p-toluidine as a side product, extensive gelation during the latter stages of the reaction was noted (7). Removal of samples just prior to gelation and spectroscopic analysis (IR and NMR) revealed significant levels of imidized units in the polymer chain. This type of gelation had also been observed for poly(amic alkyl ester) solutions which had been stored for extended periods of time in non-distilled NMP. Initially an increase in the solution viscosity can be noted, followed by gelation and eventually exclusion of the solvent, resulting in a "hockeypuck" of polymer surrounded by solvent.

The sensitivity of these poly(amic alkyl esters) toward amines is further exemplified by the attempts to transamidate one of the poly(amic alkyl esters) with an amine. Immediate precipitation of a yellow material, which turned out to be mostly insoluble, imidized polymer, occured.

Thus, the unusual behavior of these polyimide precursors described above could ultimately always be traced back to the presence of amines. It is interesting to note that only one literature reference related to this phenomenon was found (8). This reference describes the kinetics and mechanism of cyclization of poly(amic alkyl esters) to polyimides. Based on model studies, it is claimed that phthalic acid retards the imidization of the monomethyl ester N-phenyl-phthalamide model compound, whereas tributylamine has the opposite effect. However, the authors further claim that the imidization of poly(amic alkyl esters) is accelerated by the presence of acids. Based on these observations we decided to investigate this apparent base-catalyzed imidization in greater detail with the goal of utilizing this behavior to chemically imidize poly(amic alkyl esters) as well as for a new photosensitive polyimide scheme in conjunction with photogenerated bases.

EXPERIMENTAL

Materials

All of the materials were commercially available and generally used as received. N-methylpyrrolidone was purified by vacuum distillation from P_2O_5 and most of the aromatic amines were either distilled or recrystallized.

Monoalkyl hydrogen phthalate (I).   20 gm of freshly sublimed phthalic anhydride was refluxed in ca. 100 mL of dry alcohol for 10 hours. The excess alcohol was stripped under vacuum, yielding the desired product in quantitative amounts.

Monoalkyl phthaloyl chloride (II).   Compound I was reacted with a 50 % excess of oxalyl chloride at room temperature for 10 hours. The excess oxalyl chloride was stripped in vacuo to yield the corresponding phthaloyl chloride as a viscous liquid. The material was preferrably used immediately, since it has a tendency to disproportionate to phthalic anhydride.

Monoalkyl phthalamide (III).   10 gm of compound II was dissolved in 20 mL of dry THF and cooled externally via an ice/methanol bath. Next, one equivalent of the aromatic amine dissolved in 20 mL of dry THF was added dropwise. After a reaction time of 1 hour, one equivalent of pyridine was added and the reaction was allowed to proceed for an additional hour. The mixture was then precipitated in water, filtered, washed with more water and vacuum dried. Recrystallization of this material from ethyl acetate gave the desired model compound in yields ranging from 80–90%.

Aryl phthalimide (IV).   0.050 mole of aromatic amine dissolved in ca. 15 mL of dry NMP was added dropwise to a solution of 7.48 gm (0.0505 mole) of freshly sublimed phthalic anhydride in NMP contained in a three-necked flask equipped with reflux condenser, magnetic stirrer, inert gas bubbler and a liquid addition funnel. The reaction was stirred at ambient temperature for 3 hours after which 5.2 gm (0.050 mole) of acetic anhydride and 3.85 gm (0.050 mole) of pyridine were added. The reaction temperature was then raised to 120 °C and maintained for an additional 3 hours. The desired product was obtained by precipitation in 200 mL of water, thoroughly washed with water and methanol and vacuum dried at 50 °C. Two recrystallizations from ethyl acetate yielded 61% of analytically pure product.

Aryl phthalisoimide (V).   A three-necked flask equipped with magnetic stirrer, thermometer and liquid addition funnel was charged with 7.48 gm (0.0505 mole) of phthalic anhydride dissolved in 25 mL of dry NMP. Next, 0.050 mole of the desired aromatic amine dissolved in ca. 15 mL of dry NMP was added dropwise and the reaction mixture was stirred for 3 hours. The flask was cooled externally with ice/water to ca. 0–5 °C and 10.3 gm (0.0505 mole) of dicyclohexylcarbodiimide (DCC) dissolved in 30 ml of dichloromethane was added dropwise. The reaction mixture was maintained below 10 °C for 3 hours after which it was poured into 300 mL of water. After filtration and water washing, the precipitate was extracted with 100 mL of dichloromethane. The dichloromethane solution was dried over $MgSO_4$ and then evaporated under vacuum to yield a yellow solid. Two recrystallizations from ethyl acetate gave 50–60% of analytically pure product.

PMDA / ODA based Poly(amic alkyl esters).   Dialkyl dihydrogen pyromellitate obtained via the direct esterification of pyromellitic dianhydride described elsewhere (9), was converted to the diacyl chloride by chlorination with an excess of oxalyl chloride in ethyl acetate at 60 °C. Once gas evolution stopped and the reaction mixture had become homogeneous, the solvent was stripped under vacuum yielding either a crystalline residue or an amorphous mass of the desired product. In the case of crystalline products, i.e. methyl-, ethyl-, propyl- and isopropyl esters, the material was purified by recrystallization from hexane containing ca. 1–2 mL of oxalyl chloride. Amorphous products, e.g. ethyl glycolyl based esters, were thoroughly dried under vacuum and used as is. Polymerizations were then conducted at ca. − 10 °C by adding a solution of the particular diacyl chloride dissolved in dry THF to a solution of p,p-oxydianiline in dry NMP under an inert gas blanket with vigorous mechanical stirring. Upon completion of the acyl chloride addition, two equivalents of dry pyridine (relative to acyl chloride groups) were added and the temperature was allowed to gradually return to ambient temperature. Stirrring was continued overnight before precipitating the polymer in water using a blender. The precipitate was then thoroughly washed with water, methanol and finally ethyl acetate before being dried in vacuo at 50 °C for 24 hours. To control the molecular weight of the resulting poly(amic alkyl ester) a stoichiometric imbalance of r = 0.980, using the diamine in excess, was employed.

Characterization

^{1}H-NMR analyses were performed on an IBM Instruments AF250 spectrometer and IR analyses were obtained from an IBM FTIR Model 44. High temperature FTIR measurements were made in-situ on films which were solution spun onto NaCl plates. The sample cell was connnected to a temperature controller for isothermal and step-ramp temperature profiles. The relative amount of imidization was determined from the ratio of the 1776 cm^{-1} imide peak area to the peak area of the final, fully cured film (350 °C, 30 minutes) at each isothermal temperature to remove the temperature effect on the peak area.

Kinetic experiments were conducted on solutions of the desired model compound (0.30 gm in 30 gm of dry solvent) containing 1 equivalent of base. In the case of 1,8-diazabicyclo[5.4.0]undec-7-ene (DBU), reaction rates were so fast that only 0.01 equivalents of base were required. All kinetic runs were performed in a thermostatted water bath maintained at 23.3 °C. For HPLC measurements 100 μl aliquots were withdrawn and diluted to a total volume of 10 ml with acetonitrile.

HPLC experiments were carried out on a Waters Model 590 HPLC utilizing a C-18 μBondapak column. The mobile phase was acetonitrile/water = 70/30.

RESULTS AND DISCUSSION

Model Compound Imidization Studies

In order to gain a more complete understanding of the mechanistic aspects of the presumably base-catalyzed imidization reaction, initial experiments focused on model compounds derived from phthalic anhydride, i.e. monoalkyl ester monoaryl phthalamides. Although these model compounds do not necessarily represent many of the aromatic dianhydride derivatives used as polyimide precursors, they are synthetically more accessible, readily purified and soluble in a wide range of organic solvents. For these reasons we prepared a series of model derivatives where both the ester group and the amide portion of the molecule could be systematically varied. In addition, the corresponding isoimide and imide for each derivative was also prepared to facilitate the assignment of HPLC peaks and kinetic analysis of the various model compound reactions.

As shown in Table 1, the imidization of monomethyl ester p-methoxyphenyl phthalamide in NMP as a function of various organic bases tracks the apparent basicity of these materials as reported for water as a solvent. Although there is not a direct correspondence of the imidization rate and catalyst basicity in all cases, this is not too surprising since it is not uncommon for these amines to display different basicities in solvent systems other than water. For the majority of bases, stoichiometric amounts of catalyst were employed only to obtain reasonable reaction rates, while minimizing the amount of model compound required. However, as demonstrated for the case of DBU, much smaller amounts of catalyst could be quite effective. Of course, the basicity argument does not consider the differences in nucleophilic character of the various amines, particular with respect to the less basic examples, such as pyridine, N,N-dimethylaniline, and N-methylmorpholine. The question of nucleophilicity of the base catalyst turns out to be of considerable significance, since it was found that fluoride and acetate ion are also effective catalysts for the conversion of amide alkyl esters to the corresponding polyimides, with reaction rates comparable to the more basic amines.

Although we suspected that isomide could be a possible intermediate in the base-catalyzed imidization of monoalkyl ester aryl phthalamides, it was impossible to detect the isoimide by HPLC or spectroscopic means. However, it was found that aryl phthalisoimides could also be converted to the imide in the presence of base with a reaction rate significantly faster than the corresponding amide-ester. Thus, isoimides cannot be ruled out as possible intermediates in the overall kinetic scheme.

Unusual results were again observed when studying the imidization behavior of monomethyl ester p-methoxyphenyl phthalamide in various solvents using triethylamine as a catalyst, as shown in Table 2. Among these four solvents, NMP and THF show significantly lower imidization rates than chloroform and acetonitrile, with the latter examples exhibiting rate constants almost two orders of magnitude higher. The main feature distinguishing these two groups of solvents are reflected in their hydrogen-bonding power. Thus, tetrahydrofuran and N-methylpyrrolidone are classified in the medium hydrogen-bonding group with solubility parameter values of 9.1 and 11.3 $(cal/cm^3)^{1/2}$, respectively, whereas acetonitrile and chloroform are classified in the poor hydrogen-bonding group with solubility parameter values of 11.9 and 9.3 $(cal/cm^3)^{1/2}$, respectively (*11*). Within each group, the solvent exhibiting the higher solubility parameter also has the higher imidization rate. Certainly the increased basicity of triethylamine in the solvents with poor hydrogen bonding power would be consistent with this behavior.

In addition to basicity and solvent considerations, contributions from both the ester group and the amide group of the parent model compound are also to be expected. As shown in Figure 1, the relative imidization rate of the various alkyl esters with DBU decreases in going from methyl- to ethyl ester and is slowest for the isopropyl ester. This reactivity is analogous to the thermal imidization charateristics of identical poly(amic alkyl esters), which exhibit a similar trend (*5*). Thus, the susceptibility of the ester group toward imidization is related to the electro-positive character of the carbonyl group as determined by the electron-withdrawing power or relative acidity of the parent alcohol. Of course, much greater enhancements in imidization rates can be realized if esters

Table 1. Effect of Base Strength on the Imidization Rate of
Monomethyl Ester p-Methoxyphenyl Phthalamide

Base	Rate constant	pK_a[10]
DBU*	5.87×10^{-4}	-
Triethylamine	9.48×10^{-6}	10.8
Piperazine	1.49×10^{-4}	9.8
N-Methylmorpholine	2.73×10^{-6}	7.4
Pyridine	0	5.2
N,N-Dimethylaniline	4.38×10^{-6}	5.1

* Diluted 100-fold

SOURCE: Data are from reference 10.

Table 2. Imidization Rate of Monomethyl Ester p-Methoxyphenyl
Phthalamide as a Function of Solvent Medium

Solvent	Rate Constant
N-Methylpyrrolidone (NMP)	9.48×10^{-6}
Tetrahydrofuran (THF)*	1.81×10^{-6}
Chloroform*	8.55×10^{-5}
Acetonitrile*	1.04×10^{-4}

* Obtained by ^{1}H-NMR in deuterated solvent

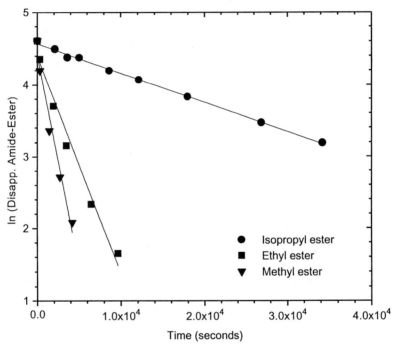

Figure 1. Imidization Rates of Various Monoalkyl Ester p-Methoxyphenylphthalamides

derived from very acidic alcohols such as allyl alcohol or trifluoroethanol are used.

Since the conversion of these amide-esters into imides involves the nitrogen of the amide group, the chemical nature of the amide group is expected to play an important role in the mechanism. As illustrated in Figure 2, the conversion of various monomethyl ester aryl phthalamides in the presence of triethylamine as a catalyst is indeed affected by the aryl residue. As compared to the phenyl amide, para-substitution of either electron-donating or electron-withdrawing groups enhances the rate of imidization. The greatest effect is exerted by the electron-withdrawing nitro group as shown for the p-nitrophenyl amide derivative. Under identical conditions, the corresponding cyclohexyl amide exhibits little or no reaction at all. Although this particular behavior is related to the acidity of the amide proton, it is not clearly understood at this time.

Based on these preliminary results we propose a reaction pathway as illustrated in Figure 3. Considering the preceeding kinetic experiments, indications are that the imidization process may involve a two-step mechanism. It is possible that this mechanism involves the formation of an isoimide-type intermediate. The first step corresponds to the complete or partial proton abstraction from the amide group with the formation of an iminolate anion. Since this iminolate anion has two possible tautomers, the reaction can proceed in a split reaction path to either isoimide- or imide-type intermediate. If the reaction proceeds via the isoimide intermediate, conversion to the imide is extremely fast as indicated by preliminary model reactions. Due to the fast rate of this reaction, the isoimide cannot be ruled out as a possible intermediate, although it has not been observed spectroscopically. However, ultimately the reaction proceeds completely to imide, which represents the thermodynamically stable species.

Imidization Studies of Polymeric Systems

Initial investigations of base-catalyzed imidization of polymeric systems has been restricted so far to PMDA/ODA based poly(amic alkyl esters). Since these polyimide precursors become highly insoluble during the latter stages of imidization (> 40%), the majority of the work was focused on IR studies of supported polymer films with amine catalysts either added to the polymer solution prior to spin-coating or via flooding the polymer film after spin-coating and soft-bake. Probably the most surprising result obtained from these initial experiments is the fact that the base-catalyzed polymer imidization reaction appears to be significantly slower at room temperature, actually requiring elevated temperatures to drive it to completion, as compared to the phthalamide model compounds. It is yet unclear whether this is a direct result of the conformational aspects associated with the polymer chain or solubility considerations arising from the less soluble, partially imidized polymer chain. To compensate for this decreased reactivity, a more activated alkyl ester precursor, such as ethylglycolyl, can be utilized. In this case, high levels of imidization at temperatures significantly lower than required for solely thermal imidization were achievable. As shown in Figure 4, up to 80% imidization could be achieved for soft-baked films which had been flooded with bis(p-aminocyclohexyl)methane for 10 minutes. Similar

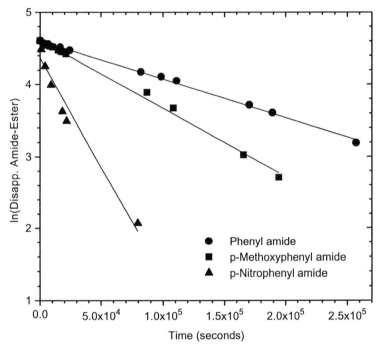

Figure 2. Imidization Rates of Various Monomethyl Ester Arylphthalamides

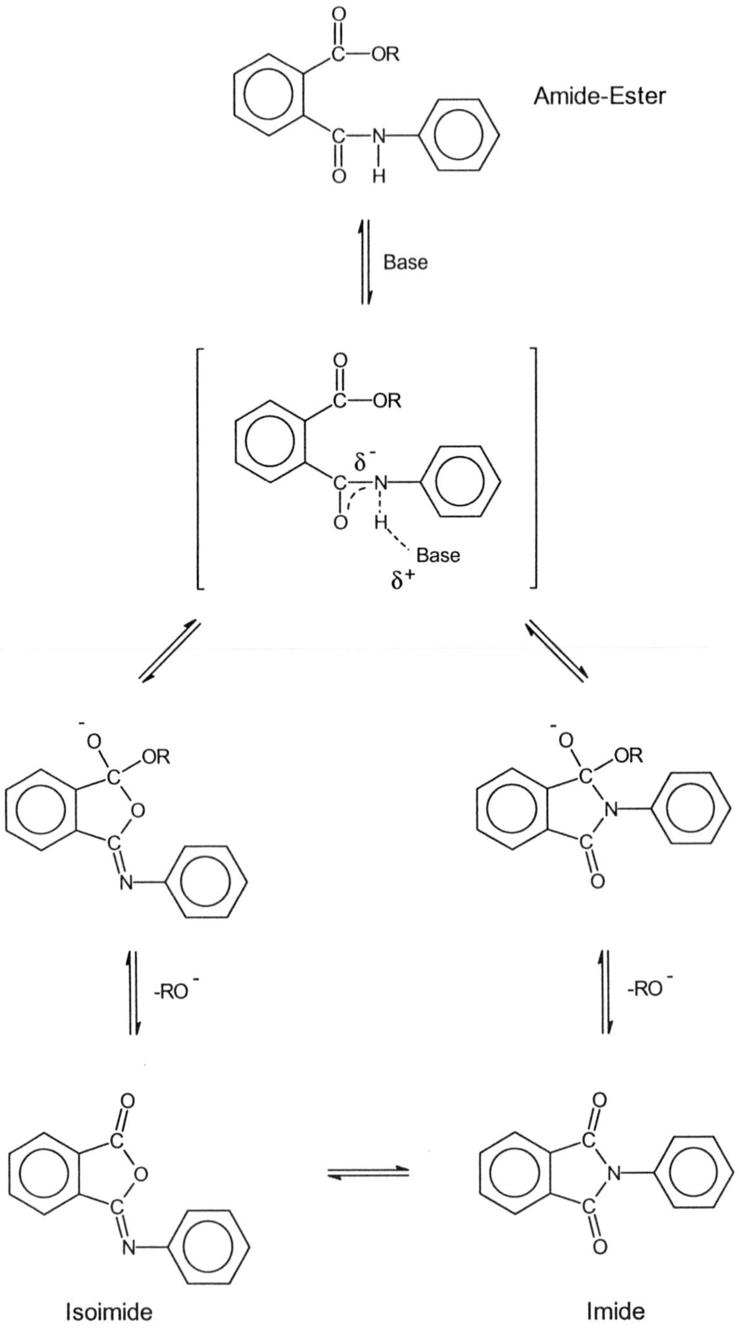

Figure 3. Proposed Kinetic Scheme for the Base-Catalyzed Imidization of ortho-Aromatic Amide Alkyl Esters

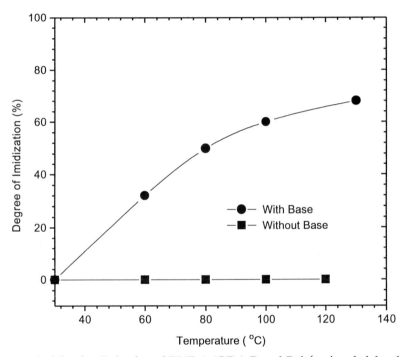

Figure 4. Imidization Behavior of PMDA/ODA Based Poly(amic ethylglycolate ester)

treatment of films which had not been exposed to base exhibited no measurable imidization.

In the case of less reactive polymer systems, such as the meta-isomer of the PMDA/ODA based poly(amic ethyl ester), mixed results were obtained when investigating the polymer imidization behavior of this polymer in the presence of various amines as illustrated in Table 3. All specimens were soft-baked at 80 °C for 5 minutes, treated with the amine for 10 minutes and finally cured at 200 °C for 15 minutes. Although imidization levels of base-treated specimens are significantly higher as compared to the non-treated specimen, no distinctive trend in the data is observable. This is most likely due to factors such as the ability of the amine to effectively diffuse into the polymer film or the actual residence time of the amine in the film at elevated temperatures as determined by volatility considerations. Another aspect not reflected by the experimental IR imidization data relates to surface effects. Since the IR measurement determines bulk imidization levels, skin-core effects are completely neglected. Thus, it is quite conceivable that in case of low imidization levels, the surface of the film could actually be completely imidized with little or no imidization in the underlying material. This skin-effect may then prevent the diffusion of catalyst into the bulk of the polymer film.

As shown in Table 4, relative degrees of imidization could be significantly improved by minimizing the evaporation of base. In this case, the neat amine, diethylamine, was reapplied to the specimen at room temperature after 5 minutes before going on to the post-bake step. Under these conditions the pyromellitic dianhydride (PMDA), p,p-oxydianiline (ODA) derived poly(amic ethyl ester) exhibits the highest level of imidization as compared to the analogous biphenyl dianhydride (BPDA) based poly(amic ethyl ester). This is somewhat expected since the PMDA based system represents the more reactive system as a result of the highly electron-deficient pyromellitate ring. It is important to remember that even after the 80 °C post-bake, these films are completely insoluble in the precursor solvent, NMP.

CONCLUSIONS

In summary, the base-catalyzed imidization of ortho-aromatic amide-alkyl esters provides a convenient pathway to the preparation of low molecular weight imide compounds. This particular reaction appears to be generally applicable to fully aromatic systems, although it is possible to achieve imidization for cycloaliphatic sytems with the use of extremely strong bases, such as DBU. The corresponding polymeric systems appear to follow the general model compound behavior, although observed imidization rates are significantly slower and require elevated temperatures. However, these temperatures are considerably lower than those required for the thermal imidization of poly(amic alkyl esters). This feature is particularly desirable for curing these polyimide precursors in the presence of materials incompatible with the otherwise high thermal excursions necessary to achieve complete imidization. Since base-catalyzed imidization of polymeric systems has been primarily focused on precursors which yield insoluble polyimides, little can be said about the effect of base-catalyzed imidization on the

Table 3. Imidization Data for PMDA/ODA Based *meta*-Ethyl Ester
Polymer in the Presence of Various Amines

Imidization Catalyst	Degree of Imidization (%)
No Amine	18
Primary Amines	
Aniline	40
Dodecylamine	41
Bis(4-amincyclohexyl)methane	41
Secondary Amines	
Diethylamine	52
Di-n-propylamine	61
N-Methylaniline	28
Piperidine	45
Tertiary Amines	
Triethylamine	30
N,N-Dimethylaniline	28
DBU	100
DBN	100

Table 4. Degree of Imidization of Poly(amic ethyl esters) in
the Presence of Diethylamine

Poly(amic ethyl ester)*	Degree of Imidization (%)	
	80 °C	120 °C
PMDA/ODA	80	90
BPDA/ODA	66	80

* Reapplication of diethylamine after 5 minutes of the standard 10 minute treatment

chemical nature of the final polyimide, such as branching and/or crosslinking. Toward this end, studies of poly(amic alkyl esters) which yield soluble polyimides may provide additional information. Finally, the base-catalyzed imidization reaction may provide a viable alternative for a photosensitive polyimide scheme using photogenerated bases, which is described in an accompanying paper.

ACKNOWLEDGEMENTS

The authors would like to express their appreciation to Dr. P.M. Cotts, Dr. D. Hofer and S. Swanson for their contributions as well as Dr. C. Grant Willson and Dr. T.C. Clarke for their support and helpful technical discussions.

REFERENCES

1. Sroog, C.E.; Endrey, A.L.; Abramo, S.V; Berr, C.E.; Edwards, W.M; Olivier, K.L., *J. Polym. Sci., Part A* **1965**, *3*, 1373.
2. Dine-Hart, R.A.; Wright, W.W.; *J. Appl. Polym. Sci.* **1967**, *11*, 609.
3. Wallach, M.L.; *J. Polym. Sci., Part A* **1968**, *6*, 953.
4. Nishizaki, S.; Moriwaki, T.; *J. Chem. Soc. Japan* **1967**, *71*, 1559.
5. Volksen, W.; Yoon, D.Y.; Hedrick, J.L.; Hofer, D.; *Proc. MRS Symposium* **1991**, *227*, 23.
6. Kim, S.; Cotts, P.M.; Volksen, W.; *J. Polym. Sci., Part B* **1992**, *30*, 177.
7. Lavrov, S.V. et. al.; *Vysokomol. Soed., Ser.B* **1978**, *20(10)*, 786.
8. Volksen, W.; Boyer, S.; unpublished results.
9. Volksen, W.; Diller, R.; Yoon, D.Y.; *Recent Advances in Polyimide Science and Techn.*; Weber, W.D.; Gupta, M.R., Eds.; Mid-Hudson Section, Society of Plastics Engineers: Poughkeepsie, N.Y. 1987; pp. 102-110.
10. Lange's Handbook of Chemistry, Dean, J.A., Ed.; McGraw-Hill: New York 1985; Chapter 5, pp. 18-60.
11. Polymer Handbook; Brandrup. J., Immergut, E.H., Eds.; Wiley-Interscience: New York 1975; Chapter 4, pp. 337-349.

RECEIVED December 30, 1992

Chapter 28

Base-Catalyzed Photosensitive Polyimide

D. R. McKean, G. M. Wallraff, W. Volksen, N. P. Hacker, M. I. Sanchez, and J. W. Labadie

IBM Research Division, Almaden Research Center, 650 Harry Road, San Jose, CA 95120-6099

A new scheme for photosensitive polyimide is described. This imaging scheme is based on the amine-catalyzed conversion of soluble polyamic esters to relatively insoluble, partially imidized polymers. The amine is generated photochemically from o-nitrobenzyl carbamate precursors which were made photochemically active at long wavelength either by using carbamate derivatives containing electron donating substituents or by using triplet photosensitizers. Imidization percentages ranging from 20-80% were obtained following irradiation and postexposure bake without any imidization in unexposed film. The imidization percentage of exposed and baked film was dependent on a number of factors including the polymer structure, the nature of the ester group, and the conditions of the postexposure bake. The differential solubility between polyamic esters and partially imidized polymer was greatest for polymers with relatively rigid polymer backbones and this resulted in significantly improved imaging properties (lithographic contrast of 3.0). A final cure of the patterned film results in fully imidized polymer.

Polyimides are widely used as dielectric materials because of a combination of desirable properties including good thermomechanical properties, the solution processability of precursor polymers (polyamic acids or esters), good planarizing properties, low dissipation factor, and a reasonably low dielectric constant ($\varepsilon = 2.7$-3.5) (1). The fabrication of electronic devices containing metal lines and vias embedded in polyimide is made considerably easier when the polyimide films can be made photosensitive and thus directly imagable. Multilayer processing of nonphotosensitive polyimide films by either dry (2) or wet (3) etch pattern transfer has been demonstrated but these imaging schemes increase the number of process steps and thus decrease manufacturing throughput compared with photosensitive polyimide (PSPI) (4). Patterning of nonphotosensitive polyimide films can also be done by laser ablation (5).

Most of the commercially available PSPI systems use either precursor polymers with pendant methacrylate functionality (Siemens technology) (6) or soluble, preimidized polymers such as Ciba-Geigy Probimide 412 (7). Both of these approaches to PSPI function lithographically in a negative mode by producing crosslinked polymers which are insoluble in a developer solvent. The photolithographic contrast is low for both systems and is typically about 1.0. Low

0097-6156/94/0537-0417$06.00/0

contrast limits the resolution that can be achieved during the imaging process which imposes a physical limitation on feature size reduction. Siemens-type PSPI has a high percentage of shrinkage on final cure (> 50%) requiring imaging of films with thicknesses greater than twice the desired final thickness. This places additional demands on the exposure step for high aspect ratio imaging and requires a larger depth of focus. Patterned image distortion frequently results from the final cure process due to the extensive shrinkage as well as higher stress.

This paper describes the development of a new approach to PSPI which is based chemically on the propensity for poly(amic alkyl esters) to imidize rapidly in the presence of catalytic quantities of amines (8). By incorporating amine photogenerators along with poly(amic alkyl esters), it has been possible to develop schemes for imaging poly(amic alkyl esters). A key aspect of this scheme is the effective photogeneration of amines, for which we utilized compounds of the type recently reported by Fréchet and coworkers (9). These compounds include 2-nitrobenzyl carbamates and 3,5-dimethoxy-α,α-dimethylbenzyl carbamates which have been applied to the imaging of other polymer films including epoxy resins (10). Polymer films containing polyamic ester and amine photogenerator are irradiated to produce an amine, which subsequently catalyzes partial imidization in the exposed portion of the film (Figure 1). The unexposed film containing only polyamic ester and amine photogenerator is more soluble in developer solvents than the partially imidized film and can be selectively removed. Subsequent to development, the negatively imaged polymer film is cured to polyimide at high temperature.

Because of the catalytic nature of the imidization step and the greatly decreased solubility of partially imidized polymer, base-catalyzed photosensitive polyimide is a chemically amplified imaging process (11). Chemically amplified imaging schemes are generally high sensitivity, high contrast processes which are advantageous for high resolution imaging. Another potential advantage of this scheme is that the imaging and curing chemistry are identical.

EXPERIMENTAL

Materials

1-(2-Nitrophenyl)ethyl N-cyclohexylcarbamate (1) and 1-(2-Nitro-4,5-dimethoxyphenyl)ethyl N-cyclohexylcarbamate (2) were prepared by reaction of the corresponding substituted benzyl alcohols with sodium hydride and cyclohexylisocyanate. 1-(4,5-Dimethoxy-2-nitrophenyl)ethanol was prepared in two steps by nitration of 3,4-dimethoxyacetophenone followed by reduction with sodium borohydride. Polyamic esters 3-5 were prepared by the standard literature procedure (12). Diglyme and N-methylpyrrolidone (NMP) were purchased from Aldrich and used as received. Coumarin 6 was purchased from Kodak.

Measurements

^{1}H and ^{13}C NMR analyses were performed on an IBM Instruments AF250 spectrometer. Infrared spectra were done on an IBM FTIR Model 44. Ultraviolet absorption spectra were recorded on a Hewlett-Packard Model 8450A UV/Visible Spectrometer. Exposure doses were measured with an Optical Associates Exposure Monitor Model 355. Film thickness measurements were done with a Tencor Alpha Step 200.

Lithographic Evaluations

Solutions were prepared by dissolving 1.0 g of polyamic ester and 0.05-0.2 g of amine photogenerator in 6.5 mL of NMP. Films were prepared on silicon substrates by spin-coating filtered solutions followed by baking to remove excess solvent. Exposures were performed through a mask using broadband irradiation. The films were then heated on a hot plate at temperatures ranging from 80 to 150 °C for 10 min. Development was done by immersion in solvent mixtures containing NMP diluted with an appropriate polymer nonsolvent such as diglyme.

RESULTS AND DISCUSSION

Short Wavelength Imaging

The initial imaging studies were conducted on poly(amic alkyl ester) **4** containing amine photogenerator **1** (Figure 2). The use of a poly(amic alkyl ester) with a glycolate ester group facilitated the base-catalyzed imidization reaction at lower temperatures. Carbamate **1** absorbs most strongly at deep-uv (230-290 nm) wavelengths and so the imaging experiments were conducted using filtered light in this spectral region. However, because of the high absorbance of poly(amic alkyl esters) at deep-uv wavelengths (Figure 3), deep-uv base-catalyzed PSPI was limited to imaging of films with a thickness less than 0.3 μm. Following exposure to broad band deep-uv irradiation, the films were baked at 100 °C for 10 minutes and gave negative images upon development with NMP. This experiment established the feasibility of the base-catalyzed PSPI scheme.

Long Wavelength Imaging

For implementation of PSPI in electronic device fabrication, thick film ($> 2\mu$m) imaging is required. For the base-catalyzed PSPI scheme, the photo-generation of amines at long wavelength ($\gamma > 350$ nm) is necessary for thick film imaging. Several methods for long wavelength amine photogeneration were explored including triplet photosensitizers and red-shifted amine photogenerators, as well as the use of more transparent polymer backbones.

Polymer Backbones with Higher Transparency

Polymer backbones with spacer groups between aromatic rings improve the transparency of these materials in the 350-400 nm range. The poly (amic alkyl ester) derived from oxydiphthallic anhydride (ODPA) and ODA (**3**) which contains oxygen spacer groups between aromatic rings was used for thick film imaging along with 20% of amine photogenerator **1**. Polymer **3** is nearly completely transparent at wavelengths longer than 360 nm and thus imaging can

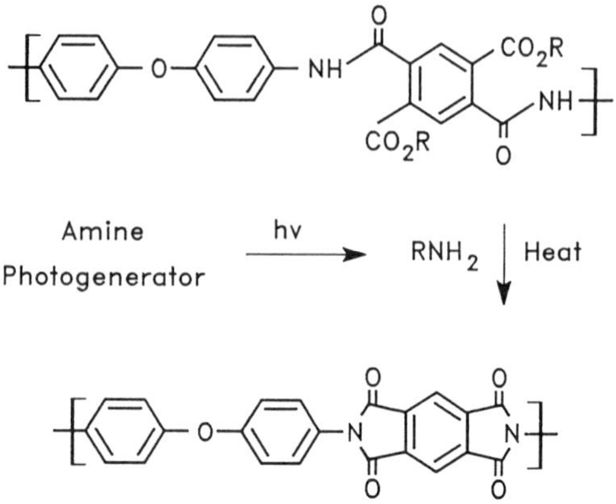

Figure 1. Chemical Scheme for Base-Catalyzed Photosensitive Polyimide.

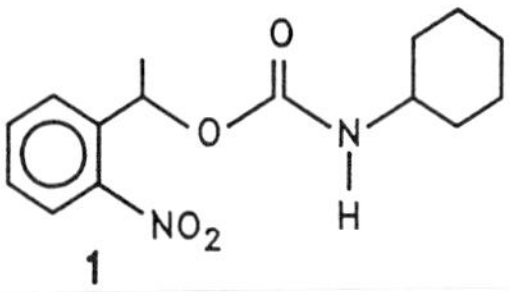

1

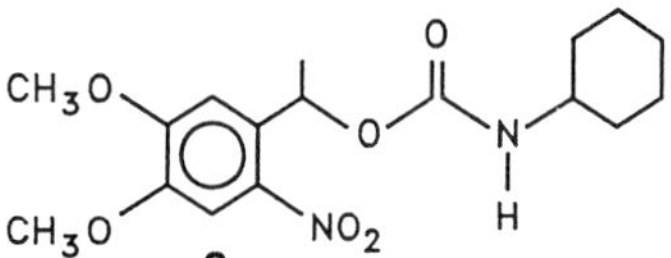

2

3

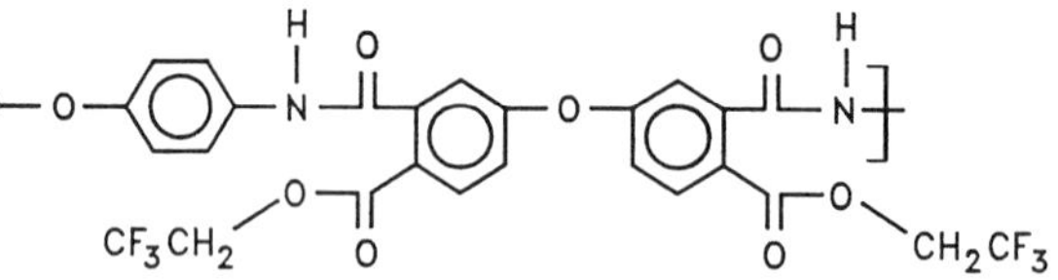

4

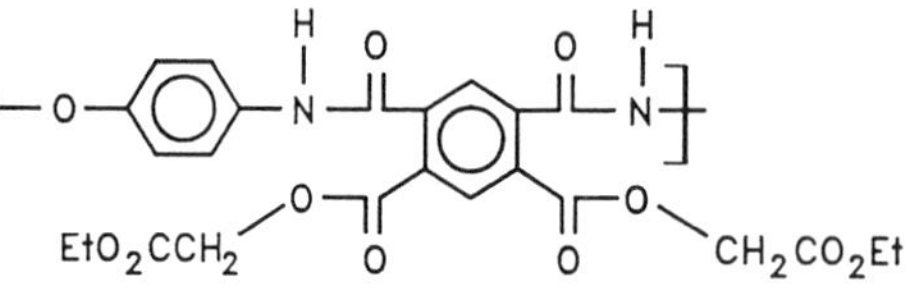

5

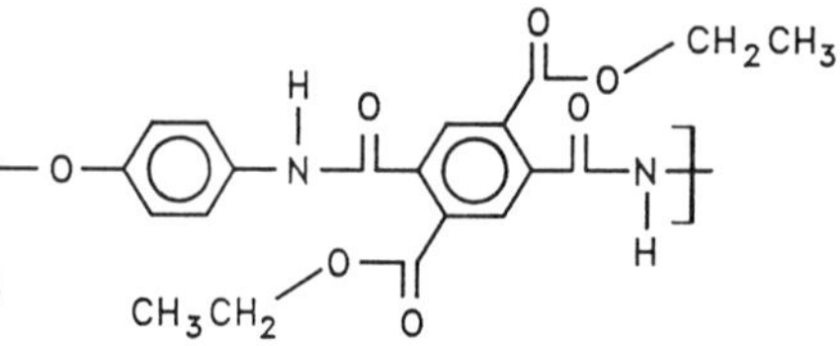

Figure 2. Structures of Amine Photogenerators and Poly(amic alkyl esters) Used for Imaging Studies.

be carried out using 365 nm light most likely due to the presence of a weak carbamate **1** absorption band which extends out to 365 nm.

The imaging of films containing **1** and polymer **3** was best done with broadband irradiation. The use of the activated ester functionality (trifluoroethyl) allowed the use of a lower postexposure bake temperature - 120 °C. The images were developed using a 1:1 solution of cyclohexanone and ethanol. Films with up to three microns of film thickness could be imaged using polyamic ester **3**; however, the solvent resistance of the exposed and postbaked films was limited. Extensive exposed film thinning (50%) was observed during development as well as swelling and solvent-induced stress cracking. The photoresist contrast (Figure 4), which was measured from the slope of a semilog plot of film thickness versus exposure time, was only 1.1 for this experiment. The extensive film thinning and low contrast is likely due to the use of the more flexible polymer backbone which diminishes the solubility differential between polyamic esters and partially imidized polyamic esters. Thus, while the introduction of oxygen spacer groups produces a polymer with greater transparency which facilitates long wavelength imaging, the solvent resistance of the partially imidized polymer is also decreased as a result of this structural change.

Triplet Sensitization of Amine Photogeneration

Thick films of more rigid polyamic esters were also imaged by using combinations of amine photogenerator **1** and a long wavelength absorbing triplet photosensitizers. The coumarin **6** (*13*), which has an absorption band extending out to 400 nm and a triplet energy of 57 Kcal/mol, was used as the triplet sensitizer.

6

Films containing 13% carbamate **1**, 15% coumarin **6**, and 72% polyamic ester **4** were imaged using filtered 404 nm light. Using the same process conditions as the deep-uv imaging experiment, negative images with greater than one micron thickness could be obtained. However, the imaging experiment required large doses (2.5 J/cm^2) at 404 nm which is probably due to the inefficiency of triplet sensitization.

Long Wavelength Sensitive Amine Photogenerators

The imaging of thicker films of poly(amic esters) could also be carried out using a long wavelength absorbing amine photogenerator. The o-nitrobenzyl carbamate derivatives when substituted with electron donating substituents have

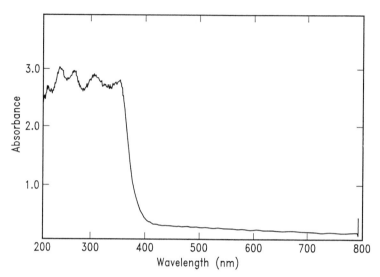

Figure 3. Ultraviolet Absorption Spectra for 3.5 micrometer thick film of PMDA/ODA based poly(amic ethyl ester).

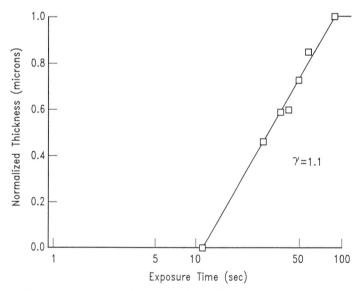

Figure 4. Contrast curve for imaging of poly(amic ester) **3** containing amine photogenerator **1**.

strong absorption bands at longer wavelength. The dimethoxy carbamate **2**, prepared in three steps from commercially available 3,4-dimethoxyace-tophenone, absorbs out to 400 nm and could be used for imaging experiments on highly absorbing poly(amic ester) backbones. Polymer films were prepared from an NMP solution containing the poly(amic ethyl glycolate ester) **4** derived from PMDA/ODA along with 20% of carbamate **2**. The films were exposed to broadband irradiation and then baked at 120 °C to affect partial imidization. Following development with 9:1 diglyme/ NMP, negative images were obtained for film thicknesses up to 6 μm; however, 20% film thinning occurred in the exposed regions of the film and a significant amount of swelling and cracking was observed. The percent imidization for these films was studied by integration of the imide carbonyl infrared absorption band, which indicated that at 100 °C approximately 80% imidization occurred in the exposed portion of the film, while no imidization took place in the unexposed film (Figure 5). Despite the large percentage of imidization, the solvent resistance of the partially imidized meta substituted poly (amic ester) **4** was not sufficient to prevent swelling and cracking during the development step.

Better imaging results for base-catalyzed PSPI were obtained by using the para substituted PMDA/ODA derived poly (amic ethyl ester) **5** which has both a more rigid backbone structure as well as a lower molecular weight ester group. The use of a smaller ester group was beneficial in producing less shrinkage after final cure. Imaging experiments were conducted on films prepared from solutions of 5% **2** and 95% polyamic ester **5**. With this low amine photogenerator percentage the dose requirement was about 500 mJ/cm^2 for imaging of 4 μm films. The use of the unactivated ethyl ester required a higher temperature for the post exposure thermolysis step (150 °C).

To define the optimum postexposure bake for poly(amic ester) **5**, films were heated to various temperatures and the imidization was monitored by infrared spectroscopy. No appreciable imidization occurred until a temperature of 170 °C was reached. Based on this analysis, a postexposure bake temperature of 150 °C was used for the imaging experiments. When 1% of cyclohexylamine was added to poly(amic ester) **5** and the resulting film heated to 150 °C for 10 minutes, infrared analysis indicated that 25% imidization had occurred which was in agreement with the imidization percentage observed when PSPI films were exposed and postbaked. Despite the low imidization percentage, the development of exposed and thermolyzed film with a solvent mixture containing 10% NMP and 90% diglyme produced negative images with minimal film thinning (10%) and no observable swelling or cracking (Figure 6). As a further consequence of high solubility differentiation, the photoresist contrast exceeded 3.0, which is an exceptionally high value for photosensitive polyimide and provides a method for overcoming many of the problems which are encountered during the lithographic exposure step (Figure 7). The higher contrast as well as better solvent resistance of the partially imidized structures is probably due to the more rigid nature of the para-substituted poly(amic alkyl ester) backbone. Furthermore, because the smaller ethyl ester group was used along with a low percentage of photoamine generator (5%), the shrinkage after final cure (37%) was less than shrinkage from Siemens-type PSPI (> 50%).

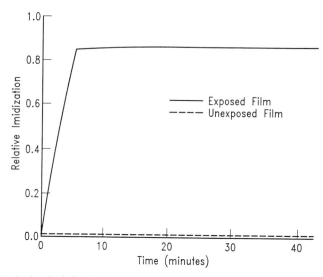

Figure 5. Relative imidization of exposed and unexposed films containing amine photogenerator **2** in poly(amic alkyl ester) **4**.

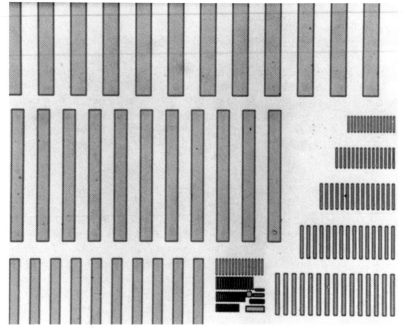

Figure 6. Optical micrograph of images obtained from base-catalyzed PSPI using amine photogenerator **2** in poly(amic ethyl ester) **5**. Smallest resolvable features are 5.0 micron lines and 5.0 micron spaces which are shown in the 5th row of features on the left.

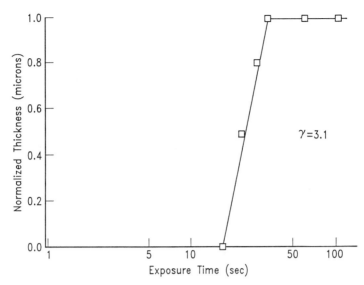

Figure 7. Contrast curve for imaging of films containing amine photogenerator **2** in poly(amic ethyl ester) **5**.

Lithographic Performance vs. Structure of Amine Photogenerators and Polyamic Esters

During the course of these studies, several important structural features were found to be important for the base-catalyzed imaging of poly(amic alkyl esters). Both amine photogenerator and poly(amic alkyl ester) structure are significant for optimal lithographic performance. Polymer structural features which are significant for imaging include the reactive ester group, polymer backbone flexibility and isomeric constitution.

Three classes of amine photogenerator were examined for this application: cobalt-ammine complexes (*14*), benzyl carbamates (*9, 15*), and o-nitrobenzyl carbamates (*9, 16*). The most effective class of amine photogenerator for this application has been the o-nitrobenzyl carbamates **1-2**.

Both the ester group and the polymer backbone structure of the poly(amic alkyl esters) (**3-5**) were important structural features for optimal performance of the base-catalyzed PSPI imaging. Studies of the imidization rate of poly(amic alkyl esters) (*8*) revealed that certain ester groups showed a higher propensity for amine-catalyzed imidization. This class of activated esters includes esters derived from ethyl glycolate and 2,2,2-trifluoroethanol and are known to undergo more facile thermal imidization relative to conventional alkyl esters (*12b, 17*). Imaging of poly(amic esters) with activated ester groups for base-catalyzed photosensitive polyimide allowed the use of a lower postexposure bake temperature to achieve partial imidization of films. However, due to relatively rapid imidization rates, the shelf life of solutions prepared from poly(amic esters) with activated ester groups was limited.

The polymer backbone structure was also important for the imaging of poly(amic esters). Poly(amic alkyl esters) derived from pyromellitic dianhydride (PMDA) and oxydianiline (ODA) are only transparent at wavelengths greater than 400 nm but the transparency at shorter wavelength can be improved by use of alkyl or heteroatomic spacer groups in the polymer backbone. The polymer backbone structure also affects polymer solubility and film development. Poly (amic alkyl esters) with more rigid backbones become insoluble at a lower percentage of imidization. Thus, although the more flexible polymer backbone may afford greater amine photogeneration, and hence, imidization, better images were often obtained with less transparent rigid backbones due to the effect on solubility and swelling properties of the partially imidized film.

CONCLUSIONS

A new scheme for photosensitive polyimide has been developed which involves the use of amine photogenerators along with poly(amic alkyl esters). Amines are generated on exposure of the films which are used to catalyze the partial imidization of the polymer. Negative images result from the lowered solubility of the partially imidized polymer. There are several advantages of base-catalyzed imaging of poly(amic esters) over other PSPI schemes. The ability to image simple ethyl or methyl esters of poly(amic esters) is an advantage in terms of cost of the polymer and also gives a lower percentage of shrinkage relative to Siemens-PSPI. The imaging and cure chemistry for base-catalyzed PSPI are the same and the process is a chemically amplified scheme which generally give higher contrast. The contrast for imaging of para-PMDA/ODA poly(amic ester) was greater than three.

ACKNOWLEDGEMENTS

The authors would like to thank Professor Jean Fréchet and Dr. James Cameron of Cornell University for providing samples of the amine photogenerators and valuable discussions regarding their use and Phil Brock of IBM for preparation of amine photogenerators. The authors also thank Dr. Grant Willson of IBM for his strong encouragement and valuable technical discussion regarding this work.

REFERENCES

1. Endo, A.; Yoda T. *J. Electrochem. Soc.* **1985**, *132*, 155; Jensen, R. J.; Vora H. *IEEE Trans. Components, Hybrids, Manf. Technol.* **1984**, *CHMT-7*, 384; Wilson A. M. *Thin Solid Films* **1981**, *83*, 145; Rothman, L. B. *J. Electrochem. Soc.* **1980**, *127*, 2216.
2. Paraszczak, J.; Cataldo, J.; Galligan, E.; Graham, W.; McGouey, R.; Nunes, S.; Serino, R.; Shih, D. Y.; Babich, E.; Deutsch, A.; Kopcsay, G.; Goldblatt, R.; Hofer, D.; Labadie, J.; Hedrick, J.; Narayan, C.; Saenger, K.; Shaw, J.; Ranieri, V.; Ritsko, J.; Rothman, L.; Volksen, W.; Wilczynski, J.; Witman, D.; Yeh, H. *Proc. Electronic Components and Technology Conference* **1991**,

362. Lichtenberger, A. W.; Lea, D. M.; Li, C.; Lloyd, F. L.; Feldman, M. J.; Mattauch, R. J.; Pan, S. K.; Kerr, A. R. *IEEE Trans. Magn.* **1991**, *27*, 3168.
3. Harada, Y.; Matsumoto, F.; Nakakado, T. *J. Electrochem. Soc.* **1983**, *130*, 129.
4. Rickerl, P. G.; Stephanie, J. G.; Slota P. *IEEE Trans. Components, Hybrids, Manf. Technol.* **1987**, *CHMT-12*, 690; Endo, A.; Takada, M.; Adachi, K.; Takasago, H.; Yada, T.; Onishi Y. *J. Electrochem. Soc.* **1987**, *134*, 2522.
5. Srinivasan, R. *J. Appl. Phys.* **1992**, *72*, 1651.
6. Ahne, H.; Domke, W. D.; Rubner, R.; Schreyer M. In *Polymers for High Technology Electronics and Photonics*; Bowden, M. J.; Turner, S. R., Eds.; ACS Symposium Series 346; American Chemical Society: Washington DC, 1987, p 457.
7. Pfeifer, J.; Rohde O. *Proc. 2nd Int. Conf. on Polyimides* **1985**, 130.
8. Volksen, W.; Pascal, T.; Labadie, J.; Sanchez, M. *Proc. ACS Polym. Mat. Sci. Engr.* **1992**, 235.
9. Cameron, J. F.; Fréchet J. M. J. *J. Org. Chem.* **1990**, *55*, 5919; Cameron, J. F.; Fréchet J. M. J. *J. Photochem. Photobiol. A: Chem.* **1991**, *59*, 105; Cameron, J. F.; Fréchet J. M. J. *J. Am. Chem. Soc.* **1991**, *113*, 4303; Beecher, J. E.; Cameron, J. F.; Fréchet J. M. J. *Proc. ACS Polym. Mat. Sci. Engr.* **1991**, 71.
10. Fréchet J. M. J. *Pure Appl. Chem.* **1992**, in print; Matuszczar, S.; Cameron, J. F.; Fréchet, J. M. J.; Willson, C. G. *J. Mater. Chem.* **1991**, *1*, 1045; Winkle, M. R.; Graziano, K. A. *J. Photopolym. Sci. Technol.* **1990**, *3*, 419.
11. Ito, H.; Willson, C. G. In *Polymers in Electronics*; Davidson, T., Ed.; American Chemical Society: Washington, DC, 1984; 11.
12. a. Nishizaki, S.; Moriwaki, T. *J. Chem. Soc. Japan* **1967**, *71*, 1559; b. Volksen, W.; Yoon, D. Y.; Hedrick, J. L.; Hofer, D. *MRS Symposium Proc.* **1991**, 23.
13. Specht, D. P.; Martic, P. A.; Farid, S. *Tetrahedron* **1982**, *38*, 1203.
14. Kutal, C.; Willson C. G. *J. Electrochem. Soc.* **1987**, *134*, 2280.
15. Birr, C.; Lochinger, W.; Stahnke, G.; Lang P. *Liebigs Ann. Chem.* **1972**, *763*, 162.
16. DeMayo P. *Adv. Org. Chem.* **1960**, *2*, 367; Patchornick, A.; Amit, B.; Woodward, R. B. *J. Am. Chem. Soc.* **1970**, *92*, 6333.
17. Volksen, W.; Yoon, D. Y.; Hedrick, J. L. *IEEE Trans-CHMT* **1992**, *15*, 107.

RECEIVED December 30, 1992

Chapter 29

Novel Cross-Linking Reagents Based on 3,3-Dimethyl-1-phenylenetriazene

Aldrich N. K. Lau and Lanchi P. Vo

Corporate Research and Development, Raychem Corporation, 300 Constitution Drive, Menlo Park, CA 94025-1164

Arylene ethers containing two or three 3,3-dimethyl-1-phenylenetriazene ($-C_6H_4$-N = N-NMe$_2$) reactive end-groups have been prepared. Differential scanning calorimetry (DSC) of these triazene functionalized arylene ethers exhibits exothermal decomposition peak temperature ranging from 263 °C to 285 °C. These arylene ethers crosslink a fluorinated polyimide at 300 °C to give high gel contents, and exert little influence on the dielectric constants of the crosslinked polymers. The crosslinked polymers exhibit improved thermal stability and resistance to organic solvents.

Aromatic polyimides find increasing use in the fabrication of multilayer interconnection systems (*1*). Aromatic polyimides containing hexafluoroisopropylidene groups -C(CF$_3$)$_2$-, exhibit low dielectric constants (*2*), but they are susceptible to organic solvents. Organic solvents swell the polymers leading to stress crazing. When the polymer crazes, it fractures the metallic circuitry that has been deposited onto the polymer surface. One method to improve the resistance to solvent-induced crazing involves the formation of a semi-interpenetrating network (*3*). Another approach involves chemical crosslinking. In our laboratories, we have been looking for sources of aryl radicals as crosslinkers that crosslink high performance polymers at or above 200 °C. It is well known that phenyltriazenes are stable sources of phenyl radicals. For example, 1,3-diphenyltriazene and 3-alkyl-1-aryltriazene arylate decompose below 160 °C to form aryl radicals. These electron deficient aryl radicals homolytically substitute aromatic nuclei to form arylated products (*4*). Arylations of benzene and pyridine with 1-aryl-3,3-dimethyltriazenes at ~ 100 °C in the presence of an acid catalyst have also been reported (*5*). The use of a catalyst, however, can potentially contaminate electronic devices.

Recently, we have discovered that arylene ethers containing 3,3-dimethyl-1-phenylenetriazene ($-C_6H_4$-N = N-NMe$_2$) end groups arylate benzene at 270 °C without introducing aliphatic moieties into the product (Scheme 1) (*6*). In addition, the arylation does not require a catalyst. We report herein the

0097–6156/94/0537–0428$06.00/0

syntheses of three 3,3-dimethyl-1-phenylenetriazene end-capped crosslinkers, **1**, **2**, and **3**, and the use of these reagents to crosslink a fluorinated polyimide **4** (Scheme 2).

RESULTS AND DISCUSSION

Preparation of Crosslinkers

In designing crosslinkers for crosslinking aromatic polymers in electronic applications, the following issues have to be considered. From the electronic point of view, the crosslinkers should not increase the dielectric constants and moisture absorption of the crosslinked polymers. From the stability point of view, stable aryl-aryl C-C bond linkages are preferred and no oxidatively unstable aliphatics should be introduced into the crosslinked network. From the processing point of view, the desired cure temperature is between 200 and 300 °C such that high boiling solvents can be used for casting polymer/crosslinker thin films. Arylene ether crosslinkers based on 3,3-dimethyl-1-phenylenetriazene meet all these requirements.

The syntheses of crosslinkers **1**, **2**, and **3** are outlined in Scheme 3, 4, and 5, respectively. The two-step, one-pot diazotization/triazenization reaction was carried out in aqueous THF. The bis-diazonium compound was not isolated before reacting with dimethylamine in the presence of an excess of sodium carbonate. DSC thermograms of crosslinkers **1**, **2**, and **3** exhibit exothermal decomposition peak temperatures (Td) at 263, 283, and 285 °C, respectively.

Crosslinking of Polyimide 4

A solution of **4** and a crosslinker in *N*-methyl-pyrrolidinone (NMP) was filtered with a PTFE filtration membrane (0.45μ). Polyimide films (10 to 20 μm) were obtained by spin-casting the filtered solution onto glass plates and subsequently "soft-baking" for 30 minutes each at 100 °C and 200 °C in a convection oven to remove the solvent. Curing is achieved by baking the polymer film at 300 °C for 30 minutes in a nitrogen oven. A free standing film of the cured polymer can be released from its substrate after soaking in 90 °C water for 1 to 4 hours. The gel content of the cured polymer film is determined after 24 hours extraction with *N,N*-dimethylacetamide (DMAc) in a Soxhlet apparatus.

Polymer films of **4** cured at 300 °C with 20 wt% of **1**, **2**, or **3** exhibit gel contents of 93, 85, and 98%, respectively (Table 1). The cured films are flexible and insoluble in polar solvents such as NMP and DMAc. Polyimide **4** cured at 300 °C without a crosslinker is fully soluble in hot DMAc and therefore exhibits gel content of 0%. Heating blends of polyimide **4** containing 20 wt% of **1**, **2**, or **3** at 200 °C for 60 minutes also showed no evidence of crosslinking.

At high concentrations of crosslinkers (20 wt%), the formation of a semi-interpenetrating network may occur, linear polymer chains of **4** physically trapped by a three-dimensional network formed by the crosslinker. To demonstrate that the triazene reagents behave as chemical crosslinkers, we studied the curing of **4**

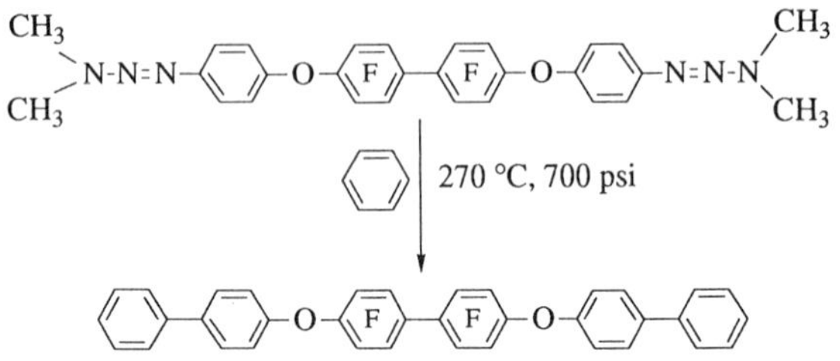

Scheme 1. Arylation of benzene with arylene ether containing 3,3-dimethyl-1-phenylenetriazene end groups.

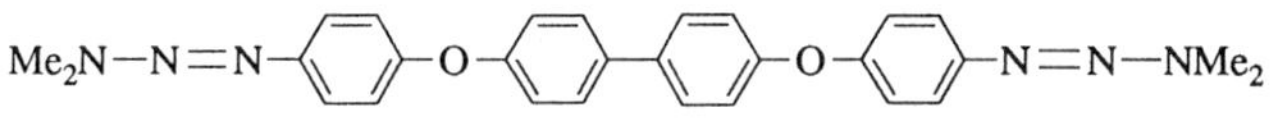

1

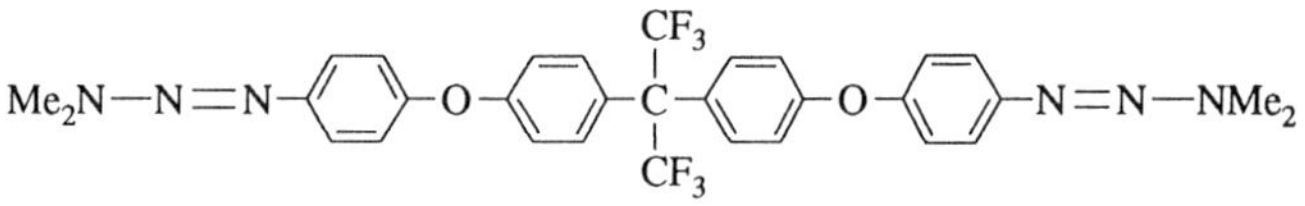

2

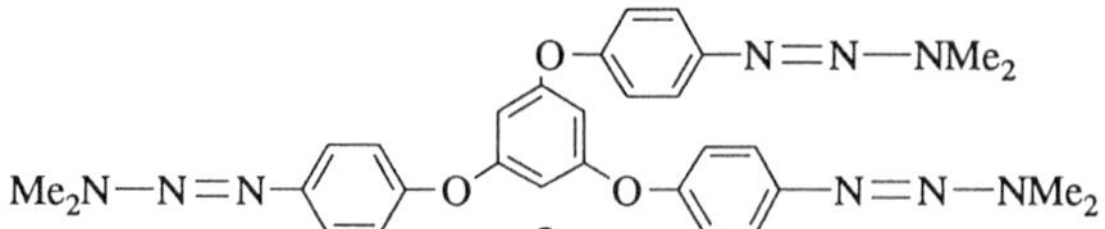

3

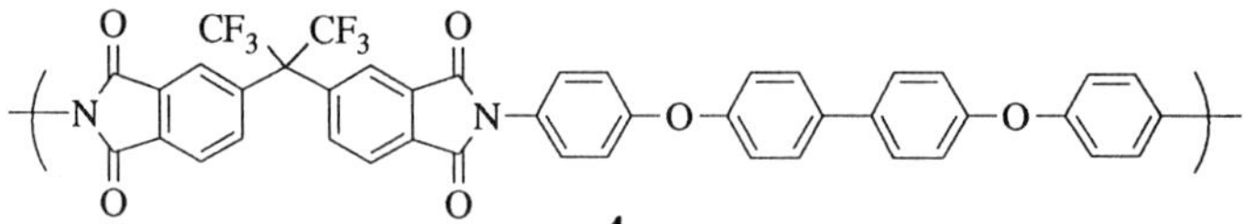

4

Scheme 2. Molecular structures.

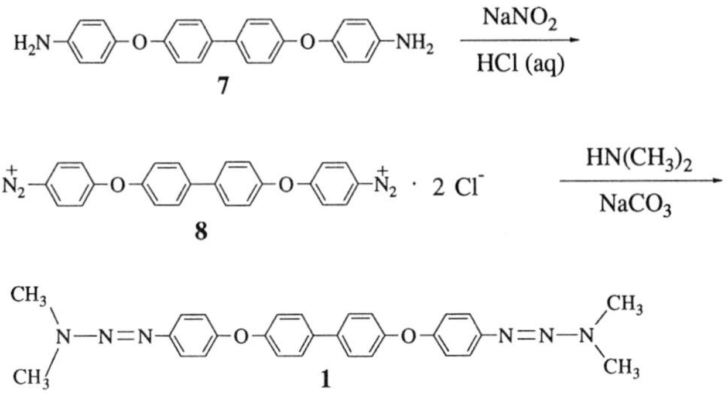

Scheme 3. Synthesis of bis-triazene crosslinker **1**.

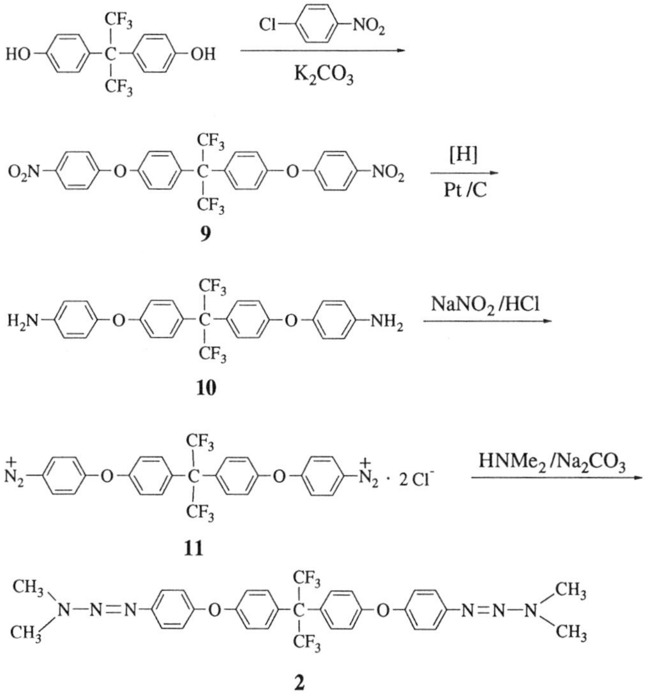

Scheme 4. Synthesis of bis-triazene crosslinker **2**.

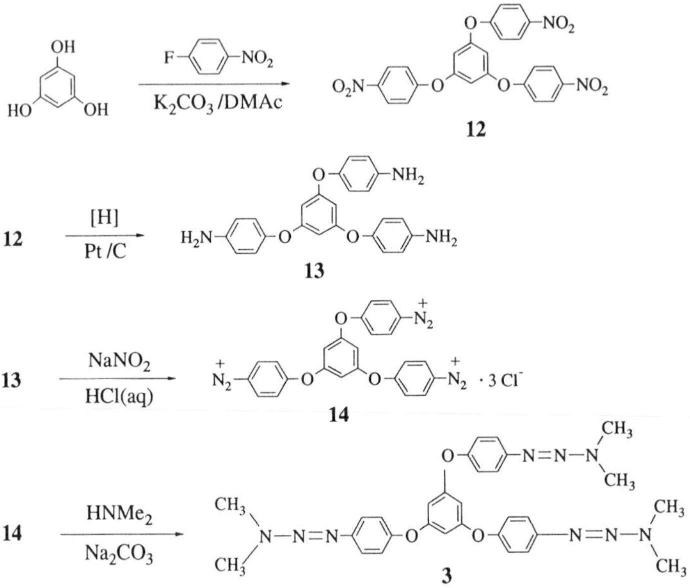

Scheme 5. Synthesis of bis-triazene crosslinker **3**.

Table 1. Gel Contents of Fluorinated Polyimide **4** Crosslinked by
Crosslinker **1**, **2**, and **3** at 300 °C

Crosslinker	1	2	3
Conc. of Crosslinker:			
None (control)	0	0	0
0.11 mmol/g of **4**	47	44	72
5 wt%	40	36	66
20 wt%	93	85	98

with lower concentrations of **1**, **2**, or **3**. We found that the fluorinated polyimide **4**, cured with 5 wt% of **1**, **2**, or **3**, exhibits gel content of 40%, 36%, and 66%, respectively (Table 1). At this low concentration, it is unlikely that the crosslinkers alone would form an effective three-dimensional network. Therefore, chemical crosslinking must be responsible for the observed high gel contents.

At equimolar concentration level of the crosslinker (0.11 mmol of a crosslinker per gram of **4**), we have found that polymer **4** cured with **1** or **2** exhibits 47% and 44% gel content, respectively (Table 1). Under the same conditions, polymer **4** cured with **3** (0.11 mmol of **3** per gram of **4**) exhibits 72% gel content. The increase in gel content is more than 50% (Table 1). It confirms that crosslinker **3**, which contains three reactive triazene end-groups, is a more efficient crosslinker than **1** or **2**, which contain only two reactive end-groups.

It is believed that polymer thin films cured at high temperature exhibit residual stress (7). Solvent swelling releases the residual stress leading to crazing upon drying. Chemical crosslinking prevents the cured polymer from swelling and, thus, improves the resistance to solvent-induced stress crazing. To determine solvent resistance, we cured a second polymer coating onto a previously cured polymer film, and looked for cracks under a microscope. We have found that polymer **4** cured with 20 wt% of **1**, **2**, or **3** does not show solvent-induced stress crazing. However, polymer **4** cured with 5 wt% of **1**, **2**, or **3** crazes (Table 2).

The dielectric constants of cured polymer films were measured at 0% and 60% relative humidities (RH). The results are shown in Table 3. Crosslinking polyimide **4** with 20 wt% of the fluorinated bis-triazene **2** does not change its dielectric constant. Polymer films of **4** cured with 20 wt% of **1** or **3** exhibit insignificant changes in dielectric constants.

The thermal stabilities of **4** prior to, and after, crosslinking were determined by TGA, and the results are shown in Table 4. Polyimide **4** cured with **1**, **2**, or **3** did not show significant change in the temperature at 5% weight loss. At 700 °C, polyimide **4** cured without a crosslinked exhibits weight loss of 85%. However, polyimide **4** cured with **1**, **2**, or **3** exhibits less than 35% weight loss at 700 °C. These results suggest that crosslinking with the triazene-containing reagents increases the thermal stability of polyimide **4**.

Mechanism of Crosslinking

A mechanism of crosslinking, which is similar to that reported for the phenylation of benzenes (5), is proposed in Scheme 6. The triazene end groups of **1 2**, or **3** decompose at elevated temperature to form σ-radicals **15**, nitrogen gas, and dimethylaminyl radicals (reaction 1). Radicals **15** homolytically substitute aromatic nuclei of the polymer chains (reaction 2) to form σ-complexes **16**. The σ-complexes **16** then aromatize by reacting with the dimethylaminyl radicals, or through other pathways, to form **17** (reaction 3) which produces a crosslinking system through the repetition of reactions 1, 2, and 3. There was no evidence that dimethylaminyl radicals were incorporated into the cured polymers.

Table 2. Solvent-Induced Stress Crazing of Polyimide **4** Crosslinked by
Crosslinker **1**, **2**, and **3** at 300 °C

Crosslinker	**1**	**2**	**3**
Conc. of Crosslinker:			
5 wt%	Yes	Yes	Yes
20 wt%	No	No	No

Table 3. Dielectric Constants of Polyimide **4** Crosslinked by
Crosslinker **1**, **2**, and **3** at 300 °C

Crosslinker	None (control)	**1**	**2**	**3**
Wt% of Crosslinker	0	20	20	15
ϵ @0% RH	2.84	2.82	2.84	2.80
ϵ @60% RH	3.21	3.30	3.21	3.25

Table 4. TGA of Polyimide **4** Crosslinked by Crosslinker
1, 2, and **3** at 300 °C

Crosslinker	None (control)	1	2	3
Wt% of Crosslinker	0	20	20	15
Onset Wt Loss in Air (°C)	544	506	510	522
Temperature@5% Wt Loss in Air (°C)	518	515	514	516
Maximum Wt Loss @700 °C in Air (%)	85	34	26	27

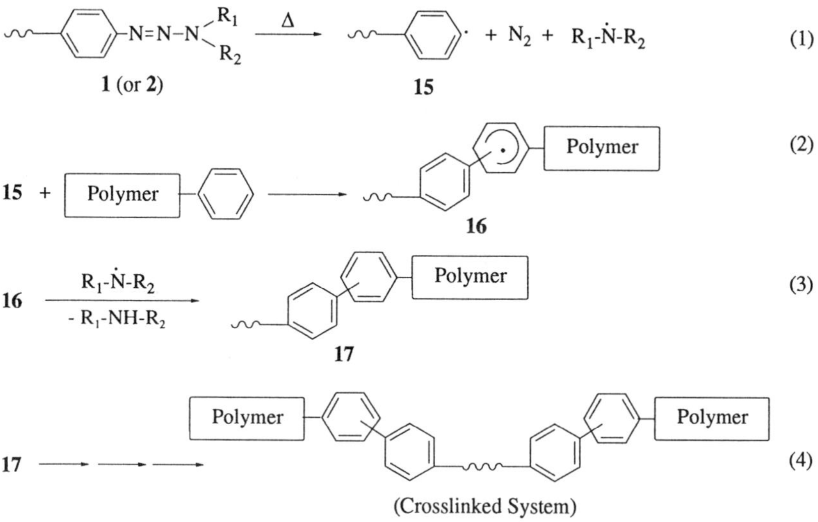

(1)

1 (or 2)   **15**

(2)

15 +   **16**

(3)

16   **17**

(4)

17   (Crosslinked System)

Scheme 6. Proposed mechanism of crosslinking.

EXPERIMENTAL SECTION

The IR spectra were recorded on a Perkin-Elmer 1420 spectrophotometer and the ^{1}H-NMR spectra were recorded on a Varian XL-300 spectrometer. Differential scanning calorimetry (DSC) and thermal gravimetric analysis (TGA) were performed on a Perkin-Elmer 7700 thermal analyzer at 10 °C/min in N_2 and 20 °C/min in air, respectively. Gel permeation chromatography (GPC) was performed with a Hewlett-Packard 1090 Liquid Chromatograph fitted with four Polymer Labs PL-Gel columns (100Å, 500Å, 103Å, and 104Å pore diameters), using tetrahydrofuran (THF) as the mobile phase and polystyrene standards. Dielectric constants were measured at 10 KHz according to the method reported by Mercer et al. (8) The polymer films (10 to 20 μm) were allowed to equilibrate in the capacitance cell for at least 30 minutes before readings were taken.

Fluorinated polyimide **4** was prepared according to the procedure reported by Jones et al (9), Tg = 260 °C, Mn = 11,570 g/mole, and Mw = 22,740 g/mole.

1,1'-[[1,1'-Biphenyl]-4,4'-diylbis(oxy-4,1-phenylene)]bis[3,3-dimethyl-1-triazene] (1)

To a stirred solution of 15.0 g (40.7 mmol) of 4,4'-bis(4-aminophenoxy)biphenyl, **7**, (Chriskev Co.) in 400 mL of tetrahydrofuran (THF) in a 1 L three-neck flask equipped with a mechanical stirrer, a thermometer and an addition funnel, a solution of 32.0 mL (384 mmol) of 12 N hydrochloric acid in 400 mL of water was added slowly. The reaction solution was then chilled to -2 °C with stirring. To this vigorously stirred mixture, a solution of 13.81 g (200 mmol) of sodium nitrite in 150 mL of water was added dropwise over a period of 30 minutes. The temperature of the reaction mixture never exceeded 0 °C. After the addition, the reaction mixture was stirred at 0 °C for an additional 30 minutes. At the end of the reaction, THF was removed under reduced pressure at 25 °C. The remaining aqueous solution was chilled to 0 °C and neutralized to pH 6-7 with an ice-cold, saturated solution of sodium carbonate. This neutralized solution was immediately poured into a 2 L beaker containing a freshly prepared solution of 15.0 g (184 mmol) of dimethylamine hydrochloride and 36.0 g (340 mmol) of sodium carbonate in 450 mL of ice water. The mixture was stirred vigorously with a mechanical stirrer for 10 minutes and then extracted with four 100 mL-portions of dichloromethane. The combined extracts were washed twice with distilled water, dried over anhydrous magnesium sulfate and decolorized with activated charcoal. The solvent was removed under reduced pressure at 35 °C and the crude product was recrystallized from a mixture of dichloromethane/acetone (1:5 v/v) to give 15.05 g (77%) of **1**, mp 167-9 °C; IR (KBr) 1492, 1250, and 1067 cm^{-1}; ^{1}H-NMR (CDCl$_3$) d 3.32 (s, 12H, CH$_3$), 6.86-7.66 (m, 16H, Ar-H). DSC of **1** exhibits an exothermic decomposition peak temperature (Td) at 283 °C. Anal. Calcd for C$_{28}$H$_{28}$N$_6$O$_2$: C, 69.98; H, 5.87; N, 17.49. Found: C, 70.74; H, 5.73; N, 16.29.

2,2-Bis[4-(4-nitrophenoxy)phenyl]hexafluoropropane (9)

To a 500 mL three-necked flask equipped with a mechanical stirrer and a reflux condenser, 17.15 g (51.0 mmol) of 4,4'-(hexafluoroisopropylidene)diphenol (Aldrich Chemical Co.), 17.57 g (111.5 mmol) of 1-chloro-4-nitrobenzene, 27.83 g (201.4 mmol) of potassium carbonate, and 130 mL of DMAc was added. The mixture was refluxed for 16 hours with constant stirring and then stirred overnight at room temperature. To the reaction mixture, 300 mL of distilled water was added with stirring. The precipitate was filtered, washed with water and recrystallized from ethanol to yield 29.0 g (98%) of **9**, mp 158-161 °C [lit. (*10*) 158-161 °C]; IR (KBr) 1590, 1510, 1340, 1240, 1170, 975, 880, 760, and 650 cm^{-1}.

2,2-Bis[4-(4-aminophenoxy)phenyl]hexafluoropropane (10)

To a solution of 23.75 g (41.0 mmol) of **9** in a mixture of 60 mL of ethanol, 60 mL of tetrahydrofuran (THF) and 40 mL of ethyl acetate, 2.0 g of platinum catalyst on activated charcoal (0.5% Pt) was added. Hydrogenation was carried out under 60 psi for 6 hours at room temperature using a Parr 3911 hydrogenation apparatus. The reaction mixture was filtered and the solvent removed under reduced pressure at about 35 °C. The residue was redissolved in 200 mL of anhydrous THF, and anhydrous hydrogen chloride gas was passed through the solution. The reaction mixture was chilled and the precipitate filtered, washed with more anhydrous THF, and vacuum dried to give 23.6 g (97.2%) of **10**, mp 270 °C; IR (KBr) 2900, 1500, 1250, and 1177 cm^{-1}; ^{1}H-NMR (DMSO-d$_6$) δ 7.0-7.8 (m, 16H, Ar-H), 10.25 (b, 6H, ammonium). Anal. Calcd for $C_{27}H_{22}Cl_2F_6N_2O_2$: C, 54.84; H, 3.75; Cl, 11.99; F, 19.28; N, 4.74. Found: C, 54.06; H, 3.93; Cl, 12.23; F, 19.57; N, 4.67.

1,1'-[[2,2-(1,1,1,3,3,3-Hexafluoropropyl)di-1,4-phenylene]-4,4-diylbis (oxy-4,1-phenylene)]-bis[3,3-dimethyl-1-triazene] (2)

Compound **10**, (10.0 g, 17.0 mmol) was converted to **2** according to the general procedure for the preparation of **1**. Recrystallization of the crude product from a mixture of dichloromethane/acetone (1:5 v/v) afforded 10.57 g (68%) of **2**, mp 125-28 °C; IR (KBr) 1495, 1247, 1198, and 1086 cm^{-1}; ^{1}H-NMR (CDCl$_3$) d 3.20 (s,12H, CH$_3$), 6.82-7.53 (m, 16H, Ar-H). DSC of **2** exhibits an exothermic decomposition peak temperature (Td) at 283 °C. Anal. Calcd for $C_{31}H_{28}F_6N_6O_2$: C, 59.05; H, 4.48; N, 13.33. Found: C, 58.50; H, 4.53; N, 12.67.

1,3,5-Tris(nitrophenoxy)benzene (12)

To a solution of 16.22 g (100 mmol) of 1,3,5-tris-hydroxybenzene (Aldrich Chemical Co.) and 44.59 g (316 mmol) of 1-fluoro-4-nitrobenzene (Aldrich Chemical Co.) in a mixture of 100 mL *N,N*-methylacetamide (DMAc) and 100 mL toluene, 44.2 g (320 mmol) of potassium carbonate was added. The mixture was heated to gentle reflux with a Dean-Stark apparatus and mechanically stirred for 24 hours. At the end of the reaction, a mixture of 100 mL methanol and 100 mL of water was added with stirring. The reaction mixture was chilled and the crystallized product filtered, washed with 1000 mL of water, followed by 300 mL of methanol. The product was redissolved in ethyl acetate, decolorized with activated charcoal, and recrystallized from ethyl acetate/methanol (1:1 v/v) mixture. The resulting product was dried in air and then in vacuo at 100 °C (5 hours) to afford 32.80 g (67%) of **12**, mp 198-200 °C [lit. (*11*) 203.5-205.5 °C]; IR (KBr) 1615, 1590, 1515, 1495, 1460, 1350, 1240, 1130, 1010, and 860 cm^{-1}; ^{1}H-NMR (DMSO-d$_6$) d 6.95 (s, 3H, Ar-H), 7.25 (d, 6H, Ar-H), 8.28 (d, 6H, Ar-H).

1,3,5-Tris(4-aminophenoxy)benzene trihydrochloride (13)

To a mixture of 100 mL ethyl acetate and 100 mL tetrahydrofuran was added 10.54 g (21.54 mmol) of 1,3,5-tris(4-nitrophenoxy)benzene. The nitro compound was only partially soluble in the mixture. To this suspension, 2.0 g of platinum (0.5%) on activated charcoal was added. Hydrogenation was carried out under 60 psi hydrogen pressure at room temperature for 6 hours. At the end of the reaction, the catalyst was filtered and the solvent was removed from the resulting solution under reduced pressure at ~ 35 °C. The residue was redissolved in 135 mL of dichloromethane at elevated temperature. The solution was dried over anhydrous magnesium sulfate. Hydrogen chloride gas was passed through the dried solution to precipitate the product. The precipitate was filtered, suction air dried and vacuum dried at 80 °C for 16 hours to give 8.52 g (77.8%) of **13**, mp 278 °C (dec.); IR (KBr), 3000(b), 1610, 1510, 1460, 1255, 1215, 1125, 1020, and 840 cm^{-1}; ^{1}H-NMR (DMSO-d$_6$) δ 6.33 (s, 3H, Ar-H), 7.30 (m, 12H, Ar-H), 8.20 (b, 9H, ammonium).

Tris-triazene Crosslinking Reagent (3)

To a solution of 5.24 g (10.30 mmol) of the hydrogen chloride salt of 1,3,5-tris(4-aminophenoxy)benzene in a mixture of 100 mL of THF and 100 mL of water was added a solution of 4 mL (48.7 mmol) of 12 N hydrochloric acid in 10 mL of water. The resulting solution was chilled to -10 °C and a solution of 3.36 g (48.7 mmol) of sodium nitrite in 20 mL of water was added dropwise with constant stirring. The temperature of the reaction mixture was kept at approximately -5 °C during the addition. After the reaction mixture was stirred at approximately -2 °C for 3 hours, the organic solvent was removed under reduced pressure at ambient temperature. The aqueous solution was chilled to 0 °C and its pH adjusted to 5 by adding saturated sodium carbonate solution. This chilled solution was poured into a solution of 1.68 g (20.6 mmol) of dimethylamine hydrochloride and 5.0 g (41.17 mmol) of sodium carbonate in 150 mL of ice water. The mixture was stirred at about 0 °C for 30 minutes and then extracted with three 80 mL portion of dichloromethane. The combined extract was dried over anhydrous magnesium sulfate and decolorized with activated charcoal. The organic solvent was removed under reduced pressure. Alumina column chromatography of the crude oil with toluene as eluent gave 1.2 g (20.5%) of **3**, mp 132-4 °C; IR (KBr) 1590, 1500, 1230, and 1082 cm^{-1}; ^{1}H-NMR (CDCl$_3$) δ 3.25 (s, 18H, CH$_3$), 6.37 (s, 3H, Ar-H), 6.83-7.60 (m, 12H, Ar-H). DSC of **3** exhibits an exothermic decomposition peak temperature (Td) at 285 °C. Anal. Calcd for C$_{30}$H$_{33}$N$_9$O$_3$: C, 63.48; H, 5.86; N, 22.21. Found: C, 63.38; H, 5.95; N, 22.09.

REFERENCES

1. For references, see, for example: (a) Shah, P; D. Laks, D.; Wilson, A. *Proc. IEDM, IEEE* **1979**, 465. (b) Wilson, A. M. *Thin Solid Films* **1981**, *83*, 145.
2. Jones, R. J.; Chang, G. E.; Powell, S. H.; Green, H. E. In *"High Temperature Polymer Matrix Composites"*, NASA Conference Publication 2385, 1983; pp 271. St. Clair, A; St. Clair, T.; Minfree, W. *Polym. Matl., Sci. & Eng.* **1988**, *59*, 28.
3. Mercer, F. W., U.S. Patent 4,920,005, 1990.
4. Hardie, R. L.; Thomson, R. H. *J. Chem. Soc.* **1958**, 1286. Vaughan, K.; Liu, M. T. H. *Can. J. Chem.* **1981**, *59*, 923.
5. Elks, J.; Hey, D. H. *J. Chem. Soc.* **1943**, 441. Buxton, P. C.; Heaney, H. *J. Chem. Soc., Chem. Comm.* **1973**, 545.
6. Lau, A. N. K.; Moore, S. S.; Vo, L. P. *J. Polym. Sci. Part A: Polym. Chem.* **1993**, *31*, 1093.
7. For references, see, for example: (a) Wachman, E. D.; Frank, C. W. *Polymer* **1988**, *29*, 1191. (b) Kochi, M.; Isoda, S.; Yokota, R.; Mita, I.; Kambe, H. In *polyimides—Synthesis, Characterization, and Applications*, Mittal, K. L., Eds.; Plenum Press: New York, 1984; Vol. 2, pp 671.
8. Mercer, F. W.; Goodman, T. D. *Proc.1990 International Electronics Packaging Conference* **1990**, 1042.
9. Jones, J. J.; Chang, G. E. C., U.S. Patent 4,477,648, 1984.
10. Jones, R. J.; O'Rell, M. K., U.S. Patent 4,111,906, 1978.
11. Takeichi, T.; Stille, J. K. *Macromol.* **1986**, *19*, 2093.

RECEIVED December 30, 1992

Chapter 30

Preparation of Novel Photosensitive Polyimide Systems via Long-Lived Active Intermediates

Takahi Yamashita[1] and Kazuyuki Horie[2]

[1]Research Center for Advanced Science and Technology, University of Tokyo, Komaba 4-6-1, Meguro-ku, Tokyo 153, Japan
[2]Department of Reaction Chemistry, Faculty of Engineering, University of Tokyo, Hongo 7-3-1, Bunkyo-ku, Tokyo 113, Japan

Polyimides are widely used in microelectronics field as passivation coatings, alpha-particle barriers, and interlayer dielectrics due to their planarization and insulating ability, as well as their good thermal and mechanical properties. Recently increasing researches have been devoted for developing photosensitive polyimides because they can eliminate a number of processing steps required for patterning usual polyimide films by using usual photoresist materials. In addition, they are expected to play a great role in manufacturing photonics devices as optical memory matrices, wave-guide materials, and non-linear optical materials in near future, when higher thermal stability and better processability would be required for designing and controlling the hyperfine structures in such materials. Several types of photosensitive polyimides have been proposed, but each system has some problems still unresolved. For example, fine patterns in polyimide precursor systems are subject to distortion due to the shrinking during imidization after exposure and development because they contain fairly large amount of additives. Solvent-soluble photoimaginable polyimides are advantageous because of their storage-stable and non-shrinking ability, but their sensitivities are rather low.

Recently we have investigated the mechanism of photoreactions of benzophenone-containing polyimide, PI(BTDA/DEDPM), which is one of the solvent soluble and intrinsically photosensitive polyimides, prepared from benzophenonetetracarboxylic dianhydride (BTDA) and 4,4'-diamino-3,3'-diethyldiphenylmethane (DEDPM). In the practical case, sensitivity of photosensitive polymers is affected by various factors, such as efficiency of light absorption, intrinsic reactivity, type of elemental reactions, characteristics of pattern formation, and so on. Quantum yield for their photocrosslinking reaction is one of the indexes which refers to the intrinsic photoreactivity. We have found that the formation of charge transfer (CT) structure of imide moieties decreases their photoreactivities, because it accelerates the deactivation rates (1). Another reason for the decrease in photoreactivity of PI(BTDA/DEDPM) is low proba-

0097-6156/94/0537-0440$06.00/0

bility of their intermolecular collision in solid-state polymer because lifetime of their triplet excited state is less than a microsecond, meaning that most of excited molecules are deactivated before they react to make intermolecular crosslinks.

Therefore, the efficiency of photocrosslinking reaction is expected to be improved by using reactions *via* long-lived active intermediates (LLAIs), which are produced by photoirradiation and react as key intermediates whose lifetime is longer enough not to be deactivated. One strategy is utilization of carbenes as LLAIs. We prepared two types of novel photosensitive polyimide systems; one is polyimide precursors containing bisacylsilanes as crosslinking reagents, and the other is diazo-containing solvent-soluble polyimide.

EXPERIMENTAL SECTION

Materials

Chemical structures of the compounds used in this experiments are shown in Figure 1. PAA(PMDA/ODA) was prepared by the condensation of pyromellitic dianhydride (PMDA) and DEDPM in dimethylformamide (DMF) solution. PI(BTDA/DEDPM) was prepared by the condensation of BTDA and DEDPM in DMF solution, followed by the chemical imidization using acetic anhydride and pyridine as dehydrating reagents. Preparation of PI(DZDA/DEDPM) and the model compounds containing diazo groups is described elsewhere (*2*). Bisacylsilane crosslinking reagents were prepared by the silylation of the corresponding thioacetals followed by the oxidative hydrolysis.

Photoirradiation of Polymer Films

Sample films of PAA(PMDA/ODA) containing bisacylsilanes were photoirradiated by using a 250 W ultrahigh-pressure mercury lamp as a light source through a combination of an interference filter (KL-40) and a cut filter (UV-39). Sample films of PI(DZDA/DEDPM) were photoirradiated by using a Xe lamp as a light source through a combination of an interference filter (KL-40) and a cut filter (UV-39). Actinometry was performed by a photometer (USHIO UIT-101).

Photoreaction of the Model Compounds

For the experiments in vacuum, sample solutions were degassed by a freeze-pump-thaw cycle whose procedure was described in detail in the previous paper (*1*). Photoirradiation of model solutions was performed by using a Xe lamp, and the changes in the absorbance were monitored by using a double beam measurement system prepared in our laboratory (*2*). An interference filter (KL-43) and a cut filter (UV-39) were used for the photoirradiation of M(DZDA/*o*-EA) and M(DZDA/*n*-BA) solutions, which were prepared from *o*-ethylaniline (*o*-EA) or *n*-buthylamine (*n*-BA), and only a cut filter (UV-29) was used for the photoirradiation of diazodiphenylmethane (DDM).

Measurements of the molecular weight of PI(DZDA / DEDPM)

GPC was measured by using a Jasco BIP-VI type HPLC with a GPC column (Shodex K-80M) using dichloromethane as eluent, and a GPC column (Hitachi kasei, GLS360 DT-5) using 1:1 mixture of THF and DMF as eluent. Calibration was carried out using monodisperse polystyrene standards. Sample films of PI(DZDA/DEDPM) were treated with acetic acid before GPC measurements in order to quench the unreacted diazo groups.

RESULTS AND DISCUSSION

Photoreactive Polyimide Precursor System Containing Bisacylsilane as Photocrosslinking Reagent

Two types of sample films of PAA(PMDA/ODA) containing bisacylsilanes, BBS and PBSK, as crosslinking reagents were spincoated on quartz plates (3). A small absorption bands were observed at 400 nm in their absorption spectra, which are due to silicon-carbonyl conjugation (4). Figure 2 shows the increase in the crosslinking densities of PAA(PMDA/ODA) during the 400-nm photoirradiation, which were calculated by the changes in their molecular weights using David's crosslinking theory (1, 5). Gelation was observed after 10-min photoirradiation, then the molecular weights did not increase due to precipitation of the higher molecular weight portions. Quantum yields for the photocrosslinking reaction of PAA(PMDA/ODA) containing BBS and PBSK were determined to be 4×10^{-4} and 3×10^{-4}, respectively.

Figure 3 shows the mechanism for the photocrosslinking reaction of PAA-bisacylsilane system. The *Si* groups in this scheme represent alkyl or aryl substituted silyl groups. Acylsilanes rearrange to give siloxy carbenes, on of LLAIs, which are inserted into poled bonds such as carboxyl groups of the polymer chains to make crosslinks. Quantum yields for the photocrosslinking reaction correspond to the number of crosslinking reagents both of whose *Si* groups were reacted. Quantum yield for the photorearrangement of a *Si* group of BBS is calculated as 1.3×10^{-3}. This relatively high reaction efficiency which is possibly due to the effect of hydrogen bonding between BBS and the carboxylic acid groups of the polyamic acid.

Photoreactivity of Diazo-Containing Polyimide

Figure 4 shows the UV spectra of PI(DZDA/DEDPM) film spincoated on a quartz plate. A shoulder at 400 nm was observed in the absorption, which can be assigned to the absorption of diazo groups. Decrease in the absorption band was observed during 400-nm photo−irradiation, showing that photo-induced decomposition of diazo groups occurred to generate carbenes. The carbenes are thought to be inserted into C-H bond of the pendant ethyl groups in PI(DZDA/DEDPM) to produce crosslinks (Figure 5). Addition reaction of carbenes to other diazo groups has been reported to give azines (6), but the ratio

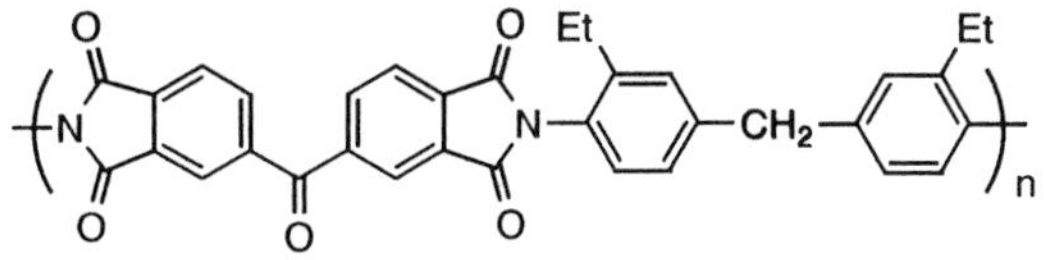

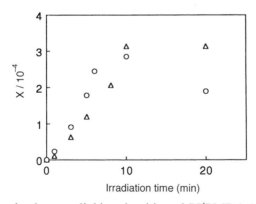

PI(BTDA/DEDPM)

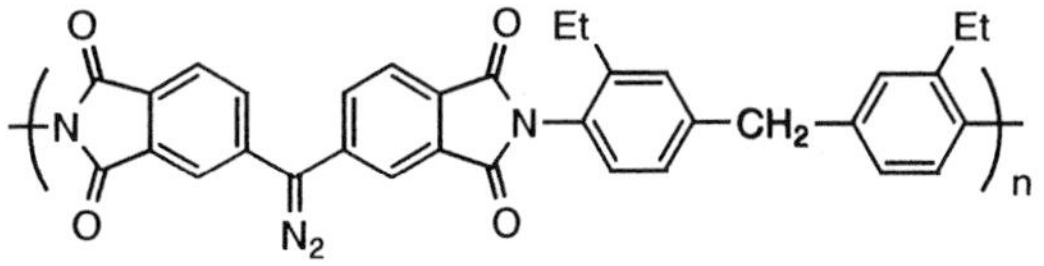

PI(DZDA/DEDPM)

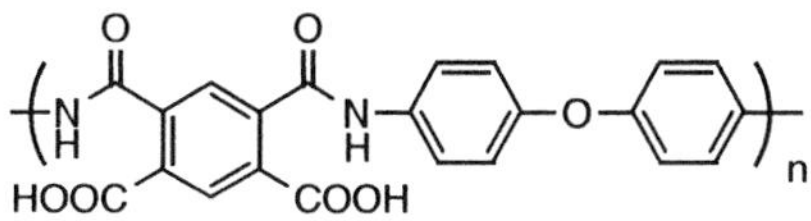

PAA(PMDA/ODA)

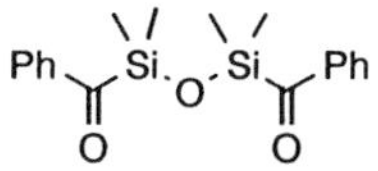

BBS 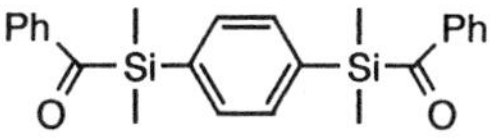   **PBSK**

Figure 1. Chemical structures of photosensitive polyimide systems.

Figure 2. Change in the crosslinking densities of PI(PMDA/ODA) containing BBS and PBSK as crosslinking reagents during 400 nm photoirradiation: BBS (○), PBSK (△).

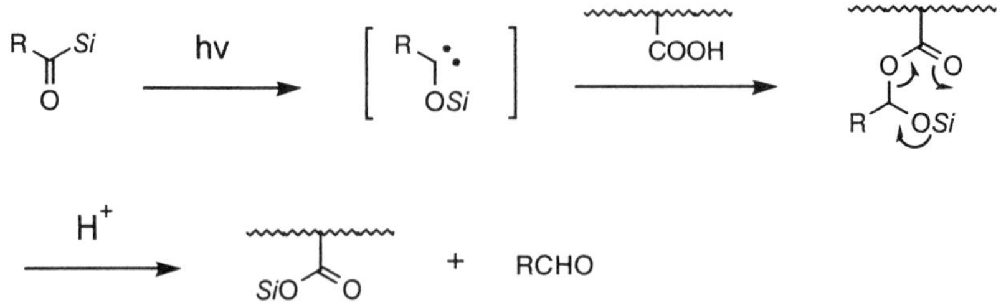

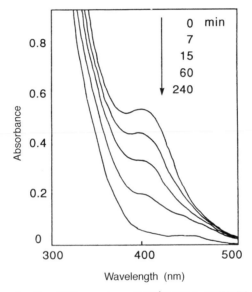

Figure 3. Mechanism for the photocrosslinking of PAA(PMDA/ODA)-Bisacylsilane system *via* Brook rearrangement.

Figure 4. Change in the UV spectra of PI(DZDA/DEDPM) during 400 nm photoirradiation.

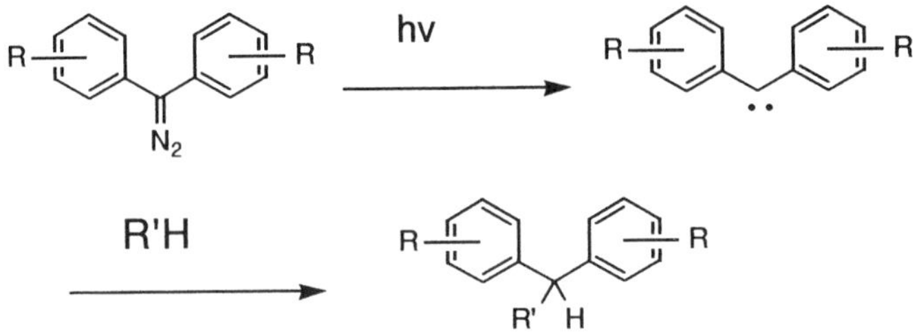

Figure 5. Photoreaction of a diazo group in PI(DZDA/DEDPM) to generate a carbene, a long-lived active intermediate (LLAI), which insert into C-H bond.

of the reaction is negligible in the present system because the concentration of ethyl groups is considerably larger than that of diazo groups. From the first-order plots of the decrease in the absorbance of diazo groups, quantum yield for the disappearance of the diazo group in PI(DZDA/DEDPM) was determined as 0.06 at room temperature in air. Quantum yield for photocrosslinking reaction of PI(DZDA/DEDPM) was obtained as 0.13 from the change in their molecular weights which were measured by the GPC during 400-nm photoirradiation. These values are about an order larger than that for PI(BTDA/DEDPM).

Photoreaction of Diazo-containing Model Compounds

Table I summarizes the quantum yields for the photoreaction of diazo-containing model compounds and those for the benzophenone-containing models,

Table I. Quantum yields for photoreaction of model compounds

Chemical Structure of the Models	$Ar\text{-}Ar$ with N_2		$Ar\text{-}Ar$ with O	
Ar \ Atmosphere:	Air	Vac	Air	Vac
(phenyl, methyl-substituted benzene)	1.1		0.3[a]	0.88[a]
N-Bu—N (phthalimide)	0.2		0.4[b]	
Et-substituted N-phenyl phthalimide (for Diazo)	0.15[c]	0.14		
Me-, MeO-substituted N-phenyl phthalimide (for Benzophenone)			0.02	0.066
Polymer Decomposition	0.06			
Polymer Crosslinking	0.13		0.001	0.002

a) measured in ethylbenzene solution. Other model compounds are measured in THF solution. b) cited from ref. 12 c) same value is obtained in THF, CH_2Cl_2, and EB solution.

whose chemical structures are shown in Figure 6. Quantum yield for the photoreaction of diazodiphenylmethane (DDM) is unity, while those for M(DZDA/n-BA) and M(DZDA/o-EA) are 0.2 and 0.15 respectively. The values are not affected by atmosphere, showing that the reactions occur from their singlet excited states. The decrease in the quantum yields for the imide-substituted models compared to that of DDM is possibly due to the relative decrease in the decomposition rates from their singlet excited states, because electron withdrawing groups such as imide groups are known to stabilize diazo groups. In the case of benzophenone derivative models, the decrease in the quantum yield for M(BTDA/DMA) compared to those for BP and M(BTDA/n-BA) is attributed to the deactivation via the intramolecular charge transfer structure which is formed by the electron donation from the aromatic amine groups to the aromatic acid anhydride groups of imide moiety. On the other hand, no remarkable difference in the quantum yields was observed between M(DZDA/n-BA) and M(DZDA/o-EA), which is a contrast to the case of photodecomposition of benzophenone derivatives, M(BTDA/n-BA) and M(BTDA/DMA).

Characterization of Charge Transfer Structures of the Model Compounds by using Fluorescence Technique

We have been developing intensive studies of the charge transfer structure of various aromatic polyimides by using fluorescence techniques (7, 8). In order to investigate the effect of charge transfer structure of the model compounds on their photoreactivities, their fluorescence spectra were measured. Figure 7 shows absorption, excitation and fluorescence spectra for M(BTDA/o-EA) in CH_2Cl_2 solution. A broad structureless emission band at 510 nm was observed when it was excited at 300 nm. This emission can be assigned to a charge transfer fluorescence because the absorption spectrum agree with the excitation spectrum and because red shift of the emission is rather large (7). On the other hand, no fluorescence at 510 nm was observed for M(BTDA/n-BA), which was not expected to form charge transfer structure (Figure 8). A very weak emission with vibration structure at 400, 450, and 480 nm can be assigned to phosphorescence from the benzophenone moiety because increase in the emission intensity was observed when the sample was degassed and cooled in order to prevent oxygen quenching, and because the carbonyl group of benzophenone is known to accelerate the rate of inter-system crossing. Indeed, the spectrum is fairly similar with that of benzophenone (9).

Figure 9 shows the absorbance, excitation, and emission spectra for M(DZDA/n-BA). No charge transfer fluorescence was observed just like the case of M(BTDA/n-BA), and a fluorescence was observed at 300 nm when it was excited at 250 nm. This observation of 300-nm fluorescence can be attributed to the lack of benzophenone-carbonyl groups. Excitation spectrum for the fluorescence indicates that the fluorescence is emitted only from the imide moiety. No fluorescence was observed if the absorption of diazo group at 400 nm was excited, showing that the electronic configuration of the diazo group in M(DZDA/n-BA) does not conjugate with the imide group and that photode-

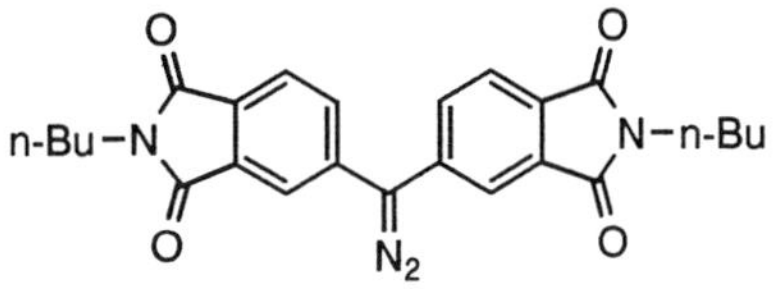

M(DZDA/*n*-BA)

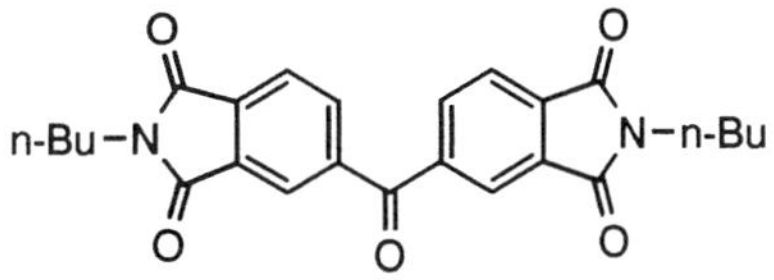

M(BTDA/*n*-BA)

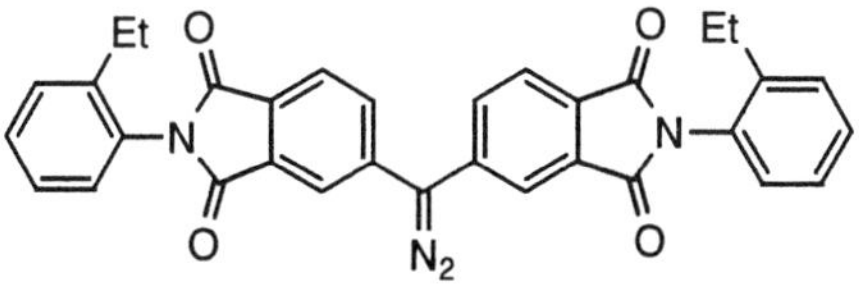

M(DZDA/*o*-EA)

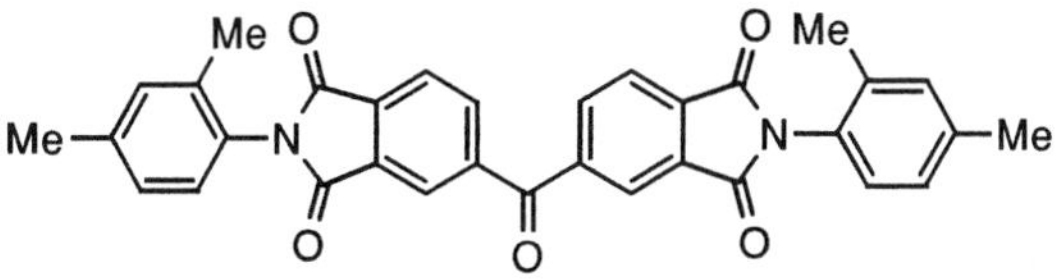

M(BTDA/DMA)

Figure 6. Chemical structures of the model compounds.

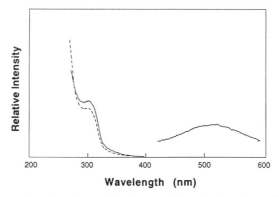

Figure 7. Absorption (-----), excitation, and emission (———) spectra for M(BTDA/o-EA) in CH$_2$Cl$_2$ solution. The emission spectrum is measured by the excitation at 300 nm, and the excitation spectrum is measured for the emission at 510 nm.

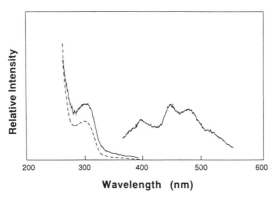

Figure 8. Absorption (-----), excitation, and emission (———) spectra for M(BTDA/n-BA) in CH$_2$Cl$_2$ solution. The emission spectrum is measured by the excitation at 300 nm, and the excitation spectrum is measured for the emission at 450 nm.

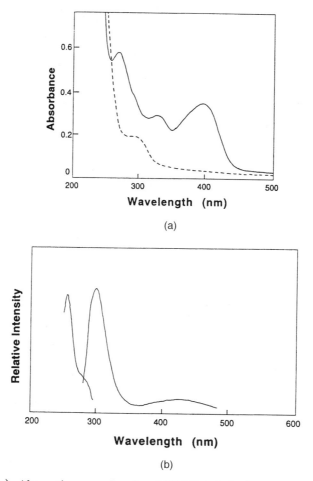

(a)

(b)

Figure 9. (a) Absorption spectra for M(DZDA/*n*-BA) in CH$_2$Cl$_2$ solution before photoirradiation (———) and after photoirradiation (-----). (b) Excitation spectrum monitored by the emission at 320 nm, and the fluorescence spectrum excited at 250 nm.

composition of diazo group occurs preferentially. Figure 10 shows absorption, excitation and fluorescence spectra for M(DZDA/o-EA) in CH_2Cl_2 solution. The absorption at 400 nm is due to the diazo group just like the case of M(DZDA/n-BA). A broad fluorescence band at 510 nm was observed when the sample was excited at 300 nm, which can be assigned to the charge transfer fluorescence just like the case of M(BTDA/o-EA). This result shows that the charge transfer structure is rapidly formed after photoexcitation of imide moiety of M(DZDA/o-EA). However, no fluorescence was observed if the sample was excited at 400 nm, showing that the electronic configuration of the diazo group of M(DZDA/o-EA) is also isolated from the aromatic imide moiety which forms charge transfer structure and accelerate the deactivation. Thus, photoexcitation of diazo group effectively produce the reactive singlet excited state, while excitation of imide moiety produce the less reactive charge-transferred singlet state.

After the photoirradiation of these model compounds new fluorescences appeared at longer wavelengths. They are thought to be the emission from photoproducts (2), because their intensities increased as the photoirradiation, and the excitation spectra were different from those before photoirradiation.

Comparison of Quantum Yields for the Model and Polymer System

Quantum yield for the photocrosslinking of PI(BTDA/DEDPM) is an order smaller than that for the model compound, M(BTDA/DMA), shown in Table I. That is partially due to the intermolecular charge transfer formation in the polymer system, but this contribution is small because the number of intermolecular charge transfer sites is estimated as not so large. Therefore the decrease in the photoreactivity of PI(BTDA/DEDPM) compared to that of M(BTDA/DMA) is explained by the restriction of molecular mobility in polymer solid state. Namely, fairly large part of the triplet excited state of PI(BTDA/DEDPM) is thought to be deactivated before they react with neighboring ethyl groups, because the lifetime of triplet state PI(BTDA/DEDPM) is smaller than a microsecond due to the fast deactivation *via* the charge transfer state, while the liberation motion of phenylene rings in polymer solid state is relatively slow (*10*). On the other hand, triplet excited state carbonyl groups in the model systems can easily abstract hydrogens of solvents because they are surrounded by a plenty of solvent molecules and because their molecular motion is free. In the case of PI(DZDA/DEDPM), quantum yield for the photocrosslinking reaction is comparable to those for the model compounds because carbenes, which are one of the LLAIs (long-lived active intermediates), are not deactivated once it is produced.

CONCLUSION

Novel photosensitive polyimide systems containing bisacylsilanes as crosslinking reagents and a polyimide containing diazo group in the main chain were synthesized and their photoreactivities were investigated. Quantum yields for the photocrosslinking of PAA(PMDA/ODA)-PBSK and PI(DZDA/DEDPM) were

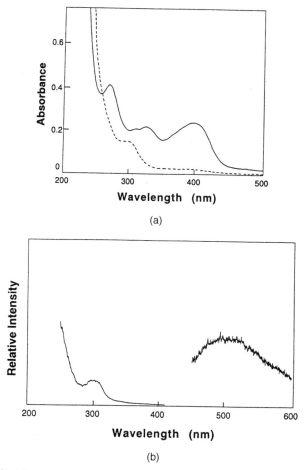

Figure 10. (a) Absorption spectra for M(DZDA/*o*-EA) in CH_2Cl_2 solution before photoirradiation (———) and after photoirradiation (-----). (b) Excitation spectrum monitored by the emission at 510 nm, and the fluorescence spectrum excited at 300 nm. Fluorescence does not observed when it is excited at 400 nm.

determined to be 4×10^{-4} and 0.13, respectively. Bisacylsilane systems are convenient because they utilize the carboxylic groups of PAA as crosslinking points and so various polyamic acids can be applicable. Quantum yield of PI(DZDA/DEDPM) was found to be much larger than that of benzophenone-containing polyimide, PI(BTDA/DEDPM). These crosslinking reagents and diazo groups in PI(DZDA/DEDPM) can be removed by the thermal curing after development without serious weight loss, which is another advantage for these systems, as well as the improvement of the photosensitivity. In addition, their photobleaching ability is of great advantage because even thick patterns can be formed by these systems.

Design of the reactions *via* long-lived active intermediate was found to be important for developing photosensitive polyimide systems, and this concept is especially effective for the reactions in solid state (*11*), because the solid-state reactions are controlled by the molecular motions (*12*). In addition, change in the electronic state in polymer solid was found to affect the efficiency of their photoreactions. Charge-transfer structure is one of the characteristic nature of aromatic polyimides, which is affected by the change in their physical properties, and which in turn controls their photoreactivities.

REFERENCES

1. H. Higuchi, T. Yamashita, K. Horie, and I. Mita, *Chem. Materials*, **1991**, *3*, 188
2. T. Yamashita, C. Kasai, and K. Horie, *Polym. Prep. Jpn.*, **1992**, *41* (3) 830; T. Yamashita, C. Kasai, and K. Horie, submitted to *Chem. Materials*
3. T. Yamashita, K. Horie, I. Mita, *Polym. Prep. Jpn.*, **1990**, *39*, 802
4. J. M. Duff and A. G. Brook, Can. J. Chem., **1973**, *51*, 2869
5. C. David, D. Baeyen -Volant, *Eur. Polym. J.*, **1978**, *14*, 29.
6. R. S. Atkinson, *Comprehensive Organic Chemistry* Vol. 2, Sutherland, I. O.; Ed.: p 219, Pergamon Press, New York, 1979
7. K. Horie, Q. Jin, T. Yamashita, M. Hasegawa, *Polym. Prepr.* **1992**, *32*, 880
8. Hasegawa, M.; Kochi, M.; Mita, I.; Yokota, R. *Eur. Polym. J.*, **1989**, *25*, 349; Hasegawa, M.; Kochi, M.; Mita, I. *Polym. Prepr. Jpn.*,**1987**, *36*, 1158, 3554; *Ibid.*, **1987**
9. K. Horie, K. Morishita, I. Mita, *Macromolecules*, **1984**, *17*, 1746
10. Cholli, A. L.; Dumais, J. J.; Engel, A. K.; Jelinski, L. W. *Macromolecules*, **1984**, *17*, 2399
11. K. Horie, I. Mita, *Adv. Polym. Sci.*, **1989**, *88*, 78
12. Q. Jin, T. Yamashita, K. Horie, *J. Photopolym. Sci. Tech.*, **1992**, *5*, 335.

RECEIVED December 30, 1992

Chapter 31

Photoregulation of Liquid-Crystalline Orientation by Anisotropic Photochromism of Surface Azobenzenes

Yuji Kawanishi[1], Takashi Tamaki[1], and Kunihiro Ichimura[2]

[1] Agency of Industrial Science and Technology, National Institute of Materials and Chemical Research, 1–1 Higashi, Tsukuba, Ibaraki 305, Japan
[2] Research Laboratory of Resources Utilization, Tokyo Institute of Technology, 4259 Nagatsuta, Midori-ku, Yokohama 227, Japan

Liquid crystals (LC) are fluid with highly ordered molecular orientation. Because of their responsiveness in orientation as well as in optical properties to an applied electric field, LCs have been materialized in production of thin displays driven by small batteries.

The LC orientation is also influenced by bringing other molecules into the system, i.e., dopants and substrate surfaces. This makes special orientation in marketed LC displays possible such as twisted nematic, super twisted nematic, surface stabilized ferroelectric, and dye doped guest-host systems, etc. Any mechanisms modifying the physicochemical nature of molecules on the surface will be available to control the LC orientation.

Introduction of photochemistry is particularly interesting since it enables us to acquire high density and fast accessible optical memories as well as new sights on molecular interactions in the LC phase. Here, photochemical approaches to regulate the LC orientation are briefly reviewed. Afterwards, our new findings on precise 3D control of the LC orientation by anisotropic surface photochromism will be introduced.

INDUCED PHASE TRANSITION BY PHOTOCHROMIC REACTIONS OF DOPED MOLECULES

Photochemical transformation of molecules with a reversible nature is termed a *photochromic reaction* (*1, 2*) or *photochromism* as represented by photoisomerization of azobenzene (Az) derivatives (1).

Direct application to optical memories based on a spectral change in photochromism seems relatively difficult, since reading light always induces the reverse photoreaction with an efficiency as long as the light is absorbed by the photoisomer.

0097–6156/94/0537–0453$06.00/0

$$\text{trans-Az} \quad \underset{\text{visible}}{\overset{\text{UV}}{\rightleftarrows}} \quad \text{cis-Az} \quad\quad (1)$$

Combining photochromism with LC makes photochemical switching of molecular orientation possible, and the event can be monitored by the light not causing either any photochemistry of photochromic systems or any damage of records. Since Sackmann (3) demonstrated elongation of helical pitch in a cholesteric mixture induced by photochromic *trans* → *cis* isomerization of Az dopant, analogous systems consisting of LC and photochromic dopants have been proposed with a view to applications in optical memories (4–8). Most of those are based on the *photoinduced phase transition* (PPT) from a mesophase to the isotropic phase. PPT is believed to be caused by a tremendous structural change in molecular shape of the photochromic dopant (Figure 1).

Although one could write images in a LC bulk by this fashion, the records easily disappear due to diffusion of dopant and flow of the LC bulk. PPT of polymer LCs (9, 10) affords long-term storage more than one month if kept below the glass transition temperature, while heating up may be required to write as fast as low-mass LCs. Long-term storage without sacrificing response properties has been demonstrated by applying the polymer dispersed LC film (PDLC) (11). PDLC is a polyvinylalcohol film in which microcapsules of a mixture of cyanobiphenyl LC analogues and 4-butyl-4'-methoxy-Az are embedded. Optical resolving power of the PDLC corresponds to the diameter of the LC capsule, that is about 1–2 μm.

INDUCED ALIGNMENT CHANGE BY PHOTOCHROMIC REACTIONS OF SUBSTRATE SURFACE

More sophisticated control of the LC orientation has been provided by photochromic transformation of surface attached molecules (12–15). When a nematic LC is placed on the substrate surface modified with a thin layer of Az molecules, the alignment (orientation-direction) of the LC can be changed reversibly between two states, homeotropic state (H, the optical axis of the LC phase is normal to the surface) and parallel state (P, the axis lies down on the surface), responding to the photochromism of the Az skeleton as shown in Figure 2. The phenomenon is not involved in PPT, but should be classified as a new kind of orientation change, *photoinduced alignment change* (PAC).

Photochemical efficiency of PAC, in which a couple of surface Az molecules is supposed to command approximately 10^4 LC molecules, is much higher than that of PPT systems requiring high concentration (> 10wt%, typically) of a

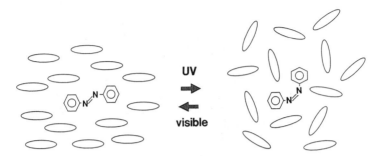

Figure 1. Photoinduced phase transition of a nematic LC by the photochromic isomerization of the doped Az.

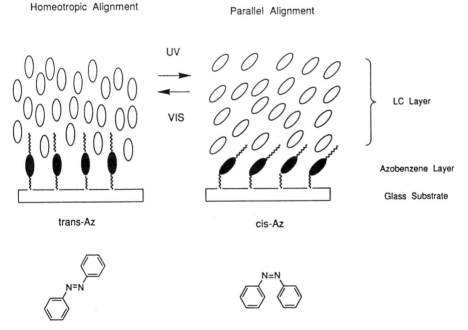

Figure 2. Photoregulated alignment change of a nematic LC by the photochromic isomerization of the surface Az.

photochromic dopant. Such Az surfaces have been prepared by silylation, Lang-muir-Blodgett method, and polymer spincoating. PAC may be monitored as a birefringence change of the LC phase or as an absorption change of a doped dichroic dye. Resolution of the system reaches 2 μm for the LC layer thickness of 8 μm. According to the laser pulse experiments (16), a typical relaxation time is about 0.1s which competes to that of electrically controlled LC devices. Since the information is written as photochemical transformation of surface molecules, no image-fading is expected even the LC phase is flowing.

Molecular mechanics of PAC has been investigated since first findings. According to *Friedel-Creagh-Kmetz* rule (17), the LC alignment to the substrate surface is predictable by taking the thermodynamic balance of the system into account. The alignment should be a function of the critical surface tension of LC (γ_{LC}) and the surface energy of the substrate (γ_S), i.e., H when $\gamma_{LC} > \gamma_S$ or P when $\gamma_{LC} < \gamma_S$. In fact, *cis*-Az surface tends to give higher γ_S than *trans*-Az surface so that the H → P alignment change seems to be explicable. However, quantitative calculation by the FCK rule did not explain all the experimental results. Deviation may be caused by the use of macroscopic values of γ. Spectroscopic investigation suggests that more specific molecular interaction must be concerned to interpret the phenomena.

Spectral changes of surface *trans*-Az moieties can be useful indices of molecular interaction at the surface. Their absorption peaks locating around ca. ~ 350nm and ca. ~ 240nm are $\pi - \pi^*$ transitions, whose transition dipoles are parallel and perpendicular to the Az long axis, respectively (18). The absorption spectrum of Az on the quartz plate is quite similar to that in solution suggesting no specific alignment of Az on the substrate in average.

It has been noticed that the H alignment of LC is never obtained right after cell preparation but requires some period to be formed from the initial P alignment. There may be a reorientation process after the LC attaches to the Az surface. As expected, the spectrum changed drastically after the Az surface was coated with a thin nematic LC layer (Figure 3). The absorption of the transition perpendicular to the molecular axis became more emphasized, which means the perpendicular alignment of the surface *trans*-Az was induced after interacting with a nematic phase. The effective interaction will probably be brought about by the rod-like shape of *trans*-Az skeleton. If so, *cis*-Az moiety with a bend structure and a higher γ_S has handicap to induce the H alignment in the LC phase.

Importance of the interaction between the Az moiety and the LC molecule has been pointed out in PAC by Az pendent methacrylic polymers (19-21). When the pendent moieties are close together and separated far from the main chain, a considerable amount of aggregates of *trans*-Az is formed in the polymer film even at the room temperature. The LC phase does not take the H alignment but the P on such a polymer surface, probably due to difficulty of insertion into paired *trans*-Az moieties. Photoirradiation to the polymer film causes isomeriza-tion of Az and affects the equilibrium on aggregation. Three interchangeable states, A, B and C, are obtained, separately.

$$aggregated\ trans\text{-}Az \underset{visible}{\overset{UV}{\rightleftharpoons}} monomeric\ trans\text{-}Az \underset{visible}{\overset{UV}{\rightleftharpoons}} cis\text{-}Az \qquad (2)$$
$$(A) \qquad\qquad\qquad\qquad (B) \qquad\qquad\qquad (C)$$

As responding to the reaction (2), the LC cell exhibits three discrete textures of the nematic LC phase (marbled parallel (Pl) on A, H for B, and then schlieren parallel (P2) for C, Figure 4). The result clearly indicates that the H alignment is highly sensitive to the amount of the monomeric *trans*-Az on the surface. To promote the H alignment, highly interactive monomeric *trans*-Az should be required.

ORIENTATION CONTROL OF LC MOLECULES BY SURFACE ANISOTROPIC PHOTOCHROMISM

The natural texture of the induced P alignment in PAC is marbled or schlieren, suggesting lack of macroscopic orientation axis in the plane. Very recently, we have found that a highly homogeneous P mono-domain can be induced by applying linearly polarized UV reaction light (*22*). We have also found that slantwise UV exposure is effective for formation of a large P monodomain even the light is not polarized (Figure 5) (*23*). Induction, rotation, and erasure of the uniaxially oriented P alignment are possible by changing light characteristics. The effects are explained in terms of the surface anisotropy brought about by the photoselective isomerization of Az moieties.

Experimental

Photoresponsive LC cells were fabricated by sandwiching a nematic mixture (DON103, K-290-N-344-I, Rodic) between a pair of Az modified glass plates (Az/LC/Az normal cell), or between a Az glass plate and a octadecylsilylated glass plate (Az/LC/ODS hybrid cell) with 8 μm spacers. For investigation of the orientation axis, guest-host (GH) cells containing 1.0 wt% of a dichroic dye (LCD118, Nippon-Kayaku) were used. Structures of chemicals and surface Az are shown in figure 6. A 500W high-pressure Hg arc lamp was used for stationary irradiation experiments (Figure 7a) with optical filters to obtain UV (365nm) and visible (440nm) light, and with a Glan-Thompson-Taylor prism polarizer for linearly polarized exposure. A pulse laser system shown in Figure 7b was used to analyse LC relaxation profiles.

Induction and Rotation of Homogeneous Orientation by Linearly Polarized UV Light

On preparation of the cell, the LC alignment was homeotropic. Ordinary UV exposure resulted parallel multi-domains with no unique macroscopic orientation axis, and subsequent visible exposure reproduced the homeotropic alignment. When the reaction UV light was linearly polarized, a large domain of homogeneous P alignment was obtained. The orientation was stable unless the visible light was applied. Results of angular dependent transmittance for the GH cell clearly indicate that the induced orientation axis is perpendicular to the electric polarization of the reaction UV light, and rotatable correlating with the light polarization (Figure 8).

Alignment relaxation was analyzed by means of laser pulse experiments. The time constants are 110 msec for the H → P (random) change by the unpolarized UV excitation, 90msec for the H → P (homogeneous) change by

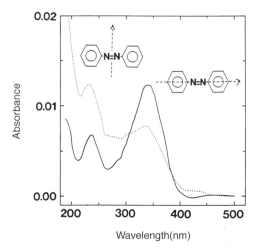

Figure 3. Absorption specta of the surface Az ($C_6H_{13}AzOC_{10}H_{20}$ $CONHC_3H_6Si(CH_3)_2O$-glass) before (——) and after (···) contact to a nematic LC. Transition vectors corresponding to two absorption peaks are perpendicular (~ 240 nm) and parallel (~ 350 nm) to the molecular long axis of Az as indicated.

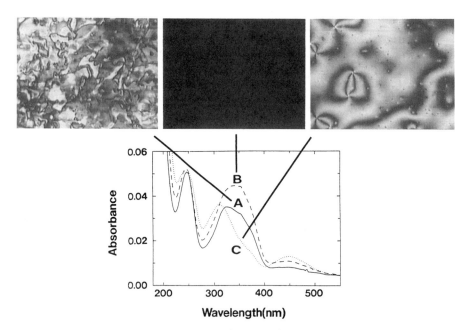

Figure 4. Textures of a nematic LC (DON103) and corresponding absorption spectra of the Az polymer $-[CCH_3(COOC_{11}H_{22}OAzC_6H_{13})-CH_2]_n-$ film.

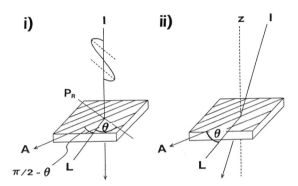

Figure 5. Photochemical induction of the uniaxially oriented parallel LC phase by the linearly polarized UV exposure (i) and the slantwise UV exposure (ii).
L: tentative axis of the LC cell,
θ: polarization angle between the incident electric polarization (P_R) and L in (i), or incident angle between the projection of the light path (I) and L in (ii).
A: induced orientation axis.

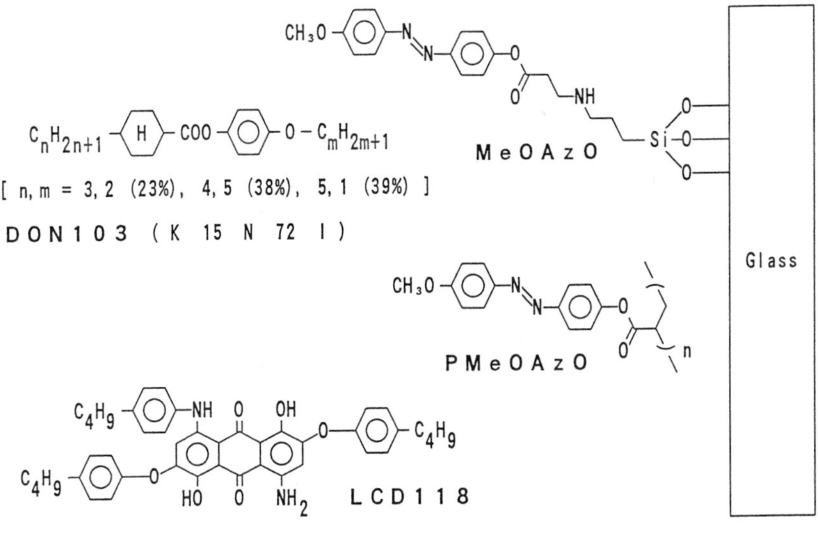

Figure 6. Chemical structures and abbreviations.

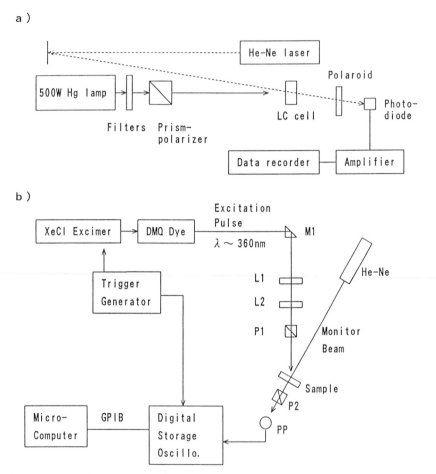

Figure 7. Experimental setup for the photoinduced alignment change in nematic LC cells.
a) Steady exposure experiments.
    Polarizer is placed in front of the Hg lamp for PAC by polarization photochromism.
b) Laser pulse experiments.
    L1,L2: lenses, Pl,P2: polarizers (Pl is used for PAC by polarization photochromism), PP: PIN photodiode.

linearly polarized UV excitation, and 240msec for rotation of the P orientation axis 45 ° to the original by changing UV light polarization, respectively. All the relaxation curves were well explained by the monoexponential function. The results show that the macroscopic orientation axis is immediately formed after the polarized UV exposure without experiencing the random P state, since no contribution from the time constant corresponding to the in-plane rearrangement was involved in the relaxation profile.

Homogeneous Orientation Induced by Slantwise UV Exposure

The homogeneous P domain can be also obtained by the alternative method, slantwise exposure to unpolarized UV light, in which the reaction light path is tilted about 10 ° to the normal to the cell surface. The results on the angular dependent transmittance of the GH cell show that the orientation axis is induced parallel to the projection of the light path on the cell surface (Figure 9). Again, the orientation axis is rotatable by changing the incident direction of the reaction light.

Dynamic conoscopic observation affords us to investigate how the orientation changes upon exposure (Figure 10). The crosspoint in the conoscopic image, which represents the optical axis of the LC phase, moves to the right when the reaction UV light comes from the left side of the microscope. By changing the amount of UV or visible irradiation, the cross point can be moved back and forth intentionally. This means that the slantwise exposure provide the uniaxial orientation axis with a tunable tilt angle.

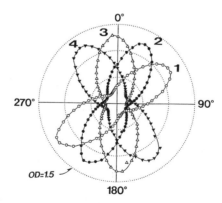

Figure 8. Absorption surfaces (angular dependence of optical density) of the MeOAzO/DON103-LCD118/MeOAzO GH cell at 633nm induced by the linearly polarized exposure at $\theta = -30$ ° (1), -60 ° (2), -90 ° (3) and -120 ° (4). Along the circle is plotted the cell rotation angle with respect to the polarization of the monitor He-Ne. Concentric circles represent 0.5, 1.0 and 1.5 of optical density, respectively.

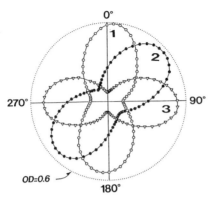

Figure 9. Absorption surfaces of PMeOAzO/DON103-LCD118/ODS GH cell at 633nm obtained by the slantwise exposure at $\theta = 0°$ (1), 45 ° (2) and 90 ° (3). The circle represents 0.6 of optical density.

Supposed Mechanism for In-plane Orientation By Anisotropic Photochromism

Present photoinduced homogeneous orientation is supposed to originate from two individual effects of the Az surface. One is an inherent ability of the *cis*-Az rich surface giving the P alignment. The other is the surface anisotropy acquired through the photoselective isomerization of Az, which may provide the inplane orientation axis. According to semiempirical LCAO-SCF-CI calculations (*18*), either the lowest $\pi - \pi^*$ transition vector of *trans*-Az or that of *cis*-Az is almost parallel to their —N≡N— bond axes. Along with the polarized UV exposure (at 365m) corresponding to the lowest $\pi - \pi^*$ transitions of both isomers, the surface becomes *cis*-Az rich, and the average alignment of Az molecules in terms of —N=N— axis should become perpendicular to the incident polarization plane due to the polarization-selective photochromism (*24*). It is not clear at present which isomer is the dominant in homogeneous orientation induction. Our preliminary results show LC molecules tend to align their long axes parallel to those of surface *trans*-Az units. On the other hand, *cis*-Az is rather known to destroy an ordered structure of LC phase as in the case of PPT.

Similar anisotropy should be produced in the Az surface by the slantwise exposure. Using Fresnel's formulas, we can expect that polarization due to reflection at interfaces of the system is negligible. To explain, we suppose that *trans*-Az lying its long axis parallel to the reaction light path hardly photoisomerize since the incident electric vector is perpendicular to the transition vector of the Az, and would give the specific orientation axis.

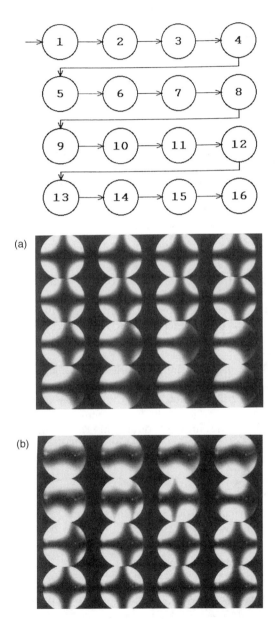

Figure 10. Conoscopic observation of induced alignment change of PMeOAzO/DON103/ODS cell by slantwise exposure to UV (a) and visible (b) lights. Exposure was made from the left of the microscope. Pictures were taken every 2s for (a) and 0.5s for (b) as the order shown above the pictures.

OTHER SYSTEMS BASED ON POLARIZATION PHOTOCHROMISM

Induction of organized structure in LC media by polarized light has recently attracted much attention. Induction of orientation axis perpendicular to the light electric field has been found in analogues of LC polymers bearing Az and mesogenic units as their side chain (*25–29*). Photoinduced reorientation depends strongly upon the driving temperature and often exhibits much slower rate below polymer's Tg. One proposed mechanism is that: thermal reverse isomerization of *cis*-Az may produce 90 ° rotated *trans*-Az with respect to the original one, which will be accumulated during photoirradiation due to photoselection.

Surface induced orientation has been demonstrated in the system consisting of two rubbed polyimide films, one of which is doped with a diazodiamine dye (*30*). Again the induced axis is found to be perpendicular to the light polarization. The system offers an enormous change in optical anisotropy, regardless of understandings how the photochemical process of the dye overcome the qualified aligning force of polyimide surface.

SUMMARY

Photochemical control of organized structures in materials is one of substantial scientific objects.

Precise control of orientation may be difficult in PPT systems, because the system is rather based on destruction of organized structure induced by a large topological change of photochromic dopant and self-organizing character of the LC phase.

When PAC systems are combined with anisotropic photochromism, precise control of the LC orientation becomes possible. Tilt angle including H-P alignment change is controlled by changing the fraction of surface photochromic molecules. In-plane orientation of the P alignment is regulated by controlling the light polarization or the incident direction. Thus, anisotropic surface photochromism will afford precise 3D control of the LC orientation by changing the light characteristics.

ACKNOWLEDGMENT

This work has been carried out under the project "Photoreactive Materials" conducted by *Agency of Industrial Science and Technology, Ministry of International Trade and Industry*.

REFERENCES

1. *Techniques of Chemistry*, Vol. 13, *Photochromism*; Brown, G. H., Ed.: Wiley Interscience: New York, 1971.
2. *Photochromism: Molecules and Systems*: Dürr, H.:Bouas-Laurent,H.,Eds.: Elsevier: Amsterdam, 1990.
3. Sackmann, E. *J. Am. Chem. Soc*. 1971, 93, 7088
4. Haas, W. E.; Nelson, K. F.; Adams, J. E.; Dir, G. A., *J. Electrochem. Soc*. 1974, 121, 1667

5. Ogura, K.; Hirabayashi, H.; Uejima, A.; Nakamura, K.; *Jpn. J. Appl. Phys.* 1982, 21, 969
6. Tazuke, S.; Kurihara, S.; Ikeda, T.; *Chem. Lett.* 1987, 911
7. Ikeda, T.; Horiuchi, S.; Karanjit, D. B.; Kurihara, S.; Tazuke, S.; *Chem. Lett.* 1988, 1679
8. Suzuki, Y.; Ozawa, K.; Hosoki, A.; Ichimura, K., *Polym. Bull.*, 1987, 17, 285
9. Ikeda,T.: Horiuchi,S.: Karanjit,D.B.: Kurihara,S.; Tazuke,S., *Macromolecules*, 1990, 23, 42
10. Ikeda,T.: Kurihara,S.: Karanjit, D.B.: Tazuke,S., *Macromolecules*, 1990, 23, 3938
11. Kawanishi,Y.:Tamaki,T.:Ichimura,K., *J.Phys.D*: *Appl. Phys*, 1991, 24, 782
12. Ichimura, K.: Suzuki, Y.: Seki, T.: Hosoki, A.: Aoki, K. Langmuir 1988, 4, 1214
13. Ichimura, K. In *Photochemical Processes in Organized Molecular Systems*: Honda,K.,Ed.: North-Holland: Amsterdam, 1991: p343.
14. Ichimura, K.: Suzuki, Y.: Seki, T.: Kawanishi, Y.: Tamaki, T.: Aoki, K. *Makromol. Chem., Rapid Commun.* 1989, 10, 5
15. Seki, T.: Tamaki, T.: Suzuki, Y.: Kawanishi, Y.: Ichimura, K.: Aoki, K., *Macromolecules* 1989, 22, 3505
16. Ichimura, K.: Suzuki, Y.: Seki, T.: Kawanishi, Y.: Tamaki, T.: Aoki, K., *Jpn. J. Appl. Phys., Suppl.* 1989, 28, 289
17. *Alignment of Nematic Liquid Crystals and Their Mixtures, Mol. Cryst. Liq. Cryst. Suppl. Ser. 1*: Cognard,J.,Ed.: Gordon and Breach: New York, 1982
18. Beveridge, D. L.: Jaffe, H. H. *J. Am. Chem. Soc.* 1966, 88, 1948
19. Kawanishi,Y.: Seki,T.: Tamaki,T.: Ichimura,K.: Ikeda,M.: Aoki, K., *Polym. Adv. Tech.* 1991, 1, 311
20. Kawanishi,Y.: Tamaki,T.: Seki,T.: Sakuragi,M.: Suzuki,Y.: Ichimura, K., *Langmuir* 1991, 7, 1314
21. Kawanishi, Y.: Tamaki, T.: Seki, T.: Sakuragi, M.: Ichimura, K., J. *Photopolym. Sci. Tech.* 1991, 4, 271
22. Kawanishi, Y.: Tamaki, T.: Seki, T.: Sakuragi, M.: Ichimura, K., *Mol. Cryst. Liq. Cryst.* 1992, 218, 153; Kawanishi, Y.: Tamaki, T.: Sakuragi, M.: Seki, T.: Suzuki, T.: Ichimura, K., *Langmuir* 1992, 8, 2601
23. Kawanishi, Y.: Tamaki, T.: Ichimura, K. *Polym. Mat. Sci. Eng.* 1992, 66, 263
24. Jones, P.: Jones, W. J.: Williams, G., *J. Chem. Soc. Faraday Trans.* 1990, 86, 1013
25. Tredgold, R.H.: Allen, R.A.: Hodge, P.: Khoshdel,E. *J. Phys. D: Appl. Phys.* 1987, 20, 1385
26. Todorov, T.: Nikolova, L.: Tomova, N. *Applied Optics*, 1984, 23, 4309
27. Anderle, K.: Birenheide, R.: Eich, M.: Wendorff, J. *Makromol. Cem., Rapid Commun.* 1989, 10, 477
28. Anderle, K.: Birenheide, R.: Werner, M. J. A.: Wendorff, J. H. *Liquid Crystals* 1991, 9, 691
29. Ivanov, S.: Yakovlev, I.: Kostromin, S.: Shibaev, V.: Lasker, L.: Stumpe,J.: Kresig, D. *Makromol.Chem., Rapid Commun.* 1991, 12, 709
30. Gibbons, W. M.: Shannon, P. J.: Sun, S.-T.: Swetlin, B. J. *Nature* 1991, 351, 49

RECEIVED May 25, 1993

Chapter 32

Factors Affecting the Stability of Polypyrrole Films at Higher Temperatures

V.-T. Truong and B. C. Ennis

Materials Research Laboratory, Defense Science and Technology Organisation, P.O. Box 50, Ascot Vale, Melbourne, Australia

The variability of electrochemically prepared polypyrrole films (with organic counterions) has been emphasised during studies of the stability of electrical conductivity at higher temperatures, up to 150 °C. Thermal analysis and changes of brittleness indicate that the films react thermally in nitrogen, but the loss of conductivity needs concurrent oxidation. The rates of oxidation and loss of conductivity are not directly related, and the descriptive kinetic equations, formally first order or diffusion, depend on the aging temperature. The permeability of the two sides of the film differ markedly and also depend on temperature. There is circumstantial evidence of T_g in the temperature range 65—95 °C. Other factors contributing to the electrical stability of these films include the particular counter anion and the thickness of the film. There is a possibility that the properties of a film can be crafted for particular applications, but the inadequacy of present chemical characterisation of these materials is a considerable limitation.

Electrically conducting polymers constitute a new class of materials. The possibility that these light-weight, inexpensive and versatile materials could replace conventional materials has stimulated much interest and presented challenging problems in applied research. In the last 15 years a large number of conducting polymers has been synthesized and, depending on dopant, polymer and polymerization conditions, the electrical properties can be varied from those of an insulator to those of a conductor like copper (1, 2). Interesting facets of their properties—redox behaviour, electrochemical effects, nonlinear optical properties and electronic junction effects for example—have suggested a number of potential applications as battery electrodes, non-metallic conductors, EMI shielding, Schottky barriers, in optoelectronic systems and as intelligent materials. Although considerable effort has been made to transform these materials into useful products and the Bridgestone Corporation has successfully commercialized lithium-polyaniline rechargeable batteries (3–5), systems or devices for household and industrial applications incorporating conducting polymers have been slow to emerge. This is mainly due to the inherent thermal and environ-

0097–6156/94/0537–0466$06.00/0
Published 1994 American Chemical Society

mental instability of conducting polymers. In this respect polypyrrole (PPy) is more stable than polyacetylene (6), and is thus a better candidate for a variety of applications. We have previously used the loss of conductivity and thermal analysis to study the effect of aging temperature, dopant and film thickness on electrochemically prepared PPy films (7–9).

In this paper we will discuss, largely on the basis of these findings, the relation between thermo-oxidative degradation and the loss of conductivity, the effect of a potential change of state—i.e. a glass transition—on the degradative mechanisms, and a possible approach to the design of PPy films with long-term thermal stability.

EXPERIMENTAL

PPy films were prepared by electropolymerization in distilled water. Pyrrole monomer was distilled under nitrogen. The polymerization solution contained freshly distilled pyrrole (0.1M) and dopant (0.1M). Sodium dodecyl sulphate (98% purity, Aldrich), sodium *p*-toluene sulphonate (*p*-TS) (98% purity, Merck) and *p*-chlorobenzene sulphonic acid (CBS) (98% purity, Tokyo Kasei) were used as dopants without further purification. Electrodeposition was performed at a constant current of 2.8 mA cm^{-2} using horizontal stainless steel electrodes under a nitrogen blanket. The film was removed from the electrode and washed with a mixture of water and acetonitrile to remove excess dopant. Dedoped films were obtained by reversing the polarity of the galvanostat. All films were stored in a vacuum desiccator before evaluation.

Thermal aging was performed in dry air in a laboratory oven. Conductivity was determined *in situ* by four-probe method at a constant current of 0.5 mA and thermal analysis was performed on Dupont equipment.

RESULTS AND DISCUSSION

Effect of the Dopant

Significant electrical conduction in PPy chains needs the presence of a dopant, the counter anion to the cationic organic polymer chain, and therefore an integral part of the structure. The term dopant is perhaps unfortunate since, in contrast to polyaniline where the conductivity is determined by the amount of acid dopant in the polymer, the level of anion present in PPy is not discretionary and cannot be varied by physical means, only by (electro)chemical reaction. It seems that PPy films with different anions, but otherwise similar, have different electrical and mechanical properties (7, 10), and it is therefore reasonable to expect that the dopant will have an effect on the thermal and environmental stability.

The conductivities of as-prepared PPy/DDS, PPy/*p*-TS and PPy/CBS films were 4, 25 and 31 Scm^{-1}. The loss of conductivity, $\Delta\sigma/\sigma_o$ ($\Delta\sigma = \sigma_o\text{-}\sigma$, where σ_o is the initial conductivity and σ is the conductivity at time t), for these films is shown in Figure 1 and the stability can be assessed by the time, $t_{0.5}$ taken for the conductivity to decrease to the half value, $\Delta\sigma/\sigma_0 = 0.5$. At 150 °C the $t_{0.5}$ values are 2, 12 and 36 h for DDS, CBS and p-TS, respectively.

Table 1 gives the elemental analyses of PPy/DDS and PPy/p-TS films. In an ideal PPy film oxygen would be only in the anion and the higher than expected value (15% excess for PPy/DDS and 30-100% for PPy/p-TS calculated on the sulphur analysis) indicates that oxidative reactions have occurred either during the electrolysis or during storage and before analysis. The closeness of the analyses of a fresh and an air-aged sample of one preparation of PPy/p-TS (Table 1, sample 4) indicates that not only are no volatile products lost during aging at 150 °C (confirming the stability indicated by thermogravimetry, TG), but also that the oxidation was complete at room temperature, which perhaps suggests that it is inherent in the preparation rather than the subsequent reaction of a very oxidisable film. On the other hand after aging at 150 °C the atomic proportions of C and H in PPy/DDS are decreased from the unaged values which implies the elimination of part of the aliphatic long chain dopant; this is not unexpected since the thermolysis of sodium DDS is seen in TG and follows melting of the salt at 200 °C. A similar loss of hydrocarbon could be expected during aging of PPy/DDS at lower temperatures. The rapid loss of conductivity in the PPy/DDS film may reflect this thermal instability of the long chain aliphatic anion. In air at 150 °C the conductivity of PPy/DDS film decayed to our limits of detection ($< 10^{-4}$ S cm^{-1}) in 20 h, whereas the conductivity of PPy/p-TS film decreased to 75% of its original value and then remained unchanged for at least 4 weeks (Figure 2).

It is an oversimplification to directly relate the conductive stability of PPy films to the thermal stability of the anion; the conductivity of both PPy/DDS and PPy/p-TS films remained constant at 150 °C for 72 hours in a dry nitrogen atmosphere, but began to decay as soon as the atmosphere was changed to air. Differential scanning calorimetry (DSC) indicated that unidentified reactions occurred from room temperature when these films were heated. When PPy/p-TS was heated in an open pan at low temperature the DSC curve (Figure 3, curve a) is dominated by a broad endotherm as very readily absorbed water is lost. However when the loss of water (consistent with the weight loss of 2-3% in TG) was curtailed by using a hermetic pan (9) the DSC curve was exothermic on the initial heating, but was flat when the sample was reheated (Figure 3, curves b and c) and the difference in slope between the two curves was attributed to the renewed exothermic reaction of intermediates or by-products (of lowish reactivity or inhibited by steric considerations) formed during the electrodeposition. It is possible that exothermic reaction at low temperature is common for electrochemically prepared materials; PPy/DDS gave similar results, and other PPy films have shown exothermic reaction immediately after the facile dehydration.

At higher temperatures (150 °C) there are obvious effects of reaction, and nitrogen aged films doped with DDS and p-TS are embrittled even though the conductivity is unaffected. (The Young's modulus and tensile strength were found to be 1.6 GN m^{-2} and 46 MN m^{-2}, respectively, for fresh PPy/p-TS film, whereas the aged samples were too brittle to test). Also, in isothermal DSC studies of the PPy/DDS oxidation, when the atmosphere was switched from air

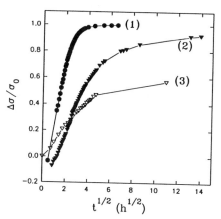

Figure 1. Effect of dopant on the stability of the conductivity of PPy films in air at 150 °C. (1) PPy/DDS; (2) PPy/CBS and (3) PPy/p-TS. (Reproduced with permission from reference (7). Copyright 1992 Elsevier Applied Science Publishers).

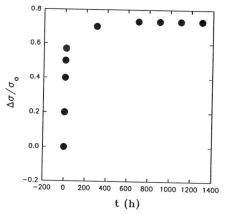

Figure 2. Long-term conductivity of 43 μm PPy/p-TS film in air at 150 °C. (Reproduced with permission from reference (7). Copyright 1992 Elsevier Applied Science Publishers).

Table 1. Empirical composition of various unaged and aged (in dry air) PPy films (7,8) (Reproduced with permission from reference (7). Copyright 1992 Elsevier Applied Science Publishers)

Film	C	H	N	O	S	O/S
p-TS counterion[a]						
1. Growth time 0.5 h	5.53	4.42	1	1.31	0.25	5.24
2. Growth time 2 h	5.68	5.18	1	0.99	0.26	3.81
aged 95 h/150 °C	5.56	4.65	1	1.35	0.27	5.00
3. Growth time 3 h	5.65	5.06	1	1.07	0.28	3.82
4. Growth time 2 h	5.59	4.67	1	1.13	0.19	5.95
aged 48 h/150 °C	5.38	4.32	1	1.14	0.20	5.70
DDS counterion[b]						
5. Growth time 2 h	7.01	9.27	1	1.20	0.26	4.62
aged 48 h/150 °C	5.20	4.97	1	1.58	0.22	7.18

In the absence of oxidation of the main chain O and S will be confined to the anion in the expected atomic ratio O:S (a), 3:1; (b), 4:1.

to nitrogen the oxidation ceased, but resumed at an increased rate when air was reintroduced (Figure 4). This is direct evidence that generation of oxidatively unstable products, which induce the initial accelerated rate, is possible in an "inert" atmosphere even though the conductivity remains constant. Table 2 shows the initial rate of oxidation, after a common holding period of 60 min in dry nitrogen, for PPy/DDS and PPy/p-TS films of different doping level and after acid treatment (DDS only). The dedoped films are more vulnerable to oxidation than the fully doped film. This tendency was also observed in undoped and doped polyacetylene (11) and PPy films (12). Quantitatively, the initial rate of oxidation increases in the following order (Table 2): fully doped film, H_2SO_4 (3M) treated fully doped film and dedoped film, whereas the rate of conductivity decay is in the reverse order (7).

Other evidence of thermal instability comes from Fourier transform infrared/evolved gas analysis (FTIR/EGA) which indicated that in PPy/DDS and PPy/p-TS carbon dioxide absorbed by amine groups was released at 100 °C and sulphur dioxide gas was evolved at 200 °C (7).

These observations suggest that the PPy films are chemically unstable and that complicated chemical processes—both endothermic and exothermic, and ranging from simple adsorption/desorption of water or carbon dioxide to evolution of sulphur dioxide and reactions leading to increased crosslink density or to chain scission—can occur to change the nature of the film in the absence of oxygen, without an inevitable loss of conductivity. It is interesting that after 1200 h at 150 °C in air the aging of PPy/p-TS film (43 µm) appears to be completed (Figure 2), at least the conductivity approaches a constant level at approximately 25% of the initial value. This suggests either a different degradation process to that in PPy/DDS film, which proceeds to completion (Figure 1), or different

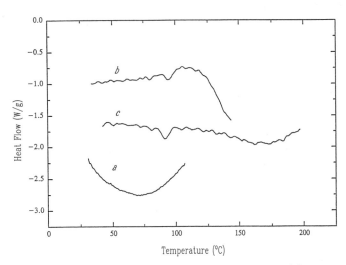

Figure 3. DSC curves of PPy/*p*-TS film at 10 °C/min: (a) open pan; (b), hermetic pan, initial heating; (c), reheat of (b). The small endotherm near 90 °C can be intensified by suitable annealing.

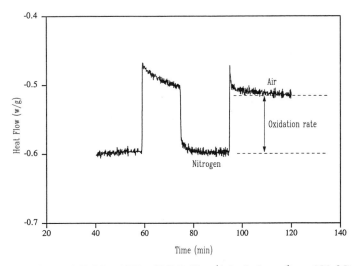

Figure 4. Isothermal DSC of PPy/DDS film (1 h dedoped) at 180 °C showing the effect of atmosphere change. Dry nitrogen was introduced to the DSC cell in the first 60 min.

Table 2. The initial rate of oxidation of PPy films at 180 °C measured by
the height of DSC signal (watt g^{-1}) (Figure 4) when the pure gas was
changed from nitrogen to air (Reproduced with permission from reference (7).
Copyright 1992 Elsevier Applied Science Publishers)

	Ppy/DDS	Ppy/p-TS
Fully doped	0.045	0.02
Dedoped	0.17	0.04
Fully doped treated with 3M H_2SO_4	0.08	N/A

conductive structures in the two materials. It is clearly different from and more
complicated than that of *trans*-polyacetylene where the loss of conductivity is
solely dependent on the loss of the conjugation (*13*). Samuelson and Druy (*14*)
have suggested that the degradation follows first-order kinetics and involves
reaction of the carbonium ion (delocalized along the polymer backbone) with
oxygen and/or water. Moss and Burford (*15*) have postulated that oxidation of
the carbon backbone results in the loss of the free radical and thus the
conductivity. Neither of these explains the residual conductivity of the present
PPy/p-TS film or the long-term stability of acid/base treated PPy films reported
by Münsted (*16*).

The reactions leading to changed electrical and mechanical properties have
not been identified, and the identity of films from different sources cannot be
assumed. In addition to the kinetic differences already noted, in contrast to Moss
and Burford (*15*) who found that carbon-oxygen species were generated at film
surface during thermal aging, no significant change in carbonyl concentration
was observed (FTIR and solid-state NMR) after aging in these laboratories (*17*).

Effect of Temperature and Film Thickness

Figure 5 shows the effect of thickness on the loss of conductivity at 150 °C.
As in Figure 1, the conductivity decay, $\Delta\sigma/\sigma_o$, is proportional to $t^{1/2}$ in the
initial stages of reaction. Attempts to treat the data as for a first order reaction,
in the manner of Samuelson and Druy (*13*), were unsuccessful, particularly at
the higher temperatures (120 and 150 °C). As with several other conducting
polymers (*15, 18, 19*), the first-order kinetic plots, $\ln(\sigma/\sigma_o)$ vs t, were nonlinear
at high temperature (> 120 °C). The inherent assumption in the first order plot
is that σ is directly proportional to the concentration of conducting species.

If it is assumed that the loss of conductivity, $\Delta\sigma$, is proportional to the
amount of oxygen absorbed by the film, M_t (*7, 8*):

$$\Delta\sigma = \sigma_o - \sigma = cM_t \tag{1}$$

then after a long time the equilibrium absorption of oxygen is achieved and

$$\sigma_o - \sigma_\infty = cM_\infty \tag{2}$$

where c is a constant, σ_∞ is the conductivity at a very large t, and M_∞ is the equilibrium sorption.

If gas sorption into a sample of thin sheet geometry follows simple Fickian kinetics, Equation 3 is an adequate description of the gas uptake (20):

$$M_t/M_\infty = 4(Dt/\pi)^{1/2}/l \qquad (3)$$

where D is the diffusion coefficient and l is the sample thickness.

If there is effectively complete loss of conductivity in the film, $\sigma_0 \gg \sigma_\infty$, then

$$M_t/M_\infty \cong \Delta\sigma/\sigma_0 = 4(Dt/\pi)^{1/2}/l \qquad (4)$$

and the reduced conductivity $\Delta\sigma/\sigma_0$ has been plotted against $t^{1/2}$ to give Figure 5 (elsewhere the parameter $t^{1/2}/l$ has been used for the several thicknesses to give a single reduced curve (8)).

It can be predicted from Figures 2 and 5 that for thick ($l \geq 43\mu$) samples the conductivity does not decay to negligible levels. The above analysis (which requires that σ_∞ is negligible) has been used for these curves on the assumption that other processes become significant only in the longer time that it takes for diffusion into a thick film. These long-term physical and/or chemical changes could affect the transport of oxygen either through the aging exterior or into the unreacted interior of the film so that electrical degradation ceases before all conductivity is lost. Explanations of the residual conductivity which require a difference in the bulk properties or composition of thick and thin films seem unlikely and have not been supported by any short term experiment or analysis. We are currently investigating the effect of thickness and temperature on the level of residual conductivity in PPy films to define this observation.

Using Fickian kinetics and $t_{0.5}$, the time required to reduce the conductivity to one-half of its initial value, i.e. $\sigma/\sigma_0 = 0.5$, D can be calculated (20) from:

$$D = 0.4919 \, (l^2/t_{0.5}) \qquad (5)$$

Thus D can be evaluated from the slope of the linear relationship of $(t_{0.5})^{1/2}$ and l. For PPy/p-TS films D was found to be 4.5×10^{-13} and 7.7×10^{-13} cm^2 s^{-1} at 120 and 150C, respectively. In contrast, this method gives D values of 10^{-9} to 10^{-10} cm^2 s^{-1} for PPy/DDS film in the same temperature range (7), which may reflect the lower density of PPy/DDS films (1.2 g cm^{-3}) compared to PPy/p-TS films (1.6 g cm^{-3}). Since, for treated PPy films, the rate of conductivity decay does not correlate directly with the rate of oxidation (Table 2) this approach to evaluation of D should be confirmed by direct measurement, but a higher diffusion coefficient could explain the lower conductive stability of DDS films.

At lower aging temperatures ($< 90\,°C$) the linearity of the $(t_{0.5})^{1/2}$ vs l plot is not obvious, and at 70 °C $t_{0.5}$ is almost independent of the film thickness (8). For a given film thickness there is a change of slope in the (Arrhenius) plot of the apparent D (from Equation 5) vs T^{-1} at 90 °C (8). These observations suggest that at lower temperatures the loss of conductivity in PPy films depends on the reactivity of the film towards oxygen, but that as the temperature is raised the rate of oxidation increases until reaction is limited by the rate of diffusion of oxygen into the film.

It would be surprising if closer examination supported the simple Fickian treatment, which should be modified to take account of chemical reaction of the diffusant in a morphological and chemically heterogeneous substrate (20). But while there is not a clear correlation between the loss of conductivity and either the rate or the amount of oxygen uptake it is clear that the diffused oxygen reacts irreversibly with the polymer leading to the loss of conjugation (loss of conductivity) and crosslinking of the polymer main chain (embrittlement), and possible decomposition of anionic dopant. The degradation mechanisms of PPy films are complex and involve both diffusion and chemical reaction.

Although the oxygen content of PPy/p-TS (growth time, t_{growth} = 2 h) can be increased by aging at 150 °C for 95 h (Table 1, sample 2) the oxygen content of this aged film was scarcely less than that of fresh (t_{growth} = 0.5 h) film (Table 1, sample 1) and from an earlier preparation aged and pristine (t_{growth} = 2 h) samples (Table 1, sample 4) with similar oxygen content were reported. This variability suggests one or all of the following possibilities:

(i) The analysis is doubtful, although duplicates on the same sample have been satisfactory;
(ii) The preparative oxidation is variable with respect to time of electrodeposition and/or position on the film—the former possibility does not necessarily imply a non-homogeneous film (but through-film variation of composition is likely) and, in regard to the latter, it has been noted that the freshly prepared film presents a mottled appearance and darkens sometime after reaction is stopped;
(iii) The post-preparative oxidation is variable and/or relatively slow;
(iv) Change in rate of oxygen uptake after 48 h.

On the basis of these observations, it is possible that, with a suitable electrodeposition technique and dopant, a thick (hundreds of μm) film could be grown that would have long term stability at high temperatures.

The Influence Of Film Structure On Degradation

It was suggested that the difference between degradation of thicker films of PPy/p-TS and of PPy/DDS could be attributed to the lower density and therefore less dense structure of the latter so that oxygen could diffuse into the film more readily. This would explain the slower reaction of the p-TS doped films, but not the eventual stabilization of the thicker film. The surface of PPy/p-TS films becomes increasingly nodular or more "cauliflower" like with increasing growth time (8), and it might be thought that this would lead to lower density material. The change is probably slight and it has not been possible to show any significant difference in the $\Delta\sigma/\sigma_0$ vs $t^{1/2}/l$ plots of thick and thin films, expected to reveal density/diffusion changes, during the initial stages of aging (8).

A more convincing example of morphological influence was provided during confirmation of the role of diffusion in PPy/p-TS film at high temperature. It was argued that if the loss of conductivity was controlled by the rate of diffusion of oxygen into the film then blocking one side of the aging film should result in a fourfold reduction in the rate of decay (*9*). At temperatures above 90 °C the expectation was satisfied; at 120 and 150 °C until equilibrium transport of oxygen through the blocking polyimide film (Dupont Kapton) was established, the predicted reduction in decay was observed (Figure 6), and at 90 °C, when chemical control was postulated, blocking one side had no effect on the rate. What was unexpected was that at 70 °C the blocking of the growth side of the film (i.e. that remote from the electrode) effectively protected the film from degradation, while the conductivity decay was unchanged when the side that had been on the electrode was protected (Figure 7). The transport properties of the two faces of the PPy/*p*-TS film are dramatically different, and must be explained by a morphological or constitutional differences.

Further evidence of anisotropy in some of these films was provided during thermogravimetry. It was noticed that after pyrolysis to 700 °C in nitrogen PPy/*p*-TS, PPy/CBS and PPy/DDS films maintain their integrity but are coiled and shiny, in particular the PPy/CBS appears quite metallic. These temperatures are much higher than those during aging, but the two sides of the film differ markedly in appearance (degree of shine) and together with the coiling, which can be quite tight, this points to a substantial differentiation between the two surfaces.

The Relevance of a possible Glass Transition to Aging Kinetics

Marked changes in the temperature dependence of the degradation of conductivity of PPy films near 100 °C have been noted in previous sections. An example of this is given in Figure 8 which shows the ratio, t_b/t_u, of the times to obtain the same conductivity decay in PPy/*p*-TS films which are either unblocked or blocked on the electrode side with polyimide film. There is a marked change at 90—100 °C from the low temperature regime where the degradation data conforms to a first order plot, for example Figure 9, to the high temperature region where the first order plot is plainly unsatisfactory and there is a linear relation between decay and $t^{1/2}$ (eg. Figure 5).

The abrupt change in the kinetic description of the decaying conductivity in PPy/*p*-TS could be explained by different dependencies for diffusion of oxygen through the film and for reaction of oxygen with the structures responsible for conduction. If the rate of chemical reaction is sufficiently slow at ambient temperatures then the rate of oxygen diffusion into the film will be fast enough to maintain the equilibrium oxygen concentration and exhibit pseudo-first order kinetics; and if the activation energy for reaction is greater than that for diffusion then at some higher temperature oxygen will be consumed faster than it can diffuse to the reaction sites and the rate of electrical decay will be determined by physical rather than chemical factors. This does not explain the protection afforded when the growth side of the film is blocked by polyimide film, or why that protection is only effective below 90 °C.

Although there are obvious differences between the two sides of the film, the sudden loss of diffusional differentiation has to be explained. One possibility

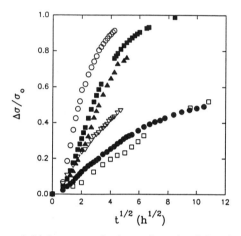

Figure 5. The effect of thickness on the loss of conductivity, $\Delta\sigma/\sigma_o$, of PPy/p-TS at 150 °C: (○) 12 μm; (■) 24 μm; (▲) 30 μm; (▽) 43 μm; (●) 57 μm and (□) 68 μm.

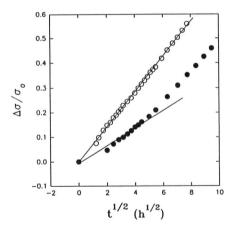

Figure 6. Loss of conductivity, $\Delta\sigma/\sigma_o$, of 43 μm PPy/p-TS film at 120 °C: (○) Unblocked film; and (●) blocked film. The change of slope at $t^{1/2} = 5$ indicates that a steady state has been established in the blocking polyimide film and thus the oxygen flux is limited by the PPy film (9).

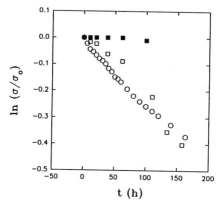

Figure 7. The first-order kinetic plot of the loss of conductivity of 43 μm PPy/p-TS film at 70 °C. (O) Unblocked film; ($\square$) initial (electrode) surface blocked; and ($\blacksquare$) growing surface blocked.

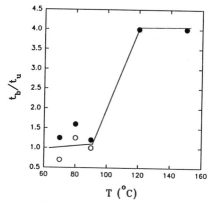

Figure 8. The temperature dependence of the relative time (t_b/t_u) for the conductivity of blocked (t_b) and unblocked (t_u) PPy/p-TS films to decay to 20% of the initial value ($\bullet$) 43 μm film and (O) 12 μm film.

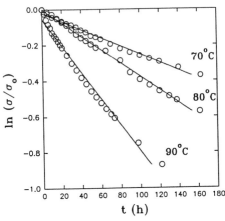

t (h)

Figure 9. The first-order kinetic plots of the loss conductivity of 43 μm PPy/p-TS film at lower temperatures.

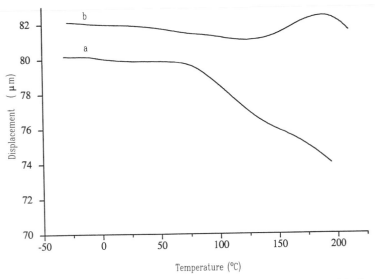

Temperature (°C)

Figure 10. TMA curves of a PPy/p-TS film, penetration probe with 5 g load, heating rate 5 °C/min: (a) initial heating; (b) reheat of (a) after overnight ambient exposure.

is that the glass transition temperature, T_g, of the PPy is in this temperature range. The diffusion of oxygen through the polymer matrix will be much easier above the T_g but it might also be expected that both sides would be equally responsive. If, however, the initial deposition is more "ordered" because of the proximity of the electrode surface so that the reacting molecules are oriented to give either a more crystalline product, or a more regular chemical structure (eg. higher degree of polymerization, less chain branching) then there could be different temperature responses for the two sides. In the former case diffusion would be largely confined to the amorphous regions and, if there was substantial degree of crystallinity, a significant lower permeability could persist beyond the T_g; and in the latter case the initial material could have a higher T_g than the bulk material.

Some evidence of a T_g in this range has been obtained and this has been detailed elsewhere (*9*). The most persuasive evidence has been found in the DSC curves of some samples where there are endotherms which could be attributed to physical aging after the T_g had been lowered because of plasticization by absorbed moisture. A typical result is seen in Figure 3; when the dehydration of PPy/*p*-TS is suppressed in hermetic pans a typical aging endotherm precedes the loss of water (Figure 3, curves *b* and *c*) and this can be enhanced by suitable annealing. Also, instead of the usual broad endotherm (Figure 3, curve *a*) obtained for fresher material in open pans, some other older samples of different PPy film have shown an additional sharp endotherm (between 60 and 90 °C) superimposed on the expected DSC curve.

Other indications have been manifest in thermomechanical analysis, TMA, where the dimensional instability of PPy films results in shrinkage and/or probe penetration as the fresh sample is heated—for example Figure 10 shows an apparent contraction of a fresh PPy/*p*-TS film as it is heated from room temperature and then marked softening/contraction at 70–100 °C, although TG showed no significant weight loss above 100 °C. After equilibration in the laboratory the previously heated sample showed evidence of increased expansion consistent with T_g before resumption of contraction at the previous higher temperature. The problem with these TMA curves rests with the several factors (expansion, penetration, stress relief etc.) which generally contribute to these curves, as well as the specific problems of loss of water and further reaction already alluded to, and the possibility of artifacts in the TMA of thin film samples. As a consequence reproducibility, in the sense of thermal cycling of TMA samples to isolate and identify specific transitions has not been possible. However, the results obtained have been consistent with drying and further reaction which raises the T_g as the sample is subjected to progressively higher temperatures.

CONCLUSIONS

The foregoing results do not admit firm conclusions, but do encourage speculation and hypothesis. In particular they present challenging and promising phenomena in electrochemically prepared PPy films and point to the need for great care in the design and the interpretation of experiments to elucidate them.

Thermal degradation of PPy films results in the loss of electrical conductivity and poorer mechanical properties. Embrittlement occurs in either dry air or in nitrogen and although this could be due to drying, chain scission or crosslinking, the last is perhaps more likely because dehydration is reversible and scission would be expected to lead to lower molecular weight and deformation temperatures. Conductivity is stable in a non-oxidising environment and even in air may not be completely lost if the film is thick enough and/or the right dopant is chosen. Although the reactions leading to the loss of conductivity have not been identified, experiment has shown that oxidation is necessary for conductivity decay but that there is no direct correlation between the rate of oxidation and the rate of decay in different PPy films. In the accessible temperature range studied (70—150 °C) the decay of conductivity seems to be determined by both diffusion and chemical reaction (oxidation). It appears that at high temperatures (above 100 °C) the diffusion of oxygen into the film is rate determining, but at lower temperature the decay follows first-order kinetics. These results indicate that it is not appropriate to extrapolate conductivity data from one temperature to another in order to predict the shelf-life at low temperature or to estimate an acceptable service temperature for PPy film. There is some evidence for a glass transition in the temperature range of interest and there is a most surprising differentiation between the growing and electrode surfaces at lower temperatures. The remarkable stability of PPy/p-TS films protected from oxidation on the growing side was confirmed for several samples of different thicknesses. This behaviour has been observed in p-TS doped film and the generality of these findings for other dopants should be confirmed. It is postulated that the stable conductivity of the thicker PPy/p-TS films after partial aging can be attributed to morphological and structural changes, brought about by thermal aging and oxidation, which progressively restrict diffusion of oxygen into the film. The stability conferred by partial oxidation was reported previously (7) and has not yet been exploited.

This study also presents a possibility of designing PPy films with long-term electrical stability at elevated temperatures, although embrittlement and loss of mechanical properties might be anticipated. Furthermore, even if the inherent stability of the electrode side cannot be achieved in the bulk polymer, the observed anisotropy is of potential importance for applications at low temperature. The more sensitive growing side may be more responsive to environmental effects and thus more useful as a chemical sensor, while for applications which require long-term conductivity stability the growing side should be preferentially protected.

PPy films are thermally and oxidatively reactive, and materials that are in the main indistinguishable by bulk analytical methods present different electrical stability, although chemical reaction does not lead inevitably to changes of conductivity. For these reasons, and since synthetic conditions can subtly vary from one laboratory to another, there is a pressing need for a better understanding of the chemical structures responsible for conduction and their distribution within the film.

ACKNOWLEDGMENTS

The authors are particularly indebted to Dr C. E. M. Morris for encouragement and to Dr P. J. Burchill for stimulating discussion.

REFERENCES

1. Naarman H., *Synth. Met.*, 1987, **17**, 223
2. Naarman H. and Theophilou N., *Synth. Met.*, 1987, **22**, 1
3. Japanese Patent (Kokai Tokkyo Koho) No. 1-150 536
4. Japanese Patent (Kokai Tokkyo Koho) No. 1-168 732
5. T. Nakajima and T. Kawagoe, *Synth. Met.*, 1989, **28** , C 629
6. N. C. Billingham and P. D. Calvert, *Adv. Polym. Sci.*, 1989, **90**, 1
7. V.-T. Truong, B. C. Ennis, T. G. Turner and C. M. Jenden, *Polym. Int.*, 1992, **27**, 187
8. V.-T. Truong, *Synth. Met.*, 1992, **52**, 33
9. B. C. Ennis and V.-T. Truong, *Synth. Met.*, in press
10. L. J. Buckley, D. K. Roylance and G. E. Wnek, *J. Polym. Sci., Part B: Polym. Phys.*, 1987, **25**, 2179
11. X.-Z. Yang, and J. C. W. Chien, *J. Polym. Sci., Chem.*, 1985, **23**, 859
12. H. Münsted, *Polymer*, 1988, **29**, 296
13. F. Ebisawa and H. Tabei, *J. Appl. Phys.*, 1985, **58**, 2326
14. L. A. Samuelson and M. A. Druy, *Macromolecules*, 1986, **19**, 824
15. B. K. Moss and R. P. Burford, *Polymer*, 1992, **33**, 1902
16. H. Münsted, *Polymer*, 1986, **27**, 899
17. M. Forsyth and R. G. Davidson, private communication
18. Y. Wang, M. F. Rubner and L. J. Buckley, *Synth. Met.*, 1991, **41–43**, 1103
19. Y. Wang and M. F. Rubner, *Synth. Met.*, 1990, **39**, 153
20. J. Crank, *The Mathematics of Diffusion* 2nd ed., Clarendon Press, Oxford, 1975

RECEIVED December 30, 1992

Chapter 33

Intrinsic and Thermal Stress in Polyimide Thin Films

M. Ree and D. P. Kirby[1]

IBM Technology Products Division, 74 Creamery Road, Hopewell Junction, NY 12533

In-situ measurements of intrinsic stress, as well as overall internal stress, were performed for three soluble preimidized polyimides as a function of temperature over the range of 25–400 °C using a wafer bending technique: Sixef-44 (6FDA-4,4′-6F), Sixef-33 (6FDA-3,3′-6F), and Probimide 412. For polyimide films with ca. 12 μm thick, intrinsic stress is 29–31 MPa at room temperature, at which the films have been prepared, indicating that at the temperature of the film preparation the intrinsic stress is not sensitive to the type of backbone chemistry among these polyimides. The intrinsic stress varied with temperature in the first heating run, where its variation with temperature was strongly dependent upon the properties of the polyimide, including mechanical properties, polymer chain stiffness, molecular order, and glass transition temperature. The measured intrinsic stress is not small enough to be neglected, as one usually does. In addition, the thermal stress was estimated from the measured intrinsic and overall stresses as a function of temperature. Its variation with temperature was dependent upon the temperature regime: regime I (above T_g), regime II (115 °C − T_g or T_f), and regime III (below 115 °C). As is well known, thermal stress is developed below T_g and, based on our results, apparently increases linearly with descending temperature only in regime II. However, the slope of the thermal stress profiles in regime II is not same with that of the overall stress profiles. In addition, T_g's of the polyimides were estimated from the overall stress-temperature profiles.

Aromatic polyimides have found wide application in the microelectronics industry as alpha particle protection, passivation, and intermetallic dielectric layers, owing to their excellent thermal stability, mechanical properties and dielectric properties (1–3). Many microelectronic devices, such as VLSI semiconductor chips and advanced multi-chip modules (3), are composed of multilayer structures. In multilayered structures, one of the serious concerns related to reliability is residual stress caused by thermal and loading histories generated through processing and use, since polyimides have different properties (i.e., mechanical properties, thermal expansion coefficient, and phase transition temperature) from the metal conductors and substrates (ceramic, silicon, and plastic) com-

[1]Current address: Barnett Institute, Northeastern University, 360 Huntington Avenue, Boston, MA 02115

0097–6156/94/0537–0482$06.00/0

monly employed. In general, residual stress in a polyimide film consists of two major components (*4*). One is the so-called *thermal stress* due to the mismatch of thermal expansion coefficients between film and substrate or metal layer, as well as the thermal history and mechanical properties. The other is *intrinsic stress* resulting from volume change due to solvent evaporation and shrinkage, from molecular structural ordering during the film formation process, and perhaps from the physical properties of the formed film. The overall internal stress of a polymer film is usually assumed to be the thermal stress, neglecting the contribution of intrinsic stress. Here, it should be noted that the intrinsic stress in a polymer film is not understood in detail yet.

In the present study we have chosen several soluble preimidized polyimides in order to understand intrinsic stress behavior and its contribution to overall stress: Sixef-44, Sixef-33, and Probimide 412 (see Figure 1). These soluble preimidized polymers are good candidate polymer systems for studying the behaviors of stress components, because completely dried films can be made without any solvent complexation and imidization which occur in polyimide precursors. For these preimidized polymers, fully dried films were prepared on silicon wafers through room temperature drying under a nitrogen flow and then followed by vacuum drying to avoid the involvement of any thermal history. For these dried polyimide films, intrinsic stress was measured in-situ during heating as a function of temperature over the range of 25–400 °C using a wafer bending technique (*5*, *6*). During subsequent cooling after baking at a certain temperature, overall stress was also measured dynamically as a function of temperature. The thermal stress component was estimated from the measured intrinsic and overall stresses.

INTRINSIC AND THERMAL STRESS

Residual stress (i.e., interfacial stress) in a polymer thin film is commonly determined by measuring the curvature (deflection) of the bilayer composite structure of the film and a substrate after film deposition. The stress of the film in equilibrium with the resultant strain can be calculated from the curvature and mechanical parameters of the substrate using the following simple plate equation (*7*):

$$\sigma_F = \frac{1}{6} \frac{E_s t_s^2}{(1 - \nu_s) t_F} \left(\frac{1}{R_F} - \frac{1}{R_\infty} \right) \tag{1}$$

Here, the subscripts F and S denote polyimide film and substrate, respectively. The symbols σ, E, ν, and t are stress, Young's modulus, Poisson's ratio, and thickness of each layer of material. R_F and R_∞ are radii of a substrate with and without a polyimide film, respectively. For Si(*100*) wafers, biaxial modulus, $E_S/(1 - \nu_S)$, is 1.805×10^5 MPa (*8*). Eq (1) has been driven under the assumption that the stress is isotropic and uniform in the film plane. The application of this equation is limited to bending displacements smaller than the thickness of the substrate. In other words, the thickness of a polymer film should be much smaller than the thickness of substrate.

The residual stress (σ_F) is known to consist of two major components: *thermal stress* and *intrinsic stress* (4). The thermal stress (σ_t) results from the mismatched thermal expansion coefficients (TECs) of the film and substrate, as well as the mechanical properties and thermal history of the film and can be estimated from the following equation (9):

$$\sigma_t = (\alpha_F - \alpha_S)(T_f - T)\frac{E_F}{(1 - \nu_F)} \tag{2}$$

with TECs (α_S and α_F), final heat-treat temperature (T_f), stress measurement temperature (T), modulus (E_F) and Poisson's ratio (ν_F).

In contrast to the thermal stress due to thermal history, the intrinsic stress arises mainly from constraints on molecular movement during film formation. A polymer film is commonly fabricated by applying the polymer in solution on to a substrate and subsequently drying it. During drying, the wet polymer film concentrates and its viscosity drastically increases due to solvent evaporation. The wet film starts to solidify when its viscosity reaches a gel point. Below the gel point, the molecules in the film are mobile enough to flow and thus residual stress can not be generated. However, above the gel point the film is extremely viscous and its glass transition temperature (T_g) increases. The increase of viscosity and T_g in the film restricts the molecular motion and results in stress. The shrinkage due to solvent evaporation takes place in the direction of film thickness but is constrained in the direction of the film plane, becuase of the interfacial adhesion between the film and the substrate. Thus, interfacial stress develops in the film plane. This stress (σ_d) generated by the polymer film deposition can be expressed by the following equation (10):

$$\sigma_d = \frac{E_F}{(1 - \nu_F)} \frac{(\phi_s - \phi_r)}{3(1 - \phi_r)} \tag{3}$$

where ϕ_s is the volume fraction of solvent at which the film solidifies and ϕ_r is the volume fraction of solvent retained in the film.

As is expressed in Eq (3), the stress (σ_d) is sensitive to the residual solvent in the film. When the polymer film is fully dried at a given temperature, the volume fraction (ϕ_r) is zero and consequently the stress is independent of residual solvent. Then, we can define that the stress (σ_d) with $\phi_r = 0$ is the *intrinsic stress* of the film:

$$\sigma_i = \frac{\phi_s E_F}{3(1 - \nu_F)} \tag{4}$$

Here, the intrinsic stress (σ_i) is only a function of the biaxial modulus ($E_F/(1 - \nu_F)$) and volume fraction (ϕ_s) which are physical characteristics of the polymer.

These parameters are of course a function of temperature. In particular, the volume fraction (ϕ_s) decreases as temperature increases.

Using Eq(4), the σ_d in Eq (3) can be expressed in terms of the intrinsic stress (σ_i):

$$\sigma_d = \sigma_i \frac{(\phi_s - \phi_r)}{\phi_s(1 - \phi_r)} \tag{5}$$

The relationship of overall stress to thermal and intrinsic stress is an interesting subject. For a given film sample, it is overall stress that is always measured. If the temperatures of heat treatment and stress measurement are the same, the thermal stress component becomes zero and consequently the measured overall stress results only from the intrinsic stress component. For this reason, we assume that overall stress is the sum of thermal and intrinsic stress:

$$\sigma_F = \sigma_t + \sigma_i \tag{6}$$

From Eq (5), this σ_F can be expressed in terms of the σ_d:

$$\sigma_F = \sigma_t + \sigma_d \frac{\phi_x(1 - \phi_r)}{(\phi_s - \phi_r)} \tag{7}$$

Using this linear additivity relationship, the thermal stress component can be estimated from the measured overall and intrinsic stress.

EXPERIMENTAL

Preimidized Probimide-412 solution (polyimide derived from benzophenone tetracarboxylic dianhydride (BTDA) and alkyl substituted aromatic diamines, ca. 12 wt% in γ-butyrolactone) was used as supplied from Ciba-Geigy Chemical Company. Two fluorinated preimidized polymers, Sixef-33 (poly(2,2'-bis(3-phenyl)hexafluoropropane hexafluoroisopropylidene diphthallimide): 6FDA-3,3'-6F) and Sixef-44 (poly(2,2'-bis(4-phenyl)hexafluoropropane hexafluoroisopropylidene diphthallimide): 6FDA-4,4'-6F), were supplied as fibers or powders by American Hoechst Chemical Company (Figure 1). These polyimides were dissolved in *n*-butyl acetate and filtered with 1 μm Fluoropore filter membranes. Their concentration was ca. 10 wt%. Double side polished Si(100) wafers (82.6 *mm* diameter and ca. 380 μm thickness) were used as substrates for stress measurements. These wafers were cleaned in a Plasmaline asher (Model 515) of Tegal Corporation before use. The curvature of these wafers was measured on a Flexus stress analyzer. An adhesion promoter solution (0.1 vol% γ-aminopropyltriethoxy silane in deionized water) was spin-applied at 2,000 rpm for 20 sec on the calibrated Si wafers. Then, the preimidized polymer solutions were spin-coated on the primed Si(100) wafers and dried in a convection oven with a nitrogen gas flow at room temperature. The partially dried samples were further dried at room tempearture in a vacuum oven with < 5 torr for two weeks. The thickness of the films was ca. 12 μm.

For fully dried polyimide films, in-situ stress measurements were performed in nitrogen ambient during thermal baking and subsequent cooling, using a double He-Ne laser beam stress analyzer (Model 2-300, Flexus Company) equipped with a hot-stage and controlled by an IBM PC/AT computer. The baking was conducted for 30 min at temperature in the range of 25-400 °C. The heating and cooling rates used were 2.0 °C/min and 1.0 °C/min, respectively.

RESULTS AND DISCUSSION

Sixef-44 films, which were dried at room temperature, exhibited 31 MPa residual stress (that is, *intrinsic stress*) on Si(100) wafers at room temperature. These polyimide films were heated to one of three different temperatures (150 °C, 250 °C, or 400 °C) and subsequently cooled to room temperature. Stress was measured in-situ during the heat-treatments. In the first heating runs (see Figs. 2-4), the intrinsic stress decreased rapidly until ca. 150 °C and thereafter decreased slowly. As given in Eq(4), intrinsic stress is a function of Young's modulus (E_F), Poisson's ratio (ν_F), and volume fraction of solvent (ϕ_s) at the film solidification. In general, for a polymer film Young's modulus (E_F) and volume fraction (ϕ_s) decrease with increasing temperature, whereas Poisson's ratio (ν_F) increases slightly with increasing temperature (6, 11). When a polymer film is fully dried at a given temperature, the volume fraction (ϕ_s) is constant. Therefore, when a polymer fully dried at a temperature is heated, Poisson's ratio (ν_F) contributes positively to the intrinsic stress, whereas modulus (E_F) contributes negatively. As shown in Figs. 2-4, the intrinsic stress of Sixef-44 films decreases with increasing temperature in spite of the positive contribution of the Poisson's ratio term, indicating that the intrinsic stress-temperature profile is dominated by the Young's modulus term.

Figure 2 shows the stress-temperature profile of the polyimide film measured during heating to 150 °C and subsequent cooling. On heating, intrinsic stress decreased from 31 MPa at room temperature to 9.3 MPa at 150 °C. During aging at 150 °C for 30 min, the intrinsic stress increased from 9.3 MPa to 13.1 MPa. This stress increase (3.8 MPa) might result from a morphological change due to molecular ordering occurred during the aging. However, stress is very sensitive to the amount of residual solvent (see Eq(5)). Therefore, this stress increase might result partially from the removal of residual solvent, if the film retained solvent even in a very small amount. On cooling after the bake, stress (*overall stress*) increased from 13.1 MPa at 150 °C to 35.4 MPa at room temperature. Using Eq(6) (or Eq(7)), the thermal stress component in the film was estimated by subtracting the measured intrinsic stress from the measured overall stress. The estimated thermal stress is compared with the intrinsic and overall stress in Figure 2. The thermal stress varied with temperature in the range of 4.2-7.1 MPa. That is, thermal stress was little sensitive to temperature over the temperature range considered and was much smaller than the intrinsic stress. Consequently, the overall stress of the film baked at 150 °C was dominated by the intrinsic stress component.

The thermal stress component became more pronounced by the elevation of bake temperature. Figure 3 presents the stress-temperature profile of the film

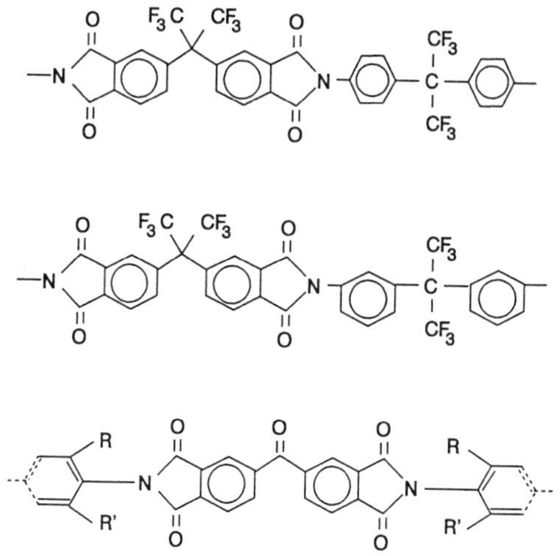

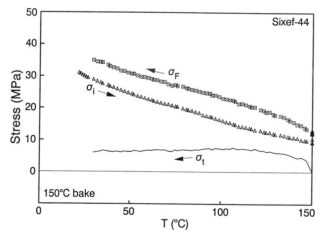

Figure 1. Chemical structures of preimidized polyimides: Sixef-44, Sixef-33, and Probimide 412.

Figure 2. Stress versus temperature behavior of Sixef-44 film dried at room temperature measured on a Si(100) wafer during baking up to 150 °C and subsequent cooling: σ_F, overall stress; σ_t, thermal stress; σ_i, intrinsic stress. The heating and cooling rates were 2.0 °C/min and 1.0 °C/min, respectively.

baked at 250 °C for 30 min. On heating, intrinsic stress decreased monotonically with temperature from 31 MPa at room temperature to 5.2 MPa at 250 °C. During aging at 250 °C for 30 min, intrinsic stress increased slightly from 5.2 MPa to 6.1 MPa. This stress increase is much smaller than that of the film aged at 150 °C. On cooling after the bake, stress increased linearly with decreasing temperature from 6.1 MPa at 250 °C to 41.4 MPa at room temperature. The thermal stress component was estimated in the same way as described above. The thermal stress exhibited two different behaviors, depending on the temperature region (see Figure 3). The thermal stress initially increased linearly with decreasing temperature from zero to 14.9 MPa in the regime of 115–250 °C and then decreased very slowly from 14.9 MPa to 12.5 MPa in the regime of 25–115 °C. The shape of the stress-temperature profile over 25–150 °C resembles that of the film baked at 150 °C. However, the stress level is 2 times higher than that of the 150 °C baked film.

Figure 4 shows the stress behavior of the film baked at 400 °C. On heating, the intrinsic stress of 31 MPa at room temperature decreased monotonically with temperature to 2.2 MPa at 400 °C. During aging at 400 °C for 30 min, stress increased slightly from 2.2 MPa to 3.0 MPa. The increase in stress on aging is only 0.8 MPa. Similar aging behavior was previously observed at 250 °C. That is, the aging effect is very small above 250 °C. In contrast, this effect was more significant at 150 °C (a highly supercooled state). Therefore, the aging effect on the stress of Sixef-44 films becomes significant at temperatures below 250 °C. On cooling after baking at 400 °C, overall stress remained at ca. 3 MPa (intrinsic stress only) until 300 °C, and thereafter increased rapidly with decreasing temperature to 50 MPa at room temperature. The final stress is much higher than that of the films baked at 150 °C or 250 °C, indicating that the thermal stress component becomes more significant in the film baked at a higher temperature.

The estimated thermal stress is compared with the intrinsic stress as well as the overall stress in Figure 4 and shows three different temperature regime behaviors. The thermal stress, which was developing during cooling, was zero or less until ca. 285 °C (regime I). On further cooling the stress increased rapidly with temperature down to ca. 115 °C (regime II) and thereafter decreased very slowly (regime III). In regime I (above T_g), thermal stress could not develop because of its rapid relaxation due to the high mobility of polymer chains causing easy deformation of the film. In regime II (below T_g), thermal stress was generated with decreasing temperature as predicted by Eq(2). In this regime, the Poisson's ratio (ν_F) term contributes negatively to the thermal stress, because it decreases slightly on cooling from the liquid state into the frozen glassy state (11). Also, the TEC (α_F) term contributes negatively to the thermal stress, due to the nature of its temperature dependency. However, despite the negative contribution of those two terms (ν_F and α_F), the thermal stress has increased overall with decreasing temperature in regime II, indicating that the degree of supercooling ($\Delta T = T_f - T$ or $T_g - T$) and Young's modulus (E_F) are predominant contributors to the thermal stress here. The increase of thermal stress in regime II was not continued into regime III where stress varied only slightly with temperature. In this regime, the modulus (E_F) and Poisson's ratio (ν_F) might be much less sensitive to temperature compared with regime II. The stress behavior

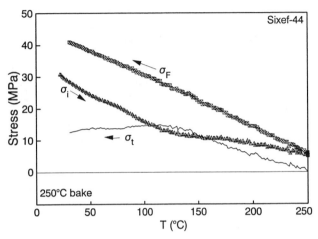

Figure 3. Stress versus temperature behavior of the Sixef-44 film dried at room temperature measured on a Si(100) wafer during baking up to 250 °C and subsequent cooling: σ_F, overall stress; σ_t, thermal stress; σ_i, intrinsic stress. The heating and cooling rates were 2.0 °C/min and 1.0 °C/min, respectively.

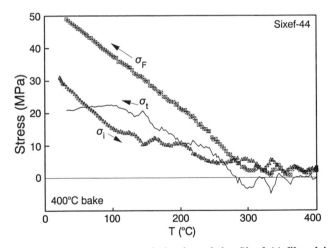

Figure 4. Stress versus temperature behavior of the Sixef-44 film dried at room temperature measured on a Si(100) wafer during baking up to 400 °C and subsequent cooling: σ_F, overall stress; σ_t, thermal stress; σ_i, intrinsic stress. The heating and cooling rates were 2.0 °C/min and 1.0 °C/min, respectively.

in *regime III* might result from either the negative contribution of α_F just cancelling the positive contribution of the ΔT or its contribution being slightly more dominant than that of ΔT.

The in-situ stress measurement was extended to Sixef-33 and Probimide 412 films. Representative results are shown in Figs. 5 and 6. The intrinsic stress of Sixef-33 films was 31 MPa at room temperature. This stress level is comparable to that of the Sixef-44 film. However, its variation with temperature in the first heating runs is different from that of the Sixef-44 films. As shown in Figure 5, on heating the intrinsic stress rapidly relaxed out with increasing temperature and then leveled off at 1.5 MPa above 150 °C. The fast stress relaxation might result from a relatively high mobility of the polymer chains. In comparison with the Sixef-44 film, the Sixef-33 film is expected to have a lower T_g. On cooling, the overall stress started to increase from 230 °C, due to the contribution of thermal stress generated in the supercooled state below T_g, and finally reached 45 MPa at room temperature. From these stress results, the thermal stress component was estimated as shown in Figure 5. The thermal stress remained at zero until 230 °C (*regime I*), increased with decreasing temperature down to ca. 115 °C (*regime II*), and then turned to decrease slowly to 16 MPa at room temperature (*regime III*). That is, the Sixef-33 film thermal stress variation with temperature showed a similar temperature regime behavior as that observed for the Sixef-44 film.

Similar stress-temperature behaviors were observed for Probimide 412 films. The intrinsic stress was 29 MPa at room temperature. In the heating run, the stress of 29 MPa decreased to ca. 4 MPa at 400 °C (see Figure 6). On cooling after the bake, the overall stress increased continuously with temperature and reached 56 MPa at room temperature. From this stress profile, one expects that the T_g of the Probimide 412 is > 400 °C. The thermal stress of the film was also estimated from the intrinsic and overall stress profiles as a function of temperature. The thermal stress was not generated above 370 °C (*regime I*), however increased with deacreasing temperature over 115–370 °C (*regime II*) and then leveled off or decreased slightly below 115 °C (*regime III*).

As described above, the three polyimides exhibited almost the same intrinsic stress (29–31 MPa) at the drying temperature, regardless of the different chemical backbone. This indicates that the mechanical properties (particularly Young's modulus and Poisson's ratio) of those polyimides are nearly the same. On heating, the intrinsic stress relaxed out with temperature. The stress relaxation was strongly dependent on T_g, that is, polymer chain flexibility. The T_g of the Sixef-33 is relatively low so that its stress relaxation is faster than that of the others (see Figs. 4-6). The stress-temperature profiles indicate that T_g is in the decreasing order of Sixef-33 < Sixef-44 < Probimide 412.

On cooling after thermal treatments, overall stress was built up again in the polyimide films by recovery of intrinsic stress and generation of thermal stress (see Figs. 4-6). The overall stress of the polyimide films was sensitive to thermal history, due to the contribution of the themal stress component. In contrast to the intrinsic stress, the overall stress of baked films varied nearly linearly with temperature below T_g. For the films baked at 400 °C, overall stress at room temperature was 45 MPa for the Sixef-33, 50 MPa for the Sixef-44, and 56 MPa

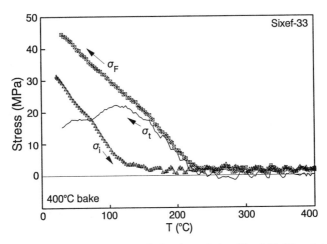

Figure 5. Stress versus temperature behavior of the Sixef-33 film dried at room temperature measured on a Si(100) wafer during baking up to 400 °C and subsequent cooling: σ_F, overall stress; σ_t, thermal stress; σ_i, intrinsic stress. The heating and cooling rates were 2.0 °C/min and 1.0 °C/min, respectively.

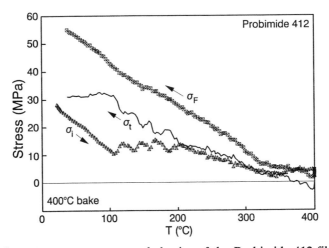

Figure 6. Stress versus temperature behavior of the Probimide 412 film dried at room temperature measured on a Si(100) wafer during baking up to 400 °C and subsequent cooling: σ_F, overall stress; σ_t, thermal stress; σ_i, intrinsic stress. The heating and cooling rates were 2.0 °C/min and 1.0 °C/min, respectively.

for the Probimide 412. Their intrinsic stress components were almost the same, 29–31 MPa. Therefore, the thermal stress component was 14 MPa for the Sixef-33, 19 MPa for the Sixef-44, and 27 MPa for the Probimide 412. In both the Sixef-33 and the Sixef-44 films, the intrinsic stress component was higher than the thermal stress component. In the Probimide 412, the intrinsic stress was comparable with the thermal stress component. The contribution of intrinsic stress to the overall stress was more significant in the polyimide films baked at low temperatures (see Figs. 2-4). This is evidence that in polymer films, particularly the polyimides studied here, the intrinsic stress component is not small enough to be neglected. Furthermore, the slope in the thermal stress variation with temperature generally is not same with that of the overall stress variation as shown in Figs. 4-6. The thermal stress variation with temperature is strongly dependent upon the temperature regime. The thermal stress varies nearly linearly only in *regime II*. However, even in this regime the slope of the thermal stress variation is not the same as that of the overall stress variation. Therefore, it may not be a good approach in the estimation of the biaxial modulus and thermal expansion coefficient of a polymer film that the overall stress measured for a thermally treated polymer film assumes to be the thermal stress itself.

The three polyimide films baked at 400 °C exhibited ca. 3.0 GPa Young's modulus at room temperature (*12*), regardless of the polymer backbone. Equivalent Poisson's ratios are expected for these polyimides. Therefore, the difference in the overall stress of the polyimide films may result from the differences in the T_g's as well as in the TECs. Considering polymer chain flexibility and T_g, TEC may be relatively high in the Sixef-33, intermediate in the Sixef-44, and low in the Probimide 412 film. However, the thermal stress was in the increasing order of Probimide 412 > Sixef-44 > Sixef-33. Consequently, the difference in the overall stress was predominantly driven by the difference in ΔT's due to different T_g's.

In addition, the T_g of the films can be estimated from the stress-temperature profiles. In general, stress is not built up above T_g because of high polymer chain mobility. However, the chain mobility is restricted in the glassy state below T_g allowing the generation of stress. In the overall stress-temperature profile measured on cooling, the temperature at which the stress started to increase was chosen as T_g. For the films baked at 400 °C, T_g was 235 °C for the Sixef-33, 300 °C for the Sixef-44, and 400 °C for the Probimide 412.

CONCLUSIONS

The intrinsic stress, as well as overall internal stress, in three different polyimide films was dynamically measured during heating and subsequent cooling as a function of temperature over the range of 25–400 °C. The thermal stress, which was generated during cooling, was estimated from the intrinsic and overall internal stresses as a function of temperature. For all the polyimide films studied here, the intrinsic stress was 29–31 MPa at room temperature. The initial level of intrinsic stress, which had been generated mainly by the solidification of cast polyimide solution through solvent evaporation, is apparently not sensitive to the backbone chemistry. However, the variation of intrinsic stress with temperature is strongly dependent upon the backbone chemistry, indicating that stress is correlated with the nature of the polyimide chain, including chain stiffness, molecular order and glass transition temperature. For these polyimides, the intrinsic stress is not small enough to be neglected. On the other hand, the thermal stress in the polyimide films was always developed below T_g on cooling after thermal treatment. Its variation with temperature was dependent upon the temperature regime: *regime I* (above T_g), *regime II:* (from 115 °C to T_g or T_f), and *regime III* (below 115 °C). That is, below T_g or the final baking temperature (T_f), the thermal stress does not vary linearly with temperature as one usually assumes. Only in *regime II* does the thermal stress apparently change linearly with temperature. However, even in this temperature regime the slope in the thermal stress profile is not same with that of the overall stress profile. There-fore, one should be carefully when estimating the biaxial modulus and thermal expansion coefficient of a film directly from the overall stress profile. In addition, the T_g's of the polyimides were estimated from the overall stress-tem-perature profiles.

REFERENCES

1.   Sroog, C. E. *J. Polym. Sci.: Macromol. Rev.* **1976**, *11*, 161.
2.   Mittal, K. L., Ed. *Polyimides:*; Plenum Press: New York, NY, 1984.
3.   Tummala, R. R.; Rymaszewski, E. J., Eds. *Microelectronics Packaging Hand-book*; van Nostrand Reinhold: New York, 1989.
4.   Hoffman, W.R. in *Physics of Thin Film*; Hass, G.; Thun, R. E., Eds.; Academic: New York, 1966; Vol.3, p 211, .
5.   Ree, M.; Swanson, S.; Volksen, W *ACS Polym. Preprints* **1991**, *32(3)*, 308.
6.   Ree, M.; Nunes, T. L.; Volksen, W.; Czornyj, G. *Polymer*, **1992**, *33*, 1228.
7.   Jaccodine, R. J.; Schlegel, W. A. *J. Appl. Phys.*, **1966**, *37*, 2429.
8.   Wortman, J. J.; Evans, R. A. *J. Appl. Phys.*, **1965**, *36*, 153.
9.   Timoshenko, S. *J. Opt. Soc. Am.*, **1925**, *11*, 233.
10.   Croll, S. G. *J. Appl. Polym. Sci.*, **1979**, *23*, 847.
11.   Brandrup, J.; Immergut, E. H., Eds. *Polymer Handbook*; Wiley-Intersci-ence: New York, 1975; Chapter V.
12.   Ree, M.; Chen, K. J. (unplublished results).

RECEIVED December 30, 1992

Chapter 34

Fluorinated, Soluble Polyimides with High Glass-Transition Temperatures Based on a New, Rigid, Pentacyclic Dianhydride

12,14-Diphenyl-12,14-bis(trifluoromethyl)-12H,14H-5,7-dioxapentacene-2,3,9,10-tetracarboxylic Dianhydride

Brian C. Auman and Swiatoslaw Trofimenko

Experimental Station, DuPont Electronics, P.O. Box 80336, E336/205, Wilmington, DE 19880–0336

Polyimides have experienced ever increasing use in the electronics industry due to their excellent combination of thermal, mechanical and electrical properties. In recent years, fluorinated polyimides have been investigated as materials with the potential for reduced dielectric constant and moisture absorption over conventional polyimides. Fluorinated polyimides based on 6FDA, have been shown to indeed yield low dielectric constant and lower moisture absorption (1), but typically are solvent sensitive and have undesirably high coefficient of thermal expansion (CTE). Recent work in our laboratory (2) has dealt with the combination of fluorinated monomers with a rigid, quasi rod-like structure as a method for producing polyimides which have not only low dielectric constant and moisture absorption, but low CTE and reduced solvent sensitivity. This work was based on new fluorinated dianhydrides, 9,9-bis(trifluoromethyl)-2,3,6,7-xanthenetetracarboxylic dianhydride (6FCDA) and its 9-trifluoromethyl-9-phenyl-analog (3FCDA) which have a rigid, *tricyclic* structure. As an offshoot of this work, we have also been investigating analogous rigid, *pentacyclic* structures as monomers for polyimides. In this paper, we describe the synthesis and characterization of 12,14-diphenyl-12,14-bis(trifluoromethyl)-12H,14H-5,7-dioxapentacene-2,3,9,10-tetracarboxylic dianhydride (PXPXDA) and polyimides prepared therefrom.

PXPXDA

0097–6156/94/0537–0494$06.00/0

EXPERIMENTAL

Starting materials

The p-phenylene diamine (PPD) and 4,4'-oxydianiline (4,4'-ODA) were high purity commercial materials and were used as received. The diaminodurene (DAD) and 2,2'-bis(trifluoromethyl)-benzidine (TFMB) were obtained in high purity from sources within Du Pont. TFMB was sublimed under reduced pressure, then recrystallized from toluene prior to use. N-cyclohexylpyrollidinone (CHP, Aldrich) was dried and distilled from calcium hydride. N-methylpyrollidinone (NMP, anhydrous 99 + % grade, Aldrich) and tetrachloroethane (TCE, Kodak) were used as received.

12,14-Diphenyl-12,14-bis(trifluoromethyl)-12H,14H-5,7-dioxa-2,3,9,10-tetramethylpentacene (TMPXPX)

A mixture of 91.4 g (0.287 mole) 1,3-bis(3,4-dimethylphenoxy)benzene prepared as in (*3*), 100 g (0.575 mole) trifluoroacetylbenzene, and 180 g (9 moles) HF was heated in an autoclave for 8 hrs at 130 °C. After venting excess HF, the contents were transferred into a polyethylene jar containing ice-water slurry and 300 ml 50% NaOH. The product was extracted with CH_2Cl_2, the extracts were filtered through alumina, and stripped. The residue was stirred with a methanol/acetone mixture, and filtered, yielding 87 g (50%) of creamy solid, which was recrystallized from toluene. The initial melting range was wide (sintering around 245 °, melting 255-260 °C), implying presence of cis-trans isomers. This mixed product was used for subsequent reactions. It was possible to isolate a single isomer by successive recrystallizations from toluene. M.p. = 287-289 °C. NMR: m 7.07, s 6.98, s 6.94, s 6.55, s 6.17, s 2.25, s 2.06 in 10:2:1:2:1:6:6 ratio. Analysis: Calc. for $C_{38}H_{28}F_6O_2$: C 72.4; H 4.44; F 18.1; Found: C 72.5; H 4.88; F 18.6 %.

12,14-Diphenyl-12,14-bis(trifluoromethyl)-12H,14H-5,7-dioxapentacene-2,3,9,10-tetracarboxylic Dianhydride (PXPXDA)

The oxidation of TMPXPX was carried out on a 31.5 g (0.05 mole) sample as described in (*4*). The tetraacid, PXPXTA, was isolated by filtration after acidification of the final oxidation filtrate with sulfuric acid, and was washed thoroughly with water. It was boiled briefly in 500 ml acetic anhydride, filtered through Celite, and stripped. The product was recrystallized from a mixture of anisole and acetic anhydride (300ml/50 ml for 35 g of crude product), using Darco, and filtering through Celite. After drying, the dianhydride exhibited a slightly broadened m.p. at about 340 °C owing to the mixture of isomers. This sample was considered pure as checked by elemental analysis and NMR and was used for polymerization studies. It was possible to isolate a single isomer by successive recrystallizations. NMR of single isomer: s 7.84, broader s 7.57, m 7.3-7.4, s 7.24, broad d 7.17 and broad s 6.37 ppm in 2:2:6:1:4:1 ratio. Calc. for $C_{38}H_{14}F_6O_8$: C 64.0; H 1.97; Found: C 64.0; H 2.25. The second isomer was difficult to isolate due to its high solubility but a fraction containing ~ 65% was obtained (additional proton peak at 6.26 ppm).

PXPXDA was also characterized as the p-tolyl, and n-butyl diimides, by stirring a solution of PXPXDA in THF with two equivalents of the appropriate amine for one hour. After evaporation of the solution, the residue was heated in vacuo at 240 ° for ten minutes. The crude product was dissolved in methylene chloride, filtered through alumina and, after evaporation of the solvent, was purified by recrystallization from toluene. Identity of these derivatives (only single isomer isolated) was established by NMR:

PXPX-bis(p-tolyl)imide: s 7.78, s (broad) 7.52, m 7.3-7.1, including a sharp spike at 7.20, s (broad) 6.35 ppm in 2:2:19:1 ratio. IR: 1725 and 1775 (vs) cm^{-1}. PXPX-bis(n-butyl)imide: s 7.67, s (broad) 7.38, m 7.30, d 7.16, with a sharp spike at 7.15, s (broad) 6.34, t 3.65, m 1.64, m 1.35 and t 0.94 ppm in the correct 2:2:6:(4 + 1):1:4:4:4:6 ratio. IR: 1720, 1775 (vs) cm^{-1}; M.p. 291-294 °C.

Polymerization

Two methods were used to synthesize the polyimides: synthesis of the poly(amic acid) with subsequent coating and thermal cure to the polyimide, or direct synthesis of the polyimide in one pot by solution imidization (5). These are illustrated in Scheme 1. Representative examples of both these methods are as follows:

Poly(amic acid), PAA. Into a 100 ml reaction kettle fitted with a nitrogen inlet and outlet and a mechanical stirrer were charged 6.2488 g (8.746 mmol) of PXPXDA and 1.7512 g (8.746 mmol) of 4,4'-ODA followed by 32 ml of NMP. Both monomers dissolved quickly and after overnight stirring under nitrogen at RT, the solution built to a moderately high viscosity. Subsequently, the solution was pressure filtered through a 1 micron filter in preparation for spin coating.

Soluble Polyimide, PI. Similar quantities of the same monomers as above were weighed into a similar apparatus followed by 48 ml of NMP and 12 ml of CHP. After stirring for several hours at room temperature under nitrogen, a Dean-Stark trap with condenser was added and the reaction mixture was raised to a temperature of 180-190 °C and maintained overnight (~ 16 hrs.) to imidize the polymer. A viscous, homogeneous solution resulted at high temperature which remained so upon cooling. This polyimide solution was pressure filtered through a 1 micron filter in preparation for spin coating.

Film Preparation

Films were prepared by spin coating the filtered poly(amic acid) or poly-imide solution onto 5″ silicon wafers containing 1000 Å of thermally grown oxide on the surface, followed by drying at 135 °C for 30 minutes in air, and then heating under nitrogen to 200 °C (2 °C/min) and holding for 30 minutes followed by heating to 350 °C (2 °C/min) and holding for 1 hour. Free standing films of about 10 μm thickness (goal) were obtained by etching the oxide layer of the silicon wafer in dilute aqueous HF to release the film.

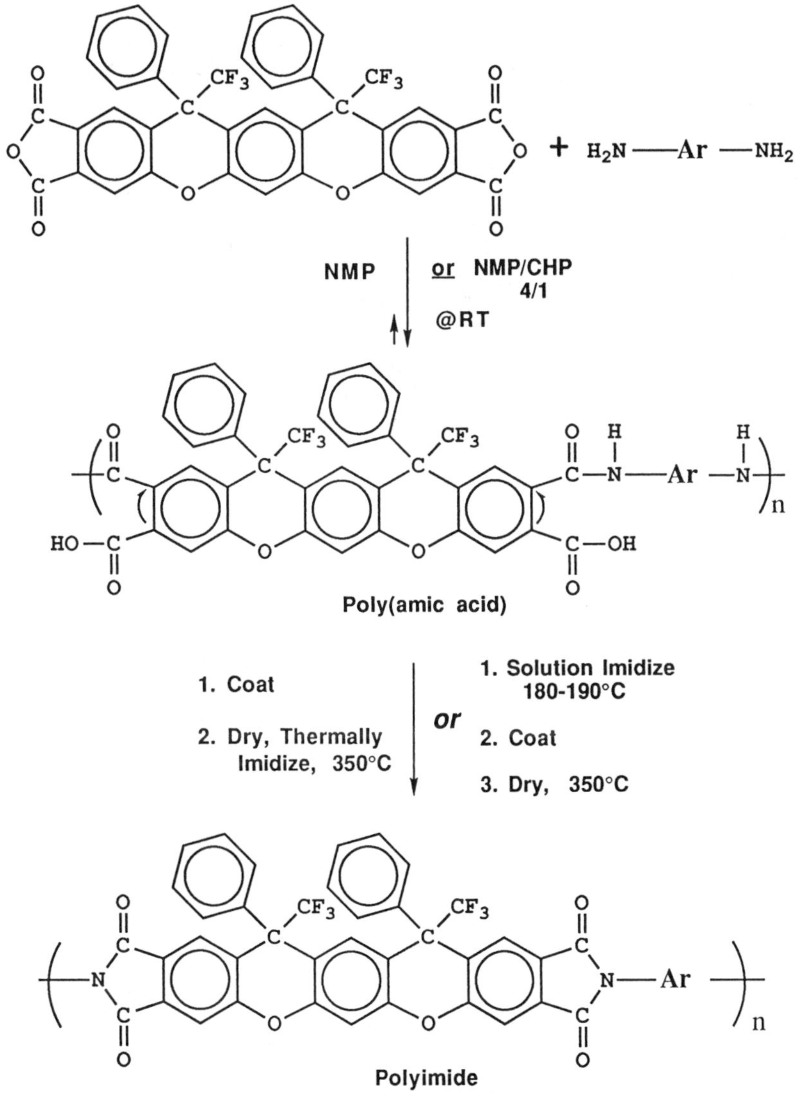

Scheme 1. Synthesis of poly(amic acid)s and soluble polyimides based on PXPXDA H$_2$N-Ar-NH$_2$ represents aromatic diamines such as those listed in Table 1

Techniques

Gel permeation chromatography (GPC) was performed either on a Waters GPC 2 at 35 °C with 4 linear Phenogel columns in the DMAC/LiBr/H_3PO_4/THF solvent system (6), or on a Waters instrument (150C) with a Zorbax TMS precolumn and 2 Shodex AD80M/S columns in DMAC at 135 °C. Flow rate was 1 ml/min, detection was by RI, and calibration was based on polystyrene standards. Mechanical properties of the films were measured in accordance with ASTM D-882-83 (Method A) on an Instron model 4501 tensile tester (crosshead speed = 0.2″/min). Linear coefficient of thermal expansion (CTE) was obtained from a P-E TMA-7 thermo-mechanical analyzer (5 °C/min, -10 to 225 °C, 30 mN tension). The value (0–200 °C) was recorded after an initial conditioning step (heat to 250 °C, hold 5 min, cool). The onset of weight loss and the temperature of 5% weight loss in air were measured on a Du Pont 951 TGA at 15 °C/min from 50 to 600 °C. The measurements were taken after an initial 150 °C/5 min. drying step. Glass transition temperatures (T_g) were obtained from a Rheometrics RSA-II dynamic-mechanical analyzer in tension (freq = 10 rads). Dielectric constant was measured by the parallel plate capacitor method in the frequency range 10 KHz–10 MHz on thin (10–20 μm) films. Gold electrodes were vacuum deposited on both surfaces of dried films, followed by thorough drying (min. 48 hrs.) at 150 °C under vacuum/N_2 prior to measurement in a sealed humidity chamber at 0% RH. Moisture absorption measurements were made by the Quartz Crystal Microbalance technique (QCM) (7, 8) on thin ($\sim$ 3 μm) films spin coated and cured (as above) onto electroded quartz crystals. Measurements were taken at various humidity settings in a controlled humidity chamber and are reported at 85% RH.

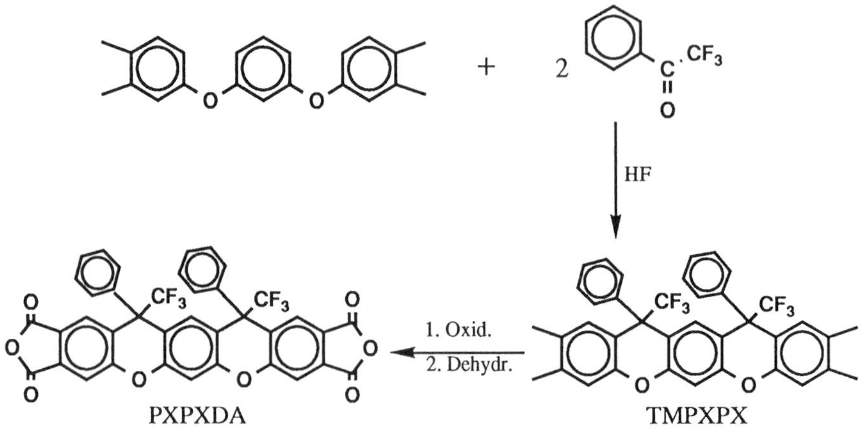

Scheme 2. Synthesis of PXPXDA

RESULTS AND DISCUSSION

The synthesis of PXPXDA was similar to that of 3FDA (*2*), involving double bridging condensation of 1,3-bis-(3,4-dimethylphenoxy)benzene with trifluoroacetylbenzene, using HF as solvent and catalyst (Scheme 2). The sequential double ring closure to form TMPXPX was more difficult than the case of a single ring closure, in the reaction of 3,4,3′,4′-tetramethylphenyl ether, as manifested by lower product yield in the present reaction. The reason for this is that after the first -C(Ph)(CF$_3$)- bridge is formed at the 4- or 6-positions of the central ring, the second ring closure may take place either at the 2- or 6-positions, only one of which yields the desired product. The by-product arising from condensation at the 2-position may have been present in the reaction mixture, but it was not isolated.

The oxidation of TMPXPX and dehydration of the resulting tetraacid, PXPXTA, to the dianhydride, PXPXDA proceeded easily, and in good yields. While the pentacyclic ring system in an analog of TMPXPX, containing -C(CF$_3$)$_2$- bridging units instead of -C(Ph)(CF$_3$)- units, was perfectly planar (and crescent-shaped), as was demonstrated by an X-ray crystallographic structure determination (*9*), TMPXPX, and PXPXDA were obtained were obtained as a mixture of two isomers. Both were crescent-shaped (due to the difference of C-O and C-C bond lengths: 1.36 and 1.54 Å, respectively), so that the imide links derived from this dianhydride would deviate by 22 ° from linearity. In addition, they exhibited folding along the O⁻C(Ph)(CF$_3$) axis . The reason for this is the steric bulk of the CF$_3$ group which strives to remain outside the molecular fold, while the planar phenyl group fits inside the fold. In this manner, the pentacyclic *trans*-system has an over-all S-shape, while the *cis*-isomer has a C-shape (schematically illustrated in Figure 1). The same *cis*- and *trans*-isomers were found in the parent pentacyclic heterocycle, 12,14-diphenyl-12,14-bis(trifluoromethyl)-12H,14H-5,7-dioxapentacene, obtained by copper-catalyzed decarboxylation of PXPXTA. According to GC/MS data, the major isomer is *trans*.

Table 1 shows the GPC molecular weight characterization of PAAs and PIs prepared from PXPXDA and several diamines. Although the molecular weight values were only relative (based on polystyrene), they were comparable on a relative basis with commercial PMDA/ODA (PI-2540) and thus indicated acceptable monomer purity. All samples listed gave good quality, creasable films with the exception of the PAA with PPD which under the chosen processing conditions gave a brittle film. Although most of the PIs were soluble in NMP at RT, TCE was sometimes chosen as coating solvent due to lower hygroscopicity vs. NMP. Interestingly, while the PI with rigid PPD precipitated from solution upon solution imidization, PIs with rigid DAD and TFMB were fully *soluble*, despite the fair amount of extended chain character of the backbone. This is likely due to the combination of bulky phenyl and CF$_3$ groups, *cis/trans* isomerism which changes the conformation, and the 22 ° bend within the PXPXDA unit, along with the solubilizing groups of the diamines. A similar solubility effect was noted with structurally similar 3FCDA (*2, 10*) containing polyimides. The solubility was even further enhanced with PXPXDA. The

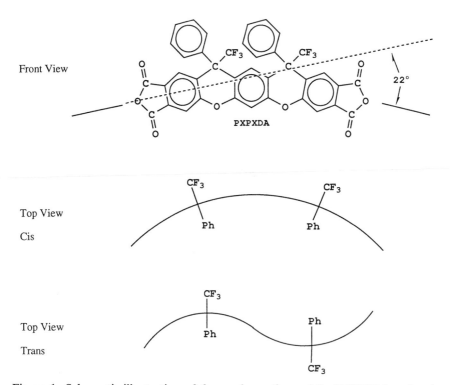

Figure 1. Schematic illustration of the conformations of the PXPXDA molecule

Table 1. Molecular Weight Characterization and Film Forming Performance
of PXPXDA-Based Poly(amic acid)s and Soluble Polyimides

Diamine	PAA or PI	%F	Mn	Mw	Mw/Mn	Film Qual./ Coating Solv
4,4'-ODA	PAAA	13.0	44000	90900	2.02	Good/NMP
4,4'-ODA	PI	13.0	68400	190000	2.78	Good/NMP
PPD	PAA	14.5	53600	108000	2.02	Brittle/NMP
PPD	PI	14.5	ppt.	—	—	—
DAD	PI	13.5	*38100	*110000	*2.87	Good/TCE
TFMB	PI	22.8	68400	181000	2.66	Good/TCE
PMDA/ODA	PAA	0.0	50800	125000	2.48	Good/NMP

*GPC in DMAC @135 °C, others in DMAC mixture at 35 °C ppt. = precipitated upon solution imidization

combination of a very stiff structure with solubility is unusual in polyimides and in polymers in general. The fact that the PXPXDA molecule had such a stiff but irregular structure allowed this unusual combiniation of properties. It is likely that other rigid diamines, even those with less bulky substituents than DAD or TFMB, would also give soluble structures.

Table 2 shows the characterization of $\sim$ 10 μm films prepared from the samples in Table 1. Mechanical properties were found to be typical of polyimides with a higher modulus and lower elongation for the stiffer DAD and TFMB materials. CTEs of the more extended chain DAD and TFMB materials were also advantageously lower than 20 ppm/°C while the ODA materials showed higher CTEs typical of increased chain flexibility. A stiff, more extended chain structure allows for better orientation during the film forming process and thereby lower in-plane CTE. CTE's lower than 20 ppm/°C are very desirable in order to match those of copper and ceramic substates often used in electronic devices. There are very few examples, if any, noted in the literature of soluble PIs which give low in-plane CTE films. The 3FCDA-based PIs were another of

Table 2. Characterization of PXPXDA-Based Polyimide Films

Diamine	Film Thick. μm	Tens. Str. MPa	% Elong.	Mod. GPa	CTE ppm/°C	%H2O abs. @ 85% RH	Dielec. Const. dry @ 1 MHz
4,4'-ODA*	12.2	123	34	1.9	46	1.3	2.6
4,4'-ODA	11.1	133	32	2.0	34	na	2.8
DAD	10.5	135	9	2.6	17	3.4	2.5
TFMB	11.2	204	18	4.5	10	na	2.6
PMDA/ODA*	12.2	168	82	1.3	31	3.5	3.2

*prepared from PAA solution, others from soluble PI solution

Table 3. Thermal Characterization of PXPXDA-Based Polyimide Films

Diamine	Tg-DMA E″ max	Td-onset in air	Td-5% wt. loss
4,4'-ODA*	439	421	485
4,4'-ODA	435	419	485
DAD	388	408	446
TFMB	436	430	486
PMDA/ODA*	425	454	565

*prepared from PAA solution, others from soluble PI solution

these examples (2, 10), while those polyimides reported by Hsu and Harris (11) may also yield low CTEs in films (film CTE data not reported). The stiff, orientable nature of these materials may also prove useful for the preparation of high modulus, high strength fibers, while the unique structure/conformations of the monomer may make useful membranes.

Dielectric constants were all notably lower than PMDA/ODA, and moisture absorption for the PXPXDA-ODA material was very low, while that based on DAD was fairly high (possibly due to the free volume effect).

Table 3 gives the thermal characterization of the various samples. The very stiff structure of PXPXDA gave very high Tg. A typical DMA trace is given in Figure 2. It should be noted that the modulus curve was very flat and that the Tg occurred in the range of the onset of weight loss, such that the modulus changed little over its entire useful temperature range. At Tg, the modulus drops very little owing to the likely onset of crosslinking reactions that occur at this temperature. On the whole, the TGA thermal stability in air of these polyimides, although high, was somewhat less than that of the conventional PMDA/ODA polyimide.

Figure 3 shows a typical ATR-FTIR spectrum of the PXPXDA-based polyimide film indicating the standard imide bands at about 1720 and 1780 cm⁻¹. In Figure 4, a typical proton NMR spectrum (CDCl₃) is presented with assignments. Interestingly, signal c was split into two resonances due to the cis/trans arrangement of phenyl and CF₃ groups of the PXPXDA unit, both isomers being present in the monomer used for polymerization. The integrals indicated that one isomer (the more insoluble one, and probably *trans*-) dominated (2/1) the monomer sample. Also of interest was the fact that the methyl groups of the DAD unit existed as an apparent doublet, possibly due to restricted mobility of the very stiff chains, leading to non-equivalent environments. A similar effect was noted for related 3FCDA polyimides (2, 10).

CONCLUSIONS

A new fluorinated rigid dianhydride has been prepared which has yielded soluble, high Tg polyimides. When this dianhydride was paired with rigid "rod-like" diamines, low thermal expansion was realized in films and an unusual combination of solubility and low thermal expansion was found. In addition to

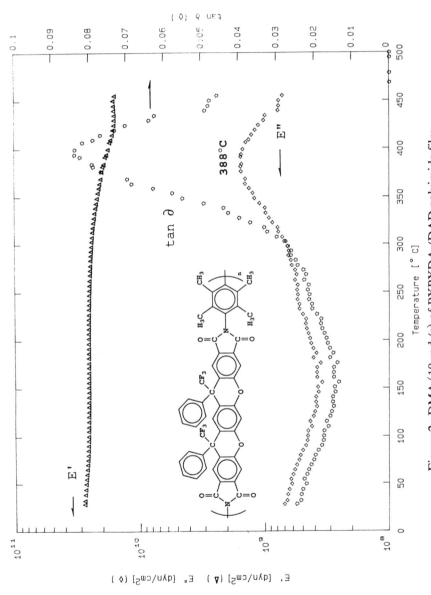

Figure 2. DMA (10 rad/s) of PXPXDA/DAD polyimide film

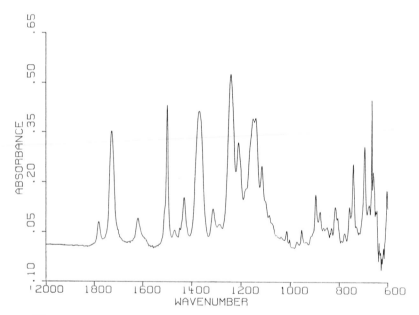

Figure 3. ATR-FTIR spectrum of PXPXDA/ODA polyimide film

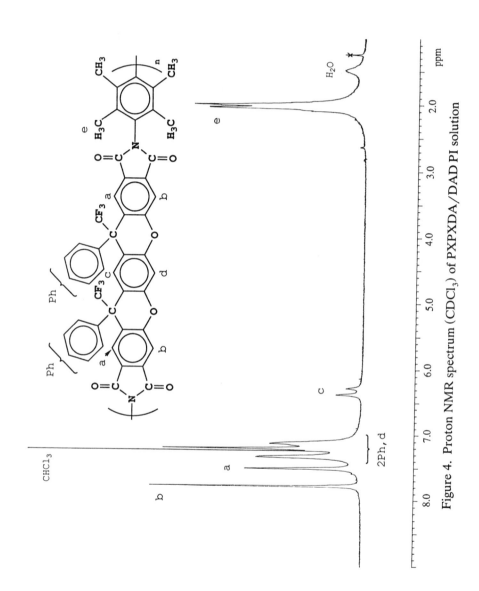

Figure 4. Proton NMR spectrum (CDCl$_3$) of PXPXDA/DAD PI solution

this, the polyimides exhibited low dielectric constant and in some cases low moisture absorption.

ACKNOWLEDGMENTS

The authors wish to thank M. J. Grovola, M. J. Enderle, M. Pottiger, R. Subramanian, J. Coburn, E. McCord, R. Pryor, M. Panco, L. Watson, G. Young, D. Oranzi, as well as Du Pont Analytical personnel for their support.

REFERENCES

1. Goff, D. L.; Yuan, E. L.; Long, H; Neuhaus H. J., In *Polymers for Electronic Packaging and Interconnection*; Lumpinski, J. H.; Moore R. S., Eds.; ACS Symp. Ser. 407; American Chemical Society: Washington, DC, 1989; pp. 93-100.
2. Auman, B. C., *Proc. of the 4th Intl. Conf. on Polyimides*, Ellenville, NY, Oct. 30–Nov. 1, **1991**. p. I-5; S. Trofimenko, ibid. p. I-3; also see *Advances in Polyimide Science and Technology*; C. Feger, M. Khojasteh, M. Htoo, Eds.; Technomic Publ.: Lancaster, PA, 1993; pp 3, 15.
3. Koton, M. S.; Florinskii, F. S., *Zh. Org. Khim.* **1968**, 4, 774.
4. Marvel, C. S.; Rassweiler, J. H., *J. Am. Chem. Soc.*, **1958**, 80, 1197.
5. Summers, J. D.; McGrath, J. E., *Polym. Prepr.*, **1987**, 28(2), 230.
6. Walker, C. C., *J. Polym. Sci., Polym. Chem. Ed.*, **1988**, 26, 1649-1657.
7. Subramanian, R.; Pottiger, M. T.; Ward, M. D. *"Proc. of the Symp. on Recent Adv. in Polyimides and Other High Performance Polymers"*, Am. Chem. Soc., Div. Polym Chem., San Diego, CA, Jan. 22-25. **1990**.
8. Moylan, C. R.; Best, M. E.; Ree, M., *J. Polym. Sci., Polym. Phys. Ed.*, **1991**, 29, 87.
9. Calabrese, J. C.; Trofimenko, S., unpublished work.
10. Auman, B. C., Trofimenko, S., Polym. Prepr., **1992**, 33(2), 244.
11. Harris, F. W.; Hsu, S. L-C., High Perform. Polym. **1989**, 1, 3; *"Proc. of the Symp. on Recent Adv. in Polyimides and Other High Performance Polymers"*, ACS, Div. Polym Chem., San Diego, CA, Jan. 22-25 **1990**.

RECEIVED December 30, 1992

Chapter 35

Processable Fluorinated Acrylic Resins with Low Dielectric Constants

Henry S.-W. Hu[1] and James R. Griffith[2]

[1]Geo-Centers, Inc., 10903 Indian Head Highway, Fort Washington, MD 20744
[2]Naval Research Laboratory, 4555 Overlook Avenue, Washington, DC 20375

The preparation of a new class of processable heavily fluorinated acrylic resins with very low dielectric constants is described. The title compounds 2 and 5 were prepared through the condensation of the respective alcohols 1 and 4 with acryloyl chloride. Unlike tetrafluoroethylene, monomers 2 and 5 are easy to process into polymers under normal conditions due to their liquid or semisolid nature. Radical polymerization of the title compounds with a trace amount of azobisisobutyronitrile or methyl ethyl ketone peroxide at 85-100 °C leads to homopolymers 3 and 6 and copolymer 7. All polymers exhibit dielectric constants around 2.10-2.24 over a frequency region of 500 MHz to 18.5 GHz; the variation of dielectric constant values over the measured frequency region is within 0.03 for each polymer. These values are very close to the minimum known dielectric constants of 2.0-2.08 for poly(tetrafluoroethylene) and 1.89-1.93 for a terpolymer of 2,2-bis-(trifluoromethyl)-4,5-difluoro-1,3-dioxole 8, perfluoropropylene and tetrafluoroethylene 9. The dielectric constants for poly(tetrafluoroethylene) measured with the same method are observed to be around 1.96-1.99 in order to validate the accuracy of our measurement.

Materials exhibiting dielectric constants below 3 are in increasing demand in the aerospace and electronic circuit industries (1). Dielectric constant is defined as a measure of the ability of a dielectric to store an electric charge. A dielectric is a nonconducting substance or an insulator. Dielectric constant is directly proportional to the capacitance of a material, which means that the capacitance is reduced if the dielectric constant of a material is reduced. For high-frequency, high-speed digital circuits, the capacitances of substrates and coatings are critical to the reliable functioning of the circuits. Present computer operations are limited by the coupling capacitance between circuit paths and integrated circuits on multilayer boards since the computing speed between integrated circuits is reduced by this capacitance and the power required to operate is increased (2). The relationship of capacitance C with dielectric constant K_s can be expressed

0097–6156/94/0537–0507$06.00/0

as $C = AK_s \varepsilon_o/d$ where A is area, d is distance, $\varepsilon_o = 8.85418 \times 10^{-14}$ F/cm and $K_s = \varepsilon_s/\varepsilon_o$. The maximum frequency of operation of the field-effect transistor can be given by the frequency corresponding to the dielectric constant in the reverse proportion (3).

Reductions in such parasitic capacitance can be achieved in a number of ways through the proper selection of materials and the design of circuit geometry. In 1988 St. Clair et al. reported a reduction of dielectric constant to 2.39 by chemically altering the composition of a polyimide backbone to reduce the interactions between linear polyimide chains and by the incorporation of fluorine atoms (4).

Poly(tetrafluoroethylene) (PTFE), which is also known as "Teflon" in trade and is a solid at room temperature, has a dielectric constant (5) in the range of 2.00-2.08 while its monomer, tetrafluoroethylene, is a gas at room temperature. Poly(tetrafluoroethylene) is exceptionally chemically inert, has excellent electrical properties, has outstanding stability, and retains mechanical properties at high temperatures. The problem with poly(tetrafluoroethylene) is that it is not processable. A family of commercial polymeric materials known as "Teflon AF" in trade is believed to be a terpolymer of tetrafluoroethylene 9, perfluoropropylene and 2,2-bis(trifluoromethyl)-4,5-difluoro-1,3-dioxole 8(a derivative of hexafluoroacetone) is reported to have a dielectric constant in the range of 1.89-1.93. It is reported to be more processable than poly(tetrafluoroethylene) (6).

A series of studies have been carried out in our laboratory to design and synthesize new fluorinated epoxy and acrylic resins. These materials have many applications because of their unique properties (7). In this paper we report the preparation of a series of new processable heavily fluorinated acrylic resins which exhibit dielectric constants as low as 2.10, very close to the minimum known values.

Experimental Section

The synthetical routes used to obtain the monomers and polymers are outlined in Scheme I. Details of the synthesis will be reported elsewhere.

Preparation of polymer "cylindrical donuts" from monomers

In order to prepare the samples for dielectric constant measurements, etherdiacrylate 2 was mixed with a trace amount of AIBN at room temperature in a cylindrical donut mold made from General Electric RTV 11 silicon molding compound. Donuts had an outer diameter of 7.0 mm, inner diameter of 3.0 mm and a thickness of 3.0 mm; the semisolid triacrylate 5 was mixed with a trace amount of liquid methyl ethyl ketone peroxide (MEKP) with some heating to obtain a clear liquid; equal masses of 2 and 5 were also mixed with MEKP with some heating.

For polymerization the filled donut molds were kept under an inert atmosphere, the temperature was raised to 85 °C over 2 hr, then kept at 85-100 °C for 20 hr. Two homopolymers 3 (from 2) and 6 (from 5), and one copolymer 50/50 (w/w) 7 were obtained.

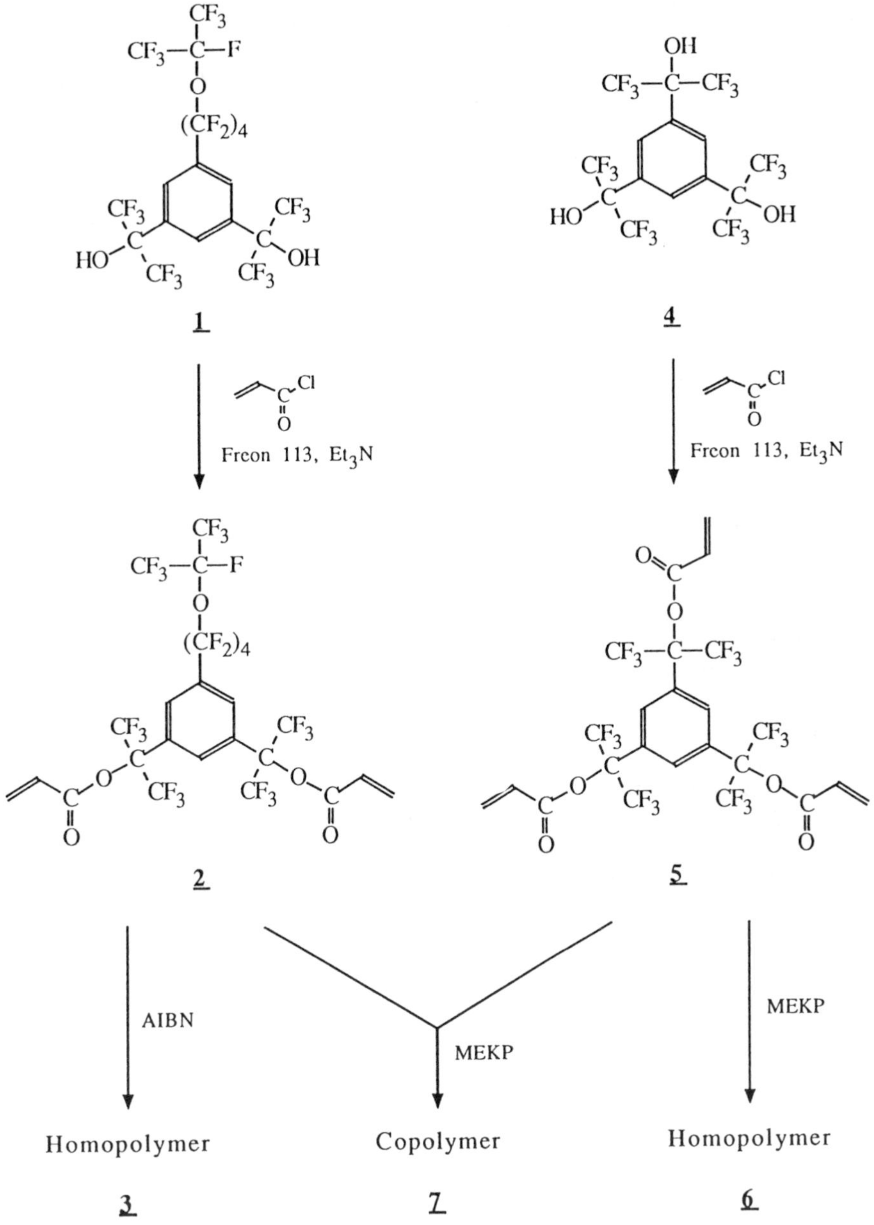

1
4
Freon 113, Et3N
Freon 113, Et3N
2
5
AIBN
MEKP
MEKP
Homopolymer
Copolymer
Homopolymer
3
7
6

Scheme I.

Dielectric constant (DE) measurements

Dielectric constant values are reported as "Permittivity" with the symbol ε or K_s (Table I). The polymer cylindrical donuts were used for the measurement of DE on a Hewlett Packard 8510 Automated Network Analyzer. The analyzer is capable of measuring 401 data points over a frequency band of 500 MHz to 18.5 GHz. Typically S11 and S21 values, which correspond to reflection and transmission respectively, are measured and then these values are used to calculate the permittivity and permeability.

Samples stood at room temperature in air prior to testing; measurements were run at room temperature and approximately 25% RH. A virgin PTFE sheet (MMS-636-2) was obtained from Gilbert Plastics & Supply Co. and cut into the same cylindrical donut size. The result for PTFE reported here is the average of DE values measured for three samples.

RESULTS AND DISCUSSION

Monomers

The preparation of etherdiol *1* and triol *4* was carried out in multistep routes according to the reported procedures (*8, 9*). As summarized in Scheme I, the synthesis of acrylic esters, *2* and *5*, was carried out in a fluorocarbon solvent such as Freon 113 by the reaction of alcohols, *1* and *4*, with acryloyl chloride and an amine acid acceptor such as triethylamine. Other attempts to esterify the fluoroalcohols directly with acrylic acid or acrylic anhydride were not successful (*10*). Product purification by distillation was not feasible because of the temperature required, but purification by percolation of fluorocarbon solutions through neutral alumina resulted in products, *2* and *5*, of good purity identified by TLC, proton-1 NMR and IR.

In proton-1 NMR spectra, both *2* and *5* showed a characteristic ABX pattern for acrylate in the region of δ 6.8-6.0 with a pair of doublet couplings for each vinyl proton; the products purified by percolation over alumina contained no detectable hydrate water or polymerized impurities. After standing in air at

Table I. Summary of Data for Dielectric Constant Measurements[a]

samples	frequency (GHz)			
	0	3.5	9.5	15.5
3	2.11	2.10	2.12	2.13
6	2.23	2.23	2.23	2.22
7	2.22	2.23	2.24	2.22
PTFE[b]	1.96	1.98	1.99	—

[a] See Experimental Section for detail.
[b] Poly(tetrafluoroethylene).

Table II. Summary of Data for Hydrate Water Relationship

samples	^{1}H NMR $\delta(CDCl_3/TMS)$ $-$ OH	$(H_2O)_n$	$(H_2O)_n$ n	functionality ratio vinyl or $-$OH vs $(H_2O)_n$
1	4.14	2.13	0.5	2:1
2		1.57	1.0	4:2
4	4.44	2.11	2.25	3:4.5
5		1.56	1.5	6:3

room temperature, both alcohols *1, 4* and acrylates *2, 5* are found to contain hydrate water as shown in Table II with the chemical shifts and integral ratios. Only the alcohol *4* was reported (*8, 9*) to be a hygroscopic solid but gave anhydrous product by distillation or sublimation. It is interesting to note that the etherdiacrylate *2* contained twice the amount of hydrate as the etherdiol *1*, while the triacrylate *5* contained only two-thirds the amount of hydrate as the triol *4* because in the case of the acrylates, there are equivalent hydrates for both *2* and *5*.

The monomers *2* and *5* are stable at room temperature, and no change can be detected by use of IR and NMR after storing in air for three months.

Polymerization

Due to their liquid or semisolid nature, monomers *2* and *5* are easy to process into polymers. For radical polymerization the use of solid azobisisobuty-ronitrile (AIBN) for liquid *2* at room temperature and liquid MEKP for semisolid *5* or a mixture of *2* and *5* with some heating is convenient. During the course of curing at 85-100 °C for 22 hours the problem of surface inhibition of free radicals by oxygen of air can be avoided by inert gas blanketing.

Since each acrylate group is difunctional, the etherdiacrylate *2* is tetrafunctional while the triacrylate *5* is hexafunctional. For polymerization of the polyfunctional monomers at sufficiently high degrees of conversion, the branching must result in the formation of cross-links to give a three dimensional network. As expected, the homopolymers *3, 6* and the 50/50 (w/w) copolymer *7* are semitransparent, hard solids, and some shrinkage in volume is observed during curing.

The fluoroacrylic polymers are high modulus, low-elongation plastics which are brittle in the sense that all thermosetting polymers are brittle. However, they are tough, rugged materials not easily damaged by impact or mechanical abuse. In an undercured state they are frangible, but when totally cured they acquire a more resilient character.

The degree of polymerization or conversion of *2* or *5* can be easily monitored with fourier transform infrared spectrophotometer, for example, by examining the intensity of the absorption frequencies at 1635 and 1410 cm^{-1} which are assigned to the acrylate functional groups.

Dielectric constant (DE)

In order to validate the accuracy of our measurements, three samples of virgin PTFE in the same cylindrical donut size were measured and found to have average DE values around 1.96-1.99, which are close to the reported values 2.0-2.08 (5). Summarized results of the DE measurement on the polymer donuts *3*, *6*, *7*, and PTFE are shown in Table I, while the observed measurement curves are shown in Figure I. All the new polymers exhibit unusually low Des around 2.10-2.24 over a wide frequency region of 500 MHz to 18.5 GHz; the variation of DE values over the measured frequency region is within 0.03 for each polymer. The 50/50 (w/w) copolymer *7*, having a molar ratio 45/55 of *2* and *5*, showed DE of 2.22 which is close to that of *6*; this indicates that the copolymer *7* was not completely homogeneous and compatible.

The wide range frequency independence of these low Des and the processability of these monomers suggest many potential applications. In particular, it is known that the addition of a fluorine-containing group to the polymer backbone will reduce the polymer chain-chain electronic interactions, resulting in a reduction of DE as reported by St. Clair *et al.* (*4*), but the DEs for *3* (fluorine content 57%) are at most 0.13 below those for *6* (fluorine content 46%), which may indicate that a minimum value might have been nearly reached regarding the structure/property relationship.

In 1990 Resnick (*6*) reported that the lowest dielectric constant (1.89-1.93) of any polymer thus far was obtained from copolymers of 2,2-bis(trifluoromethyl)-4,5-difluoro-1,3-dioxole *8* and tetrafluoroethylene *9* with structures shown in Scheme II. In addition to the effect of the fluorine content, the existence of similar perfluoroalkyl ether linkage on the double bonds in both monomers *8* and *2* seems to play a important role in further reducing the dielectric constant values.

It has been reported (*11*) that in the preparation of polyimides the more flexible *meta*-linked diamine systematically gave lower DE values than the corresponding *para*-linked system and that this may be related to free volume in the polymer since the *meta*-substituted systems should have a higher degree of entropy. This *meta* effect may be somewhat responsible for the low DE values for the fully 1,3,5-substituted *3*, *6* and *7*.

St. Clair *et al.* (*4*) also observed that the low dielectric constant polyimides containing the -CF$_3$ group show excellent resistance to moisture, exhibiting almost PTFE-like behavior. Since PTFE contains 76% fluorine, it appears that the DE values for the heavily fluorinated *3*, *6* and *7* with fluorine content around 46-57% will be humidity-independent. It is known that water has a very high DE of 79.45. The low DE values for *3*, *6* and *7* suggest a very low water content after polymerization of the hydrated monomers.

Processibilty and Applications

The liquid or low melting solid monomers *2* and *5* can be cured to the solid state by incorporating therein a curing catalyst and heating the mixtures below the decomposition temperature thereof. Moreover the cured solids *3*, *6*, and *7*

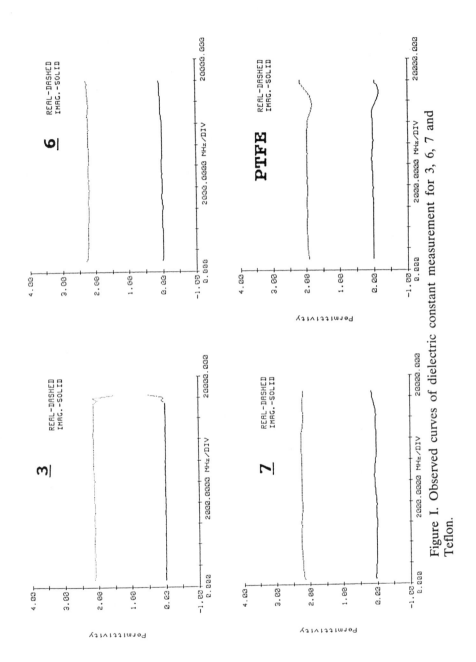

Figure I. Observed curves of dielectric constant measurement for 3, 6, 7 and Teflon.

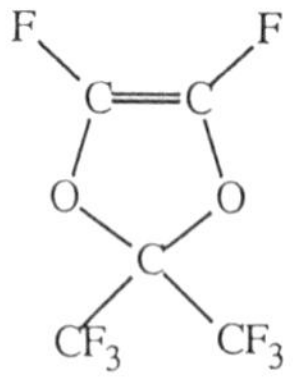

8    **9**

Scheme II.

are transparent and hard polymers formed of three dimensional networks and reasonable thermostability.

In the course of curing, the viscosity varies from thin to syrupy liquids which is favorable for use to impregnate reinforcing materials, such as fiber glass scrim, used in making wiring boards or circuit boards or other electronic components used in electric or electronic applications.

Dielectric constants of these materials can be further lowered by known means such as by incorporating air bubbles in the materials, or by inhibiting the crystallization. A difference of a couple of hundredths in the dielectric constant value may be important when one is at the low extremes thereof. Recently Singh et al. calculated the dielectric constants of polyimide films from the measured free volume fraction and found that the calculated values for the dielectric constants are close to the experimental results (12). The calculation was based on the relation: $1/\varepsilon = (1 - f)/\varepsilon_R + f/\varepsilon_{Air}$ where ε_R is the value of ε for zero free volume fraction which can be obtained from the plot of ε vs f.

Although a composite electronic board may contain a fibrous reinforcement for dimensional stability and strength, the electrical interactions between conductors occur on such a small scale that only the resin needs to be between the interacting units. Thus, only the dielectric constant of the impregnating resin would be of consequence and not that of the full composite which may be high if glass fiber reinforcement is employed.

A compound containing two or three unsaturated groups would be expected to polymerize to a crosslinked, solid, and non-linear thermosetting polymer. Fluorinated compounds which have one unsaturated group can be used in a mixture with 2 or 5 to polymerize to a solid polymer of a thermoset nature.

Lower Dielectric Constants?

In 1991 Groh and Zimmermann (13) estimated the theoretical lower limit of the refractive index of amorphous organic polymers by using the Lorenz-Lorentz equation and reported the lower limit to be very close to 1.29, while in 1979 Dislich (14) proposed a lower limit of about 1.33 from a screening of published polymer data.

The amorphous terpolymer of 9, perfluoropropylene and 8 is reported with the lowest refractive index in the range of 1.29-1.31 and also with the lowest dielectric constant in the range of 1.89-1.93. Furthermore, Groh and Zimmermann reported that functional groups with a high fluorine content, like CF_3 and CF_2, have the lowest refractive index contribution, the value for the ether group is also remarkably low, while the values for the carbonyl and carboxyl groups are high. In view of the good agreement between the refractive index and the dielectric constant on amorphous organic polymers, the opportunity to obtain the dielectric constants in the range of 1.89-2.10 by modifying our synthesis is high.

CONCLUSIONS

In this work we have demonstrated that a new class of heavily fluorinated acrylic resins can be efficiently synthesized and then cured to solid form with a catalyst at elevated temperatures. These cured resins were found to have unusually low dielectric constants, which are close to the minimum known values for PTFE and its terpolymers. In contrast to tetrafluoroethylene, our monomeric fluorinated aromatic acrylates are processable under normal conditions due to the fact that they are liquids or low melting solids, and moreover are soluble in common organic solvents. A practical manipulation in making network polymers is also established. We foresee many applications of these new fluoropolymers in aerospace and electronic circuit industries.

ACKNOWLEDGEMENT

We are indebted to Mr. Jonas K. Lodge of SFA, Inc. for the dielectric constant measurements. Partial funding support from the Office of Naval Research is gratefully acknowledged.

LITERATURE CITED

1. Dorogy, W. E. Jr.; St. Clair, A. K. *Polym. Mater. Sci. Eng.* 1991, *64*, 379.
2. Licari, J. J.; Hughes, L. A. *Handbook of Polymer Coating for Electronics*; Noyes Publications: Park Ridge, NJ, 1990; p 114.
3. Grove, A. S. *Physics and Technology of Semiconductor Devices*; John Wiley & Sons: New York, NY, 1967; pp 254-255.
4. St. Clair, A. K.; St. Clair, T. L.; Winfree, W. P. *Polym. Mater. Sci. Eng.* 1988, *59*, 28.
5. see Ref. (2), pp 378-379, Table A-13: Dielectric Constants for Polymer Coatings (at 25 °C).
6. Resnick, P. R. *Polym. Prepr.* 1990, *31(1)*, 312.
7. (a) Griffith, J. R.; Brady, R. F. Jr. *CHEMTECH*, 1989, *19(6)*, 370. (b) Hu, H. S.-W.; Griffith, J. R. *Polym. Prepr.* 1991, *32(3)*, 216.
8. Griffith, J. R.; O'Rear, J. G. *Polym. Mater. Sci. Eng.* 1985, *53*, 766.
9. Soulen, R. L.; Griffith, J. R. *J. Fluorine Chem.* 1989, *44*, 210.
10. Griffith, J. R.; O'Rear, J. G. In *Biomedical and Dental Applications of Polymers*; Gebelein, C. G., Koblitz, F. F., Eds.; Plenum: New York, NY, 1981; pp 373-377.
11. St. Clair, T. L. In *Polyimides*, Wilson, D.; Stenzenberger, H. D.; Hergenrother, P. M., Eds.; Blackie: Glasgow, U.K., 1990; pp 58-78. see p 74.

12. (a) Singh, J. J.; Eftekhari, A.; St. Clare, T. L. *NASA Memorandum* 102625, 1990. (b) Eftekhari, A.; St. Clare, A. K.; Stoakley, D. M.; Kuppa, S.; Singh, J. J. *Polym. Mater. Sci. Eng.* 1992, *66*, 279.
13. Groh, W.; Zimmermann, A. *Macromolecules* 1991, *24*, 6660.
14. Dislich, H. *Angew. Chem., Int. Ed. Engl.* 1979, *18*, 49.

RECEIVED February 4, 1993

Chapter 36

Enhanced Processing of Poly(tetrafluoroethylene) for Microelectronics Applications

Charles R. Davis and Frank D. Egitto

Technology Products Division, IBM, 1701 North Street, Endicott, NY 13760

Polymers having low-dielectric constants (ε_r) are of strategic importance to the electronics industry since this physical property largely determines a device's overall signal speed and wiring density (1). In general, a lower value of ε_r allows higher wiring density and greater signal speeds. Although Teflon®-type fluoropolymers, such as poly(tetrafluoroethylene) (PTFE), offer the lowest dielectric constant values ($\varepsilon_r \simeq 2.1$) of commercially available materials, they have not been viewed as attractive candidates for microelectronic applications. This is due in part to processing difficulties, e.g., generation of high resolution features (μm range).

One means of microstructuring materials is by laser ablation. Kawamura, et al. (2) and Srinivasan and Mayne-Banton (3) were first to identify the use of high-energy ultra-violet (UV) excimer laser radiation to photo-ablate polymers. These early investigations were performed using materials that inherently absorbed radiation at the wavelength studied. For instance, ablation of poly(ethylene terepthalate) (PET) (1) and poly(methyl methacrylate) (PMMA) (2) was investigated at 193 nm. Subsequent work included ablation of materials which absorbed at longer wavelengths, such as polyimide at 308 nm (3–6).

Not all polymers have the requisite chemical functionality in their molecular structure to inherently absorb the high-energy UV photons as required for excimer laser ablation. For instance, PMMA and PTFE are not readily structured at 308 nm. However, several researchers have reported successful excimer ablation of UV-transparent materials using a technique known as "doping" (7–11). In these investigations, a low-molecular weight, highly conjugated organic compound that absorbs strongly at the excimer laser wavelength of interest is incorporated into the non-absorbing host matrix. These studies have been limited to polymers and dopants that have similar solubility charactecteristics, for example, PMMA doped with pyrene using 308 nm exposure (9).

0097–6156/94/0537–0517$06.00/0

The chemical inertness, intractability and extreme processing requirements of PTFE present a significant challenge in identifying and incorporating a suitable sensitizing dye. For some time, clean ablation of neat PTFE had only been achieved using very short pulses (e.g., femtosecond pulse width) (*12*) to provide high-intensity radiation, or using shorter wavelengths than commonly produced with commercially available excimer lasers (*13*). It is the purpose of this paper to report the successful formation of high resolution features in PTFE, using an excimer laser at 308 nm and a pulse duration of 25 ns, by doping with an aromatic polyimide. In addition to structuring behavior, the electrical characteristics, specifically in terms of dielectric constants, ε_r, of the polymer blends are discussed.

EXPERIMENTAL

A. Materials

Biphenyl tetracarboxylic acid dianhydride/phenylene diamine (BPDA-PDA) polyamic acid, 14.5% solids in N-methylpyrrolidone (NMP), TEF 30B, 60%

solids PTFE in an aqueous dispersion from E.I. du Pont de Nemours Co. and 40% dimethyl amine in water from Aldrich were used as received. Films of 50 μm Upilex-S® polyimide (BPDA-PDA) were obtained from Imperial Chemical Industries, Wilmington, DE.

B. Film Formation

An organic salt of the BPDA-PDA polyamic acid was formed by partially reacting the acid functionalities on the polyimide precursor with a suitable organic base, such as dimethyl amine. Once formed, the polyamic acid salt was freely added to the aqueous PTFE dispersion. Polymer films, $\simeq 100\mu$m thick, of PTFE containing various amounts of polyimide were prepared using a draw-down bar coater to apply the solution onto a metal substrate. Once applied, the coatings were baked in a Blue M high-temperature convection oven to remove solvent and dispersion medium and to thermally imidize the polyamic acid. Table 1 illustrates the bake/cure cycle utilized for film formation. Following the bake, the samples were laminated at high temperature and pressure to provide material sintering.

C. Excimer Laser Ablation

Ablation of polymers was performed using a Lambda Physik Model 203 MSC XeCl excimer laser that emits at 308 nm with a pulse width (FWHM) of 25 ns. A uniform spot intensity profile was obtained using a beam homogenizer built by XMR, Inc. The rectangular homogenized spot was focused onto a stainless steel mask with a 1.27 mm diameter aperture. Fluence was controlled by adjusting the homogenizer which changed the cross-sectional area of the rectangular beam. This did not alter the total energy in the beam at the mask. Fluence was determined by measuring the energy of the beam exiting the mask using a Laser Precision RjP-700 Series energy probe. The beam was then directed through a 10X set of reducing optics and focused at the polymer surface. Resulting hole diameters were $\approx$ 127 μm.

D. Analysis

5 × 10 arrays of holes were formed in the polymer and used for analysis. Ablation rates were determined from measurements of etching depth using a Sloan Dektak IIA profilometer. The reported rates are an average of five measurements obtained at randomly chosen positions. Scanning electron microscopy (SEM) was used to evaluate material structuring quality. Dielectric constant values were determined using a high frequency sweep generator coupled with an oscilloscope in a range of 500 to 1000 MHz.

RESULTS AND DISCUSSION

Dopant-induced ablation is of great significance because it provides the opportunity to process materials that do not exhibit desirable excimer laser structuring characteristics, i.e., micro-feature formation, but offer other outstanding properties, e.g., low ε_r. Like the matrix polymers of the initial doping investigations, neat PTFE is transparent to conventional excimer laser emissions. PTFE, whose chemical repeat unit is

$$-(CF_2-CF_2)-$$

has only $\sigma \rightarrow \sigma^*$ and n $\rightarrow \sigma^*$ excitations available which are high-energy (short wavelength) transitions. Sorokin and Blank reported that PTFE is highly transparent at wavelengths greater than 140 nm (*14*). This is consistent with observations of Egitto and Matienzo (*15*), Takacs et al., (*16*) and Egitto and Davis (*17*).

Aromatic polyimides are well known in advanced-technology industries as having a number of desirable properties, notably including high thermal stability (*18*). In addition, these materials have significant absorptions in the UV energy spectrum, including the longer wavelength regions, e.g., $\alpha_{308} \approx 1 \times 10^5 \mathrm{cm}^{-1}$ (*19*,

20), where α_{308} is the absorption coefficient of the material at a wavelength of 308 nm. The material's absorption coefficient, $\alpha_\lambda = A_\lambda/b$, is a convenient method to describe the interaction of emitted laser energy with the polymer, where A_λ is the wavelength-dependent absorption (dimensionless) and b is the thickness of the polymer. Thus, aromatic polyimides have several critical characteristics required to behave as an effective dopant for PTFE. However, in processing coatings of PTFE, aqueous dispersions of the fluoropolymer are frequently employed. Water is a non-solvent for aromatic polyimides and their respective precursors. However, by partially reacting the precursor's acid functionalities with a suitable organic base, e.g., an amine, Figure 1, aqueous compatibility is quickly achieved. This permits its introduction into the PTFE dispersion, resulting in an UV-sensitized fluoropolymer coating. Once processed, uniform polyimide-doped PTFE films are obtained that exhibit excimer laser structuring behavior substantially different from its neat fluoropolymer counterpart. For instance, Figure 2 shows a representative structure formed in a blend containing 5% (wt/wt) polyimide and etched at a fluence (energy per unit area per pulse at the substrate) of 12 J/cm^2. The excellent hole quality, minimal hole-entry deformation and smooth wall profile are clearly seen and are characteristics typically associated with strongly absorbing homopolymers. Figure 3 illustrates that successful ablation of PTFE occurs at several laser fluences and dopant concentrations. That is, the etch rate, μm/pulse, of 0.5%, 1.0% and 5% (wt/wt) polyimide are shown at several laser fluences. For comparison, the etch rate of Upilex-S polyimide is included. Under the conditions investigated neat PTFE was not observed to ablate. Etch rate is dependent on the depth of absorption in the polymer. This depth of absorption is typically governed by Beer's law, $A_\lambda = \varepsilon_\lambda \cdot b \cdot c$ where A_λ and b are as previously defined and ε_λ and c are the material's wavelength-dependent extinction coefficient and chromophore concentration, respectively. If α_λ is low, little or no ablation occurs due to the lack of photon/material interaction (except at high-radiation intensities (*12, 13*)); if α_λ is high, strong, but shallow, absorption will result in low etch rates. A detailed discussion of the relationship between absorption coefficient and etch rate can be found in references (*9, 17*).

Since there is no electronic interaction between the excited dopant and the host polymer (*9*) a "rule of mixtures" relationship is used to predict the effect of incorporating polyimide into PTFE on the absorption coefficient of the blend. Figure 4 shows predicted values of α_{308} for PTFE containing 0% to 5% weight fraction of polyimide. Clearly, only a small quantity of dopant is required to alter the materials' interaction with photons.

Since performance of an electronic device is directly impacted by the ε_r of the insulating material, the effect of incorporating the sensitization agent on the ε_r of PTFE must be known. ε_r for polyimide is greater than that for PTFE, about 2.90 and 2.10, respectively. Table 2 shows predicted values of ε_r for several polyimide-PTFE blends, using a "rule of mixtures" relationship, along with their respective experimentally determined values. Successful structuring of PTFE can occur with minimal additions of polyimide sensitizer, while the electrical characteristics of the resulting blends can be maintained at essentially

Table 1. Bake/Cure Cycle

Temperature (°C)	Time (Minutes)
100	60
200	60
320	120

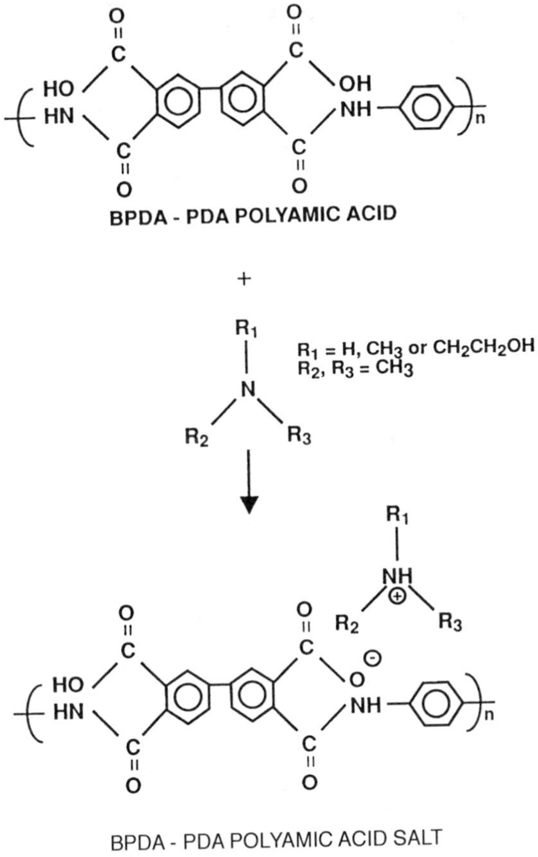

BPDA - PDA POLYAMIC ACID

+

R_1 = H, CH_3 or CH_2CH_2OH
R_2, R_3 = CH_3

BPDA - PDA POLYAMIC ACID SALT

Figure 1. Polyamic acid salt formation.

Figure 2. SEM micrograph (cross section) of an ablated feature in a blend containing 5% polyimide. Ablated hole diameter, measured at the upper surface of the polymer, is ≅ 127 μm. Reprinted with permission from reference 19. Copyright 1992 Springer-Verlag.

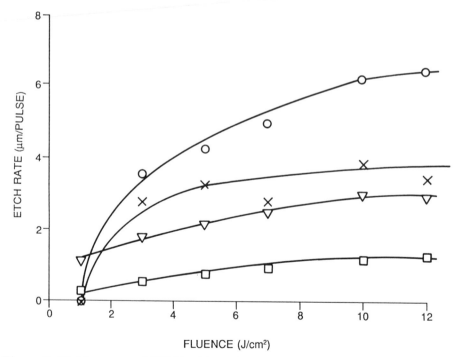

Figure 3. Etching rate of PTFE with various doping levels as a function of fluence; 0.5% (○), 1.0% (×), 5.0% (▽), and Upilex-S film (□).

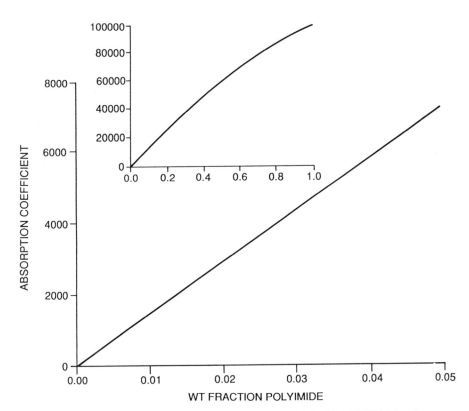

Figure 4. Absorption coefficients at 308 nm for polyimide/PTFE blends predicted using a rule of mixtures relationship for polyimide dopant concentrations (weight fraction of polyimide) up to 5.0%. Calculated values of α blend over the full range of polyimide concentrations, from neat PTFE to neat polyimide (Upilex-S film), are given in the inset.

Table 2. Predicted and Measured Values of ϵ_r

% Polyimide	Predicted	Measured
0	2.10	2.10
0.5	2.10	2.12
1.0	2.11	2.15
5.0	2.16	2.15

that of neat PTFE. In addition, from the table it is clearly seen that the experimental values closely follow those predicted.

SUMMARY

Homogeneous films of excellent quality comprised of PTFE and polyimide were prepared using an aqueous dispersion of the fluoropolymer and an organic salt of the BPDA-PDA polyamic acid. Polyimides, due to their high thermal stability, strong absorption coefficients and ability to be made water soluble (when in the polyamic acid form), act as effective and efficient excimer laser sensitization agents for PTFE, both in terms of structuring quality and etch rate. Since PTFE sensitization requires only minimal quanitites of polyimide, the final blends have electrical properties (ε_r) essentially equivalent to those of the neat fluoropolymer. Thus, a Teflon-like polyimide-PTFE blend with excimer laser micro-structuring capability is readily achievable.

ACKNOWLEDGEMENTS

The authors are grateful to D.A. Koehler and F.D. Curry for assistance with laser ablation experiments and etch depth measurements, D.L. Dittrich for film preparation, L.P. Wilding for performing dielectric constant measurements and Dr. S.L. Buchwalter for helpful discussions.

REFERENCES

1. Y. Kawamura, K. Toyoda, and S. Namba, Appl. Phys. Lett. **40**, 374 (1982).
2. R. Srinivasan and V. Mayne-Banton, Appl Phys. Lett. **41**, 578 (1982).
3. J.H. Brannon, J.R. Lankard, A.I. Baise, F. Burns, and J. Kaufman, J. Appl. Phys. **58**, 2036 (1985).
4. G.D. Mahan, H.S. Cole, Y.S. Liu and H.R. Philipp, Appl. Phys. Lett. **53(24)**, 2377 (1988).
5. R. Sauerbrey and G.H. Pettit, Appl Phys. Lett. **55(5)**, 421 (1989).
6. R. Srinivasan, E. Sutcliffe and B. Braren, Laser Chem. **9**, 147 (1988).
7. H. Hiraoka, T.J. Chuang and K. Masuhara, J. Vac. Sci. Technol. B **6(1)**, 463 (1988).
8. R. Srinivasan and B. Braren, Appl. Phys. A **45**, 289 (1988).
9. T.J. Chuang, A. Modl, and H. Hiraoka, Appl. Phys. A **45**, 277 (1988).
10. H. Hiraoka and S. Lazare, Applied Surface Science **46**, 342 (1990).
11. M. Bolle, K. Luther, J. Troe, J. Ihlemann, and H. Gerhardt, Appl. Surf. Sci. **46**, 279 (1990).
12. S. Kuper and M. Stuke, Appl. Phys., B **44**, 199 (1987).
13. D. Basting, U. Sowada, F. Voss, and P. Oesterlin, Proc. SPIE **1412**, 80, (1991).
14. O.M. Sorokin and V.A. Blank, Shurnal Prikladnoi Spectroskopii, **9**, 927, (1968).
15. F.D. Egitto and L.J. Matienzo, Polym. Degrad. and Stabil., **30**, 293, (1990).

16. G.A Takacs, V. Vukanovic, D. Tracy, J.X. Chen, F.D. Egitto, L.J. Matienzo and F. Emmi, Polym. Degrad. and Stabil., **40**, 73 (1993).
17. F.D. Egitto and C.R. Davis, Appl. Phys. **B**, **55**, 488 (1992).
18. J. H. Jou and P.T. Huang, Macromolecules, **24**, 3796 (1991).
19. C.R. Davis, F.D. Egitto and S.L. Buchwalter, Appl. Phys. **B 54**, 227 (1992).
20. Y. Liu, H. Cole, H. Philipp and R. Guida, Symp. Proc. SPIE, 1987.

RECEIVED February 4, 1993

Chapter 37

Fluorinated Poly(arylene ethers) with Low Dielectric Constants

Frank W. Mercer, David W. Duff, Timothy D. Goodman, and
Janusz B. Wojtowicz

Corporate Research and Development, Raychem Corporation, Mail Stop
123/8512, 300 Constitution Drive, Menlo Park, CA 94025

High performance polymer films and coating materials are increasingly being required by the electronics industry for use as interlayer dielectrics and as passivation layers. Aromatic polyimides are generally the polymers of choice for these applications because of their unique combination of chemical, physical, and mechanical properties (1). Another class of polymers which have been investigated for these applications are the poly(arylene ether)s that can be prepared by the nucleophilic displacement of activated aromatic dihalides by alkali metal bisphenoxides. Heterocycles such as benzoxazoles (2), imidazoles (3), phenylquinoxalines (4), and 1,2,4-triazoles (5) have been incorporated within poly(arylene ethers) utilizing this synthetic procedure.

Aromatic poly(oxadiazoles) (POX's) are a class of high temperature resistant heterocyclic polymers that show excellent thermal stability. POX's can be prepared by the cyclodehydration of polyhydrazides containing the -CON-HNHCO- group. This can be done thermally near the glass transition temperature of the polyhydrazide or in solution by using dehydrating agents such as polyphosphoric acid, sulfuric acid, or phosphorous oxychloride (6). Until recently, the use of POX's has been limited since they are soluble only in strong acids and could not be processed from organic solvents.

The synthesis of poly(arylene ether oxadiazoles) from bis(hydroxyphenyl oxadiazoles) and commercially available activated aromatic dihalides has recently been reported (5). As part of an effort to develop high performance, high temperature resistant polymers for microelectronic applications, we have prepared a series of poly(arylene ether-1,3,4-oxadiazoles) and characterized their thermal, mechanical, and electrical properties. The poly(arylene ether-1,3,4-oxadiazoles) reported herein were prepared by the reaction of a bisphenol with the bis(4-fluorophenyl)-1,3,4-oxadiazoles, 1 and 3, using potassium carbonate in N,N-dimethylacetamide.

0097–6156/94/0537–0526$06.00/0
© 1994 American Chemical Society

EXPERIMENTAL

Starting Materials

N,N-dimethylacetamide (DMAc), methanol, phenol, pyridine, cyclohex-anone, N-methyl-2-pyrrolidinone (NMP), 4,4-(hexafluoroisopropylidene)diphe-nol, dimethyl isophthalate, hydrazine hydrate, 4-fluorobenzoyl chloride, phospho-rous oxychloride, and potassium carbonate were obtained from Aldrich and used without purification. 9,9-bis(4-hydroxyphenyl)fluorene and 1,1-bis(4-hydroxy-phenyl)-1-phenylethane were obtained from Kennedy and Klim and used without purification. 2,5-Bis(4-fluorophenyl)-1,3,4-oxadiazole (1) was prepared as previ-ously described (7) and its synthesis is depicted in Scheme 1.

The 2,2'-(1,3-phenylene)bis[5-(4-fluorophenyl)-1,3,4-oxadiazole] (3) was pre-pared as depicted in Scheme 2 and as follows: To a 1 L round bottom flask was added 50.0 g (0.258 mol) of dimethyl isophthalate, 50.0 g (1.0 mol) of hydrazine hydrate and 500 mL of methanol. The mixture was heated at reflux for 4 hours, allowed to cool to room temperature, and stirred for an additional 20 hours. The mixture was filtered, washed with water and dried affording 42 g (84% yield) of isophthalic dihydrazide. In a 250 mL round bottom flask containing 11.0 g (0.0565 mol) of isophthalic dihydrazide and 100 mL of pyridine was added 18.6 g (0.117 mol) 4-fluorobenzoyl chloride over about 5 min. The mixture was refluxed under nitrogen with stirring for 45 min. The mixture was allowed to cool, water was added to precipitate the product, and the product was filtered, washed with water, and dried affording 20 g (86% yield) of the difluoro-bis(hydrazide) (2).

In a 500 mL round bottom flask a mixture of 17.5 g (0.0427 mol) of the difluoro-bis(hydrazide) (2) and 175 g of phosphorous oxychloride was refluxed overnight. Most of the phosphorous oxychloride was removed by distillation from the reaction mixture. The residue was allowed to cool to room temperature and poured slowly into 500 mL of ice water with stirring. The solid was filtered, washed with water, and dried affording 14.8 g (86% yield) of 3. GC/Mass. Spec. M + /e = 402. M.P. = 249-251 °C.

Model Reaction

2,5-Bis(4-phenoxyphenyl)-1,3,4-oxadiazole (4) was prepared as depicted in Scheme 3 and as follows: To a 100 mL round bottom flask was added 2.5 g (0.00968 mol) of 1, 2.18 g (0.0232 mol) of phenol, 3.2 g (0.0232 mol) of potassium carbonate, and 40 g of DMAc. The mixture was heated to 150 °C for 17 hours and the reaction was followed by GC/MS. Isolation by aqueous work up followed by recrystallization (toluene/hexane) afforded 3.4 g (86% yield) of 2,5-bis(4-phenoxyphenyl)-1,3,4-oxadiazole (4) as a white crystalline solid. M.P. = 161-163 °C. M + /e = 406. Key features in the FT-IR spectrum of 4 include the following absorptions: Aromatic C-H, 3062 cm^{-1}, Oxadiazole C = N, 1614 cm^{-1}; Aromatic C = C, 1594 and 1491 cm^{-1}; and Aromatic ether Ar-O-Ar, 1247 cm^{-1}. Calculated for $C_{26}H_{18}N_2O_3$:C, 76.83; H, 4.46; N, 6.89. Found: C, 76.74; H, 4.36; N, 6.92.

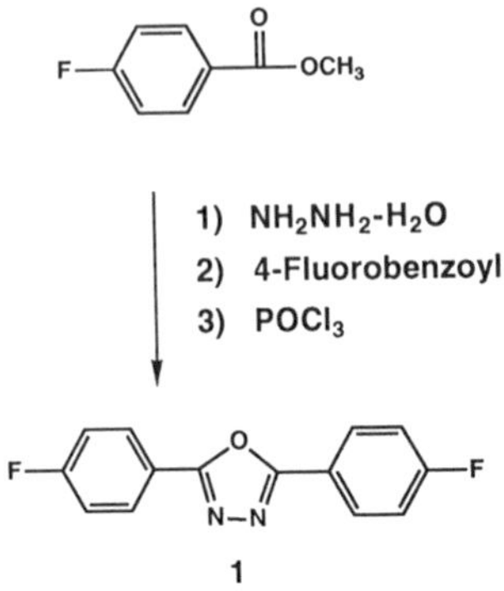

Scheme 1.

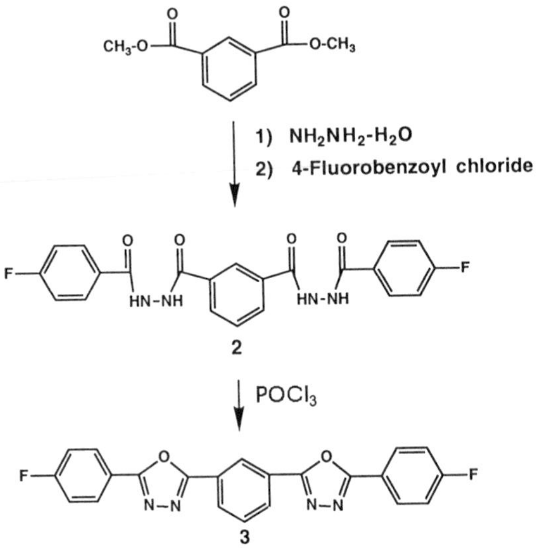

Scheme 2.

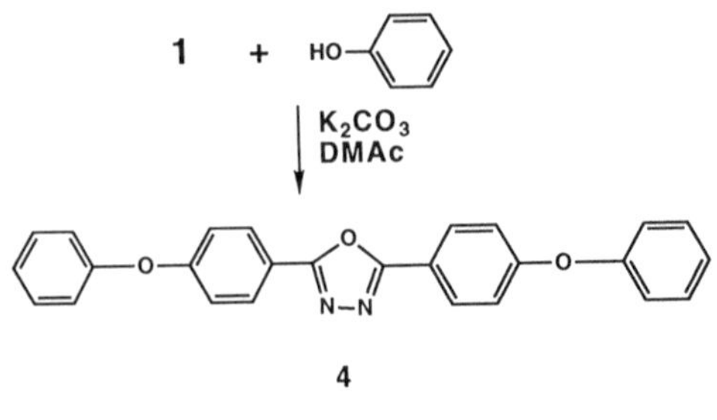

Scheme 3.

Polymerizations

Poly(arylene ether oxadiazole)s were prepared by the reaction of a bisphenol and either **1** or **3** in the presence of potassium carbonate in DMAc at 160 °C as depicted in Schemes 4 and 5. A typical polymerization was carried out as follows: To a 100 mL round bottom flask was added 2.58 g (0.010 mol) of **1**, 3.36 g (0.010 mol) of 4,4-(hexafluoroisopropylidene) diphenol, 31.2 g of DMAc, and 3.1 g (0.022 mol) of potassium carbonate. The mixture was heated to 160 °C with stirring under nitrogen for 17 hours. The mixture was allowed to cool to room temperature. The polymer was precipitated by pouring the reaction mixture into a blender containing about 100 mL of water, filtered, washed three times with water and dried to yield 5.1 g (92% yield) of a white powder. Anal. Calcd for $C_{29}H_{16}N_2O_3F_6$: C, 62.82; H, 2.91; N, 5.05. Found: C, 62.48; H, 3.19; N, 5.07.

Films

Solutions of the polymers (15-25 wt.% solids) in NMP or a 1:1 mixture of DMAc and cyclohexanone were spin coated onto a glass substrate. The coatings were dried 1 hour at 100 °C, 45 min. at 200 °C, and 15 min. at 300 °C.

Measurements

Dielectric constants were measured using a previously described method (*8*). Glass transition temperatures (Tg) reported in this paper were determined using differential scanning calorimetry (DSC) using a heating rate of 10 °C/min. Thermal gravimetric analyses (TGA) were determined using a heating rate of 20 °C/min. Both DSC and TGA were performed on a Perkin-Elmer Series 7 DSC/TGA. Gas chromatographic/mass spectral (GC/MS) analysis was performed on a Hewlett-Packard 5995 GC/MS. Gel permeation chromatography was carried out on a Hewlett-Packard 1090 liquid chromatograph fitted with four Polymer Labs PL-Gel columns (500Å, 100Å, 103Å, and 104Å pore diameters), using tetrahydrofuran as the mobile phase. ^{13}C and ^{1}H decoupled NMR results were obtained on a Varian XL-300 operating at 75.4 and 299,9 MHz for ^{13}C and ^{1}H, respectively. Tensile testing of compression molded films was conducted on an Instron tensile tester (ASTM D-882-64T).

RESULTS AND DISCUSSION

The difluoro oxadiazole monomers **1** and **3** were readily prepared from commercially available starting materials in high yield. A model reaction between **1** and potassium phenoxide was carried out in DMAc and followed by GC/MS. The reaction was near completion after 6 hours at 150 °C. After 18 hrs. the reaction was worked up and the expected 2,5-bis(4-phenoxyphenyl)-1,3,4-oxadiazole (**4**) was isolated.

The difluoro oxadiazole monomers **1** and **3** were subsequently reacted with aromatic diphenols in DMAc using potassium carbonate to give poly(arylene ether oxadiazole)s (PAEO). PAEO's **5a**, **5b**, and **5c** were soluble in DMAc, NMP, and chloroform, whereas, **6a** and **6b** were only soluble in NMP. Solutions of the

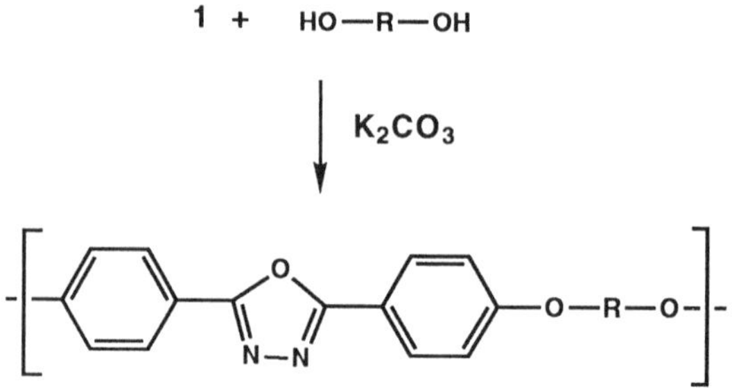

1 +   HO—R—OH

↓ K₂CO₃

5

Scheme 4.

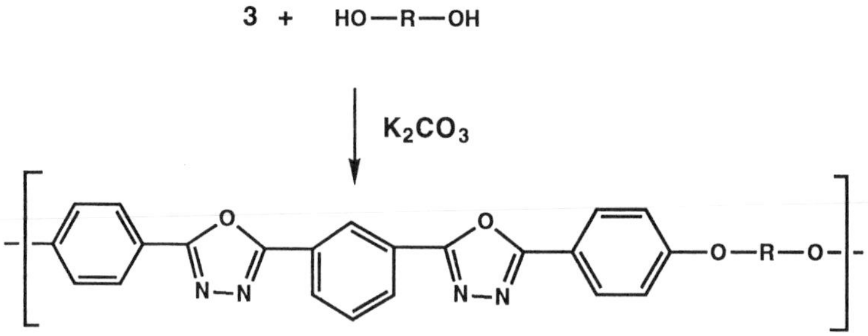

3 +   HO—R—OH

↓ K₂CO₃

6

5, 6 a: R =

b: R =

c: R =

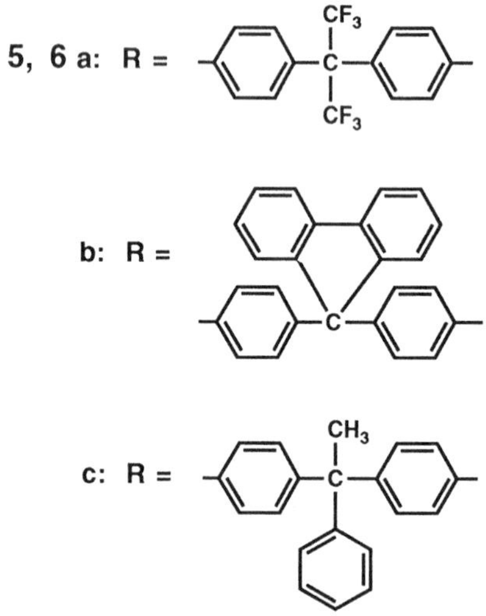

Scheme 5.

fluorinated polyarylethers **5a**, **5b**, **5c**, **6a**, and **6b**, containing up to 25 wt.% polymer, were prepared and spin-coated onto glass substrates and dried. The resulting tough, flexible films were about 10 microns thick.

At 0% relative humidity (RH), **5a**, **5b**, and **5c** have dielectric constants of 3.1, 3.0, and 3.0 respectively. At 60% RH, the dielectric constants of **5a**, **5b**, and **5c** all increase to 3.4. At 0% RH polymers **6a** and **6b** have dielectric constants of 3.0 and 3.1, respectively. At 60% relative humidity, the dielectric constants of **6a** and **6b** increase to 3.8 and 3.7, respectively. The relationship of dielectric constant to relative humidity for these PAEO's is depicted in Figure 1.

For comparison, one of the most common polyimides used as a dielectric in electronics applications is PMDA-ODA (**9**), made by reaction of pyromellitic dianhydride (PMDA) and 4,4′-diaminodiphenylether (ODA). For PMDA-ODA, the dielectric constant at 0% RH is 3.10 and at 58% RH the dielectric constant increases to 3.71. The smaller change in dielectric constant as a function of relative humidity for 5a, 5b, and 5c compared to PMDA-ODA may offer a performance advantage in microelectronic applications where low dielectric constant at high relative humidities is required.

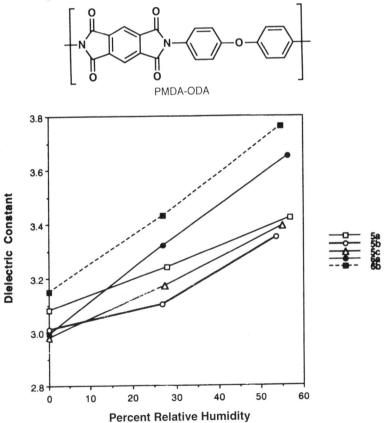

Figure 1. Dependence of dielectric constant on relative humidity.

Table 1. Properties of Selected PAEO's

Polymer	Tg (°C)	Mn	Mw	Tensile Strength (psi)	Elongation (%)	Modulus (Kpsi)
5 a	205	11,890	165,400	8,910	50	271
5 b	277	9,670	51,840	—	—	—
5 c	216	10,390	62,580	9,401	74	185

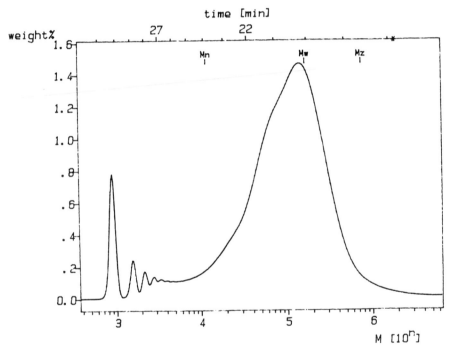

Figure 2. Molecular weight distribution of Sa.

Table 2. ^{13}C Chemical Shifts for Poly(arylene ether oxadiazole) **5a**

Carbon	**5a** ^{13}C Chemical Shift (ppm)[a]
1	164.0
2	119.3
3	128.9
4	119.4, 118.6[b]
5	159.3
6	156.7
7	119.4, 118.6[b]
8	132.0
9	128.7
10	63.8
11	129.8, 126.0, 122.2, 118.4[c]

[a] Chemical shifts referenced internally to CDCl$_3$ (77 ppm).
[b] Unambiguous assignment not determined.
[c] Quartet due to JCF coupling.

The Tg's for the poly(arylether oxadiazole)s **5a**, **5b**, and **5c** were 205 °C, 277 °C, and 216 °C, respectively, and for **6a** and **6b** were 234 °C and 288 °C, respectively. Thermal gravimetric analysis of **5b** reveals that the polymer exhibits initial weight loss in air at 475 °C (scan rate = 20 °C/min.) whereas **6b** exhibits initial weight loss in air at 405 °C. Therefore, PAEO's prepared from **1** are predicted to show improved thermal stability compared to the PAEO's prepared from **3**.

Results from GPC analysis of **5a**, **5b**, and **5c** are listed in Table 1. In each case, low molecular weight peaks were observed at a retention time of about 30.5 minutes. These peaks are thought to be the result of cyclic oligomer formation during the polymerization process. The number of repeat units in the cyclic oligomers is not known. The GPC spectrum of **5a** is shown in Figure 2.

PAEO's **5a** and **5c** were compression molded at 285 °C to yield tough, transparent films. The mechanical property measurements for **5a** and **5c** are shown in Table 1. PAEO's **5a** and **5c** have elongations of 50% or greater. This was unexpected since the PAEO's previously reported by Hergenrother (5) exhibited elongations of less than 10%.

The chemical shifts for structure **5a** are given in Table 2. ^{13}C chemical shift assignments were made based on comparisons with the model compounds

4,4'-(hexafluoroisopropylidene) diphenol, 2,5-bis(4-fluorophenyl)-1,3,4-oxadia-zole (1), 2,5-bis(4-phenoxyphenyl)-1,3,4-oxadiazole (4) and from calculations based on substituted benzenes (10).

CONCLUSION

Oxadiazole groups are effective activating groups for nucleophilic aromatic substitution. We have demonstrated that aryl fluorides para to 1,3,4-oxadiazoles groups were readily displaced with phenoxides. A series of new PAEO's were synthesized by the nucleophilic displacement reaction of bis(4-fluorophenyl)-1,3,4-oxadiazoles with a bisphenol. High molecular weight was readily achieved yielding polymers that could be processed from solution or by melt processing. The polymers had Tg's ranging from 205 °C to 288 °C and thermal decomposi-tion temperatures in excess of 400 °C. The polymers yielded tough, thermally stable films. The properties of these polymers may make them useful as di-electrics in microelectronics applications. Additionally, this polymerization pro-vides a general method for preparing poly(aryl ether 1,3,4-oxadiazole)s where the structure of the polymer can readily be controlled by varying both the bisphenol and the bis(aryl fluoro) substituted 1,3,4-oxadiazole.

REFERENCES

1. a) Denton, D.D.; Day, D.R.; Priore,D.F.; Senturia,S.D.; Anolick, E.S.; Schei-der, D. *J. of Elec. Mat.*, **1985**, 14, 119-136. b) Larsen, R.A. *IBM J. Res. Develop.*, 24, 268 (1980). c) Wilson, A.M. *Thin Solid Films*, 83, 145 (1981).
2. Hilborn, J.G.; Labadie, J.W.; Hedrick, J.L. *Macromolecules*, 23, 2845 (1990).
3. Connell, J.W.; Hergenrother, P.M. *Polym. Matl. Sci. Eng. Proc.*, 60, 527 (1989).
4. Hedrick, J.L.; Labadie, J.W. *Macromolecules*, 21, 1883 (1988).
5. Connell, J.W.; Hergenrother, P.M.; Wolf, P.W. *Polym. Matl. Sci. Eng. Proc.*, 63, 366 (1990).
6. a) Bach, H.C.; Dobinson, F.; Lea, K.R.; Saunders, J.H. *J. Appl. Polym. Sci.*, 23, 2125 (1979). b) Unishi, T.; Hasegawa, M. *J. Polym. Sci. Part A*, 3, 3191 (1965). c) Sato, M.; Yokoyama, M. *J. Polym. Sci. Chem. Ed.* 18, 2751 (1980). d) Ueda, M.; Sugita, H. *J. Polym. Sci. Part A: Polym Chem.* 26(1), 159 (1988).
7. Hayes, F.N.; Rogers, B.S.; Ott, D.G. *J. Am. Chem. Soc.*, 77, 1850 (1955).
8. Mercer, F. W.; Goodman, T. D. *High Perf. Polym.* 3, 297 (1991).
9. Economy, J. In *Contemporary Topics in Polymer Science*; Vandenberg, E. J., Ed.; Plenum: New York, 1984; Vol. 5.
10. Ewing, D.F. *Org. Magn. Reson.*, 12, 499 (1979).

RECEIVED May 4, 1993

Chapter 38

Microstructural Characterization of Thin Polyimide Films by Positron Lifetime Spectroscopy

A. Eftekhari[1], A. K. St. Clair[2], D. M. Stoakley[2], Danny R. Sprinkle[2], and J. J. Singh[2]

[1]Department of Physics, Hampton University, Hampton, VA 23668
[2]National Aeronautics and Space Administration, Langley Research Center, M/S 235, Hampton, VA 23665-5225

Positron lifetimes have been measured in a series of thin aromatic polyimide films. No evidence of positronium formation was observed in any of the films investigated. All test films exhibited only two positron lifetime components, the longer component corresponding to the positrons annihilating at shallow traps. Based on these trapped positron lifetimes, free volume fractions have been calculated for all the films tested. A free volume model has been developed to calculate the dielectric constants of thin polyimide films. The experimental and the calculated values for the dielectric constants of the films tested are in reasonably good agreement. It has been further noted that the presence of bulky CF_3 groups and meta linkages in the polyimide structure results in higher free volume fraction and, consequently, lower dielectric constant values for the films studied.

Polyimides are an important class of polymers for high temperature aerospace applications. Thin polyimide films are ideal candidates for protective coatings on antenna reflectors and other electronic applications. Their properties, both physical and electrical, are expected to be strongly influenced by their morphology. We have developed a novel technique for monitoring microstructural characteristics of thin polymer films. It is based on the sensitivity of the positron lifetimes to the molecular architecture of the polymers. Specifically, positron lifetimes can be used to calculate free volume hole radii and free volume fractions in the test polymers. A free volume model has been developed to calculate dielectric constants of thin polyimide films. It has been tested on a series of special purpose polyimide films developed for aerospace communication networks. The results are described in the following sections.

EXPERIMENTAL TECHNIQUES

Materials

The aromatic polyimide films (1–3) used in this study are listed in Table I. The dianhydride monomers 2,2-bis(3,4-dicarboxy-phenyl)hexafluoropropanedian-

0097–6156/94/0537–0535$06.00/0

Table I. Physical and Electrical Properties of the Test Films

Sample	Density (gms/cc)	Sat. Moist Content, volume % $(V/o)^{(*)}$	Dielectric Constant ε(at 10 GHz)
Kapton (Reference)	1.431	2.02	3.20 + 0.03
BFDA + ODA	1.384	0.74	2.63 ± 0.03
BFDA + 4-BDAF	1.400	1.38	2.44 ± 0.03
6FDA + DDSO$_2$	1.486	0.74	2.86 ± 0.03
BFDA + DABTF	1.440	0.49	2.55 ± 0.03
6FDA + APB	1.434	0.53	2.71 ± 0.03
BTDA + ODA	1.380	1.21	3.15 ± 0.03

(*)Saturation Moisture Volume Percent = V/O
$$= 100 \text{ (Saturation Moisture Fraction by Volume)}$$
$$= \frac{100\,\rho(W/O)}{100 + \rho(W/O)}$$
where W/O = Saturation Moisture Weight Percent

hydride (6FDA) and 3,3',4,4'-benzophenone tetracarboxylic dianhydride (BTDA) were also obtained from commercial sources. The diamine monomers 4,4-oxydianiline (ODA), 3,3'-diaminodiphenylsulfone (DDSO$_2$) and 1,3-bis(aminophenoxy)benzene (APB) were also obtained commercially . The bis[4-(3,4-dicarboxy-phenoxy) phenyl] hexafluoropropane dianhydride (BFDA) was an experimental material obtained from TRW Inc. The 2,2-bis[4(4-aminophenoxy)phenyl]hexafluoropropane (4-BDAF) and 3,5-diaminobenzotrifluoride (DABTF) were experimental monomers synthesized by Ethyl Corporation and NASA Langley Research Center, respectively. Kapton H film was obtained from duPont de N'emours and used for comparison. Figures 1(a) and 1(b) show the chemical structure of the dianhydride and diamine monomers investigated.

Polymer and Film Synthesis

Polyamic acid precursor solutions were prepared in closed vessels at ambient temperature by reacting stoichiometric amounts of diamine and dianhydride in dimethylacetamide at a concentration of 15-20% solids by weight. The resulting high viscosity polyamic acid solutions were cast onto glass plates in a dust-free chamber at a relative humidity of 10%. Solutions were spread with a doctor blade and gaps were set so as to produce a 25 μm thick film. The polyamic acid films were thermally converted to the corresponding polyimide films by heating in a forced air oven for one hour each at 100, 200 and 300 °C.

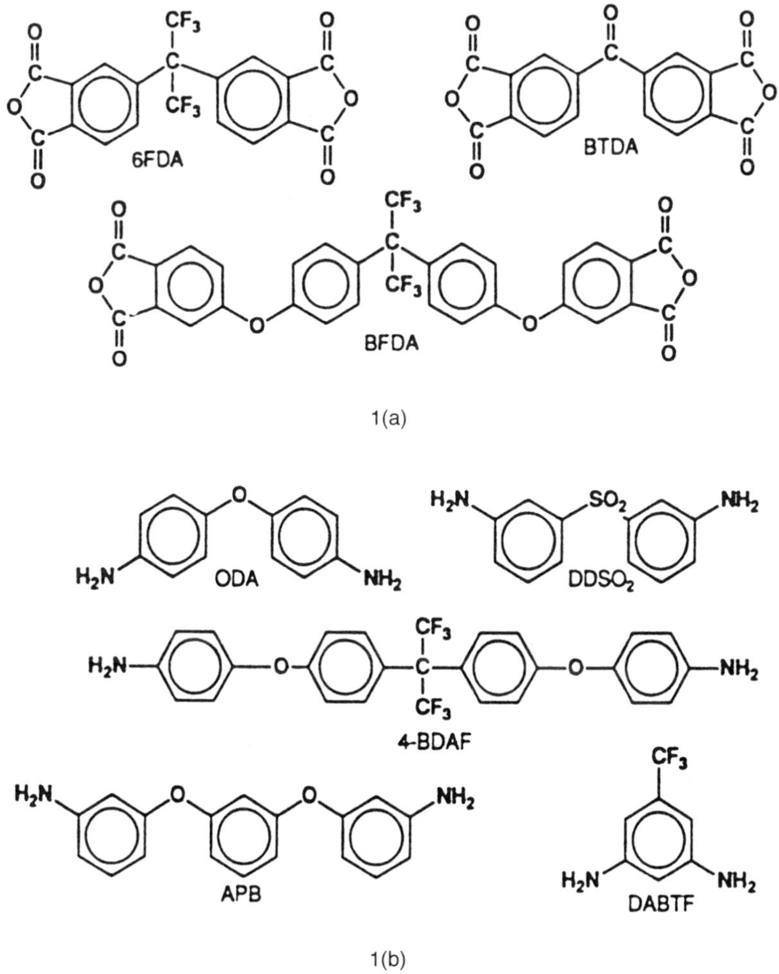

1(a)

1(b)

Figure 1. (a) Chemical structure of dianhydride monomers. (b) Chemical structure of diamine monomers.

Density and Dielectric Measurements

Densities were determined for the fully cured polyimide films in a density gradient tube prepared with aqueous $ZnCl_2$ solutions according to ASTM D1505-60T. Dielectric constants of the polyimide films were determined using a Hewlett Packard 8510 Automated Network Analyzer over the frequency range of 8-12 GHz. All films were desiccated in a heated vacuum oven prior to positron lifetime measurements.

Saturation Moisture Content Measurements

The samples were first desiccated by heating them to 120 °C in a vacuum oven till their weights became constant. They were then submerged in water at 90 °C till their weights stabilized. The saturation moisture content by weight was determined as (saturated sample weight- desiccated sample weight)/(desiccated sample weight). The saturation moisture fraction by volume was then calculated from the measured weight fraction as follows:

(Saturation Moisture Fraction by Volume)

$$= \frac{\rho(\text{Saturation Moisture Fraction by Weight})}{1 + \rho(\text{Saturation Moisture Fraction by Weight})}$$

where ρ is the density of the polymer sample.

Positron Lifetime Measurements

Positron lifetimes in the test films were measured using a recently developed low energy positron flux generation scheme (4). Briefly, the procedure involves the use of a thin aluminized mylar film as the positron source holder sandwiched between two well-annealed polycrystalline, high purity, tungsten moderator strips. The test films are introduced between the source holder and the moderator strips, thus serving an additional purpose of electrically insulating the source holder from the moderator strips. When a small negative potential $(-V)$ is applied between the source holder and the moderator strips, the thermalized positrons diffusing out of the moderators are attracted to the source holder thus forcing them to enter and annihilate in the test films. A positive potential, $(+V)$, on the other hand, forces the outdiffusing positrons back into the moderator strips. The difference between the positron lifetime spectra with the source at $\mp V$ volts with respect to the moderator strips is thus entirely due to the positrons annihilating in the test films. The lifetime data were acquired using a 250 microcuries Na^{22} positron source and a standard fast-fast coincidence measurement system with a time resolution of approximately 225 picoseconds. Attempts were made to analyze the lifetime spectra into 2- and 3-components using POSITRONFIT-EXTENDED (5) program. The best least squares fit to the data were obtained with 2-component analysis. Typical positron lifetime spectra are shown in Figure 2(a) and 2(b). It required about 6 hours to

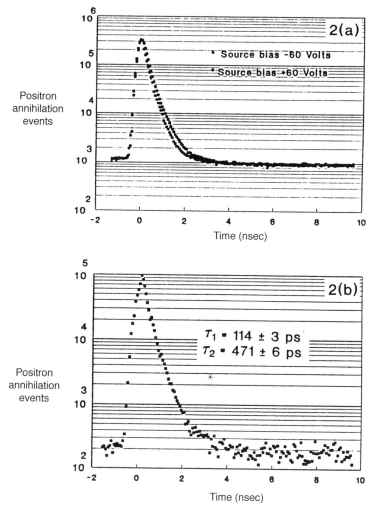

Figure 2. (a) Comparison between positive and negative source bias lifetime spectra in Kapton films. (b) Positron lifetime spectrum in Kapton films obtained by subtracting normalized positive bias spectrum from negative spectrum.

obtain 10^6 counts in the total spectrum in each case. The measurements were made at room temperature and ambient pressure.

RESULTS AND DISCUSSION

The physical and electrical properties of the films investigated are summarized in Table I. The positron lifetime component values, their respective intensities, and the microvoid volumes are summarized in Table II.

The longer positron lifetime in each test film corresponds to the positrons trapped in the potential defects (microvoids) in the test films. These microvoids result from fluctuations in the packing density of the macromolecular chains. The sizes of these microvoids are too small for the formation and localization of positronium atoms. However, free positrons can be trapped at these sites with subsequent annihilation. The radii of the microvoids (R) in nanometers and the trapped positron lifetime (τ_2) in nanoseconds are related as follows:

$$\frac{1}{2.5\tau_2} = \left(1 - \frac{R}{R + \Delta R} + \frac{1}{2\pi}\sin\frac{2\pi R}{R + \Delta R}\right) \tag{1}$$

where $\Delta R = 0.1659$ nanometers.

This equation differs from the conventional model (6) for positronium-forming media in having $1/2.5\tau$ instead of $1/2\tau$ as the left hand side term. This form has been dictated by the following considerations: (a) The positron annihilation in polyimides reportedly (7) differs considerably from that observed in most polymers. It proceeds from the free or trapped positron states without the formation of positronium atoms; (b) Positron lifetime spectra in all of the polyimides (PMDA, BFDA, BTDA and 6FDA-based polyimides etc.) investigated in this laboratory exhibit only two lifetime components. The shorter lifetime (τ_1) ranges from 100 to 300 picoseconds and arises from free positron

Table II. Positron Lifetimes and the Free Volume Fractions in the Test Films

Sample	τ_1/I_1 (ps/%)	τ_2/I_2 (ps/%)	Microvoid Volume $V_f(A^3)$	Free Volume Fraction f(%)
Kapton (Ref)	114 ± 3/65.3 ± 1.	471 ± 6/34.7 ± 1.	1.54 ± 0.19	2.02 ± 0.27
BFDA + ODA	223 ± 4/73.2 ± 1.	699 ± 17/26.8 ± 1.	12.26 ± 1.10	12.43 ± 1.09
BFDA + 4-BDAF	131 ± 2/71.8 ± 1.	790 ± 12/28.2 ± 1.	18.36 ± 0.85	19.56 ± 0.91
6FDA + DDSO$_2$	170 ± 3/72.5 ± 1.	623 ± 14/27.5 ± 1.	7.94 ± 0.75	8.26 ± 0.79
BFDA + DABTF	135 ± 4/68.6 ± 1.	653 ± 15/31.4 ± 1.	9.57 ± 0.96	11.41 ± 1.14
6FDA + APB	254 ± 4/86.2 ± 1.	867 ± 39/13.8 ± 1.	23.93 ± 2.83	12.5 ± 1.48
BTDA + ODA	124 ± 3/60.9 ± 1.	531 ± 7/39.1 ± 1.	3.7 ± 0.25	5.48 ± 0.37

(*)τ_1 and τ_2 are the lifetimes in picoseconds of positrons annihilating promptly and after trapping in microvoids, respectively. I_1 and I_2 are their respective intensities in percent.

annihilation. The longer lifetime (τ_2), whose value ranges from 400 to 900 picoseconds, has been reportedly (8, 9) associated with positrons trapped at defect sites (microvoids). Recently, Deng et al. (10) have reported that $1/2\tau_3$ for localized positronium can be paralleled by $1/n\tau_2$ for localized positrons for determining the microvoid radius. The maximum value of $1/n\tau_2$, which corresponds to the minimum radius of the free volume hole, is ≈ 1. Thus for $\tau_2 \geq 0.4$ nanoseconds, $1/2.5\tau_2$ becomes the appropriate form. It implies that the radius of the smallest free volume hole (microvoid) in polyimides corresponds to a trapped positron lifetime of 400 picoseconds. The microvoid volume (V_f) is given by $4/3\pi R^3$. The free volume fractions in the test films were calculated as follows:

$$f = CI_2V_f \qquad (2)$$

where C is a structural constant and I_2 is the intensity of the trapped positron lifetime component. The structural constant was calculated by equating the free volume fraction in Kapton with its saturation moisture fraction by volume. This is justifiable since Kapton has no hydrophobic atoms in its molecular structure. If, however, the space accessible to positrons is not identical to the volume accessible to water, the value of the structural constant C would be slightly different. It has been assumed that the structural constant C has the same value for all of polyimide films tested. The free volume fractions of all the films tested are summarized in Table II.

Even though the chemical structures of the films investigated are not identical, their polarizabilities are not expected to be significantly different. Recently, Misra et al. (11) have reported similar conclusions about the effects of fluorine substitutions on the dielectric constants of similar aromatic polyimides. Thus, the differences in the dielectric constants are expected to arise mainly from the differences in the free volume fractions of the films studied. The effect of free volume fraction (f) on the dielectric constant (ϵ) of the films can be calculated as follows. As illustrated in Figure 3, the aggregate of the free volume cells in the film of thickness d can be represented by an empty(air) strip of thickness df. The effect of the air strip on the dielectric constant of the film can be calculated by considering a parallel plate condenser with the test film as its dielectric medium. The dielectric medium is thus made up of a resin strip of thickness of d(1-f) and an airstrip of thickness df in series. The capacitance (C) of such a parallel plate condenser can be written as:

$$\frac{1}{C} = \frac{1}{C_1} + \frac{1}{C_2} \qquad (3)$$

where $C = \epsilon(A/4\pi d)$
$C_1 = \epsilon_R(A/4\pi d(1\text{-}f))$
$C_2 = \epsilon_{Air}(A/4\pi df)$
and A = Area of the condenser plates.

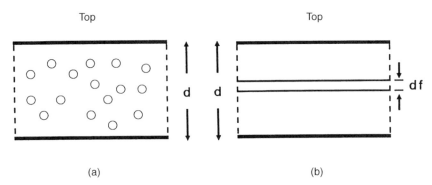

(a)   (b)

Figure 3. Cross sectional view of the test film. (a) Randomly distributed free volume cells. (b) Equivalent free volume strip.

Substituting the values of C, C_1, and C_2 in equation 3, we obtain:

$$\frac{1}{\varepsilon} = \frac{1-f}{\varepsilon_R} + \frac{f}{\varepsilon_{Air}} \tag{4}$$

The value of ε_R, which corresponds to the value of ε for zero free volume fraction, was obtained from the ε vs f data illustrated in Figure 4 and has been found to be 3.55. It should be emphasized that $\epsilon_R = 3.55$ is the predicted value of the dielectric constant of a free-volume-hole-free polyimide film. This value should not be compared with experimental values of ϵ_R for any polyimide film unless it is known to have zero free volume fraction. Furthermore, if the saturation moisture content of Kapton is not equal to its free volume fraction, the value of ϵ may be slightly different. The measured and calculated values of the dielectric constants for various test films are summarized in Table III and illustrated in Figure 5. It is noted that the test samples with the largest free volume fractions have the lowest values of dielectric constants as suggested by equation 4. The large free volumes in these samples are due to the presence of meta linkages and bulky CF_3 in their molecular architecture. A comparison of the saturation moisture data in Table I and the free volume fraction values in Table II shows that the saturation moisture fractions are much less than the free volume fractions in all samples, except Kapton in which they have been assumed to be equal. This is presumably the result of hydrophobicity of fluorine atoms present in high concentration in the backbones of the test samples other than Kapton.

CONCLUSIONS

Positron lifetime spectroscopy provides a sensitive technique for characterizing

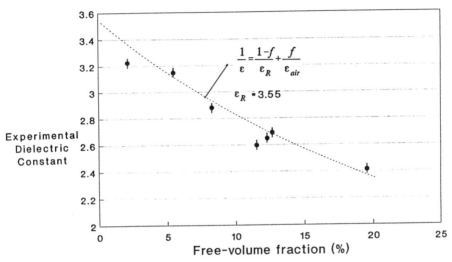

Figure 4. Experimental dielectric constant vs. free-volume fraction.

Table III. Comparison Between the Experimental and Calculated Values of the Dielectric Constants

Sample	f(%)	ε(Expt)	ε(calc)
Kapton (Ref)	2.02 ± 0.27	3.20 ± 0.03	3.31 ± 0.03
BFDA + ODA	12.43 ± 1.09	2.63 ± 0.03	2.69 ± 0.05
BFDA + 4-BDAF	19.56 ± 0.91	2.44 ± 0.03	2.37 ± 0.04
6FDA + DDSO$_2$	8.26 ± 0.79	2.86 ± 0.03	2.93 ± 0.05
BFDA + DABTF	11.41 ± 1.14	2.55 ± 0.03	2.74 ± 0.06
6FDA + APB	12.5 ± 1.48	2.71 ± 0.03	2.70 ± 0.07
BTDA + ODA	5.48 ± 0.37	3.15 ± 0.03	3.12 ± 0.03

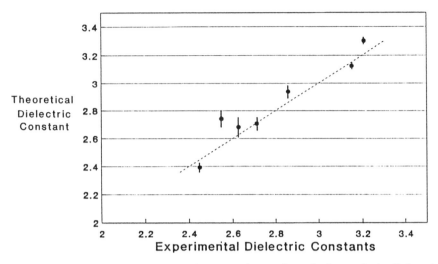

Figure 5. Comparison between the experimental and theoretical dielectric constants.

thin polymer films in terms of their free volumes and dielectric constants. When combined with saturation moisture determination, it also provides useful information about their moisture susceptibility. It is apparent that the presence of fluorine atoms in the polymer architecture reduces their moisture pick up, and hence the moisture-induced degradation in their physical and electrical properties. Also, the bulky CF_3 groups and meta linkages in their structure enhance their inter-molecular spacing resulting in higher free volume fraction and, consequently, lower dielectric constant.

REFERENCES

1. St. Clair, A. K.; St. Clair, T. L., U.S. Patent 4,603,061, 1986.
2. Stoakley, D. M.; St. Clair, A. K. *Proc. of the Mat. Res. Soc. Symp.*, 1991, 131-136.
3. St. Clair, A. K.; St. Clair, T. L.; Winfree, W. P. *Proc. of the ACS Division of PMS & E.* 1988, 59, 28.
4. Singh, J. J.; Eftekhari, A.; St. Clair, T. L. *Nucl. Instrum. Methods. Phys. Res.*, 1991, B53, 342-348.
5. Kirkegaard, P. *Comput. Phys. Commun.*, 1974, 7(7), 401-409.
6. Nakanishi, H.; Wang, S. J.; Jean, Y. C. *Proc. Int'l Symp. on Positron Annihilation Studies of Fluids.* 1988, 292-298.
7. Askadskii, A. A.; Tushin, S. A.; Kazantseva, V. V.; Kovrins, O. V. *Polymer Science U.S.S.R.*, 1990, 32 (No. 12) 2560-2568.
8. Hautojarvi, P.; Vehanen, A. In *Positrons in Solids*; Hautojarvi, P., Ed.; Springer-Verlag: New York, 1979; pp 1-23.

9. Stevens, J. R. In *Methods of Experimental Physics*; Fava, R. A. Ed.; Academic Press: New York, 1980; Vol. 16 (Part A) pp 371-403.
10. Deng, Q.; Sundar, C. S.; Jean, Y. C. *J. Phys. Chem.*, 1992, 96, 492-495.
11. Misra, A. C.; Tesoro, G.; Hougham, G.; Pendharkar, S. M. *Polymer*, 1992, 33, 1078-1082.

RECEIVED December 30, 1992

Chapter 39

Synthesis and Characterization of New Poly(arylene ether oxadiazoles)

Frank W. Mercer, Chris Coffin, and David W. Duff

Corporate Research and Development, Raychem Corporation, Mail Stop
123/8512, 300 Constitution Drive, Menlo Park, CA 94025-1164

Considerable attention has been devoted to the preparation of fluorine-containing polymers because of their unique properties and high temperature performance (1). Recently we reported the preparation and characterization of novel fluorine-containing polyimides and polyethers which exhibit low moisture absorption and low dielectric constants (2, 3). Fluorinated polyimides absorb 1 wt% water and have dielectric constants of about 2.8 (all dielectric constants reported in this paper were measured at 10 kHz) whereas their non-fluorinated analogs absorb as much as 3 wt% water and have dielectric constants of about 3.2. Fluorinated polyarylethers, which are free of polar groups such as ketones, imides and sulfones, absorb as little as 0.1 wt% water and have dielectric constants less than 2.8.

In our continuing effort to develop polymers with low dielectric constants and which exhibit low moisture absorption, we have prepared and characterized six new fluorinated polyarylethers (FPAE). FPAE 1, 2, 3, 4, 5, and 6 (depicted in Scheme 1) were prepared by reaction of decafluorobiphenyl with 4,4'-(hexafluoroisopropylidene)diphenol (Bisphenol AF), 9,9-bis(4-hydroxyphenyl)fluorene, 1,1-bis(4-hydroxyphenyl)-1-phenylethane (Bisphenol AP), phenolphthalein, fluorescein, and methyl 3,5-dihydroxybenzoate, respectively. The properties of FPAE 1, 2, 3, 4, 5, and 6 make them useful for electronic applications.

EXPERIMENTAL

Reagents

All reagents were reagent grade and were used without purification. Decafluorobiphenyl, phenolphthalein, fluorescein, 3,5-dihydroxybenzoic acid, cyclohexanone, dimethylacetamide (DMAc), g-butyrolactone (GBL), and potassium carbonate were obtained from Aldrich Chemical. Methyl 3,5-dihydroxybenzoate

0097–6156/94/0537–0546$06.00/0

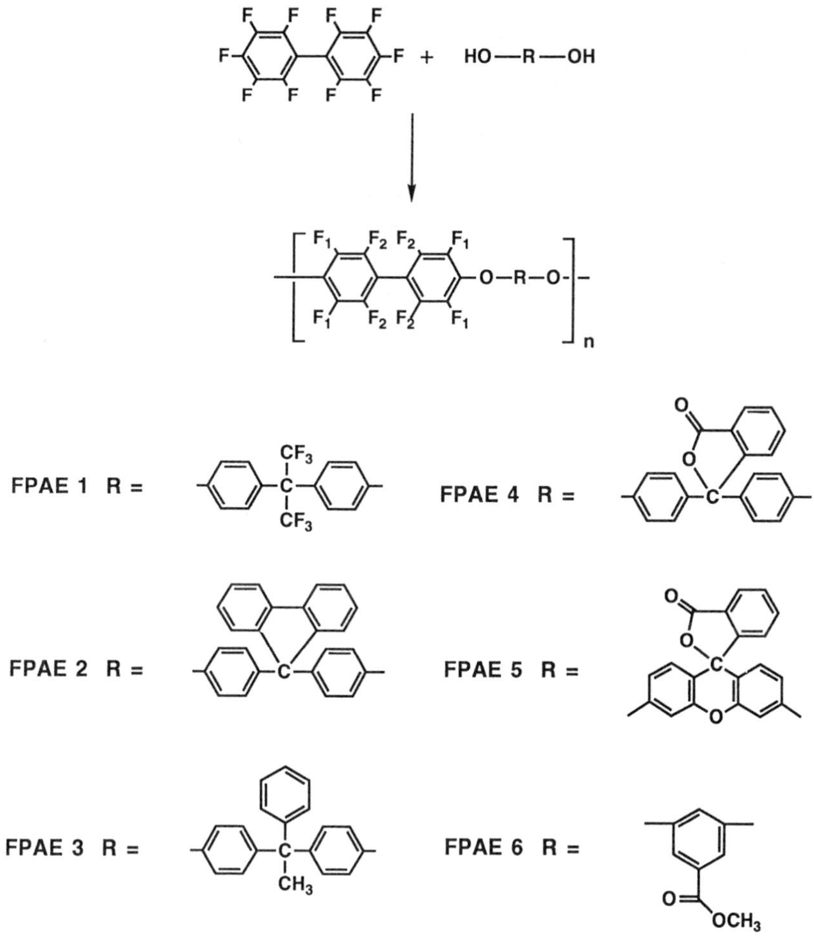

FPAE 1 R =

FPAE 2 R =

FPAE 3 R =

FPAE 4 R =

FPAE 5 R =

FPAE 6 R =

Scheme 1.

was prepared by esterification of 3,5-dihydroxybenzoic in methanol using sulfuric acid catalysis. Bisphenol AF, Bisphenol AP, and 9,9-bis(4-hydroxyphenyl)fluorene were obtained from Kennedy and Klim and used as received.

Measurements

Dielectric constants were measured at 10 KHz using our previously described method (2). Tensile testing of compression molded films was conducted on an Instron tensile tester (ASTM D-882-64T). Moisture absorption was calculated following immersion of solution-cast films of the polymers in water for 16 hrs at 90 °C. Glass transition temperatures (Tg) reported in this paper were determined using differential scanning calorimetry (DSC). Both DSC and thermal gravimetric analyses (TGA) were performed on a Perkin-Elmer Series 7 DSC/TGA. Gel permeation chromatography was carried out on a Hewlett-Packard 1090 liquid chromatograph fitted with four Polymer Labs PL-Gel columns (500Å, 100Å, 103Å, and 104Å pore diameters), using tetrahydrofuran as the mobile phase and polystyrene molecular weight standards. Nuclear magnetic resonance analysis was performed on a Varian XL-300 NMR spectrometer. Gas chromatographic/mass spectral (GC/MS) analysis was performed on a Hewlett-Packard 5995 GC/MS. The percent gel in crosslinked polymers was determined by Soxlet extraction with DMAc for 16 hrs.

Polymer Synthesis

Polymerization of decafluorobiphenyl with the bisphenols was carried out using the following general procedure: to a 250 mL round bottom flask was added 11.36 g (0.034 mol) of decafluorobiphenyl, 11.54 g (0.033 mol) of 9,9-bis(4-hydroxyphenyl)-fluorene, 105 g DMAc, and 12.2 g (0.090 mol) of potassium carbonate. The mixture was stirred at 120 °C under nitrogen for 17 hrs. The mixture was allowed to cool to room temperature and poured into a blender containing 300 mL of water to precipitate the polymer. The polymer was isolated by filtration, washed with water and dried to yield FPAE **2** as a white powder.

RESULTS AND DISCUSSION

A model reaction between 1 mole of decafluorobiphenyl and 2 moles of phenol was carried out in DMAc and followed by GC/MS. The reaction, depicted in Scheme 2, was near completion after 6 hrs at 150 °C. After 18 hrs, the reaction was worked up and the expected 4,4′-diphenoxyoctafluoro-biphenyl (**4**) was isolated as the exclusive product by ^{19}F-NMR and GC/MS.

Decafluorobiphenyl was subsequently reacted with aromatic diphenols in DMAc using potassium carbonate to give the fluorinated polyarylethers. Solutions of FPAE **1**, **2**, **3**, **4**, **5**, and **6** containing up to 25 wt% polymer, were prepared in DMAc, tetrhydrofuran, bis(2-ethoxyethyl)ether, GBL, cyclohexanone, methyl isobutyl ketone, or mixtures of the above. Solutions of the polyethers in a 50/50 mixture of GBL and cyclohexanone (w/w) were spin-coated

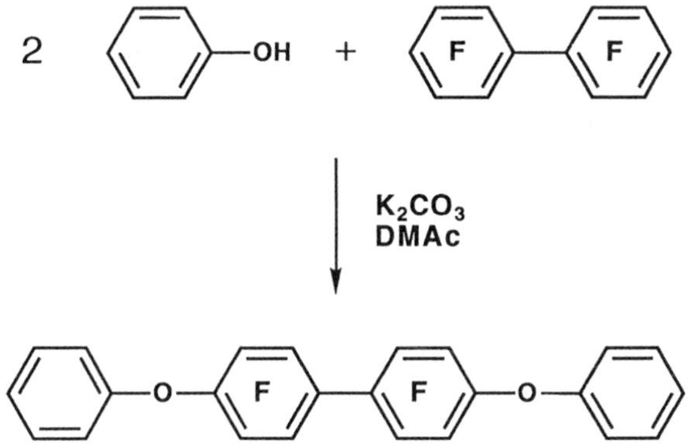

Scheme 2.

onto glass and dried 15 min at 100 °C, 15 min at 200 °C, and 30 min at 350 °C. The resulting tough, flexible films were about 10 microns thick.

The dielectric constants of the FPAE polymers were measured at 0% relative humidity (RH) and at 60% RH and are listed in Table 1. Moisture absorption of the FPAE polymers was measured and is also listed in Table 1.

FPAE 1, 2, 3, 4, 5, and 6 have glass transition temperatures of 189 °C, 260 °C, 208 °C, 243 °C, 285 °C, and 148 °C, respectively, and show excellent thermal stability.

Thermal gravimetric analysis of FPAE 1 and FPAE 2 reveals that the polymers exhibit initial weight losses in air at 500 °C (scan rate = 20 °C/min). FPAE 2 has improved thermal stability over FPAE1 and FPAE 4, revealing only a 3.6% weight loss after 3 hr in air at 450 °C, whereas FPAE1 and FPAE 4 exhibit 19.8% and 50% weight loss, respectively, after similar treatment. The results of TGA analysis for FPAE1, FPAE 2, and FPAE 4 are listed in Table 2.

FPAE 1 and FPAE 2 can be compression molded at 260 °C and 315 °C, respectively, to yield transparent, flexible films. Tensile specimens were prepared from from the resulting films of FPAE 1 and FPAE 2. The mechanical properties of these films are listed in Table 3.

The molecular weight of FPAE polymers was controlled by addition of an excess of decafluorobiphenyl. The GPC analysis of several FPAE polymers having differing levels of excess decafluorobiphenyl are listed in Table 4. Attempts to prepare FPAE polymers with an excess of bisphenol always led to crosslinked polymer gels and no soluble polymers were obtained.

The 282.3 MHz ^{19}F spectrum for FPAE 2 prepared using a 3% molar excess of decafluorobiphenyl was recorded. The peaks in the spectrum were referenced externally to a,a,a-trifluorotoluene at -63.7 ppm. The spectrum is dominated by the peaks centered at -139.0 and -153.7 ppm corresponding to the F2 and F1 fluorine atoms, respectively, in FPAE 2 (Scheme 1). These assignments were confirmed by examining the ^{19}F spectrum of the model compound, 4,4'-diphenc

Table 1. Dielectric Properties and Moisture Absorption of FPAE Polymers

Polymer	Dielectric Constant		Moisture Absorption (wt. %)
	0% RH	60% RH	
FPAE 1	2.50	2.60	0.10
FPAE 2	2.60	2.70	0.15
FPAE 3	2.60	2.70	0.10
FPAE 4	2.75	3.00	0.45
FPAE 5	2.80	3.10	0.50
FPAE 6	3.10	3.35	—

Table 2. TGA Analysis of FPAE 1, 2, and 4

Property	FPAE 1	FPAE 2	FPAE 4
TGA Weight Loss			
Onset in Air (°C)	500	500	450
Onset in Nitrogen (°C)	510	540	—
Maximum Weight Loss			
@1000 °C (%)	60	37.5	98
Isothermal Weight Loss in Air			
3 Hours at 400 °C (%)	2.5	2.7	—
3 Hours at 450 °C (%)	19.8	3.6	50

Table 3. Mechanical Properties of FPAE 1 and 2

Property	FPAE 1	FPAE 2
Thermal Coefficient of Expansion (ppm/°C)	76	65
Ultimate Tensile Strength (Kpsi)	8.3	10.7
Elongation at Break (%)	85.0	36.0
Elastic Modulus (Kpsi)	245	295

Table 4. GPC Analysis of FPAE Polymers

Polymer	Mole % Excess Decafluorobiphenyl	Mn	Mw
FPAE 1	2	24,250	955,000
FPAE 1	3	18,560	56,140
FPAE 2	3	17,090	78,970
FPAE 4	4	13,990	49,190

xyoctafluorobiphenyl, mentioned above. The predominance of these two fluorine environments suggests the FPAE **2** is linear with little or no branching. The ^{19}F NMR spectrum also contains three spectral features centered at -138.3, -151.0 and -161.4 ppm corresponding to the F3, F5 and F4 fluorine atoms, respectively, in the nonafluorobiphenyl endcapper (Figure 1). Peaks corresponding to the fluorine atoms F_1' and F_2' of the ether-substituted ring of the end cap have essentially the same chemical shifts as the fluorine atoms F1 and F2 in FPAE **2**. These assignments were confirmed by examining the ^{19}F spectrum of the model compound 4-phenoxynonafluoro-biphenyl (*5*). Expansion of the spectral region centered at -151.0 ppm in the ^{19}F NMR spectrum of FPAE **2** corresponds to the F5 fluorine atoms in 4-phenoxy-nonafluorobiphenyl. Integrating this ^{19}F NMR spectral region of FPAE **2** and either the spectral region centered at -139.0, or that centered at -153.7 ppm, of FPAE **2** yields a DP of ∼ 29 which is consistent with the DP of 32 calculated using the stoichiometric ratio of the reactants.

FPAE polymers can be crosslinked by heating in air between 300 and 450 °C or heating in nitrogen using a peroxide crosslinker, such as dicumyl peroxide. Crosslinking leads to a themoset material having improved solvent resistance, and is necessary to prevent solvent induced stress cracking of the FPAE polymers when coatings or films of the polymers are exposed to polar aprotic solvents. Table 5 lists the gel content of FPAE **1** crosslinked by heating in air. The results of crosslinking of FPAE **1** and FPAE **2** with peroxides are listed in Table 6.

CONCLUSION

FPAE polymers were prepared by reaction of decafluorobiphenyl with bisphenols. These polymers exhibit low dielectric constants, low moisture absorption, and excellent thermal and mechanical properties. Tough, transparent films of the polymers were prepared by solution casting or compression molding. These polymers may be useful in electronic applications. Synthesis and characterization of other polymers containing perfluoroaryl units is continuing.

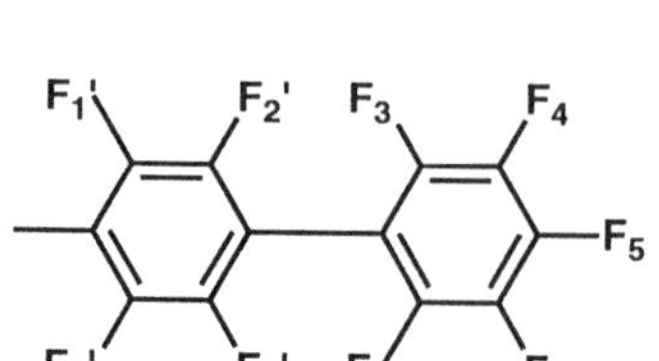

Figure 1. Nonafluorobiphenyl endcapper.

Table 5. Crosslinking FPAE 1 by Heating in Air

Cure Temperature (°C)	Cure Time (min)	% Gel
300	30	55
400	15	79
400	30	80
400	60	86
400	105	92

Table 6. Peroxide Crosslinking of FPAE Polymers*

Polymer	Peroxide	wt % Peroxide	% Gel
FPAE 1	—	—	3.0
FPAE 1	dicumyl peroxide	5	68.8
FPAE 1	dicumyl peroxide	10	69.2
FPAE 1	benzoyl peroxide	5	31.9
FPAE 1	benzoyl peroxide	10	50.7
FPAE 2	dicumyl peroxide	10	94.0
FPAE 2	benzoyl perodixe	10	49.4
FPAE 2	cumene hydroperoxide	10	81.3
FPAE 2	t-butyl perbenzoate	10	75.8

* All samples cured at 400°C for 30 min. under nitrogen.

REFERENCES

1. Cassidy, P.E.; Aminabhavi, T.M.; Farley, J.M. *J. Macromol. Sci.–Rev. Macromol. Chem. Phys.*, **1989**, **C29**(2 & 3), **365**.
2. Mercer, F.M.; Goodman, T.D. High Perf. Polymers, **1991**, 3(4) 297.
3. Mercer, F.M.; Goodman, T.D., Polymer Preprints, **1991**, 32(2), 188.
4. Richardson, G.A.; Blake, E.S. Ind. Eng. Chem. Prod. Res. Dev., **1968**, 7, 22, .
5. The model compound, 4-phenoxynonafluorobiphenyl, was prepared upon treatment of decafluorobiphenyl (15 mmol) with phenol (15 mmol) in the presence of K2CO3 (43 mmol) in DMAc. The product was isolated following treatment of the reaction mixture with water. Subsequent preparative thin-layer chromatography of the crude product on silica gel with heptane elution afforded 4-phenoxynonafluorobiphenyl which crystallized upon standing, mp 110.5-111.5 °C. The ^{19}F NMR spectrum of the purified model compound consisted of resonances centered at -138.3, -139.1, -151.1, -153.8 and -161.4 ppm.

RECEIVED December 30, 1992

INDEXES

Author Index

Affiliation Index

Subject Index

Production: Meg Marshall, Paula M. Bérard, and C. Buzzell-Martin
Indexing: Deborah H. Steiner
Acquisition: Anne Wilson

Printed and bound by Maple Press, York, PA

Bestsellers from ACS Books

The ACS Style Guide: A Manual for Authors and Editors
Edited by Janet S. Dodd
264 pp; clothbound ISBN 0–8412–0917–0; paperback ISBN 0–8412–0943–X

The Basics of Technical Communicating
By B. Edward Cain
ACS Professional Reference Book; 198 pp;
clothbound ISBN 0–8412–1451–4; paperback ISBN 0–8412–1452–2

Chemical Activities (student and teacher editions)
By Christie L. Borgford and Lee R. Summerlin
330 pp; spiralbound ISBN 0–8412–1417–4; teacher ed. ISBN 0–8412–1416–6

Chemical Demonstrations: A Sourcebook for Teachers,
Volumes 1 and 2, Second Edition
Volume 1 by Lee R. Summerlin and James L. Ealy, Jr.;
Vol. 1, 198 pp; spiralbound ISBN 0–8412–1481–6;
Volume 2 by Lee R. Summerlin, Christie L. Borgford, and Julie B. Ealy
Vol. 2, 234 pp; spiralbound ISBN 0–8412–1535–9

Chemistry and Crime: From Sherlock Holmes to Today's Courtroom
Edited by Samuel M. Gerber
135 pp; clothbound ISBN 0–8412–0784–4; paperback ISBN 0–8412–0785–2

Writing the Laboratory Notebook
By Howard M. Kanare
145 pp; clothbound ISBN 0–8412–0906–5; paperback ISBN 0–8412–0933–2

Developing a Chemical Hygiene Plan
By Jay A. Young, Warren K. Kingsley, and George H. Wahl, Jr.
paperback ISBN 0–8412–1876–5

Introduction to Microwave Sample Preparation: Theory and Practice
Edited by H. M. Kingston and Lois B. Jassie
263 pp; clothbound ISBN 0–8412–1450–6

Principles of Environmental Sampling
Edited by Lawrence H. Keith
ACS Professional Reference Book; 458 pp;
clothbound ISBN 0–8412–1173–6; paperback ISBN 0–8412–1437–9

Biotechnology and Materials Science: Chemistry for the Future
Edited by Mary L. Good (Jacqueline K. Barton, Associate Editor)
135 pp; clothbound ISBN 0–8412–1472–7; paperback ISBN 0–8412–1473–5

For further information and a free catalog of ACS books, contact:
American Chemical Society
Distribution Office, Department 225
1155 16th Street, NW, Washington, DC 20036
Telephone 800–227–5558